The Natural History of Blenheim's High Park

The Natural History of Blenheim's High Park

edited by
Aljos Farjon

With contributions from A. Martyn Ainsworth, Keith N.A. Alexander, Alona Yu. Biketova, Peter Chandler, Anthony S. Cheke, Graham A. Collins, Brian Coppins, Sandy Coppins, Martin Corley, Phillip Cribb, Aljos Farjon, Richard Fortey, Ray Heaton, Rosemary Hill, Angela Julian, Jovita F. Kaunang, Alison Moller, Torsten Moller, David M. Morris, Benedict John Pollard, Neil Sanderson, Caroline Steel, Laura M. Suz, Peter Topley, Thomas Walker, Rosemary Winnall, Pat Wolseley, Ivan Wright, Jacqueline Wright

PELAGIC PUBLISHING

First published in 2024 by
Pelagic Publishing
20–22 Wenlock Road
London N1 7GU, UK

www.pelagicpublishing.com

The Natural History of Blenheim's High Park

https://doi.org/10.53061/QXBT1148

British Library Cataloguing in Publication Data
A catalogue record for this book is available from the British Library

ISBN 978-1-78427-430-6 Hbk
ISBN 978-1-78427-431-3 ePub
ISBN 978-1-78427-432-0 PDF

Front cover image: Aerial view of High Park from the south (photograph
by Damian Grady), copyright Historic England Archive.

Frontispiece: A particularly impressive ancient oak in High Park. This one
has a girth of 9.4 metres. (Photograph by Aljos Farjon)

Every effort has been made to trace copyright holders and obtain permission to reproduce
the material included in this book. Please get in touch with any enquiries or information.

MIX
Paper | Supporting
responsible forestry
FSC
www.fsc.org   FSC® C013056

Printed and bound in Great Britain by
TJ Books Limited, Padstow, Cornwall

Contents

Foreword

This wonderful book owes its origins to Aljos Farjon's decades-long fascination with ancient English oaks. His initial research led to the publication of his landmark *Ancient Oaks in the English Landscape* and a number of articles (several of which appeared in *British Wildlife* magazine, of which I am proud to be the publisher). From this starting point it was inevitable that Aljos would be drawn to High Park, a largely unaltered ancient pasture woodland adjacent to the world-famous 'Capability Brown' designed landscape in the other, larger part of Blenheim Park with its Palace, Lake, vistas and gardens.

Inevitable because High Park (on the western edge of the estate) has the greatest number of large unplanted ancient oaks of any site in England and, indeed, in all of Europe. They form an ecological continuum from a distant past, for several of the present living trees date back to the Middle Ages and through only a few tree generations all the way back to the prehistoric 'wildwood'.

These oaks would warrant a book in themselves; their presence (together with some rare saproxylic beetles dependent on dead or decaying wood) was the rationale for the notification of High Park as a Site of Special Scientific Interest as early as 1956. But *The Natural History of Blenheim's High Park* goes beyond the oaks to present the first comprehensive assessment of all the biodiversity of this extraordinary place, including non-lichenized fungi, lichens, bryophytes, ferns, woodland plants, snails, slugs, spiders, flies, wasps, bees, ants, butterflies, moths, beetles, amphibians, reptiles, birds and mammals. In doing so, the book highlights how ancient oaks support an abundance of biodiversity – an aspect of their contribution to nature that has been neglected until relatively recently.

The work to identify and record the large array of species – alongside the compilation of a wealth of information on both self-seeded and planted oaks – was carried out by many specialists over several years. This large undertaking was entirely voluntary, a tribute to the dedication and talents of those involved – and an honourable instalment in the long tradition of contributions by naturalists toward a better understanding of the wildlife of the British Isles.

Bernard Mercer, Chairman NHBS, Publisher British Wildlife

Acknowledgements

The survey of the biodiversity of High Park, as well as the writing of this book, were dependent on the assistance and contributions of many people. I am greatly indebted to the many recorders and specialists who joined me to contribute their time and expertise to build the volume of data on which this book is based: Martyn Ainsworth, Keith Alexander, A.J. Allen, R.B. Angus, R.J. Barnett, Lawrence Bee, Alona Yu. Biketova (fruit body DNA), J. Blincow, BNHS, Marc Botham, J.M. Campbell, Ben Carpenter, Peter Chandler, Anthony Cheke, Ryan Clark, Keith Cohen, J. Cole, Martin Collier, Graham Collins, Jon Cooter, D. Copestake, Brian Coppins, Sandy Coppins, Martin Corley, Bob Cowley, Phil Cribb, Andy Cross, Penny Cullington, Chloe Dalglish, Jonty Denton, Molly Dewey, Mary Elford, ESB from Thames Valley Environmental Records Centre (TVERC) data, Jim Fairclough, G.L. Finch, R.N. Fleetwood, Richard Fortey, Steve Gregory, Sara Gregory, S. Grove, Peter Hall, Clive Hambler, Ray Heaton, Russell Hedley, Wil Heeney, Alick Henrici, Rosemary Hill, Emily Hobson, Julian Howe, Caroline Jackson-Houlston, Stuart Jenkins, Angie Julian, Jovita Kaunang, Keith Kirby, Ian Lewington, Wendy MacEachrane, Linda Losito, Ceri Mann, Darren Mann, Alison Moller, Torsten Moller, David M. Morris, R. Morris, Sylfest Muldal, T. Newton, OBRC (= TVERC), Steve Nash, Oxfordshire County Museum Records, Bill Parker, Chris Perrins, Pedro Pires, Benedict Pollard, Keith Porter, Mark Powell, Margaret Price, G. Prior, Paulo Salbany, Neil Sanderson, Kate Sharma, Paula Shipway, Helen Simmons, Brian Spooner, Caroline Steel, Laura M. Suz (ectomycorrhizal root DNA), Peter Topley, Martin Townsend, TVERC Grassland Survey 2015, Tom Walker, Rosemary Winnall, Pat Wolseley, Ivan Wright, Jacqueline Wright, Wytham Survey.

In the Appendices, the recorders of each taxonomic group are listed with those who contributed most of the records named first. Many people listed above contributed to several taxonomic groups, and they are mentioned in the relevant Appendices.

Without permission from the Blenheim Estate to conduct the High Park Biodiversity Survey (HPBS), neither that survey nor this book would have been possible. This was granted on behalf of Blenheim Palace by Roy Cox, Managing Director of Estates. The specifics of our visits were managed by Rachel Furness-Smith, Head of Estates, while individual visits by surveyors were managed by Sally Tustian at the Estate Office. It was Sally's job to tell us if a visit was possible on the requested day. Her always prompt communication made visits run smoothly, and we are grateful to her for that assistance with the HPBS. Other staff at Blenheim (if not mentioned in the chapters) were helpful with information or publicity; I mention Nick Baimbridge (Head Forester) and Rachel Leach (Marketing) in particular. All at Blenheim Palace are thanked for their help and support with the HPBS.

I signed a Data Agreement with TVERC, so that we were able to incorporate earlier records of species held by TVERC in the chapters and in the Appendices. Caroline Coleman signed the agreement on behalf of TVERC. Henrietta Pringle, Biological Recording Coordinator, and Kate Prudden, Administration Officer, are thanked for their advice and help with this aspect of the HPBS. Administrative assistance was given by the Wychwood Forest Trust

(Wychwood Project), I thank Neil Clennell, Project Director, for his support. Other organizations involved with wildlife conservation in Oxfordshire expressed an interest in the project; I mention Natural England (Thames Valley Area), Berks, Bucks & Oxon Wildlife Trust and Trust for Oxfordshire's Environment. Through them I was able to inform conservationists and naturalists in the region about our project through presentations. In some instances this led to new participants in the HPBS.

Although participation was entirely voluntary, I considered it appropriate to be able to offer compensation for certain expenses made. Funds to enable this were generously provided by the Finnis Scott Foundation, the Trust for Oxfordshire's Environment (TOE2) and TVERC. Thanks to this financial support it was possible for those living at a distance from High Park to travel, and it also paid for some equipment costs. Many waived these costs, which could otherwise have risen above the available budget. I am grateful to the funders as well as to the generosity of the recorders.

In order to plan this book, a meeting was convened with the (key) participants in the HPBS. Blenheim Palace kindly hosted this meeting on 20 April 2022 in the Marlborough Room at the Palace. Rachel Furness-Smith attended to answer questions for the Blenheim Estate that might arise. Our records of the HPBS were handed to the Blenheim Estate represented by Rachel and to TVERC represented by Henrietta. My thanks go to all involved in making this meeting a success.

People who have given assistance or information to authors of chapters are acknowledged at the end of these chapters. Photographs and illustrations are all credited in the captions. Most were contributed by authors and surveyors, some by others not otherwise involved in the project, and were nearly always made available free, for which generosity I am grateful.

There were technical issues with the chapters submitted to me with which I found I needed help. Torsten Moller, author and co-author of several chapters, helped me with these to such an extent that I asked him to act as co-editor of this book, and he accepted the task. He created the various graphics for technical illustrations, helped authors with their presentation of statistics and made up the pages to guide the publisher to our preferred format. As two pairs of eyes always see more than one, he also helped to eliminate errors in the text. I am most grateful for all the work Tosh has put into this book. Proofreading was done by Sylfest Muldal, a participant in the project but not an author. I thank him for his thorough vetting of the entire book.

The publisher requested a contribution to the production costs of this book. The Blenheim Estate and TOE2 made donations towards these costs, for which I thank them warmly. Finally, I found Bernard Mercer prepared to write a foreword. Bernard is well known among our recorders of the biodiversity of High Park as the supplier of field guides and equipment for their surveys.

Aljos Farjon

Abbreviations

AOD	Acute Oak Decline
ATI	Ancient Tree Inventory
AWI	Ancient Woodland Indicator
BAS	British Arachnological Society
BBOWT	Berkshire, Buckinghamshire and Oxfordshire Wildlife Trust
BENHS	British Entomological and Natural History Society
BLS	British Lichen Society
BMS	British Mycological Society
BWARS	Bees Wasps & Ants Recording Society
CAI	Current Annual Increment [of wood]
DAFOR	Dominant–Abundant–Frequent–Occasional–Rare
FIT	Flight Interception Trap
GIS	Geographical Information System
HLM	Historic Landscape Management Ltd
HPBS	High Park Biodiversity Survey
ICOMOS	International Council on Monuments and Sites
IEC	Index of Ecological Continuity
IUCN	International Union for the Conservation of Nature
JNCC	Joint Nature Conservation Committee
LiDAR	Light Detection and Ranging
LSU/ha	Livestock Unit per hectare
MPR	Management Plan Review [Blenheim Palace]
NHBS	Natural History Book Service
NR	Nationally Rare
NS	Nationally Scarce
NT	Near Threatened (IUCN)
NVC	[British] National Vegetation Classification
OBRC	Oxfordshire Biological Record Centre (=TVERC)
OCC	Oxfordshire County Council
OCM	Oxford County Museum
OPM	Oak Powdery Mildew
OxARG	Oxfordshire Amphibian and Reptile Group
PLI	Pinhead Lichen Index
PMP	[Blenheim Palace] Parkland Management Plan
RDB	Red Data Book
RSPB	Royal Society for the Protection of Birds
s. lat.	*sensu lato* (in a wider sense)
s. str.	*sensu stricto* (in a narrower sense)
SOWI	Southern Oceanic Woodland Index
SQI	Saproxylic Quality Index

SQS	Saproxylic Quality Score
SSSI	Site of Special Scientific Interest
TOE2	Trust for Oxfordshire's Environment
TVERC	Thames Valley Environmental Records Centre
UNESCO	United Nations Educational, Scientific and Cultural Organization
UNITE	Database for molecular identification of fungi
VC 23	Vice-County 23 [Oxfordshire]
VCH	Victoria County History
VU	Vulnerable (IUCN)
WGP	Windsor Great Park

Ancient oaks in High Park. (Photograph by Aljos Farjon)

Introduction

Aljos Farjon

High Park, a relatively small section of Blenheim Park in Oxfordshire, has long been known to contain large numbers of ancient and veteran oaks – it was notified as a Site of Special Scientific Interest as early as 1956 on the basis of the ancient oaks and some rare saproxylic beetles dependent on dead or decaying wood that were already known to occur there. A 'veteran tree' survey was conducted in the period 2001–6, and this included the ancient and veteran oaks. Apart from one very large oak, known on the Blenheim Estate as the King Oak, none had been recorded for the Ancient Tree Inventory (ATI). It was for this reason that I sought permission from Blenheim Palace to record the ancient oaks for the ATI, using my association with the Royal Botanic Gardens, Kew, as the academic background for my request. It was granted, and in 2014–18 I recorded all oaks in High Park, dead and alive, with a minimum girth of 5m. It was soon found that no other site could match the numbers of large ancient oaks recorded in High Park (see Chapter 2). This density of ancient trees meant that the biodiversity of High Park should turn out to be comparable to the other sites with ancient trees.

The significance of ancient trees for biodiversity has been recognized more widely only relatively recently (Green 2010), in particular for saproxylic invertebrates (Alexander 2002; Chapter 15 in this volume), for non-lichenized fungi (Ainsworth 2005) and for lichens (Rose 1993; Wolseley 2022). Where these ancient trees occur in large numbers in a relatively small area, their significance is multiplied, especially for slow-dispersing species. Among Britain's native trees, the two oaks (Pedunculate Oak *Quercus robur* and Sessile Oak *Q. petraea*) support a greater number of species than any other trees (Tyler 2008). Published surveys of the biodiversity of a location with many ancient trees extending beyond mere checklists of selected species groups are rare. The best example is Moccas Park in Herefordshire, a site both ecologically and historically comparable to High Park, Blenheim (Farjon 2022). The study of Moccas Park (Harding and Wall 2000) was the inspiration for the High Park Biodiversity Survey (HPBS) and this book.

Two sites with ancient woodland in Oxfordshire, Wytham Woods (Savill et al. 2011) and Shotover (Wright and Wright 2018) for which the results of biodiversity surveys are described in a book, are ecologically different and with only a few ancient trees. Other old deer parks and royal forests or chases with high numbers of ancient or veteran trees, and cited as having had more or less complete biodiversity surveys, have not been treated in publications attempting to record all their biodiversity. These locations include Windsor Forest and Great Park (mostly for beetles and fungi), Epping Forest, New Forest, Sherwood Forest, Richmond Park, Staverton Park, Langley Park and Yardley Chase (Ainsworth 2022; Alexander 2022; Blincow 2023; Chandler, Chapter 9). Colin Tubbs gave a summary for the ancient woodland of the New Forest (Tubbs 1986). Biodiversity of the New Forest was treated in more detail by Newton (ed.)

Aljos Farjon, 'Introduction' in: *The Natural History of Blenheim's High Park*. Pelagic Publishing (2024). © Aljos Farjon.
DOI: 10.53061/YKRB6101

in 2010. For Staverton Park, Paul Harding compiled a preliminary list of the fauna in three parts (Harding 1972–5), stating that at the time '[t]he fauna of Staverton Park seems not to have been studied in any detail'.

The task of exploring the biodiversity of High Park was formidable from the outset and required the participation of many specialists over a number of years. Ours was a very different situation from sites that are more open to the general public. High Park only has a public footpath along its north-western periphery and one permissive path leading through it; it is closed to the public everywhere else. The Blenheim Estate had complete control over access and permissions were required to visit, set traps and sample specimens, all the while avoiding interfering with the Estate's business, whatever it was. With these conditions, permission was again given and in 2016 I could begin organizing this new project, with surveying starting in February 2017. I had planned a survey period of four years, but this was extended by a further year in 2021 because of the travel restrictions imposed by Government in 2020 related to the COVID-19 virus outbreak, which prevented those living further afield from reaching High Park. The number of visits we could make was further limited by a closed period, usually lasting from August to February, to facilitate an annual pheasant shoot. The cancellation of the shoot in 2021 owing to the same COVID-19 restrictions – no pheasants had been released – gave us one full year of unlimited access; in other years we had to stay within 5m either side of the permissive path during the shooting period. Recruiting volunteers for the survey was fairly easy, despite the fact that a small budget I had raised could only pay for some travel expenses. Many participants waived these; interest to visit this 'forbidden' site remained strong and was apparently for many a sufficient reward. In effect, we constituted a successful 'citizen science' team. I am very grateful for that goodwill, without which the project could have faltered.

Prior to the HPBS undertaken in 2017–21, several visits and biological surveys in High Park had been made, the records of which are held by the Thames Valley Environmental Records Centre (TVERC). These records go back to 1968 (e.g. fungi, higher plants) – with some incidental reports earlier still – and end in 2016 (e.g. beetles). All groups of organisms (except lichens) reported here are represented in these earlier records, with varying levels of completeness. For example, the TVERC records for butterflies observed in High Park between 1982 and 2015 contain 25 species; the present survey found 22 species. Between 1985 and 2002 just four amphibians and three reptiles were recorded; this survey found the same species. However, the more speciose the group, the less complete the older records are likely to be in comparison with those of the present surveys. Many of the older records are 'field records', which usually means observations during walks rather than by more systematic methods such as the use of various traps. For most of the invertebrates, the HPBS made use of a variety of trapping devices, left *in situ* throughout the Park for a period or operated during the night with light traps. Although GPS-based locations are not given in the older TVERC records, it is likely that many of them were restricted to the immediate vicinity of the footpath through High Park. However, the earlier records often span a much longer period, for instance for beetles they range from 1946 to 2016. The differences in results are nevertheless very notable. As an example, the records of beetles prior to 2017 held by TVERC contain 225 species; the present survey brought the total to 932 species. However, for most groups the recorded biodiversity is still incomplete. Certain species were found prior to 2017 that did not again turn up in the recent surveys; in almost all cases the correct interpretation should be that they were missed, not that they have disappeared.

It is in this light that the results of our surveys must be compared with and judged against the more impressive results for selected taxa from places such as Windsor or the New Forest. Is it possible to extrapolate? Could we estimate probable numbers of species based on what we have found and the similarity of habitats? When I asked, the specialists involved with surveying High Park were reluctant to guesstimate numbers. 'More than

1,000 species' for the most speciose groups was all that could be said with a modicum of confidence. The only way to find out would be to continue recording; this may in future become possible as and when the significance of biodiversity surveys and their detailed information for sites such as High Park become more widely recognized. The consent and support we obtained from Blenheim Palace for this project are a clear indication of this growing appreciation.

For some groups of organisms, our surveys will have been closer to reaching completion. This applies to sedentary organisms such as lichens – but not to non-lichenized fungi that do not move but have cryptic 'lifestyles' – and vascular plants, and to more conspicuous animals such as birds, butterflies, bees and wasps, or species-poor groups such as amphibians and reptiles. Mammals, although poor in number of species and often conspicuous, pose another problem: they are often hiding and disappear before they can be seen properly, or they are nocturnal while the recorders are diurnal. Bat detectors and interpretation of signs can to some extent alleviate these obstacles, but it is much work for often minimal returns. Despite this, we could increase the number of mammal species recorded in High Park to 26 from just 9 prior to our surveys.

For vascular plants we could apply the so-called DAFOR notation (Dominant–Abundant–Frequent–Occasional–Rare) in High Park for each species. Molluscs were counted and mapped for each designated location on repeated visits. In most cases, numbers of individuals found were noted for each species on each visit. Records were collated in Excel spreadsheets including minimally the basic information of species name, numbers of individuals found, site name, GPS-based locality (preferably as an OS grid reference), name of recorder, date of record, name of identifier and habitat note or substrate, all as required by TVERC. One person in each taxonomic team compiled the spreadsheet for that group and sent updated files to me for safekeeping. Under a data exchange agreement signed with TVERC, the HPBS project had access to all records from High Park made prior to 2017. These records were additional to our data and could inform the accounts of the relevant groups about observations in the past.

Although the HPBS included all the commonly surveyed taxonomic groups on land-based sites, a few were excluded, either because I did not find an expert field surveyor or because of their expected low diversity or cryptic nature. Of plants, we did not survey algae, a group that has probably a low diversity in High Park. Of the enormously diverse Invertebrates, we did not survey earthworms (Annelida) and crustaceans systematically; those that were encountered are listed in an appendix. Micro-organisms were excluded as being in the cryptic category. Butterflies and moths (Lepidoptera) are here treated in separate chapters although they belong to the same taxonomic group.

The two maps presented here show the area of High Park included in the HPBS. In the north, near the end of the north-west arm of the Great Lake, an area with tree plantations was excluded as being different in its ecological characteristics from High Park and more like other parts of Blenheim Park. This reduced the area of High Park for the purpose of the HPBS from 129ha, as cited in the High Park Management Plan, to 121ha. This includes the grounds around High Lodge, also not surveyed, in the total area here defined as High Park.

The total number of species recorded in High Park is about 4,100. This number includes species found prior to the HPBS and not found again during this project. This number is therefore the result of a five-year survey with imposed access limitations, supplemented with a limited number of earlier records.

How does this compare with Moccas Park in Herefordshire? Harding and Wall (2000) presented checklists for most groups but did not give totals of species found, so I had to tally their lists. Amphibians, reptiles and mammals had not been systematically surveyed but are mentioned in the text. The total number of species I could count for Moccas Park is 3,571, or

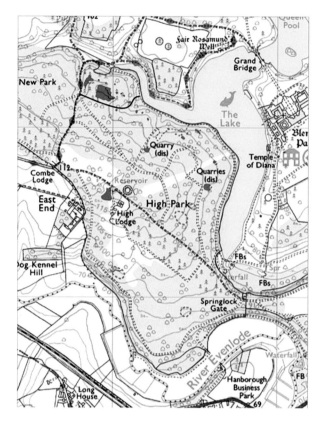

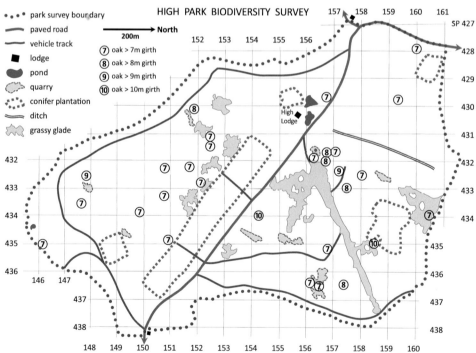

Blenheim's High Park environs (Ordnance Survey) *(top) and HPBS survey area grid* (Diagram by Torsten Moller) *(bottom).*

with a rounded figure 3,600. Thus, the numbers are of a similar order for the two parks and can be broken down into taxonomic groups as follows:

Species group	High Park	Moccas Park
Bryophytes	120	128
Vascular plants	279	261
Fungi	459	654
Lichens	231	235
Molluscs	76	48
Arachnids	106	121
Diptera	679	714
Hemiptera	86	60
Hymenoptera	233	62
Butterflies	28	18
Moths	684	150
Beetles	932	960
Other invertebrates	44	33
Amphibians and reptiles	7	5
Birds	74	106
Mammals	26	16
Totals:	**4,064**	**3,571**

Some groups, such as moths, were apparently under-recorded in Moccas Park, as is acknowledged by Paul Harding (Harding and Wall 2000: 195). Of the invertebrates, only beetles (Coleoptera) and flies (Diptera) were surveyed comprehensively. Considering this, both groups compare closely, especially the beetles. The bryophyte and vascular plant diversities are similar. More non-lichenized fungi have been found in Moccas, while lichens are equally diverse in High Park. With fungi, time spent on forays makes all the difference. Birds are more numerous in Moccas Park, but with this group it much depends on what birds are considered resident; in High Park we included only these and excluded birds seen on the Lake (noted in Appendix 14, but not counted).

In conclusion, it may be said that both sites have been surveyed for their biodiversity at a comparable level of detail; a longer but less taxonomically complete survey in Moccas Park as opposed to a shorter, but more complete survey in High Park. We can only speculate about the number of species that are actually present and would have been found eventually. The graphic curves would attain an asymptotic shape, which means fewer and fewer additional species will be found with the same effort over time. However, the figures for Diptera in Burnham Beeches, Epping Forest, and Windsor Forest and Great Park cited in Chapter 9 strongly indicate that High Park also has over 1,000 species of fly. An even higher estimate is likely true for beetles. These and similar places with many ancient trees are true biodiversity hotspots. This realization is important in an age in which negativity about the state of nature, in the media (good news is no news), but also among conservationists, is currently setting the tone. This is not to deny that there are serious environmental problems. It only serves to demonstrate that not all is 'crisis' or 'loss', and that we can still discover the extraordinary diversity of England's nature when we make a serious effort to find it.

Overview of the chapters

Chapter 1 'Historical Review' provides a framework within which the various accounts of the biodiversity of High Park in the ensuing chapters can be interpreted. It casts its net wider than the location of High Park on which the surveys of biodiversity were concentrated. Geology, landscape, archaeology and history are reviewed, leading to the present description of High Park as a largely unaltered ancient pasture woodland, in contrast with the more artificial 'Capability Brown'-designed landscape in the other, larger part of Blenheim Park with its Palace, Lake, vistas and gardens.

Chapter 2 'The Ancient Oaks in High Park' describes its most salient feature: the large ancient oaks, with the greatest number of any site in England and, indeed, all of Europe. These oaks determine the high level of biodiversity on this approximately 121ha site, which would be unremarkable for most groups of organisms without them (Chapter 15). These unplanted ancient oaks form an ecological continuum from a distant past, for several of the present living trees date back to the Middle Ages and through only a few tree generations all the way back to the prehistoric 'wildwood'.

Chapter 3 'Planted and Self-Seeded Trees' accounts for all the tree species present, including the younger native oaks. Self-seeded trees, including non-ancient oaks, are ubiquitous, but there are also many trees planted by people, most notably the oaks planted by the 9th Duke of Marlborough at the turn of the twentieth century. We decided to map these oaks as they were intended by the duke to be the successors of the ancient oaks, which he believed were all dying. Natural tree succession has been an important factor in a development from half-open pasture woodland to a more closed-canopy woodland in many places. In the recent past, this development was aggravated by local forestry plantations; this ecological mistake is now being rectified.

Chapter 4 'Flora and Vegetation' gives an account of all naturally occurring and some introduced plants other than the trees, including bryophytes, ferns, herbaceous plants and shrubby or climbing woody plants. Characteristic woodland plants and notable plants are listed and the (rather limited) diversity of native species is discussed. The vegetation is analysed and explained in relation to the various drivers, both natural and artificial, that have determined the various vegetation types present in High Park. Among these factors is a change from more acidic to more basic soils marked across the Park by the disappearance of Bracken towards the basic soils along an undulating line of transition.

Chapter 5 'Fungi (Excluding Lichens)' deals with the non-lichenized fungi found in High Park during fungal forays in the period 2017–21 and through the analysis of DNA extracted from mycorrhizal root samples of some of the oak trees. Fungi are by and large cryptic organisms; for detection in the field and identification in the field or the lab the fruit bodies ('mushrooms') are required, or the sequencing of DNA from fungal hyphae hidden in the soil to be matched with the sequences of identified samples in the laboratory. By both methods the results are slow to accumulate and, perhaps more than for any other organisms we tried to find, it would have taken many years before a more or less complete inventory was achieved. Despite the limited number of forays, many notable species were found, concentrating on saproxylic fungi on the oaks but not ignoring other types.

Chapter 6 'Lichens' describes the rich 'flora' of lichenized fungi present in High Park. This diversity is mostly associated with trees, and in particular with the ancient and veteran oaks. Those species specifically associated with these old oaks are highlighted in the text and listed in tables and detailed accounts were made of the lichen associations observed on several ancient or veteran oaks, illustrated by an example with photograph and drawing. It is concluded that High Park is one of the richest habitats for lichens in Middle England. The survey has led to the discovery of some species new to Britain.

Chapter 7 'Snails, Slugs and Bivalves (Mollusca)' deals with the 76 species – mostly snails and slugs plus a few mussels – found at 113 locations distributed throughout the site. The emphasis of the survey was on the ancient oaks and their immediate surroundings. Other locations yielding many molluscs were some of the larger old limestone quarries. The difference in habitat determined by pH of the soil was found to be significant, with the more acidic soils of the highest parts of High Park yielding fewer species and lower numbers than elsewhere. The presence or absence of Bracken was, as for the flora, a good indicator of this distinction. Snails with large shells such as the Roman Snail, of which there is a viable population in High Park, need easy access to exposed weathered limestone to build their 'houses'.

Chapter 8 'Spiders and Relatives (Arachnida)' gives an account, arranged by family, of arachnids recorded in High Park. Most are spiders and some 100 species of these are listed in Appendix 6, including the results of surveys under the HPBS for 2018–2020 and earlier records held by TVERC. Species of interest are mentioned and discussed in the context of their occurrence in Britain and/or Oxfordshire. Some rare or uncommon species are highlighted in text and illustrations.

Chapter 9 'Two-Winged Flies (Diptera)' focuses on one of the highly diverse groups of invertebrates. In High Park, many species are saproxylic, either living as larvae directly in decaying wood, or indirectly via the fungi that grow in and decompose this wood. The ancient oaks are particularly important in this respect, and this chapter therefore concentrates on the species with these associations. Other notable species of fly are also mentioned, especially when they represent first records for the county. Comparisons are made with other sites with ancient trees and the saproxylic or fungi-associated flies recorded there. Despite large differences in recording activity between High Park and these other sites, a high number of species was found and many of these are noted under their respective families.

Chapter 10 'Sawflies, Wasps, Bees and Ants (Hymenoptera)' describes the familiar bees, wasps and ants, as well as sawflies and Parasitica, some of the latter causing galls on oaks and other plants. High Park is a prime site for those species that nest in the cavities of dead wood, but less suitable for species dependent on flowers other than the abundant Elder and thorns. Interestingly, many solitary-nesting bees have their own 'cuckoo' – a cleptoparasite bee to be expected and indeed often found in the surveys. A particularly rare 'cuckoo', only known from seven localities in Britain, was found together with its presumed host in a trapping device set by Ivan Wright. There are at least 50 colonies of honey bees nesting in the hollow oak trees (surveyed by Filipe Salbany independently from the HPBS) and it is under investigation to establish whether or not these represent a separate population from domesticated bees.

Chapter 11 'Butterflies (Lepidoptera)' examines the 22 species recorded in High Park in 2017–21. A few are woodland habitat specialists, but most are more generally distributed in the wider countryside. Most of the species that could be expected were recorded, but a few that could occur in this mixed grassland/woodland habitat were absent or were missed. Previous records amounted to 25 species, and in total 28 species were found over the years. High Park has a good butterfly presence, but is rather species-poor compared with some other nearby sites that have a greater variation in habitats.

Chapter 12 'Moths (Lepidoptera)' shows that the diversity of species is far higher than that of butterflies. Moths have a strong connection with plants because the larvae almost exclusively feed on these, often on particular species only. The species are therefore listed and described in terms of their feeding plants and/or feeding habits. Relatively few species are saproxylic; those found are also mentioned, with their preferred substrates. Many are active at night and can only be found with light traps; the methods used are fully described to give information relevant to what has been recorded. As with other very speciose groups, recording remained incomplete by the end of the project owing to constraints, but many notable species were nevertheless found and are discussed.

Chapter 13 'Beetles (Coleoptera)' focuses to a large extent on describing and illustrating the various methods employed to trap and collect beetles. It shows that with different contraptions new species can be found that did not appear with other methods. Beetles are an inordinately diverse group of insects, and High Park is very rich, mostly owing to the high density of ancient oaks and their dead, decaying wood. The survey concentrated on this woody habitat, but did not ignore the other types of vegetation, although sampling there was less intensive. One specialist substrate is animal dung, much increased with the introduction of cattle halfway through the survey period. The involvement of top specialists from the Oxford University Museum of Natural History and elsewhere boosted the species list, so that High Park now ranks among the most species-rich sites, behind the two top sites New Forest and Windsor Great Park. There are numerous new records for Oxfordshire, and some species found are only known from a few specimens elsewhere in England.

Chapter 14 'Bugs (Hemiptera) and other Insect Orders' gives brief accounts of those insects that belong to orders with few species, either in High Park or in the whole of Britain, and which were therefore not given a full chapter of their own. One group, the Hemiptera, is larger than the others but reported on here to give the chapter more substance. In order to find them, many of the insects discussed here require special methods beyond general observation or the trapping methods employed for such groups as beetles and moths. As none of these methods was employed targeted at these insect orders, it is noted that there are likely many more species in High Park than were recorded during the HPBS.

Chapter 15 'Assessing the Importance of High Park for Saproxylic Beetles' provides a comparison of High Park with other sites with ancient trees in lowland England. Two measures are used to compare the ecological quality of sites using saproxylic beetles: the Index of Ecological Continuity and the Saproxylic Quality Index. Sites are ranked in tables for each index according to their scores. As a result of intensive surveying, High Park now ranks among the five top sites in the country, making it of European significance for its saproxylic beetles. All 268 scoring saproxylic species, with their scores, are listed in a table.

Chapter 16 'Amphibians and Reptiles' discusses Great Britain's limited number of native amphibians and reptiles. All but one of the eight species that can be expected to occur in High Park were recorded; the eighth is probably present but no adult of this newt – needed for certain identification – was found. Some species have been found often enough to conclude that a viable population exists; among these is the highly protected Great Crested Newt.

Chapter 17 'Birds' divides birds into the categories breeding birds and visitors, with emphasis on those that are known or assumed to breed in High Park: birds are highly mobile and, unlike with most of the organisms recorded in High Park, it is difficult to say – unless they are breeding there – whether they are resident or passers-by. The artificial lake in Blenheim Park borders on High Park, so birds that make use of the shore on this side are also mentioned.

A third category deals with the deliberately introduced game birds, here pheasants, which are annually reared and released in High Park for the shoot. As they are birds, it is appropriate that this practice, its history on the Blenheim Estate and its impact on other birds through gamekeeping, are discussed here.

Chapter 18 'Mammals' gives an account of the 26 species found in High Park arranged by groups following the Mammal Society list. These are Insectivores (5 species), Lagomorphs (2 species), Rodents (4 species), Carnivora (4 species), Ungulates (3 species) and Bats (8 species). All are common in the UK and only one species of bat has a high affinity and some reliance on the ancient oaks of High Park. Owing to the cryptic habits of many species of mammal it is noted that some species not found are nevertheless very likely to be present. The apparent absence of several mustelid carnivores is perhaps linked to gamekeeping activity for the pheasant shoot.

Chapter 19 'Review of Management Practices' gives a brief overview of the more recent management carried out in High Park in relation to documented management plans. Management of game birds (pheasants) in High Park is discussed in Chapter 17. The environmental management is seen to be increasingly focused on maintaining and where appropriate restoring the pasture woodland landscape in which the ancient oaks including deadwood are the major component in the ecosystem. Recently, grazing with cattle was introduced and this is discussed in a wider context, comparing High Park with similar sites elsewhere in England, some of which have had a longer time in which the effects of this type of land management may be evaluated. The purpose of conservation grazing and its likely influence on the landscape, specifically in High Park, are also discussed.

CHAPTER 1

Historical Review

Alison Moller and Torsten Moller

The aim of this overview is to provide a framework within which current patterns of species distribution, ecology and related findings in Blenheim Park can be evaluated in relation to changing environmental conditions, land usage and management practices since pre-Roman times. In order to provide context in space and time, the net is cast wider than the scope of the modern High Park, with a focus on aspects of the history, archaeology, geology, topography and climate of the whole Park in its setting deemed relevant to the overarching purpose of the report.

The ancient and veteran oaks have long played a central role in the ecology of Blenheim High Park and have also to some degree been part of the designed landscape first created in the eighteenth century in Lower Park. Complementary datasets will need to be compared in order to estimate the likely age and maturity of the oak-inhabited community; a task made more complicated by the way in which oaks grow and change over time. To this end, the history of land utilization and management can serve to supplement the analysis of species composition, biodiversity and other ecological signposts reported in later chapters.

1.1. The ancient landscape and early settlements

The twisting course of the River Evenlode (Figure 1.1), is an example of a 'misfit stream' flowing in a broad and gently meandering channel carved by a once-larger river system swelled intermittently by post-glacial meltwaters, but with a catchment truncated by the Severn & Avon (Dury 1958: 107; Whiteman and Rose 1997: 330). Before it was beheaded around 450,000 years ago (Anglian Ice Age), the proto-Thames occupied the whole of the Evenlode valley and built gravel terraces along its banks containing Triassic quartzose conglomerates from the West Midlands 'Bunter Pebble Beds' (Shotton et al. 1980: 84). Falling sea levels during repeated ice age cycles allowed rivers to deepen their valleys, and gravel remobilized by erosion was distributed into multiple fluvial terraces at successively lower levels down-stream. The original beds survive as 'Northern Drift' on flat high ground such as the Combe Plateau where a thin surface layer of rounded quartzite pebbles is testament to the winnowing of the former sediment matrix by drying winds across an unwooded tundra landscape (Hardaker 2004: 30). The LiDAR image of the Combe Plateau and River Evenlode adjoining High Park (Figure 1.1),[1] reveals a drainage pattern of numerous dry valleys cut into the bedrock in wetter

---

1   LiDAR is an acronym for Light Detection and Ranging. It employs airborne laser scanning of the ground to generate a digital high-resolution 3D representation of surface features that are revealed by artificial lighting from one side.

Alison Moller and Torsten Moller, 'Historical Review' in: *The Natural History of Blenheim's High Park*. Pelagic Publishing (2024). © Alison Moller and Torsten Moller. DOI: 10.53061/ZZBX6109

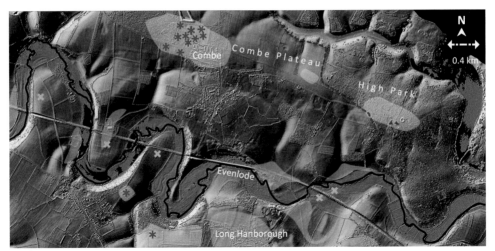

Figure 1.1.   LiDAR image (https://houseprices.io/lab/lidar/map?ref=SP43041554) of the Combe Plateau and Evenlode valley. Palaeolithic finds (✳ – Hardaker 2004); Neolithic (✖ – Hardaker 1992); Northern Drift cover (blue); Debris fans and Gravel beds (green); Blenheim Park wall (red dashed line).

post-glacial climates and shaped by a fissured limestone aquifer, periglacial weathering and soil creep (solifluction), creating fans of gravelly debris.

A flint hand-axe (Acheulian) found in an unrolled condition in a river terrace gravel bed at Long Hanborough (Arkell 1947: 3) is an early pointer to human habitation and perhaps attributable to the Early Wolstonian cold stage, some 350,000 years ago (Early/Middle Palaeolithic). Although more easily worked than quartzite, good knappable flints are rare in proto-Thames/Evenlode gravel beds and it is arguable that it was the higher isolated patches of quartzite-rich Northern Drift that attracted the greatest attention of early hominids. Clusters of their tools, knapped flakes and spoil have been found among stones in ploughed fields at Combe and near Long Hanborough. The weathered condition of these artefacts varies greatly and suggests the sites were visited repeatedly and exploited for much of the early Palaeolithic period (Hardaker 2004: 32). In High Park the quartzite pebble layer is largely hidden by vegetation, but there is good reason to suspect the presence of Palaeolithic artefacts among accumulations of hand-size stones in the portion of the Park covered by Northern Drift (Figure 1.1).

Over 2,000 Neolithic artefacts have been recorded during systematic field-walking in the Evenlode valley (Hardaker 1992: 49). There are several Neolithic long barrows, Bronze Age round barrows and ring ditches found in the vicinity of Blenheim Park. Signs of Iron Age occupation within the Great Park include a settlement, a possible hill fort and a section of Grim's Ditch.[2] The river gravel terraces of the Upper Thames Valley held the main pre-Roman population centres and two Iron Age hill forts are situated to the south, at Eynsham Camp and Bladon Camp (Figure 1.2).

The purpose of Grim's Ditch is uncertain (Copeland 1988: 283), but it encompassed a large area (64km²) either side of the River Evenlode, at the junction between three principal tribes, Dobunni in the west, Catuvellauni in the east and Atrebates to the south (Lambrick 2013: 47). Such a key position suggests a region of prosperity and relative independence, which may have led to this land also being favoured by early (first century CE) Roman settlers (Booth 1999: 47; Salway 1999: 6). It has been proposed that the gaps in the Ditch indicate where woodland once stood. This interpretation would suggest that the northern portion of Great Park was under cultivation at the time. The Roman road known as Akeman Street crosses the northern

---

2   https://historicengland.org.uk/listing/the-list/map-search?clearresults=True, accessed 25 July 2019.

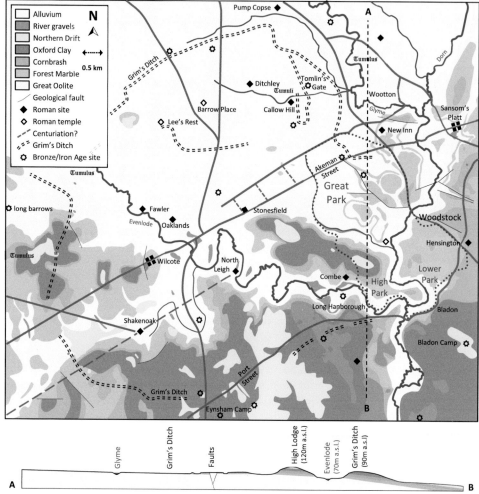

Figure 1.2.   Geological map of Jurassic bedrock (162–7Ma) with surficial drift (<3Ma) and a cross-section (A–B) with elevations from 70 to 120m above sea level. Traces of early settlement history shown as symbols; Blenheim Park marked in red. (Authors' illustration)

tip of the Great Park and cuts the older Grim's Ditch. A small excavation carried out in 1936 at the intersection of the two features indicated that the Ditch was created just before the Roman occupation of early first century CE (Harden 1937: 91). The Ditch was thought to have been about 5m wide and 1.5m deep with a bank almost 2m high. The bank covered older ploughed soil containing Belgic (Catuvellaunian) pottery sherds, which showed that this part of the Great Park had indeed been under cultivation before the Romans arrived. The evidence pointed to the Ditch being deliberately filled in, since all the sherds in the backfill were of Roman date.

Akeman Street, which connected St Albans and Cirencester, provided further opportunities to cultivate areas on both sides – most probably for cereal and wool production. The climate was relatively warm in this period and terraces at North Leigh may have been used for growing grape vines (Baggs et al. 1990: 76). The road would have provided easy access to garrison towns and markets along the route. Fragments of the distinctive carrot-shaped amphorae used by the Roman army have been found at Ditchley, Fawler, Shakenoak and Wilcote (Figure 1.2), suggesting a close connection with military supply lines (Booth 1999: 48).

Larger nucleated settlements at regular ten-mile intervals along Akeman Street include Wilcote and Sansom's Platt.

There is a parallel trackway connected to the Street by straight field boundaries about half a mile to the north of Akeman Street. A second parallel trackway, almost a mile to the south of Akeman Street and also connected by field boundaries at right angles, has been observed in aerial photographs (Emery 1974: 45). These could be signs of *centuriation*, the geometric grid pattern of land layout used by the Roman army when awarding land to its retiring troops (Rivet 1964: 101).

The Roman occupation brought greater emphasis on converting land use to arable farming. Several high-status Roman villas have been discovered around Blenheim Park, notably those at Ditchley, North Leigh and Stonesfield. They are generally situated close to rivers and streams, or to springs emanating from the Cornbrash substratum. The concentration of villas in the Glyme-Evenlode interfluve suggests intensive cultivation on free-draining soils of Forest Marble and Great Oolite Limestone, which weather to loam (Bond and Tiller 1987: 11).

The villas were working farms with up to 1,000 acres of cultivated land (Emery 1974: 44). The villa at Ditchley is surrounded by a 2.7-acre (1.1ha) enclosure with a large granary and supposed threshing floor (Radford 1936: 28), which were probably associated with production from several nearby dependent farms run by tenants (Rivet 1964: 122). The Combe dwelling is the closest to Blenheim High Park and was inhabited from pre-Roman to late Roman times.[3] It is of more modest proportions than nearby villas and was perhaps part of the Stonesfield estate, occupied by a tenant farmer growing cereal and rearing sheep, cattle and pigs (Speake 2012: 8, 90).

The broadened horizons and increased trade brought by the Roman occupation were reflected in a growing network of long-distance paved roads and local trackways for movement of people, bringing animals and goods to markets and avoiding the trampling of growing crops. Wheeled transport became increasingly important, and factors such as the more efficient drainage afforded by ditched tracks and a convenient source of a river gravel substrate are likely to have influenced the selection of routes and the method of trackway construction (Booth 2011: 7).

The study area straddles the boundary between Cotswold Jurassic limestones in the north and the Thames Valley Oxford Clay Vale to the south (Figure 1.2). The contrasting soil types may partly explain the large number of Roman villas in the Cotswolds and the relative scarcity of corresponding settlements on the poorer and heavier clay grounds to the south, more suited to a thriving pottery industry (Salway 1999: 15). In addition to providing suitable agricultural land, the concentration of Roman settlements in the Cotswolds may also reflect social status and the proximity to good hunting country (Salway 1999: 16). Figure 1.3 illustrates a Roman passion for deer hunting, captured in Sicilian mosaics – a theme that was to become a prominent feature of Oxfordshire land management for more than a millennium.

While the distribution of Roman villas appears to reflect the underlying geology and associated soil types, there is a noticeable gap in the area of Great Park. Bond and Tiller (1987: 20) suggest this could have been filled by a villa on the site of the medieval manor/palace (Figure 1.7), where Roman coins have indeed been found (Baggs et al. 1990: 431). Other Roman finds in Great Park include burial objects just north of Akeman Street and a possible temple site by

---

3   'The ceramic evidence would indicate that there is continuity and transition from earlier middle Iron-Age occupation to a later Iron Age farmstead, with its enclosure ditches and suspected roundhouses. This appears to be followed, without any break in the ceramic sequence, by the rectangular plan and more sophisticated building techniques of a Romano-British villa with its ancillary structures' (Speake 2012: 2). 'A record of prehistoric activity on the site, from the Mesolithic to the Neolithic, [is] evidenced by flint and stone artefact scatter' (ibid.: 10).

Figure 1.3.   Deer hunting scene, Villa del Casale, Sicily. (Photograph by T.H. Moller)

the River Glyme, 2.5km south-east of the Street. Temples provided both spiritual refreshment and opportunities for meetings to conduct secular business (Salway 1999: 13).

Much of the present High Park is covered by Northern Drift deposits and Oxford Clay, making it much less suitable for cereal cultivation. The clay forms soil of the Denchworth Association (Chapter 3, Figure 3.16), and as in other areas of the country, this is good for pasture woodland (PMP 2014: 8), although poor drainage encouraging shallow tree root development can lead to increased wind-throw. Cornbrash occurs sporadically in both High Park and Lower Park, and although rich in phosphates and lighter than the Oxford clay, it too becomes difficult to work when wet (Bond and Tiller 1987: 11). While the relatively poor soil quality in High Park favoured hunting, it is possible that farm animals were grazed in these upland areas during the summer when the lower areas of Great Park and Lower Park were given to corn. The woodlands may have been managed to some extent since timber was needed for fuel, building material and joinery. On this basis, the areas of the present Great Park and Lower Park were in all probability cleared in the pre-Roman and Romano-British periods and then at least partially recolonized by forest after the Romans had left.

In the fifth-century, post-Roman period, it is likely that mass-produced pottery and large-scale grain and wool production would have been early victims of a fragmenting road network and less secure access to distant markets. This may have encouraged a gradual contraction, consolidation and diversification of estates into more economically viable units scaled to local demand and self-sufficiency. A shift in the balance between arable and pastoral farming is also conceivable (Thomas 1971: 120). It is likely that other aspects of Roman civilization and innovation introduced over nearly four centuries survived, albeit in different forms. Latin persisted as the language of learning and religion. In a sense, the Roman civic administration morphed into the early Christian organization of ecclesiastical provinces, dioceses and parishes, which increasingly provided a structure and code at all levels of society.[4] The impact of this process can be seen in the parish system, which changed little over time and was already old by the time of the Norman Conquest (Rackham 1997: 19).[5] The ancient parishes predated the first known boundaries of Woodstock Park since the earliest Park was carved out of the ecclesiastical parishes of Bladon and Wooton (Baggs et al. 1990: 14, 443). It is perhaps no coincidence that the Park expanded over time to incorporate all the elements of nominal self-sufficiency that characterized a typical parish unit: arable, meadow, pasture and woodland.

---

4   'Christianity, from Roman to Viking times, was not just a religion that happened to survive. It was a code of ethics, a way of life, literature, education, the commemoration of the dead, a major import customer, a tenurial power, a social system, a patron of pure and applied arts' (Thomas 1971: 71).

5   'It is likely, that ministers had for centuries provided pastoral care to big "proto-parishes" around them. By 1100 these were breaking down into the small local parishes, centred on parish churches, that have existed ever since' (Blair 2010: 26).

Figure 1.4.   Scenes from the Bayeux Tapestry showing Harold Godwineson with the text 'SOLDIERS RIDE TO BOSHAM: THE CHURCH: HERE…'. (After Shapland 2012)

1.2. The hunting forest

In the longer term, the power vacuum left after the Roman withdrawal led to the emergence of Anglo-Saxon kingdoms, growing in size from the sixth century onwards. The area around Woodstock may have become sparsely populated since this was disputed land between the kingdoms of the West Saxons and Mercia (Bond 1981: 202) – much as seen in the south-east corner of Mercian territory vis-à-vis the East Saxons (Williamson 2010: 97). Saxon settlements in Long Hanborough and North Leigh stem from the seventh century. By this time, the climate was turning wetter and cooler, and it is likely that the forest had become re-established in most of the present-day Blenheim Park. The Old English words for 'wood' (*wudu*) and 'clearing or wood-pasture' (*lēah*) are conspicuous in early local place names (Bond 1981: 201).

The re-afforested land was probably prized for the opportunities it offered to royalty and the aristocracy for hunting. An early depiction of the Anglo-Saxon love of hunting and its high status is shown on the Bayeux Tapestry (Figure 1.4). Harold's hunting lodge/hall is depicted in the Tapestry as if made of timber.

Wodestoch is a Saxon name meaning a woody place (Taunt 1914: 63), or a wooden stockade, alluding to the links between the constructed wooden world of the Anglo-Saxons and the natural environment surrounding them (Bintley and Shapland 2013: 14). Woodstock appears to have become royal property by the ninth century (Bond 1966: 5) since successive witans (high councils of wise men) were held there to choose Kings of Wessex.[6] Ethelred II (978–1016) held witan at Woodstock,[7] presumably in his hunting lodge, which may have been on the site of a later palace, just within the forest and the royal estate (Baggs et al. 1990: 435). This was not yet a royal palace, so was probably of timber construction since 'timber from living trees was perceived as suitable for everyday life' (Bintley and Shapland 2013: 6) and stone was usually reserved for ecclesiastical buildings and royal palaces. It is possible that the area around the king's manor at Woodstock was defined by a boundary or enclosure before or very soon after the Conquest, since it was described as 'in defensione regis' – a restricted woodland space (Jørgensen 2010: 125) – in the 1086 Domesday survey.

There is a Woodstock charter of 1005 that referred to a hunting lodge and enclosure, but no clue is revealed as to its location or extent (Woodward 1982: 2). Further circumstantial evidence of the importance of the hunting lodge is supplied by Dornford Lane (Figure 1.5). This is a

6   '[St] Aethelred King of Wessex [c. 848–871] – married to a Mercian princess – held Witenagemot (or Witan) at Woodstock in 866. His younger brother King Alfred (849–899), who had also married a Mercian noblewoman, is said to have translated Boethius' De Consolatione Philosophiae at Woodstock in 888' (Banbury et al. 2010: 4).

7   'This is the ordinance which King Ethelred and his Witan ordained as "frithbot" … for the whole nation, at Woodstock in the land of the Mercians' (McLean 1989: 99).

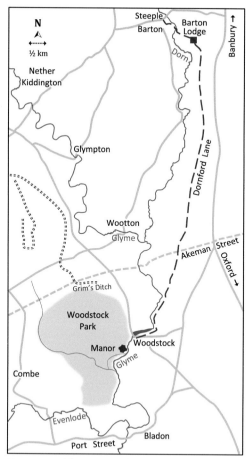

Figure 1.5.   The course of Dornford Lane in the Middle Ages. (Authors' illustration)

green lane running parallel to the Banbury to Oxford road and connecting Barton Lodge in Steeple Barton with Woodstock. Barton can mean a barley farm, but in later times meant a demesne farm (Hoskins 1977: 239). The lane runs for 7.2km and does not follow any known boundaries, and so has been interpreted as a track established no later than the tenth century to bring supplies from a royal demesne farm to the hunting lodge.

Woodstock was recorded in Domesday as having been part of the king's (Edward the Confessor, 1003–1066) demesne forest, and measuring 9 leagues in both length and breadth. The area included Scotorne (Shotover), Stauuorde (Stowood), Wodestoch (Woodstock), Corneberie (Cornbury) and Huchewode (Wychwood) and was still sparsely populated with just six villagers (villeins) and eight small-holders (bordars) recorded with three and a half plough teams (Baggs et al. 1990: 400). The locations of these forests as they waxed and waned are shown in Figure 1.6.

Hunting was the main attraction of the forest: the quarry would have been both Red Deer (browsers/grazers), of which the females are gregarious and the males solitary and territorial; and Roe Deer (browsers, suited to dense forest), which are territorial, but non-gregarious (Fletcher 2011: 97–8). Both species therefore need a large area in which to roam (Creighton 2013: 126). The deer were hunted 'par force de chien'; that is, a single deer was chased down with hounds over the course of a day (Liddiard 2007: 5). This suggests that the forest was not enclosed but, to some extent, managed through permitted animal grazing, which facilitated the passage of horse and rider.

King William I (1028–1087) had reinforced the concept of hunting being an elite privilege by laying claim to all the beasts of the chase (deer and Wild Boar). By the time Henry II (1133–1189) succeeded Stephen in 1154, one-quarter of England was declared royal 'forest' and subject to separate laws. Forest Law had no specific relationship to woodland since it derives from the Latin *foris* meaning 'outside'; that is, outside the common law.[8] These expansions were reversed in Magna Carta and in the Forest Charter of 1217. Two inquisitions in 1219 and 1229 identified the areas to be disafforested. Perambulations conducted in Wychwood in

---

8   While agreeing that 'In medieval parlance, forest was an organizational term rather than a geographical one,' Jørgensen (2010: 115, 126–8) noted that 'Numerous [Domesday] entries indicate that "the woodland is in the forest" (*silva est in foresta*), stressing the distinction between the two terms'; and that 'Game never appears as a forest product in Domesday Book; when hunting is mentioned, it is always in reference to woodlands (*silvae*).' She argued on the basis of a strict analysis of contemporary texts that 'forest had changed by the late twelfth century, and had become . . . a legal entity centred on royal hunting rights'.

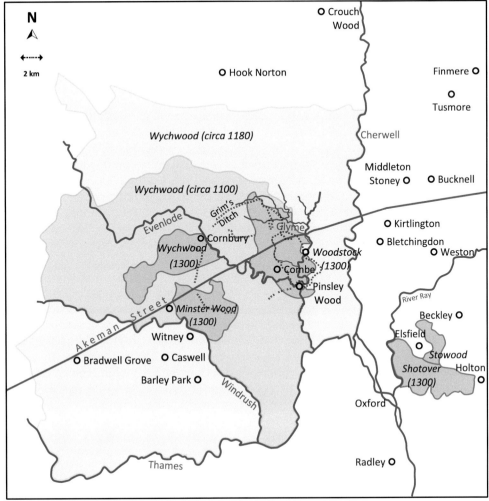

Figure 1.6. Medieval forests and deer parks in Oxfordshire. The name of Wychwood was probably derived from Hwicce, the kingdom that succeeded the Dobunni and was subsumed into Mercia (Blenheim Park outline in red). (Adapted from Bond and Tiller 1987: 24 and Schumer 2004: 2)

1298 and 1300 further reduced the ambit of Forest Law to just the royal manors, although the boundaries were not finally ratified by Parliament until 1641 (Schumer 2004: 24). Nevertheless, the records of the perambulations serve to delineate the proscribed area within which the part known as Woodstock Royal Park was enclosed.

Enclosing the habitat of the deer was a logical progression. The term 'park' is derived from the Old English '*pearroc*', meaning an enclosed field or paddock with a fence, but by the Middle Ages applied to deer enclosures, especially in royal forests (Woodward 1982: 2). At the time of the Conquest there was probably little difference between the deer hunting forests in England and those in Normandy (Liddiard 2003, Jørgensen 2010: 128). However, the Normans soon introduced Fallow Deer from their newly conquered territories in Sicily (Creighton 2013: 125). Their small size and less aggressive nature made them easily transportable (Fletcher 2011: 100). Native to the Levant, grazing Fallow Deer needed enclosed wood-pasture, meaning woodland with laundes or clearings in which the grass could grow well. Once enclosed, and the deer introduced, the forest would have been transformed. Grass in the cleared laundes would have rapidly replaced the understorey and new saplings would have to be protected from the grazing deer, thus resulting in a much more managed landscape (Creighton 2013: 133).

Henry I (1068–1135) visited Woodstock most years, often for several weeks at a time (Schumer 1984: 46). In around 1110 he had a lodge built and a seven-mile wall constructed around his Park. Outside he founded the village of Old Woodstock on 40 acres of waste ground to provide lodging for his retinue (Trinder 1983: 3). An 'enclosure' that was probably the wall is first mentioned by the monk William of Malmesbury in his chronicle of about 1125.[9] It would have encircled an area of about 2,500 acres (1,000ha), making it a similar size to the royal parks at Windsor and Clarendon (Lasdun 1991: 12). The position of the wall is uncertain (Figure 1.5), but is likely to have taken in most of the modern Great Park and High Park (Baggs et al. 1990: 441, 443). Woodstock Park was identified as an extra-parochial entity separated from neighbouring parishes and with tithes granted by Henry II to Godstow Abbey, Oxford, where his mistress Rosamund Clifford was buried in around 1176 (Baggs et al. 1990: 430). Evidence from the Pipe Roll of 1165 shows that the stone walls were repaired at that time (Baggs et al. 1990: 441). They were repaired again in 1232 and 1250, when the section near New Woodstock was raised. In 1251 Henry III (1207–1272) ordered a deer leap to be constructed at Woodstock (Woodward 1982: 7), allowing deer to enter the Park, but not to escape.

Contrary to a report from 1677, the wall around Woodstock Park was probably designed primarily to contain the newly introduced Fallow Deer rather than the king's collection of exotic animals.[10] It seems likely that the beasts were not allowed to roam free. No specific information has been found on how these animals were kept at Woodstock, but once the menagerie was moved to the Tower of London in about 1204 (Grigson 2016: 1) there are some clues, albeit from a later date, that the animals were confined and largely hand-fed (Turner-Bennett 1829: xiv).

At about the same time as founding New Woodstock in the late twelfth century, Henry II extended the area of Woodstock Park by incorporating Hensgrove (Figure 1.7), maybe thereby seeking to control its woodland in the face of a growing population in the new borough. Bond and Tiller (1987: 48) have presented evidence for 'emparkment of the open fields at Hensington' before about 1200, but how Woodstock Park was managed in the longer term is not known. In 1274 Edward I (1239–1307) ordered the felling of trees 'in the park and in the wood of Hensigrove', probably to make rides and to sell the timber. The twelfth/thirteenth century is probably when some of the most ancient oaks in High Park and Lower Park germinated. There are records of oaks being planted in this period, but there is no such evidence pertaining to Woodstock Park. The Bishop of Coutances (1049–1093), a nephew of William I, is recorded as planting acorns in his newly formed park in Normandy and importing deer from England (Harvey 1990: 8). Instances of acorns bought and planted in England in the thirteenth and early fourteenth century are recorded from Winchester, Wiltshire, Northamptonshire, Chertsey (Surrey) and Cuxham (Oxfordshire).[11]

---

9   'Paul, Earl of Orkney, although subject by hereditary right to the king of Norway, so looked up to the friendship of the king that he repeatedly used to send him little gifts. For indeed he took exceptional pleasure in the wonders of foreign lands – begging foreign kings with great enthusiasm for lions, leopards, lynxes, camels, which cannot be bred in England. And he had an enclosure called Wudestoche in which he cherished the delights of such things. And he kept there an animal which is called a porcupine . . . covered completely with bristly spines, which naturally stick in dogs chasing them' (Hardy 1790; Latin translation by Jan Dicks).

10 'King Henry the First enclosed the parc at Wudestoc with a wall, though not for deer, but for all foreign wild beasts, such as lions, leopards, camels and linxes which he procured abroad from other Princes' (Grigson 2016: 1).

11 'Also belonging to St Swithun's was Alton Priors, Wiltshire where ploughed land was planted with 19 quarters of nuts costing £1 5s 4d and 8½ quarters of acorns costing 8s 6d in 1260–1261. Henry de Bray of Harlestone, Northants is recorded sowing acorns in 1305–1306 and Abbot John Rutherwyle of Chertsey had young oaks and acorns sewn at Hardwick Grange, Chertsey on four occasions from 1307' (Harvey 1990: 13–17).

Figure 1.7. Woodstock Park (red dashed line), c. 1400. Signs of ridge and furrow may indicate the site of a medieval settlement assimilated into the Park. (Adapted from Bond and Tiller 1987: 29)

Forest Law gave the king sole right to hunt, only allowing others to do so as a mark of favour. Clearing woodland for agriculture (assarting) was forbidden. To the extent that it interfered with hunting, local people were banned from grazing their animals (agistment) in the Park or allowing their pigs to forage for acorns (pannage) or take anything that would sustain the deer – collectively known as 'vert' (Lasdun 1991: 15). They were not allowed to protect their crops by fencing, and could not use timber from the woodland for their buildings. The underwood was also protected and there were restrictions on gathering wood for fuel (estover).[12] The harshest enforcement of such constraints could be expected in smaller deer parks of the nobility that were used more frequently for hunting, whereas royal parks were larger and used less often. There is evidence at Woodstock that some of these restrictions were relaxed – for a fee – and the Forest Charter of 1217 during the reign of Henry III (1216–72) allowed free men to graze their domestic animals within the forest, except when the deer were fawning (Watkins 2014: 57). Tenants engaged in mowing the king's meadow (Figure 1.7) were allowed to turn their hobbled horses out into it. There is a record of 70 oxen, 12 cows, 6 horses and 164 pigs admitted to Woodstock Park in 1254, rising to 600 pigs in 1279 (Ballard 1908: 446–7).

A keeper was in overall charge of the administration of the Royal Park. Parkers or rangers were responsible for maintaining the deer herds and killing the deer to supply venison during the periods when royalty was not using the Park for hunting (Rowe 2019: 3).

---

12 http://www.foundationforcommonland.org.uk/rights-of-common.

Theirs was a high-status position, often held by second sons of landed families. Verderers were court officials dealing with poaching, regarders were responsible for the condition of forest boundaries and agisters dealt with pannage and the number of pigs allowed in the forest (Watkins 2014: 59).

The forester was responsible for wood, timber and charcoal production (Schumer 1984: 47). Evidence from the Calendars of Close and Fine Rolls shows that the trees that made up the enclosed forest were used for many purposes. Oak for building was highly prized (330 trees were needed to construct a house) and other timber by-products are recorded, such as laths for walling, pegs for stone roof tiles and shingles for wood roofing, piles for the causeway, paling for fencing, sluice gates for the fish ponds, hurdles for temporary fencing and various wooden implements and items of furniture. Reed for thatching was harvested from Woodstock Park along with wood for fuel, although demands to heat the Palace were met by supplies of firewood and charcoal from woods outside the Park (Bond 1981: 204).

It is possible that some areas of Woodstock Park could have been used for coppicing, which by the thirteenth century was widely practised on a seven-year cycle (Rackham 2011: 47). Such areas could not be used for deer or any other livestock since the new growth of the coppice would be grazed. In contrast, pasture woodland and pollarded trees are compatible with grazing animals and early commentators assumed this had been the practice at Woodstock.[13] Signs in support of pollarding come from the use of terms in the Rolls such as browse wood, robur, coupon (severed branches) and escaet (strippings) (Bond 1981: 207). Yet most of the ancient trees in the present High Park are maidens and show no signs of ancient pollarding. Their misshapen appearance is a consequence of the oak's propensity to shed limbs (Farjon 2017: 246).

Parks were multifunctional, providing grazing, fuel, recreation and social display (Rowe 2019: 24). There are records of an eyrie for falcons in 1250, and from 1254 to 1361 the Royal Stud was situated within the Park. As shown in Figure 1.8, there were fish ponds along the Glyme and its tributaries, reedbeds for thatching and meadows in the wide valley bottom providing pasture for grazing animals, once haymaking for winter deer fodder had been secured (Baggs et al. 1990: 442). The different functions led to subdivisions within the Park, for which there is evidence in illustrations (Figure 1.8) and in references to repairs to paling. Railing was also used to restrict livestock selectively, but allowed deer to leap over and fawns to creep under the rails – 'leap and creep' (Langton 2014: 125). Bond and Tiller (1987: 30) suggest some internal divisions may have been of the usual type – a ditch with a bank topped by a wooden fence, rather than a stone wall.

The 31 parks mentioned in Domesday had become 1,900 by the thirteenth century (Lasdun 1991: 5). The enclosed parks led to a new style of hunting, with the quarry flushed out of the covert by specially trained hounds and chased by horsemen towards stationary bowmen at a standing (Liddiard 2007: 5). There is documentary evidence from the sixteenth and seventeenth centuries of money being spent on standings at Woodstock (Baggs et al. 1990: 444). The deer were often hunted in the summer when they carried the most meat and were 'in grease' (Rowe 2019: 22), but the Park was also simply a larder providing important winter protein.[14] Deer could be fed and fattened more efficiently in parks, and records for Woodstock – thought to be conservative – show over 1,000 deer were taken in the period 1230–1300.[15] They were,

13 'We are indebted for some of the most picturesque trees in our oldest parts to a practice that once extensively prevailed of pollarding for "verte" – or firewood – boughs of oak and beech being lopped off for the deer to gnaw the bark, of which they are excessively fond: but no bough was permitted to be cut larger than a buck was able to turn over with his horns' (Marshall 1873: 23).
14 In 1250, 100 does from Woodstock were killed, salted down and sent to Winchester for Christmas (Bond and Tiller 1987: 25).
15 It has been calculated that 1,100 acres of park yielded 44 deer annually in the period 1234–63, whereas an unenclosed forest area would have needed to be ten times that size to yield the same number (Lasdun 1991: 6, quoting Oliver Rackham).

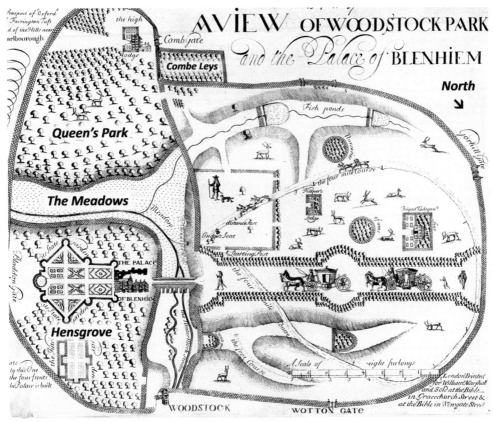

A VIEW OF WOODSTOCK PARK and the Palace of BLENHIEM

North

Figure 1.8.   Woodstock Park and the Palace of Blenheim, *c.* 1710; print for William Marshall, London, showing boundary walls and internal divisions. (Authors' annotations to Gough Additions, Oxon. a.79, Weston Library, Oxford)

however, expensive to keep through the winter when they were fed hay, for which pasture had to be reserved,[16] as well as ivy and browse-wood – cut branches from which the deer ate the leaves, or cut branches stored as 'leafy hay'. Deer could not be sold since they all belonged to the Crown; they used grazing land which could not then generate an income from agistment; they were difficult to contain and the pales and walls were expensive to maintain. Generally, where deer parks did persist it was because they were near the main mansion and seen as being its aesthetic adjunct (Rowe 2019: 27).

As a Royal Park, Woodstock survived the economic hardships in the fifteenth century when there was civil unrest and the climate was worsening. Many deer parks were disparked, but arrow heads from the fourteenth and fifteenth centuries found in the present High Park attest to its continued use for hunting (Bond and Tiller 1987: 38). Wild Boar were kept in the Park in the mid-fourteenth century, and hare and partridge were hunted in the sixteenth century (Baggs et al. 1990: 445). The Tudor and early Stuart monarchs appreciated Woodstock Royal Park and supported it with substantial investments for its upkeep. However, Elizabeth I (1533–1603) was less favourably disposed, having been held prisoner there in 1554.

Strategically positioned lodges were usually provided in deer parks, and at Woodstock these stood in laundes or clearings. In 1337 'a House called "Logge"' built of stone and maintained

---

16 'In 1577 Henry Lee the Ranger of Woodstock maintained a herd of 2,000–3,000 deer in the Park which was fed on hay from the Royal Meadows in the winter' (Bond and Tiller 1987: 25).

by the keeper/bailiff to provide accommodation for the parker was recorded for Woodstock Park. This may have been the original High Lodge, situated just west of the medieval manor/palace and so named because what is now Great Park was then known as High Park (Baggs et al. 1990: 442). Over the years other lodges were added,[17] and in 1586/7 a new one was built in 'The Straights' (now High Park) by the ranger, Sir Henry Lee.[18] After the Civil War, in 1649, this lodge was refitted by Sir William Fleetwood, and probably it was then it became known as High Lodge.

It was also Sir Henry Lee who reintroduced Red Deer to Woodstock Park (Baggs et al. 1990: 448). To contain these larger browser/grazers he imparked Bladon Wood and Heynes Close, which, together with The Straights, then formed Queen's Park (Figures 1.8, 1.9) and surrounded it with an eight-foot-high wall. The walled Red Deer park was later extended in Charles I's reign when large sums were spent in 1633–5 on merging Queen's Park with Hensgrove and the Meadows in the Glyme river valley (Baggs et al. 1990: 443). Extensive walling is depicted schematically in Figure 1.8.

Woodstock fared badly under the Commonwealth, and after Charles I (1600–1649) was executed, the manor was leased for £300 per annum and 1,000 deer were sold in 1650 for £1,000 (Marshall 1873: 209). Woodstock Manor House and Park were sold in 1652 to Lt-Col. Charles Fleetwood, Cromwell's son-in-law, for £20,665 12s 11½d (Taunt 1914: 69).

After the Civil War, the Parliamentary Surveys stated that most of the timber at Woodstock was only suitable for fuel, but 2,500 trees were reserved for the navy (Baggs et al. 1990: 447). The trees in 'The Straights' and Hensgrove were valued at £500 whereas those in the northern park, which seems to have been largely cleared of trees, were valued at just £110. While some felling took place,[19] there is a suggestion it was incomplete given the well-wooded appearance of High Park in Figure 1.14. Probably the ancient oaks in the southern Parks survived because they were already hollow and of little commercial value. However, the 'King's Oak' in High Park was destroyed by the Parliamentarian Commissioners, and so 'that nothing might remain that has the name of the king attached to it, they digged [it] up by the roots' (Taunt 1925: 22–3).

Around 1670, Lord Lovelace lived in the old gatehouse of the Palace – the only habitable building after the Civil War since many of the buildings had been demolished and the material sold off (Baggs et al. 1990: 439). He laid out a four-mile racecourse in High Park (present Great Park), which suggests that this part of Woodstock Park was not heavily wooded (Figure 1.8). High Lodge was either enlarged or rebuilt between 1674 and 1680 by John Wilmot, Earl of Rochester (Baggs et al. 1990: 337–8). Wilmot wanted to be keeper so he could use the lodge for private writing and entertaining (Cooper 1999: 110).

Although stag hunting was losing its popularity with the court, Woodstock still fulfilled its function as a venison larder, as shown by the Privy Council's demand in 1696 for a 'Brace of fatt Bucks of this Season' (Banbury et al. 2010: 53). Royal interest in Woodstock waned, and although Queen Anne (1665–1714) continued to enjoy hunting by pony chariot after her health declined, Woodstock was not one of her favourite sites (Lasdun 1991: 61).

---

17 High Lodge, Bladen Lodge, Gorrell Lodge and New Lodge (Bond and Tiller 1987: 32). The Parliamentary Survey of 1649 recorded that three of the lodges had more than six rooms (Cooper 1999: 10).

18 This could be the 'neat lodge for a ranger, sweetly seated on a hill' described in 1634: 'On the large high leads, which is over the goodlye fayre gatehouse, I had full prospect of that grand and spacious walled park, ye brave lawnes and waters' (Green 1951: 296).

19 'June 1660 A warrant to seise and to stay all timber cut down or stocked, grass or other product of Woodstock Park and Manor . . . and in spite of the order of Parliament May 17th 1660 against spoils, or cutting down of timber in the royal woods' (Marshall 1873: 223).

1.3. Buildings, landscaping and garden design

Maps through the ages showing the outline of Woodstock/Blenheim Park include several right-angle notches along the perimeter (Figure 1.7). It is an unusual feature for a deer park needing a costly enclosure, which was usually of a more economical circular shape, but there is evidence for this anomaly also in park pale traces in Windsor Great Park (Farjon 2017: 264). Such rectilinear elements suggest field boundaries that predate the Park enclosure. It is tempting to suppose that New Woodstock could lie atop a Celtic or Roman field complex and that the rectangular shapes and parallel alignments of parcels of land in High Park (Combe Leys, Figure 1.12; New Park, Figure 1.17) might correspond to lands originally managed from the adjacent Roman villa at Combe. In the same vein, the northern perimeter of Great Park is defined by the Roman track running parallel to Akeman Street.

Stone for buildings, boundary walls and roads was sourced locally. The locations of selected buildings, quarries, gravel and clay pits and brick works in relation to the underlying geology are shown in Figure 1.9. The fossiliferous Great Oolite was used in walling and lime production (Great Park), but was prone to frost damage and unsuitable for building, although

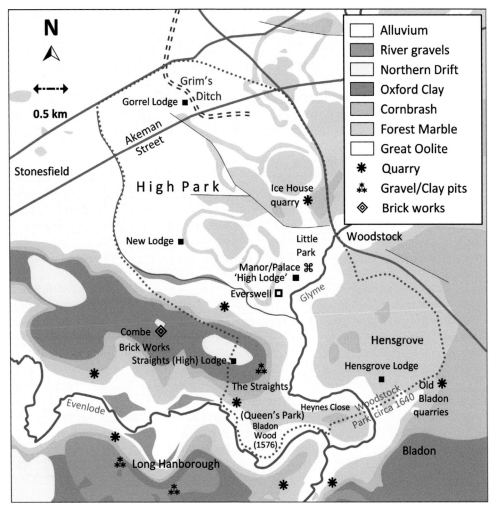

Figure 1.9. Geology and sources of building material in and around Woodstock Park, outlined in red, c. 1640. (Authors' illustration)

frost-split oolite 'slates' quarried in Stonesfield were a prized Cotswold roofing material; Forest Marble was quarried for building and paving (Bladon quarries, Ice House quarry); Oxford Clay was dug for bricks (Combe); drift deposits and river gravels were extracted for roads (Long Hanborough).

An early example of landscaping was the construction of a causeway across an artificially dammed portion of the Glyme in the twelfth century, creating an impressive approach to the Palace from Woodstock (Figure 1.7). Areas in the vicinity of the King's Manor/Palace served specific aesthetic functions. Henry II made a garden at nearby Everswell (Rosamund's Well), begun in 1166, which included 'three baths in trayne' fed by water from an existing spring (Henderson 2005: 76). At this period a lead-roofed alley and covered walkway joined the Woodstock Palace gardens to the Everswell gardens and spring. A herb garden was made in about 1230 (Bond and Tiller 1987: 38), and when 100 pear saplings were planted in 1264, the gardens were described as hedged and palisaded (McLean 1989: 111). In 1240, Henry III had walls and a fountain repaired at the Palace and an iron trellis made, presumably for vines, roses or other climbing plants giving shade (McLean 1989: 109, 111). Further work in 1248 created gardens on either side of the king's chamber and one stew (probably a water pleasance). His queen, Eleanor of Provence, was reported to delight in gardens and made several of her own, but of unknown design. Matthew Paris, writing in the mid-thirteenth century, describes how most castles, manors and monasteries had enclosed quadripartite gardens divided by gravel walks with a fountain in the middle and a shaded arbour from which to view the garden.

By the thirteenth century, in the time of the Plantagenets, the enclosed park, where the complex etiquette of the hunt played out in the managed laundes and woodland, was being seen as an interface between the safe inhabited world of house and garden, and the wildwood forest (Creighton 2013: 150). Aesthetic considerations meant that the Park as a setting for the residence became increasingly important – and parks set on rising ground, such as the present High Park, were particularly valued (Creighton 2013: 130). Edward III (1312–1377) and his queen, Philippa of Hainault, favoured their Palace at Woodstock, overlooking the Queen's Pool, named after Philippa (Taunt 1925: 67). In 1354 a balcony was constructed for their daughter Princess Isabella to give her an unrestricted view of the Park (Taigel and Williamson 1993: 33).

Henry VII (1457–1509), the first Tudor king, spent lavishly to bring piped water from High Park via a vaulted tunnel and wooden trough on stone piers across Combe Bottom to Woodstock Palace (Baggs et al. 1990: 438). In the following centuries the Palace slipped into gradual decline. When Princess Elizabeth was held prisoner in 1554 there was a wardrobe court with a fountain, a privy garden and a larger garden called Lockley Green. In Elizabeth's reign it is likely that Lodge Green, the gardens of the adjacent High Lodge kept by the queen's ranger, Sir Henry Lee, were in better condition than those of the Palace, which sprawled over a 3-acre site (Baggs et al. 1990: 439).

After the Restoration in 1660 – when Woodstock had reverted to the Crown – 60 acres of 'furze ground' called Combe Leys was added to Woodstock Park (Figure 1.10). It cost £616, which included £80 for a new wall to integrate it and £50 in compensation to the tenant for the loss of common rights (Bond and Tiller 1987: 61). At this time the principal character of Combe Leys was likely to have been pasture woodland, since ancient oaks are present there now.

The western boundary of Blenheim Park had further additions in the eighteenth century (Figures 1.10, 1.18) most notably in about 1780 when Workhouse Close, Furzy Leys and parts of Old Assart Furlong were added, which together became known as New Park. Reflecting its history of long-term cultivation, there are no ancient oaks in New Park. A rectangular portion of 22 acres (9ha) enclosed by a ditch and bank contains about ten pillow mounds, constructed to accommodate rabbits in a sizeable grazing area (Figures 1.11, 1.21). Such an elaborate enclosure cutting across the three parcels taken into the Park may signify a late eighteenth-century expansion of an earlier enterprise, and perhaps centred on rabbit fur as much as meat. Warrens or coneygarths were often enclosed in this way, for protection from

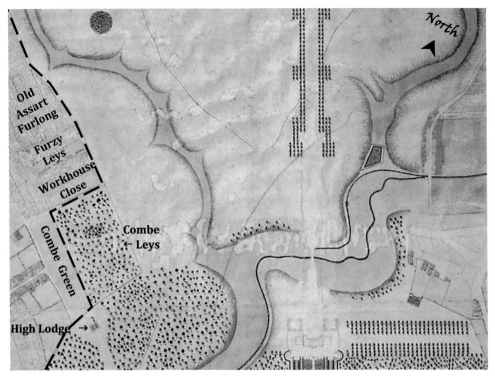

Figure 1.10. Detail of a 1709 survey map attributed to Charles Bridgeman showing wooded portions of High Park and fields outside the Park boundary at that time – dashed line (authors' annotations). Blenheim Palace Heritage Foundation

Figure 1.11. Two eroded pillow mounds up to 8m × 3.5m × 1m high, protruding through leaf litter on sloping ground in New Park and forming part of a scheduled monument (https://historicengland.org. uk/listing/the-list/list-entry/1009418). (Photograph by T.H. Moller, 7 April 2019)

poachers and predators and to discourage the rabbits from straying (Williamson 2007: 65).[20] Rabbit burrows were also a hazard for horse and rider, adding a further incentive to their containment. Even so, in the nineteenth century, crop damage in Combe parish from animals escaped from Blenheim Park was principally caused by rabbits (Baggs et al. 1990: 76).

As with Fallow Deer, rabbits were introduced on a small scale by the Romans but then reintroduced by the Normans who prized them for their meat and fur; the first documentary evidence is from 1135 (Williamson 2007: 11). The rabbits at that time were poorly adapted to the damp English climate, which made them prone to foot rot – hence the need for pillow mounds. These are typically low, flat-topped structures about 1.5 m high, 9 m long and 4–7 m wide, created on a sloping site in light soil so that the excavated earth would fall away (Williamson 2007: 32). In some waterlogged areas stones were first scattered on the land surface to aid drainage. Some mounds were simple dumps of earth with no internal features; others had a burnt layer at ground level from clearing the vegetation. More elaborate examples had artificial rabbit runs made of stone tunnels laid out systematically and then covered with earth. Dispatching the rabbits was achieved by introducing a ferret on one side of the mound while netting the other side, making it easy to club the creatures as they emerged.[21]

1.4. Ducal park management

After defeating the army of King Louis XIV at Blindheim, Bavaria in 1704, John Churchill, 1st Duke of Marlborough (1650–1722) was granted the Royal Manor of Woodstock in 1705 and Parliament voted funds for a new residence on behalf of a grateful nation. The gift included the royal demesnes of Hordley, Wooton, Stonesfield, Old Woodstock, Handborough, Combe and Bladon (Ballard 1908: 424). John Vanbrugh was appointed architect/surveyor and a grand design was put forward, set on a north-west to south-east orientation to give the longest possible axis. The original plan had an imposing entrance from the Ditchley gate with a two-mile approach to the Palace. The axis continued through the house, out through the walled Great Parterre and Bastion Garden to Bladon church (Figure 1.12) (Lasdun 1991: 86). The hexagonal Bastion Garden, also known as the 'Wood-work', covered approximately 36 acres and referenced Marlborough's military background. It gave views into the Park, but was not finished until 1723, a year after the duke's death. Although the axial layout of the garden was probably conceived by Vanbrugh, the duke trusted the royal gardener Henry Wise to oversee its execution, as expressed in a letter to his wife: 'for the Gardening and Plantations I am at ease being very sure that Mr Wise will be diligent' (quoted in Bapasola 2009: 19).

Lower Park, where building work started on the Palace in June 1705, must have been wooded, since Vanbrugh suggested Mr Wise should 'open ye Ground, cut down Trees etc in order to laying the Foundations' (Green 1951: 315). The felling was compensated for by the massive tree planting that Wise then embarked upon. Tree planting was seen as patriotic, prestigious and necessary after the depletions of the Civil War and a series of disastrous storms.[22] The northern and eastern avenues alone were planted with 1,600 elms (Bapasola 2009: 16). In the Wood-work Henry Wise supplied 9,357 hedge yews as well as '5,900 hornbeam,

---

20 Coneygarth rather than 'warren' was used to describe rabbit housing until the fourteenth century, and the term 'Freewarren' does not refer to rabbits but to land over which the individual is granted the right to take game (Williamson 2007: 17).

21 Two other possible mound sites are known in Blenheim Park, near to Ice House Quarry (Figure 1.9); this is probably the scene of the aborted 1649 poaching of coney-burrows when the fleeing men left their ferret behind (Bond and Tiller 1987: 62).

22 In 1665 Pepys recorded a severe storm and Evelyn another in 1690; on 2 November 1703 the Great Storm uprooted rows of elm, oak and Beech and whole orchards (Lasdun 1991: 52).

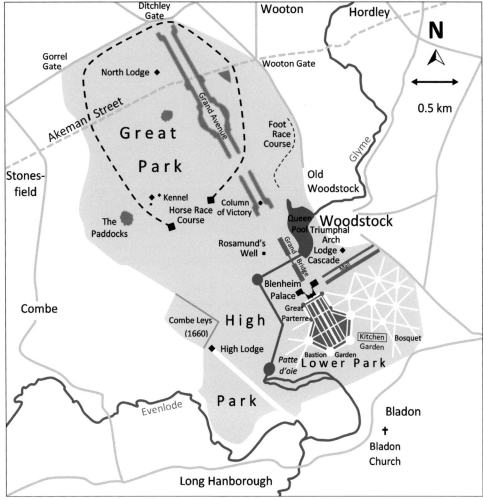

Figure 1.12. The formal layout of Blenheim Park on a Ditchley Gate – Bladon church axis, *c.* 1730. (Adapted from Baggs et al. 1990: 462)

privatt and sweet bryer'. Stephen Switzer, who worked for Wise, later said that 'out of about ten thousand Hedge-yews etc. that were planted under my direction at Blenheim in 1706, there were not two hundred that fail'd' (Bapasola 2009: 21). There is good evidence that the existing forest trees were preserved where possible.[23] An ancient oak still survives on the site of the Bastion Garden, and the 1719 plan (Figures 1.13, 1.20) shows such trees in the Bosquet; there is also a contemporary account from Viscount Percival that decribes 'private walks cut thro an old wood to which a great deal of new plantation has been added'.[24] Percival's remarks probably apply to the Bosquet, which was laid out with paths cut in a star pattern; existing trees were incorporated into the design wherever possible, or even left encroaching on paths. As in the Wood-work, some survive to this day. As well as bringing large trees from his Brompton

23 Vanbrugh himself was concerned about what the Clerk of Works, Tilleman Bobart, was doing 'About the Forest Trees in the Quarters of the Woodwork I mean as to bring in Lines of 'em every where behind the Hedges, for I take that to be the grand point of all' (Dalton 2012: 113).

24 *Letter Book of Lord Egmont*, 1724, Viscount Perceval to Daniel Dering (August 1724) British Library Add Mss 47030 (quoted in Dalton 2009: 17).

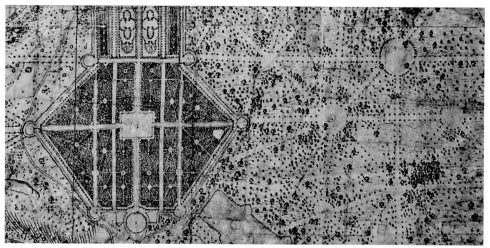

Figure 1.13.   Detail of a 1719 plan of the Bastion Garden (left) and Bosquet (right) in Lower Park, from 'A Plan of Blenheim House and Gardens 1719.' (Blenheim Palace Heritage Foundation)

nursery,[25] Wise may have moved some mature trees to accommodate the formal pattern.[26] He was responsible for the gardens until 1716 and his total charge was £14,559, of which £3,050 was for plants and trees.

The early eighteenth century was the time when the ambitions of man imposing order upon Nature were being called into question and not everyone was admiring of the formal gardens at Blenheim (Lasdun 1991: 85). Alexander Pope wrote:

> 'Grove nods at grove, each alley has a brother
> And half the Platform just reflects the other'

To the west of the Bastion Garden the three-pronged *patte d'oie* was probably of newly planted trees, but the 1719 plan suggests a few of the trees of the ancient deer park were preserved around it. It focused the gaze onto the steep river bank and the densely wooded slope of High Park, which can be seen from further upstream beyond Vanbrugh's bridge across Col. John Armstrong's canalization of the River Glyme (Figure 1.14).

The 1st Duke died in 1722 and his title passed to his daughter Henrietta (1681–1733), who became the 2nd Duchess, but her son pre-deceased her, so she was succeeded by her nephew, Charles Spencer. He took up residence at Blenheim after his grandmother Sarah's death in 1744.

The 3rd Duke (1706–1758) was interested in gardening and, on the evidence of the John Spyer survey of 1763 undertaken for Lancelot Brown, was probably responsible for removing the – by then – unfashionable Grand Parterre and the western bastion walls around the Wood-work, thus merging the Park and garden on that side. The 3rd Duke was also drawn to arboriculture, and the first known planting of Lombardy Poplar in England was at Blenheim in the 1750s (Bapasola 2009: 48).

---

25 The building accounts mention '52 standards lawrells, 18 cedars of Lebanon, 186 large elm and sycamore. 2,219 large espalier limes, 2,599 smaller size ditto, 2,000 small hedge hollies' (Bapasola 2009: 22).
26 'his Grace bad him consider, he was an old man and could not expect to live til the Trees were grown up; and therefore he expected to have a garden as it were ready made for him. Accordingly, Mr Wise transplanted thither full-grown Trees in Baskets, which he bury'd in the Earth; which look and thrive the same as if they had stood there thirty or forty years' (Macky 1722: 117).

Figure 1.14. Vanbrugh's Grand Bridge spanning Armstrong's canalization of the Glyme. 'A North East view of Blenheim House and Park in the County of Oxford' published by John Boydell, 1752. (Gough Additions, Oxon. a.79., Weston Library, Oxford)

The 4th Duke (1739–1817) had engaged Lancelot Brown – the most famous exponent of the Landscape Movement – to work at his house at Langley Park, Buckinghamshire, but prioritized his work at Blenheim. Although Brown completed the removal of the Bastion Garden, he retained the *patte d'oie* and most of the Bosquet – the western end was made into a Flower Garden. He considerably increased the density of planting along the Lake margins of High Park; its wild nature acted as a foil to the smooth landscape he was creating in the foreground (Bond 1981: 201). The sweep of grass resulting from the levelling of the Bastion Garden was framed by cedars and separated from grazing animals by a ha-ha (Figures 1.15, 1.17). Tree planting in the wider Park was an important part of Capability Brown's scheme. The Northern Avenue was cut into aesthetically pleasing clumps and new groves, such as the Fourteen Acre Clump, were planted to form coverts for game birds and foxes (Figure 1.17). This led to the development of tree nurseries on the estate, and trees would later become an important source of income (Bapasola 2009: 78).

Under the day-to-day direction of Brown's project foreman, Benjamin Read, Armstrong's canal (Figures 1.14, 1.18) was flooded, along with the lower portions of Vanbrugh's bridge and the tributary of the Glyme, which forms the north-eastern perimeter of High Park. The river banks near Rosamund's Well, which had already been reshaped by Townesend and Peisley for Sarah, Duchess of Marlborough, were further sculpted to make the slopes less severe. However, it was the 4th Duke who lowered the valley rim to allow a view of the water from the Palace (Baggs et al. 1990: 466). In damming the river to make the Great Lake, Brown was able to create the Cascade (Figure 1.16). The wildness of High Park and Brown's carefully placed rocks on the far shore provided a suitably 'Picturesque' background for the tumbling white water, and a striking visual contrast to the smooth grassed slopes leading down to the tranquil lake waters in other parts of Blenheim Park. Benjamin Read continued to implement the scheme in the English Landscape Style after Brown's contracts ended in 1774. 'But, after all, the water is the capital feature and principal object of Blenheim: it adorns, enriches, enlivens, and connects the whole' (Combe 1794: 89).

Brown's landscapes were designed to be enjoyed on a circuit – on foot close to the Palace and to the Cascade, but in a carriage or on horseback for the wider estate. As shown in Figure 1.17, High Park was an important part of a circuit route that gave selected views of the Palace and its setting, framed by the woodland. Far from cutting them down, Brown would

Figure 1.15.   A view of Brown's landscape from Blenheim Palace towards Bladon. (Authors' image)

Figure 1.16.   Brown's Cascade against the dark backdrop of High Park. (Authors' image)

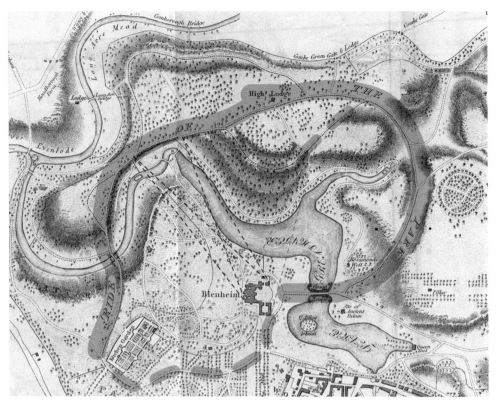

Figure 1.17. A map of High Park published in 1806 featuring a 'RIDE OF THE PARK' past High Lodge and giving views of Blenheim Palace (authors' annotations). (Blenheim Palace Heritage Foundation)

have valued the ancient trees for supplying variety – their ancient and irregular forms giving the scene a frisson of 'terror', which was part of the emerging 'Picturesque' ideal.[27] For this reason, High Lodge was probably Gothicized either by Brown or at his suggestion, since this style of architecture was seen as suitable for a wild setting and creating a foil to the smooth classicism of the rest of the parkland. The Lodge would probably have been used as a destination in which to take refreshment on the circuit.

From a management perspective, the earlier story of Woodstock Park has much to do with royal whim and favour, whereas the common thread running through the more recent history of Blenheim Park is one of financial constraints and the challenge of wealth creation.

The 4th Duke (1739–1817) was ranger of Wychwood Forest and thus responsible for the deer – approximately 100 were dispatched each year and six sent to the king. Deer were still valued for sport as well as a food source, and in 1781 *The Diary of Silas Neville* recorded over 3,000 head of deer in the Park. Deer hides and sheepskin were the raw materials for the flourishing glove-making industry centred on Woodstock and with hundreds of outworkers

---

27 Another reason for preserving the ancient trees is suggested by William Gilpin's comment on another Brown project – Cadlands in Sussex. Here Brown retained old trees to give a 'spurious air of antiquity' to a newly created scene. Gilpin goes on to say 'abundance of old timber gives the house, tho' lately built, so much of the air and dignity of an ancient mansion that Mr Brown, the ingenious improver of it, used to say "it was the oldest new place he knew in England"' (Brown and Williamson 2016: 104). These remarks could equally well apply to Blenheim.

in surrounding parishes (Schulz 1938: 148). Thus, High Park would have been an important part of the management of deer at this time, giving another reason for its preservation.

In 1767 the 4th Duke bought the Lince (Figure 1.18) as an addition to the Park and a new pheasant ground was established, densely planted with trees by Capability Brown's foreman, Benjamin Read (Bapasola 2009: 80–1, 86). This purchase also allowed the Glyme to be widened with the construction of an embankment and a second cascade where the Glyme joins the Evenlode. Other large acquisitions of lands in nearby parishes were steadily added to the ducal estate – not to be taken into the Park, but to boost income for its upkeep. Following a string of parliamentary enclosure awards between 1767 and 1804, the consolidated estate included most of Bladon/Hensington and Combe parishes and substantial holdings in surrounding parishes (Baggs et al. 1983: 186, 272; Baggs et al. 1990: 25, 87, 125, 169). The combined acreage far exceeded that of Blenheim Park, which covers about 2,400 acres (970ha).

The 4th Duke made few alterations to the serene park landscape Brown had created for him, and it is essentially the scene we look upon today. He did, however, develop the area on

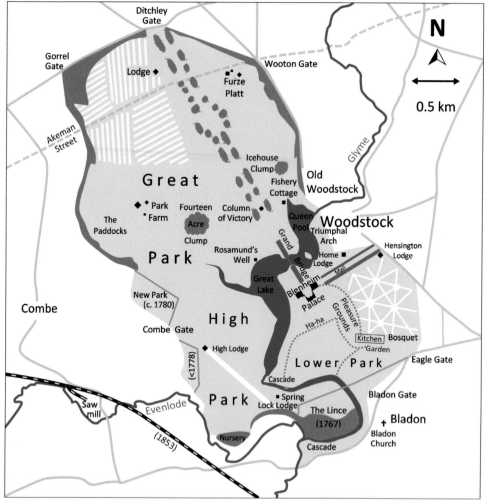

Figure 1.18. Imprint of the Landscape and Picturesque styles on the layout of Blenheim Park, *c.* 1820. (Adapted from Baggs et al. 1990: 468)

both sides of the Cascade as a Pleasure Garden and made an aviary in his flower garden in 1812. He passed on his love of gardening to his son – but the extravagant spending by the 5th Duke (1766–1840) on plant collecting almost ruined the dynasty. He transformed the lawns east of the Palace into 'a harlequin jacket of little clumps and beds' (Bapasola 2009: 106) and further extended the Pleasure Gardens near the Cascade, making separate specialist gardens (Rock Garden, Botany Bay Garden, Chinese Garden). The Rock Garden is the main one that impacts High Park, being on the slope down to the Cascade and Great Lake – an area first developed by the 4th Duke. The ingenious pivoting boulder that guarded the entrance to this garden is the derivation of the name of the nearby Springlock Lodge. The 5th Duke also created a 12-acre arboretum in front of the aviary of 'choicest and most beautiful Forest Trees' (Bapasola 2009: 112). This did not dissuade him from felling existing trees for their timber value, to such an extent that his son and heir tried to stop him through court action.

Despite his engagement of the courts to curb his father's depredations in the Park woods, it was ironically the 6th Duke (1793–1857) who authorized the biggest tree felling operation in Blenheim's history. He had been granted a mortgage by Parliament, and to meet the repayments he was allowed to fell timber up to the value of £10,000 (Bapasola 2009: 115). The corn mill on the Evenlode was converted to a saw mill, fitted with steam-driven machinery and provided with a siding to the new railway line passing through Combe parish (Figure 1.18) (Baggs et al. 1990: 91). Many of the trees left in Henry Wise's bosquet to the south-east of the Palace as well as one of the three rows of elm trees that made up the Eastern Avenue fell to the axe.[28]

The 7th Duke (1822–1883) created formal gardens east and west of the Palace, a rose garden on the site of the Chinese garden and had a menagerie and a large orchid collection, but seems to have left High Park alone. His son, the 8th Duke (1844–1892), was also passionate about orchids and extended the hot houses, and, like his father, shored up his finances by selling off family heirlooms rather than timber.

As well as employing the French garden designer and architect Achille Duchêne to remodel the north, west and east gardens at the Palace, the 9th Duke (1871–1934) planted over half a million trees and shrubs throughout the estate between 1893 and 1931 (Bapasola 2009: 148). Some in High Park were planted in rows as if intended as a cash crop, but in the wider estate he made numerous new clumps and replanted the eastern and northern avenues, in which about 2,300 trees were planted. Having made this heavy investment, he sought to protect his legacy from premature demise.[29]

The First World War saw a sale of timber from around the estate and ended the managed existence of deer in Blenheim Park – 'in the interests of food supply' (Bapasola 2009: 149). Deer numbers had already declined steeply – in 1867 there were 779 Fallow Deer recorded at Blenheim (Woodward 1982: 5). However, this was still a substantial number, and the 1897 guide says 'The Park is well stocked with herds of fallow deer, while the number of rabbits is legion' (Vincent 1897: 14).

The 10th Duke (1897–1972) introduced softwood (conifer) plantations from the 1950s, and some of these can be seen in High Park. The 11th Duke (1926–2014) was keen to retain the Brown parkland and to conserve the ancient woodland in High Park.[30] To this end, he drew

28 A poster in the Blenheim Archives advertises the sale of 220 maiden oak timber trees, and a record shows that the sale of 122 elm trees brought in about £1,000.
29 The 9th Duke did not mince his words in a stark message to his successors: 'Any man who cuts these trees down for the purpose of selling the timber is a scoundrel and deserves the worst state that can befall him' (Spencer-Churchill 1892–1931).
30 In this book, we accept the current definition of ancient woodland adapted from Peterken (1993: 12) by the Woodland Trust and the Forestry Commission: 'Ancient woodland is defined as land that has been under trees since at least 1600 CE.'

up a Heritage Management Programme on a 200-year cycle. Dutch elm disease led to clear-felling the avenues and replanting with lime for the Northern Avenue, and plane and lime trees for the Eastern Avenue.

1.5. The present day

'The present day' equates to 'in living memory', a time that is neither this moment, nor yet history. It is the period which provides us with the most tangible links to the landscape. Thus, the aerial photographs of High Park from 1961 (Figure 1.19) together with the laser scanning image of the twenty-first century (Figure 1.20) can serve as bookends bracketing the last 60 years. These images provide us with information that can potentially be verified on the ground or by personal corroboration. The observation of Rackham (1997: 25) is pertinent at this juncture: 'I cannot analyze the historic landscape without noticing how much of almost every aspect of it has been lost since 1945.'

Figure 1.19.   Composite aerial photograph of High Park from June 1961, with the parallel rows of the 9th Duke's oak plantation (1898–1900) top left. (Oxfordshire County Council imagery)

Figure 1.20.   A 1719 map of the Parterre (P), the Bastion Garden (BG) and the Bosquet (B) with individual trees marked as dots is superimposed on a Google Earth image of Lower Park *c.* 1945. Post-war tree canopies appear as dark patches; 71 extant ancient and veteran oaks are marked (⊙).

The grid pattern of the 9th Duke's oak plantation in High Park shows up clearly in Figure 1.19, as does the mosaic of pasture woodland to the south. The trees tell their own story, whether they be alive or present only as stumps or as gaps in a regular matrix of planting. Thus, the vista that gives a view of High Lodge from Blenheim Palace was not cut until after the Second World War since it is not present on aerial photographs from 1945, and a few large oak stumps still visible in the grass are further confirmation of relatively recent felling.

The 1st Duke occupied High Lodge while the Palace was being constructed in the early eighteenth century, and it seems that rather than providing a view from High Lodge, the Marlboroughs' purpose for a narrow vista created towards the south-east was to impress visitors approaching from Bladon, across the Glyme. The narrowness of the vista in the aerial photograph also shows that the adjacent long plantation of conifers encroached upon the ancient pasture woodland (Chapter 3, Figure 3.1). Confirmation on the ground comes from a search in the long plantation, which has revealed the existence of many large stumps of ancient oaks (Chapter 2, Figure 2.6). Although High Park together with the Brownian lakes were designated as a Site of Special Scientific Interest in 1956, these oaks must have been felled much later since they are still seen standing south-west of the vista in the aerial photograph taken in June 1961 (Figure 1.19). The measurements of tree girth and estimations and trunk size from surviving stumps allow a glimpse of the likely sequence of events in the past and the probable growth pattern and age of the ancient oaks in High Park. These questions and their implications are fully explored in Chapter 2.

The superimposed image of a 1719 plan of the formal gardens onto the relatively modern post-war RAF photograph of Lower Park (Figure 1.20) serves to illustrate how the precision of seventeenth-century surveying, and the present-day distribution of ancient oaks,[31] match up as faithfully as a latter-day parch mark.[32] Judging by the low density of extant ancient and veteran oaks (71 trees), many more specimen trees were evidently alive there in the 1940s.

---

31   Locations determined and girths recorded by Aljos Farjon.
32   Actual parch marks revealing the position of the Grand Parterre were prominent in July 2018 (https://www.bbc.co.uk/news/uk-england-oxfordshire-44980314).

Remnants of Henry Wise's Bosquet (Figure 1.13) can just be traced as stunted alignments. The Eastern Avenue of elms at the top of Figure 1.20 was planted in 1901 by the 9th Duke, but had succumbed to Dutch elm disease by 1976.

In the modern LiDAR image shown in Figure 1.21, the geomorphology of High Park is laid bare, highlighting the drainage pattern of ancient dry valleys and the 'misfit stream' of the Evenlode. Another relic is a sink hole in High Park caused by a collapse in the underlying limestone aquifer and creating an incipient dry valley. Karstic dissolution is also consistent

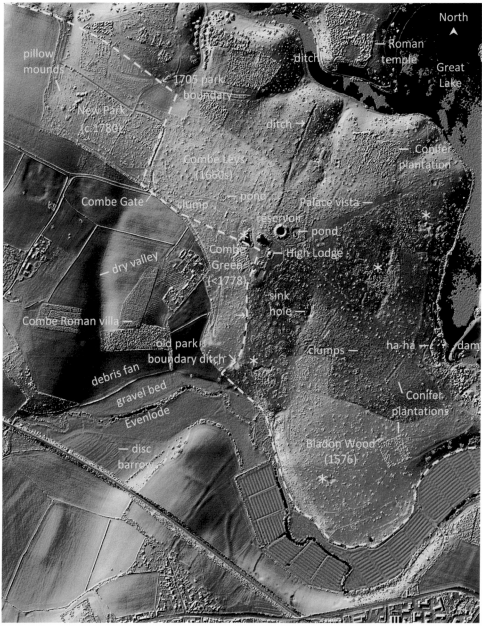

Figure 1.21.   LiDAR image (https://houseprices.io/lab/lidar/map?ref=SP43041554) of High Park with four superimposed colour-washed parcels of land acquired from 1576 to 1780. Dashed line shows the boundary of Blenheim Park in 1705. (✱ quarries). (Authors' annotations)

with the cover of Northern Drift deposits on the Combe Plateau that is rich in transported quartzite pebbles, but containing no limestone clasts; they were once present, but have since dissolved (decalcified). Ancient and modern human-made features can also be seen in the image, including a Bronze Age disc barrow (Harden 1947: 175), a Roman temple and various earthworks.

More recent signs in the landscape include a prominent depression bisecting High Park from south to north. Field-walking has revealed that the middle section, each side of High Lodge, is occupied by two streams running north and south. The two distal sections become human-made dry ditches, of which the southern is an old park boundary. The northern ditch is more prominent and ends abruptly as a result of the valley sides being softened during the creation of the ornamental Great Lake. This ditch therefore

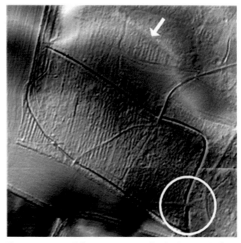

Figure 1.22.   Colour-contoured LiDAR Digital Terrain Model (DTM) of New Park. A break in slope (arrow) marks the line of the 1705 park boundary. (Environment Agency Survey Open Index Catalogues)

predates the eighteenth-century landscapers and has been suggested to match historical descriptions of a water conduit once supplying the old Palace in the reign of Henry VII. However, the northern ditch runs along a spur rather than following a more logical course in the adjacent valley. The ditch may more plausibly be interpreted as an old internal boundary and a continuation of the known old boundary ditches separating Combe Leys and Combe Green from the Park (Figures 1.8, 1.21). Brown's sculpting of the slopes down to the Lake on both sides erased the ditch, but a short segment can still be seen at the top of the opposite slope. There are many historical examples of internal boundaries, such as in the description of Henry Lee's retaining wall of over 2.4km around 'Queen's Park' constructed in the sixteenth century to confine Red Deer. All internal walls were probably removed and the stone reused in park perimeter walls in the early eighteenth century. According to a 1727 memorandum of changes at Blenheim, walling was pulled down and Combe Leys 'laid all open together' with the Park (Baggs et al. 1990: 441).

More subtle signs of old boundaries can also be deduced from the LiDAR imagery. In Figure 1.22 a small misalignment in the ditch just north of Combe Gate (circled) may suggest that the gate was repositioned when New Park was acquired in about 1780, and that the ditches on either side of the discrepancy are of different ages. The individual parcels of land indicated in Figure 1.10 that made up New Park can still be seen, as well as the position of the previous (1705) park boundary (arrowed).

1.6. Summing up

High Park is essentially a fragment of the medieval Wychwood that was relatively undisturbed until the 1960s. The presence of ponds around High Lodge and parallel short drainage trenches visible in Combe Leys all reflect the moist conditions of a perched water table formed by the relatively impermeable cap of Oxford Clay. It was partly the poor agricultural potential of High Park that favoured the creation and maintenance of pasture woodland for hunting and an environment conducive to the long-term survival of oak-rich woods and glades. Moreover, the contribution of woodland towards the economic viability and self-sufficiency of the Royal Park and the later ducal estate had long been recognized.

Figure 1.23. '[H]ere and there a huge oak-tree, so incredibly ancient that it may well have seen the Plantagenets riding to the chase . . . and every spring these monsters put forth a few pale leaves among their stiff, dead, scarecrow branches' (Shelmerdine 1951: 19). (Drawing by Iain Macnab)

In contrast, much less survives of the community associated with the few widely scattered ancient and veteran oaks remaining in the open grassland of Lower Park (Figure 1.23). A natural understorey has been absent for more than three centuries, and in consideration of the visiting public, any dead, unsafe or unsightly trees are felled, removed and replaced by replanting, in accordance with the guidelines of the Parkland Management Plan (PMP 2014: 78) and the Management Plan for High Park (Mottram and Kerans 2014: 37). Beginning with the appointment of consultants to advise on dealing with the impact of elm and Beech disease in the 1970s, the task of reconciling conflicting interests and setting management policy for a multi-functional Park has passed from the dukes to a team of professional experts reporting to a Steering Group (Historic Landscape Management Ltd 2017: 72).

Acknowledgements

We thank Dr Alexa Frost, archivist at Blenheim Palace, for her generous support, and Dr Anthony Cheke for drawing our attention to the online availability of high-quality aerial photographs of High Park.

CHAPTER 2

The Ancient Oaks

Aljos Farjon

2.1. The survey

The native species of oak (*Quercus robur* and *Q. petraea*) are considered in this chapter. Of these, only *Q. robur* or Pedunculate Oak has been found in High Park. A number of oaks have been recorded in this survey simply as 'oak' but this is done because the two species can only be separated with confidence when there is foliage or acorns on (or under) the tree, making winter identification more difficult. Tree survey data for High Park by the Blenheim Estate (see below) do not mention *Q. petraea*, so it is reasonable to assume that all native oaks on this site belong to *Q. robur*.

In February 2014 I began to record all oaks with girths of 6m and larger, following the Ancient Tree Inventory (ATI) guidelines as closely as possible. Later, I added oaks with girths between 5 and 6m as well as dead oaks, standing or fallen, with a minimum size of 5m girth, occasionally assisted by Benedict Pollard. (All oaks with girths of ≥5m in High Park appeared to be ancient or veteran, while smaller oaks may or may not fall in these categories; see below). This work was considered completed in May 2018. Halfway through my efforts I was given the results of a comprehensive survey concerning 'veteran trees' of High Park carried out from 2001 to 2006 (Blenheim Estate data), which had recorded oaks from 1.67m girth to the largest found with a girth of 10.32m. As this survey included oaks of smaller size as well as those that I was interested in, I decided to continue my own survey but using the Blenheim Estate's tree survey as an additional dataset for analysis. Permission was given by Roy Cox to upload my records, including photographs, to the ATI website (https://ati.woodlandtrust.org.uk/) where they can be viewed, marked as 'Private, not visible from public access' or 'Public, partial access' where they stand close to the road alongside and through High Park that is partly public footpath and partly permissible access for walkers. I had also noted large old oak stumps cut low above ground level in the vista from Blenheim Palace towards High Lodge. In April 2019, with Alison and Torsten Moller, a search was made there and in the long plantation, where more such stumps were found and recorded. With these, only a rough estimate could be made to establish their girth when still standing. All recorded oaks were georeferenced using GPS readings with a Garmin *etrex* 12 channel GPS device to a horizontal accuracy of 5–7m whenever possible. Girth in cm was taken by measuring tape at approximately 1.50m above the ground where possible, avoiding gross irregularities if feasible. In addition to the ancient and veteran oaks in High Park, I also completed the ATI records by Owen Johnson in May 2008 of oaks with girths of 5m and larger in adjacent Lower Park, a part of the landscaped Blenheim Park.

Aljos Farjon, 'The Ancient Oaks' in: *The Natural History of Blenheim's High Park*. Pelagic Publishing (2024).
© Aljos Farjon. DOI: 10.53061/WPSJ5341

2.2. The numbers

The total of recently recorded oaks in High Park, alive and dead, with a minimum girth of 5m is 266. Of these, 219 were recorded as alive and 47 as dead (standing or fallen). Of the dead oaks 9 were recorded as 'dead-sawn' (all fallen) and these are assumed to have been moved from Lower Park to High Park in a clean-up action of the landscaped park by order of the duke (Nick Baimbridge, Head Forester, pers. comm. 2018). There were, with the single exception of a smaller oak (Figure 2.15), no *in situ* stumps near any of the sawn oak trunks found in High Park. The total of recorded large ancient and veteran oaks originating in High Park by May 2018 was therefore 266. There has been a long-standing policy, probably dating from the first notification of the site as (part of) a Site of Special Scientific Interest in 1956, to leave dead oak wood *in situ*. The large oaks that have left stumps in the Palace Vista were removed between 1945 and 1961 (see Figure 1.19, page 34), while those in the long plantation were felled after 1961. This historical evidence gives us an approximate date for these cuttings, for which no saws were used. Although now partly decomposed, these stumps also provide a point of reference for establishing how long large oak wood can remain as a physical entity on this site. In April 2019 there were 8 oak stumps found in the Palace Vista and 17 oak stumps in the long plantation. The girths of these oaks when still standing could only be estimated roughly from the diameters measured across the basal remains, ranging from a girth of >4m to <6.50m. Given the uncertainties about the girth estimates, it was decided to record these stumps but not to add them to the database of ≥5m girth veteran and ancient oaks. Their locations are shown in Figure 2.6.

In Table 2.1 numbers are presented for different characteristics of all 266 recorded large ancient and veteran 'indigenous' oaks in High Park. The vast majority are maidens and only five are true pollards, oaks that were cut repeatedly well above the ground and allowed to regrow limbs from that level. Storm damage can have similar effects, often on ancient or veteran oaks with hollowing trunks, and when this happened a considerable time ago, with large regrown branches present, it can be difficult to make out the difference. However, it is rare for this regrowth to recur at the same level under natural circumstances. A total of

Table 2.1.   The 266 recorded ancient and veteran oaks in High Park, divided into 11 size classes (type 1 in the database), with some of their characteristics. 'Maiden' includes 'pollard form (natural)' in the records. 'Dead' includes standing and fallen in situ oaks and excludes trunks transported to High Park as well as in situ stumps of oaks cut in the early 1960s.

Size class	Maiden	Pollard	Living	Dead	Ancient	Veteran
1 (5–5.5)	65	0	42	23	13	52
2 (5.5–6)	58	0	46	12	13	45
3 (6–6.5)	63	2	61	4	26	39
4 (6.5–7)	41	0	37	4	30	11
5 (7–7.5)	14	3	15	2	17	0
6 (7.5–8)	8	0	8	0	8	0
7 (8–8.5)	6	0	5	1	6	0
8 (8.5–9)	1	0	1	0	1	0
9 (9–9.5)	2	0	2	0	2	0
10 (9.5–10)	1	0	0	1	1	0
11 (>10m)	2	0	2	0	2	0
Total	**261**	**5**	**219**	**47**	**119**	**147**

16 'pollard form (natural)' oaks were recorded, and these are considered maiden trees because we are interested in the proportion of human-made pollards, which is a mere 2% of ≥5m oaks in High Park.

The distinction between 'ancient' and 'veteran' is not clearly defined because we are dealing with a gradual process in which dead wood, hollowing and the scars of damage increase over time, together with crown reduction compensated (or not) by regrowth, leading from a state of mature optimum through over-mature and senescence to death and decay. Although the ATI gives guidelines and the ageing phenomena have been discussed at length in the literature (Shigo 1985; Read 1999; Green 2010; Farjon 2022) the distinction between ancient and veteran is a matter of judgment, in the ATI recordings left to the verifier, whereby the recorder provides observational data only. This author is both a recorder and verifier of oaks in the ATI. Not surprisingly, all oaks in High Park with girths exceeding 7m were found to be ancient; they would invariably be centuries old and would therefore have moved into that category unless exceptional preservation had been maintained for that long. Below size class 5 one sees an increase in the number of veterans, which at size class 3 in Table 2.1 already outnumber the ancients. This shift can also be predicted; were we to include oaks with smaller girths than I have, fewer and fewer would be considered ancient or veteran as their size decreases (see Figure 2.2). Although there is a correlation with size and this trend, a smallish oak can still be found to be ancient and a much larger one veteran. In general terms, though, oaks move from veteran to ancient as their girth and thus their age increases, as is shown by the records. Since the majority of oaks (186 or 70%) fall in size classes 1–3 (5–6.50m) there are more veteran than ancient oaks in High Park. There are 47 'indigenous' large dead oaks in High Park, making 13% of the total of oaks with a minimum girth of 5m. Of these, 14 were fallen oaks and 33 were still standing when recorded. Although the proportion of dead oaks seems higher than the average found across England (Farjon 2022), this may be related to the non-removal policy applied to High Park for at least 65 years, combined with the long period large dead oak wood persists before it has completely disintegrated. There is also a likely national recording bias towards living trees. We will come back to this topic later on as it has implications concerning our understanding of the dynamics of the population of ancient and veteran oaks in High Park.

In Figure 2.1 the data from Table 2.1 for High Park are shown graphically together with corresponding data for Lower Park. Both histograms show a decline in numbers (y-axis) as size increases beyond 6m (size class 3), but more steeply in the oaks in High Park. The markedly higher numbers in size classes 1–4 (5–7m girth) could be the consequence of a warrant to stop timber cutting in 1660 (Chapter 1, fn. 19) and more or less continuous 'benign neglect' of the High Park area since then, allowing more mature oaks of that vintage and younger to survive to the present. Before 1660, it is likely that mature oaks (>100 years) would have been subject to felling for their timber. The more irregular distribution of size classes in Lower Park could be a function of the smaller sample size, but is more likely influenced by the fact that this 'population' is managed. All trees deemed to be dangerous to the public – whether dead or alive – are removed from this landscaped part of Blenheim Park.

In Figure 2.2 the results of a survey of 'veteran trees' conducted by the Blenheim Estate in the years 2001–6 are shown together with the 266 oaks of Table 2.1 and Figure 2.1. All native oaks, living and dead, considered veterans are included, regardless of girth size. No distinction between veteran and ancient was made in this survey. We can here see a steep rise in numbers from the smallest veteran to a girth of 4m. From 4m to 5.50m the numbers of oaks fall gradually and beyond that size steeply to the largest oaks beyond 7.50m girth. These largest oaks linger on in small numbers (Table 2.1), down to the 'champion oak of Blenheim' at 10.37m girth (Figure 2.7). The rise in numbers of smaller veteran oaks is caused by the fact that few oaks with girths under 4m will have been considered veterans in the tree survey. It does not mean that oaks smaller than 4m were less abundant. The dip at size class 2 (5.5–6m) in the dataset presented in Table 2.1 and Figure 2.2 is an artefact caused by a change of protocol

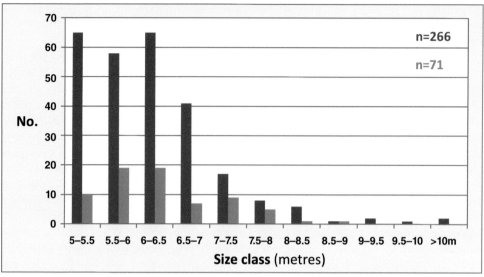

Figure 2.1.   Size distributions of 266 ancient and veteran oaks in High Park (red) and 71 in Lower Park (green) according to 11 size classes with 50cm intervals.

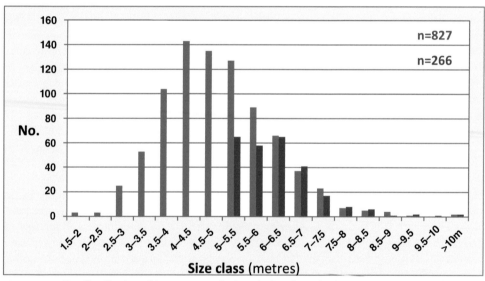

Figure 2.2.   Size distribution of 827 veteran oaks (*Q. robur*) with girths from 1.67 to 10.32m in the veteran tree survey, 2001–6 (blue), compared with the 266 oaks in Figure 2.1 (red).

in my surveying. It does not signify a gap in the age distribution of ancient or veteran oaks in High Park. One reason is probably that I followed the ATI guidelines and measured girths at 1.5m from the ground, while the veteran tree survey, 2001–6, used the standard forestry height of 1.3m ('breast height'). This would have given more oaks with girths of ≥5m in the latter survey, especially in the 5–5.5 m range. More important is that the two surveys were conducted about 15 years apart, so that many oaks not attaining 5m girth in the earlier survey had reached or exceeded that size in the later survey. Having begun the survey of 5–6m oaks after completing the larger trees, I may not have found all oaks that could have fallen in this size range (Figure 2.2, red columns for 5–5.5m and 5.5–6m oaks). However, comparative analysis of both datasets shows that errors of measurement and missed oaks also occurred in the veteran tree survey, 2001–6 (T.H. Moller, pers. comm.). From girths of 6m upwards the two graphs are consistent. We can therefore assume that for ancient and veteran oaks in High Park the blue histogram in Figure 2.2 presents a natural distribution, confirmed by the later survey of oaks with girths from 6m. This natural distribution of size classes in High Park has implications for the understanding of population dynamics of the ancient and veteran oaks, to which we will return later.

High Park has the highest number of very large (≥6m girth) ancient and veteran oaks of any site in England (130), exceeding the total numbers in Savernake Forest (84), Cranbourne Chase (74), Windsor Deer Park (71) and Chatsworth Old Park (63). Adding Blenheim Lower Park (42) to Blenheim High Park gives a total of 172 for Blenheim Park; the total for Windsor Great Park (Deer Park + Cranbourne Chase) is 145.

2.3. The age of the oaks

There is one question that invariably comes from the audience when I give lectures on the ancient oaks: how old are they? This is the most difficult question to answer with confidence; all we are capable of is estimates. All ancient and veteran oaks are hollowing, and conse-quently the wood that could provide the evidence has nearly always disappeared and merged into the soil. We are left with proxy data, and the most useful and convenient of these is the girth of the tree, since there is a relationship between this and the number of annual rings of wood laid down upon it. That relationship is complex, influenced as it is by many factors, some variable and difficult to quantify, and there are many pitfalls to be avoided (Farjon 2022). However, a new model to estimate the age of native oaks has recently been developed (Moller 2019), improved from the model developed by White (1998). Moller essentially used White's formula, but calibrated this beyond the first 100 years (when an oak reaches its full crown) with growth calculations obtained from repeated girth measurements of 12 large oaks reaching back in some cases to >200 years. It appears that White's assumption of a constant current annual increment of wood on the trunk (CAI) after the first 100 years, resulting in ever narrower growth rings on an expanding bole, is incorrect. In almost all ancient oaks the crown contracts, reducing the total amount of wood to be made, leaving more to be put on the bole but with a smaller crown of foliage. White erroneously assumed that the crown would remain as expansive as at maturity into old age, requiring wood to be added to all the branches. So the CAI on the bole does increase with time, resulting in a more constant ring width (but with variations owing to external factors that impact growth). The very few sawn oak trunks of large girths with intact wood from the earliest years to the last, one of them deposited in High Park, bear this out. This implies that White's model (and tables based upon it) overestimates the age, with an increasing discrepancy the larger in girth the oak has become. Moller's revised model (Moller 2019: Table 4) appears to avoid this bias and is here used to estimate the ages of the ancient oaks in High Park. Given the predominant soils (sand with coarse gravel, loam and Oxford clay) and an annual precipitation of 130cm, High Park is considered an 'average site' for oaks, which is also the baseline for Moller's calculations in his Table 4. However, it should be noted that markedly different age estimates can be derived from different assumptions regarding soil quality and other variable factors that have governed growth rates spanning many centuries. No precision can be claimed for any age calculation and a realistic margin of error is probably a century – or even more for the oldest oaks.

From the numbers of living oaks in Table 2.1 the oaks that are heavily burred and those that are severely fragmented have been excluded in Table 2.2. Their measured girths no longer have relevance to estimates of age, as in the first category the girth has expanded in an abnormal way, while in the second category it has been reduced by an unknowable amount. Dead oaks are also excluded because in most cases we do not know how long they have been dead. This leaves 203 ancient and veteran oaks with girths of 5m and above for which an age

Table 2.2.   The ancient and veteran oaks in High Park divided in 11 size classes, with the corresponding age estimates using the revised model by Moller (2019).

Size class	Girth range (m)	Girth average (cm)	Age range (years)	Age average (years)	Numbers (living)
1	5–5.5	525	305–350	327	41
2	5.5–6	575	350–400	375	44
3	6–6.5	625	400–445	422	58
4	6.5–7	675	445–495	470	34
5	7–7.5	725	495–535	515	14
6	7.5–8	775	535–585	560	7
7	8–8.5	825	585–625	605	2
8	8.5–9	875	625–675	650	1
9	9–9.5	943	675–715	695	2
10	9.5–10	975	715–755	735	0
11	>10	1035	755–800	777	2

can be estimated. The average age given in Table 2.2 may apply to any tree in its size class. Conversely, the age range is calculated for the lowest and highest girth value in its size class and gives a minimum and maximum age within the class, although with an unknown margin of error. In size classes 8–11 there are very few 'eligible' oaks, but I have maintained range and average figures for these; calculating an age based on a single measurement would give a falsely precise figure. We can therefore deduce that the large ancient and veteran oaks in High Park may have ages between 300 and 800 years. From size class 6, living oaks in High Park are likely to date back to the Middle Ages, with the two largest probably having been young trees in the fourteenth or one even in the thirteenth century. As noted previously, oaks in size class 4 with an average age of 470 years will have reached maturity by 1660 and together with their successors may have enjoyed greater survival as a result of the warrant in that year prohibiting timber cutting in Woodstock Park.

2.4. Distribution of oaks

The area of High Park as defined in the Management Plan (Wychwood Project 2014) is 129ha. The distribution of 266 ancient and veteran oaks with ≥5m girth within this area is shown in Figure 2.4. It appears that the old oaks are randomly distributed where they still occur. Absence is particularly notable in the long plantation, stretching from High Lodge to the south-east. Other voids occur in the northern part, where the 9th Duke planted many new oaks around the turn of the twentieth century, and west and directly

Figure 2.3.   Stump of a large oak in the long plantation. Alison Moller recording. (Photograph by Aljos Farjon, 7 April 2019)

south of High Lodge, areas that were also replanted by the 9th Duke (Chapters 1 and 3). The land sloping down to the Lake in the eastern part of High Park and to the meadows along the River Evenlode in the south-east is nearly devoid of large old oaks. There are no ancient or veteran oaks in the four small conifer plantations (c. 0.7, 1, 1.2 and 3ha). Finally, a strip of grassland making up the vista from Blenheim Palace towards High Lodge is now empty of ancient or veteran oaks but the oak stumps mentioned earlier are the remaining witnesses of their demise, sometime between 1945 and 1961 (Figure 2.6). The south-western half of the long plantation was cleared in similar fashion, but after 1961 (Figures 2.3, 2.6). In the other smaller plantations established after 1981, no old oak stumps were found and must have been entirely removed when the sites were cleared for planting since aerial photographs show mature trees present in June 1961.

Direct evidence (stumps) for a past presence of large ancient or veteran oaks in the other empty areas is also lacking. Given the presence and distribution of existing ancient oaks in High Park and Lower Park, it is almost certain that most if not all these areas have had at least one but probably several large oaks, either ancient or veteran. The reasons for clearance of the north-east half of the long plantation can be surmised from the fact that here a vista was created from Bladon Heath across the valley of the River Evenlode to High Lodge in the eighteenth century. The large clearance in the north of High Park similarly appears on an eighteenth-century map (Figure 1.10), but its reasons have not been discovered.

The western, south-eastern and lake fringes are characterized by mature plantings of Beech, Common Lime, Ash, Horse Chestnut and Sycamore, perhaps serving as wooded landscaping as seen from outside (the trees are situated on sloping ground and on top of slopes). This tree planting with (in part) non-indigenous trees that grew faster than the indigenous oaks appears to have been initiated by Lancelot ('Capability') Brown in the eighteenth century, but it was maintained long after (with plantings especially in 1893, 1898, 1926 and 1928), and few

Figure 2.4.   Google Earth map showing the distribution of 266 ancient and veteran oaks with girths of ≥5 m in High Park, divided into three size classes and all indigenous dead oaks.

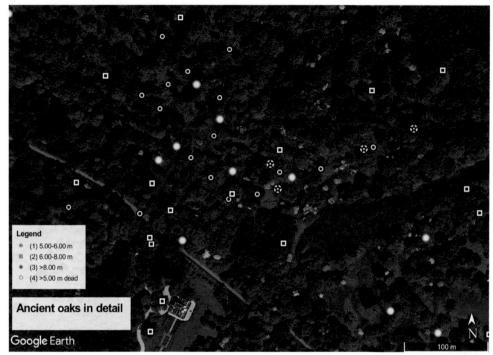

Figure 2.5.   Detailed view on Google Earth of part of High Park, showing the distribution of ancient and veteran oaks with girths of ≥5m. High Lodge, part of the road and the high end of the Palace Vista (grassland) are seen. The Google Earth image was created in mid- to late May with hawthorns in flower and tree crowns in full leaf.

remaining trees seem to date from this earlier period (Chapter 3). It is possible that many remaining old oaks were simply out-competed for light by the faster growing and more shade-tolerant planted trees. Some evidence of this competition leading to the demise of veteran oaks may be present on the western slope in High Park above the valley of the Evenlode, where many standing dead oaks are present among Sycamore and Ash (recently thinned). But caution should be applied; most of the oaks are long dead, perhaps dying before the infill trees had grown to maturity. Smaller 'veteran' oaks are still present here and there in these marginal areas (Figure 2.6). Judging by aerial photographs from 1961 and 1981, the absence of oak stumps in the four small conifer plantations is probably owing to their removal when the plantations were created.

Density of large ancient and veteran oaks is highest in the southern part of High Park, but there are areas with high density also in the northern section (i.e. north of the road from Combe Lodge via High Lodge to Spring Lock Lodge), such as directly north of High Lodge (Figure 2.5). There are more exceptionally large ancient oaks (>8m) in the northern section and three of the four living >9m oaks also occur here. The distribution of the three size classes on the map also appears a random mix. Large dead oaks seem to be absent from an area north of the Palace Vista, but this partly coincides with the areas of dense oak planting under the 9th Duke where ancient oaks are absent or scarce. These random distribution patterns are consistent with a history of natural regeneration that goes back many centuries. Most of the larger gaps seem to be due to losses, most likely for the reasons mentioned.

The distribution of all 'veteran' oaks recorded in the veteran tree survey, 2001–6 (Blenheim Estate data), gives a similar pattern to that of the larger oaks in Figure 2.4. The absence of data points in the long plantation is here more clearly marked, as are the two smaller conifer plantations surrounded by veteran oaks. The area to the south of the long plantation has again

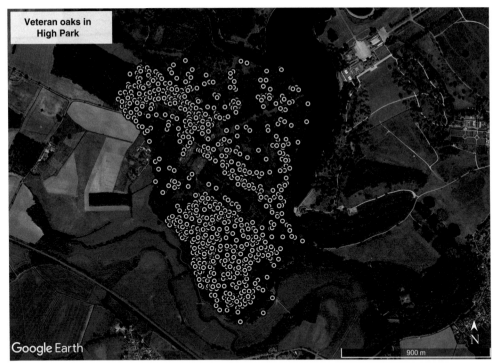

Figure 2.6. Google Earth map showing the distribution of 827 'veteran' oaks in the veteran tree survey, 2001–6 (Blenheim Estate data), and 26 oak stumps in the Palace Vista and the long plantation (yellow).

a greater density of ancient and veteran oaks than that to the north. The area to the north of the road near High Lodge is also very dense with 'veterans', but including oaks smaller than 5m girth extends this density much further towards New Park. This area partly coincides with an extension to Blenheim Park known historically as Combe Leys (Chapter 1). A gap in the 'veteran' oaks here that is not a plantation of trees becomes more evident. Some other areas lacking ancient and veteran oaks with minimum girths of 5m are confirmed in Figure 2.6, but the slopes to the Lake and to the River Evenlode in the south-east of High Park do have some smaller 'veteran' oaks that may have survived tree planting, or perhaps grew up with the earlier planted trees.

2.5. The largest ancient oaks

There are five ancient oaks in High Park with a girth in excess of 9m (Table 2.3). This is the highest number of >9m oaks at any one site in Europe. Four of these are still alive and healthy, one of them is standing dead; the dead oak could not be identified but is most likely also a Pedunculate Oak. Two are verified as 'pollard form (natural)' and three are 'maidens'; the distinction here being only that the apparent loss of limbs and regrowth that give the tree an appearance of a pollard have natural causes.

The largest tree (no. 136563, Figure 2.7) has formed two more or less ascending stems; a large gap was probably left by a third, but no remains are present on the ground. These stems are dying back towards the top and instead new branches, some large, have been formed from epicormic buds lower down on the tree. However, some large limbs are likely first growth, originating on a smaller ancient or veteran oak. Large burrs occur near the ground on one side and similar irregular growth is present elsewhere. Reiteration, that is, the formation of new shoots, mostly emanates from these burrs, and where they succeeded they have formed some of the branches that carry the foliage-bearing crown. Extensive hollowing is open to one side through the gap, which a person could enter, but hollowing also extends well into the two remaining old stems and probably the oldest branches. The girth of this oak has been reduced by the gap, but it is also increased by the burrs that cannot entirely be avoided in the measurement.

The division into two (formerly three) large stems is by some interpreted as originating from 'bundle planting', whereby three acorns (or saplings) were put together in a planting hole. Later, the stems would then have fused to form a multi-stemmed tree. There are several arguments to be made against this notion. This oak, like the other ancient oaks in High Park, was planted by a bird, most likely a Jay. It dates back from a time well before people planted trees here, as far as we can tell. Jays do not normally put more than one acorn down at one time; the birds tend to spread their caches. The split occurs too high up to have originated on the ground; more likely it is a case of early loss of the leader, which was then replaced by three new shoots. This can result from browsing and is often seen in young oaks that grew up in the open, where the formation of a low spreading crown is favoured by the absence of

Table 2.3.   Largest five oaks in High Park, with some characteristics. Tree numbers are ATI identities. (https://ati.woodlandtrust.org.uk/)

Tree number	Species	Tree form	Condition	Veteran status	Girth
136563	Q. robur	Pollard form	Alive	Ancient	10.37m
144530	Q. robur	Maiden	Alive	Ancient	10.34m
142735	?	Maiden	Dead (2015)	Ancient	9.62m
144526	Q. robur	Pollard form	Alive	Ancient	9.47m
12181	Q. robur	Maiden	Alive	Ancient	9.40m

Figure 2.7.   Ancient oak no. 136563 (10.37m girth). (Photograph by Aljos Farjon, 5 March 2014)

Figure 2.8.   Ancient oak no. 144530 (10.34m girth). (Photograph by Aljos Farjon, 28 April 2015)

competition and increased light reception. The tree stands on a hillock that has partly been formed by ancient digging (for clay and/or gravel?) forming a ditch around one side of the oak. If the root system was affected, it has probably regrown because the oak's bark and branches are living on that side. In wet winters the ditch can hold some water, indicating the presence of Oxford Clay deposits in the vicinity of this great oak. The root systems of large open grown oaks can extend far beyond the drip line of the crown (Tyler 2008; Green 2010) to distances in excess of 30m from the trunk. Roots that connected with parts of the tree that are now dead will presumably also be dead, but this is impossible to know for this oak.

The second largest oak (No. 144530, Figure 2.8) has almost the same girth, and the difference would fall within a margin of error that is unavoidable with ancient oaks of this size. However, its shape and growth form are different. In this tree a central trunk exists, although it is dead at least in its upper part. In the lower half massive burr growth surrounds the trunk, which is hollow and presumably mostly dead. This has given this ancient oak its girth of 10.34m; a rough estimate from where bits of normal bark or wood can still be seen indicates an 'original' girth of no more than 7m. However, it is the burr growth, unavoidable in measuring this tree, that now makes up much of the living tree above ground. All branches are reiterations from this burr growth, up to the largest limbs present. Many hundreds of buds produce shoots, almost all of which abort very soon, but some succeed to grow into branches bearing foliage. Where these are shaded by canopy above, they also die, but those that grow enough leaves, presumably when connectivity via sapwood, however contorted, and living roots is well established, can grow into the spreading branches that make up a secondary, healthy crown. This oak is one of the best examples in High Park demonstrating the compartmentalized nature of tree growth and death emphasized by Francis Hallé (Hallé 2002). A gap near the ground shows the extent of hollowing in the central trunk, which serves as a prop holding the burr growth and its branches in an aerial position. We could be looking at a relatively young oak being propped up by an old hulk. Burr growth does not produce wood laid down in concentric layers (annual growth rings) but instead in an irregular, contorted accumulation. As some

Figure 2.9.   Dead ancient oak no. 142735 (9.62m girth). (Photograph by Aljos Farjon, 30 December 2014)

areas cease to grow, others may pick up, which may account for the fluted appearance of this tree. Any correlation between girth and age of this undoubtedly ancient oak cannot be determined, and estimates of age on such grounds are spurious.

The dead oak (No. 142735, Figure 2.9) is the third largest in girth and is still standing. There was still a bit of life left when I first saw it in the winter of 2014, in the form of a few small epicormic shoots with live buds. These had failed the following spring. I could see no other cause than that decay had finally engulfed the entire tree, parts of which had been dead for a long time while foliage branches still grew on other limbs (not seen but inferred from thin dead branches still present). A large oak tree like this does not live or die as an entire organism; in this it is very different from an animal (Hallé 2002). Instead, large parts would have been dead, while other parts were alive. A tree is a compartmented organism (Shigo 1985), or even a colony, as Hallé argues, in which dead compartments have no other relation to the living parts than a mere physical connection, or at most by default serving as a prop to hold living parts up above the bracken to access sunlight. An ancient compartmented tree is no longer 'obliged' to add wood around the trunk or onto dead branches; so long as a strip of bark (with cambium) connects foliage to some living roots able to take up water there is life in the tree. New wood is only formed under that cambium. We consider the oak tree as 'alive' as long as this takes place, but instead we may be dealing with one or more 'young' oaks propped up by an old carcass. The bark on the trunk is now (2019) coming loose and beginning to fall. The tree is of course hollow, although there are no visible large gaps or holes (inspection from above could reveal these but has not been attempted). Elsewhere in this chapter we shall revisit the question of how long large dead oaks can remain before decomposition has finally returned all woody matter to the soil. This exceptionally large standing dead oak will provide habitat for many organisms for a long time, even when it breaks up and falls apart.

The fourth largest oak (no. 144526, Figure 2.10) also has massive burr growth around most of the trunk, from which numerous epicormic shoots arise. Several of these have grown into large branches, forming a low canopy. The central leading trunk has been long dead and is

Figure 2.10.   Ancient oak no. 144526 (9.47m girth). (Photograph by Aljos Farjon, 28 April 2015)

much broken, giving the oak some resemblance to a pollard, so it was classified as a 'pollard form (natural)', but its original 'maiden' shape is still discernible in the dead upper parts. There is a gap on one side of the tree and several dead limbs have fallen; most of these are likely to have come from the now dead and much reduced upper trunk. The epicormic growth of shoots on the burrs is vigorous, but most of these do not seem to ultimately succeed owing to shading from the larger branches, which produce a full crown of leaves in summer. This oak is, owing to its burr growth, very squat, approaching a 'baobab-like' habit. Ancient oaks like this may have a longer future life than those of comparable girth without burrs, such as the next big oak here described. The short, squat trunk supporting a low spreading crown is less prone to wind throw. When limbs are broken by storm, epicormic shoots, suddenly exposed to light, can quickly take over. Gaps are less likely to appear and expand in actively growing burr wood; the gap in this oak and the one found in no. 144530 probably predate most of the burr growth, and they give access to the hollow bole inside. This ancient oak is surrounded by young oaks, most of which were planted in one of the more recent attempts to replenish the stock of oaks in the Park (Chapter 3). Unlike the other >9m girth oaks in High Park, this tree stands in the southern end of the site where there is no bracken and the abundance of the grass False Brome (*Brachypodium sylvaticum*) indicates the proximity of limestone under a thin sandy soil. Other oak sizes do not seem to differ between the two areas (bracken/no bracken) and this is probably coincidence.

The fifth largest oak (no. 12181, Figure 2.11), although it has the smallest girth, is actually the largest oak in High Park by volume. It is a tall oak, unlike the other four, with a bole almost devoid of burrs and with massive primary limbs and a very large nearly full crown. Limbs have been lost, in part compensated by new branches. There are two very large main limbs arising 5–6m above the base at ground level. From these grow secondary limbs, but there is also much evidence that more limbs arose at around 6–8m; these are either sawn or broken off. When these are taken into account, this oak might have had the appearance of a pollard many years ago, but pollards were not cut at these heights above ground. Unlike the other

Figure 2.11.   Ancient oak no. 12181 (9.40m girth, tree on the right). (Photograph by Aljos Farjon, 25 March 2019)

four largest oaks, in this tree there is no retrenchment to a smaller crown, with concomitant dieback of the upper trunks or limbs, although there are some dead branches in the crown. Evidence for substantial hollowing comes from broken limbs and from extensive digging by Badgers under the trunk that has brought up sand and coarse gravel as well as bits of decaying wood. This tree has been managed (unlike the other four) to prevent collapse or breakage, by spanning steel cables between the limbs at various points ('bracing') and by sawing off other limbs. The sawn wounds have been mostly closed by the tree forming callous bark and wood, showing that this treatment was performed many years ago. There is heavy dead wood on the ground from a broken limb that fell after these incursions. This oak is locally known as the 'King Oak', presumably as a name replacement for an oak destroyed by the Parliamentary Army during the Civil War (Chapter 1). This oak is well known on the Blenheim Estate and is often shown to visitors; it was the only tree recorded for the ATI before my records (hence the lower ATI number). Recently, the oak has been 'haloed' by cutting many Ash trees that had sprung up nearby and were reaching canopy height.

2.6. Other ancient and veteran oaks

Many aspects and characteristics of the ancient oaks in High Park are exemplified in the five largest oaks described above in some detail. However, one growth form, pollard, is not. As mentioned, there are few convincing human-made pollards in High Park, but a larger number of 'pollard forms (natural)'. Here I mention and describe two examples of these.

Ancient oak no. 136568 (Figure 2.12) is a lapsed pollard with many limbs originating at about the same height, 2–3m above the ground. The tree is leaning heavily (literally!) as a likely consequence of the limbs growing larger every year. Old oak pollards have a 'bolling' (trunk) that widens upwards in consequence of repeated cuts and the reaction of the tree to this damage. Callous growth surrounding the cuts builds up, while new branch bases also add to the bulk. This is clearly seen in this oak, so it was cut several times. We do not know

Figure 2.12.   Ancient pollard oak no. 136568 (7.20m girth). (Photograph by Aljos Farjon, 22 February 2014)

when it was last cut, but it is possible that this happened at least 150–200 years ago, as was the case in many parts of England (Rackham 1997; Farjon 2022). In Richmond Park, Surrey, I found in 1997 a sawn section of a limb from a pollard oak with 285 annual rings. The limbs on this leaning pollard are now becoming heavy, and there is a danger that they will tear the undoubtedly hollow bolling apart or keel the tree over, rather than just breaking off. Why this and a few other human-made pollards exist here is a conundrum: 98% of large oaks in the Park are technically maidens. Pollarding was done by peasants (commoners, tenants) and their access to the oaks for timber in Blenheim Park must have been very limited or non-existent throughout most of its history.

Ancient oak no. 145093 (Figure 2.13) is verified as 'pollard form (natural)' and its truncated bole with many branches emerging at a height above the ground commonly seen in human-made pollards is accidental. A storm most likely broke off the trunk and major limbs, several of which are still lying under the new canopy. Such breakage in an oak with abundant healthy bark and cambium often initiates sprouting from dormant buds, some of which will succeed to branches as there is suddenly more available light. On this oak several now quite large branches arise from the bole under the breakage area and are curved upwards towards the light. A tree with >6m girth would have been hollow when the major damage occurred, as has been found when inspecting the tree at this level. However, the remaining 'shell' shows no major gaps and, so far, easily supports a regenerated healthy crown, just as a human-made oak pollard would have done. In comparison with tree no. 136568, it is easy to see the difference with a human-made pollard: the 'bolling' is not widening upwards and both 'cut' (breakage) and regrowth of limbs were one-off events. This happened relatively recently, when the tree was already >6m in girth; not an age or size at which pollarding would have been started by users of the timber.

The 16 'pollard form (natural)' oaks of ≥5m girth recorded in High Park are examples of the extraordinary capacity of veteran and ancient oaks to restore major damage. In some cases,

Figure 2.13. 'Pollard form (natural)' no. 145093 (6.78m girth). (Photograph by Aljos Farjon, 4 June 2015)

the tree has almost collapsed, with large dead wood lying around but with one upright part still connected to a root system developing a secondary crown. In such trees, however, there is a precarious mechanical balance between life and death.

The two ancient oaks in Figure 2.14 are examples of trees abundantly seen in High Park. These oaks have large low branches, though often broken away, leaving gaps to the hollow interior of the trunk (farthest oak in Figure 2.14). Their height is limited relative to girth of the trunk when compared with forest oaks, which in England seldom reach 5m girth, but tall forest oaks with girths of 5–7.30m occur in the Strict Reserve of Białowieża National Park in Poland (Keczyński 2017). More wood is present in limbs and branches relative to the trunk than in forest oaks and the crown spreads wider. The ancient and veteran oaks in High Park have grown up in parkland (pasture woodland or wood-pasture) where the trees are more or less widely spaced, not forming a closed canopy. This situation still prevails in many areas, but in other parts of High Park the open areas have been gradually filled with younger trees, mostly Ash, oak and locally birch, Sycamore, Wych Elm (rare) or other trees, all except many of the oaks from spontaneous regeneration. In recent years many of the overgrown areas have been thinned, mostly of Ash and Sycamore where the latter was abundant. When this secondary tree growth matures and closes the canopy, the 'architecture' of the old oaks cannot adapt to this new situation by growing taller. Aerial photography of High Park taken in around 1945 by the RAF (Figure 1.18) shows a more open parkland almost everywhere with well-spaced trees; most of these would have been oaks. The infill with predominantly Ash has not yet led to the suppression of existing large oaks. Although Ash grows faster than oak, it comes as late or later in leaf and does not intercept as much sunlight as do other competitors such as Beech and Sycamore. In a comparison of light requirements of common tree species in the UK (Green 2010), Ash and Pedunculate Oak are neighbours, with Ash only slightly more shade tolerant. Much depends on the size and vigour of the crown of the oak, and I have seen many ancient and veteran oaks that appeared unaffected or little affected by the proximity of one or more Ash trees. The competition between Ash and oak is more likely to favour Ash in earlier stages of growth, from seedlings onward.

Figure 2.14.   Ancient oaks in size classes 1–3 (5–6.50m girth) in High Park. (Photograph by Aljos Farjon, 9 May 2018)

2.7. The process of hollowing

Oaks have a well-marked division of sapwood and heartwood. As the tree grows, sapwood cells cease to function, die, and are added to the heartwood. All heartwood is dead wood and, when conditions are right, will be slowly decomposed, foremost by fungi but also by microbial organisms, boring insect larvae and other invertebrates such as woodlice. This decay often starts from the base upwards where the taproot has died off, providing an entry point for fungi. In High Park, many fallen oaks have hollowing trunks with a cavity that is widest near the base (Figure 2.15). Broken limbs are other entry points for fungi. The most common wood rot in oaks is brown rot, but white rot can also be present. These differ in the components of the wood that are primarily decomposed, and different fungi are responsible for each type (Alexander 2017). The brown rot leaves the lignin component, while white rot fungi decompose cellulose and lignin. This partially broken-down wood is fragmented further by invertebrates and composted by fungi and microbes. Field observation here and elsewhere indicates that almost all oaks with girths of ≥5m will have hollowing trunks, even if this cannot be observed from the outside in the absence of holes or gaps. In standing live oaks, openings can often be found between root bases that lead under and inside the trunk. This decay

Figure 2.15.   Oak hollowing from the base of the trunk. (Photograph by Aljos Farjon, 28 April 2015)

Figure 2.16.   Fallen dead oak no. 158755 (5.65m girth). (Photograph by Aljos Farjon, 26 March 2017)

progresses with time while the tree still grows new wood. In ancient oaks that are found to be completely hollow in the bole and large branches, an outer layer of very dense and hard wood forms a resistant shell that keeps branches aerial as long as it remains standing. This wood, the most recent wood produced, can be found to be more than 150 years old, with very narrow annual rings. High Park, with its large number of oaks in all phases of hollowing, is an excellent research site to learn more about this process, with obvious connections to the questions posed in the next section of this chapter.

2.8. Survival of ancient oaks

Mature and veteran oaks may be felled by storms, killed by drought or pathogens or by what often seem to the casual observer to be unknown causes (Figure 2.16). Young oaks become mature; they develop further to become veterans; and then become ancient oaks until senescence leads to death. This sequence presumes no catastrophic causes of death, such trees are assumed to have survived all the detrimental influences and contingencies they have met in the course of many centuries. In High Park, as we have seen, there is a high number of large dead oaks, 13% of a total of 360 recorded oaks (veteran tree survey, 2001–6) with girths of ≥5m. When we analyse the size distribution of these 47 dead oaks (Figure 2.17) we find a similar receding curve as in Figure 2.1 for all 266 large oaks. Doing the maths only for the living oaks produces the same curve (not shown). There is no size class with a disproportionate number of dead oaks; this indicates that there is a more or less constant rate of death over time, with no peaks or troughs at certain periods. In an analysis of all recorded 'veteran' oaks in High Park (Wychwood Project 2014) it was found that 'there does not appear to be a particular age at which the proportion of dead to live trees increases substantially'. The dead oaks with girths of ≥5m recorded during the recent survey period (2014–18) were mostly standing (33 of 47). Decay is slower in standing trees because the wood can dry out in summer, slowing the process of wood decomposition by fungi since dry wood (up to about

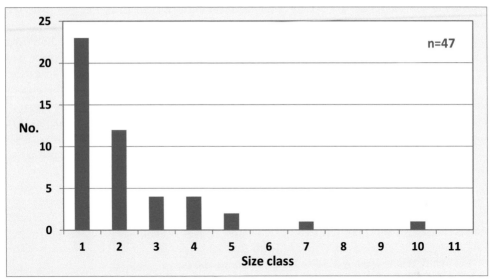

Figure 2.17.   Size distribution of 47 dead 'indigenous' oaks in High Park with girths of ≥5m according to 11 size classes with 50cm intervals.

28% moisture content) does not decay. Standing dead oaks will disintegrate, dropping bark and branches, and will usually fall before decay is complete, although short, squat trunks may remain standing despite being hollow almost to the end. Fallen trunks, especially when being overgrown by Bracken and brambles, will remain moist longer and can decay faster.

It would be of great interest in this context to have an estimate of the time large dead and hollow oak trunks remain before disappearing into the soil. As noted previously, we have a minimum estimate for the low-cut stumps in the long plantation and the Palace Vista (60 to 65 years). These are very low to the ground, partly overgrown, and may decay faster than large fallen trunks. While several of these trunks superficially seem to have reached comparable levels of decay, their bulk (although hollow) indicates that it will take many years for them to disappear, while we have no knowledge of when these trees died and/or fell. The dates of more recently fallen oaks can be determined; a few of these I have seen standing and alive. It would be possible to monitor the process and progress of decay, but this will obviously take much more than 60 to 65 years and probably more than a century, a project unlikely to be brought to fruition. I have not found nearly decayed remains of fallen oaks that could be estimated to have been of ≥5m girth, but that can of course be difficult when little rotting wood remains. Smaller fallen oaks would take less time to disintegrate, so these, while abundantly present (veteran tree survey, 2001–6; AF's unrecorded field observations) cannot serve to estimate 'residence time' for dead oaks in the database of ≥5m oaks. Most other kinds of dead wood, while sometimes present in large sizes, Beech, for example, are less durable than oak.

Streeter (1974) presented a table for oaks in Britain in which the number of dead trees per 100 trees was calculated 'on theoretical grounds' for two assumed life spans. For a life span of 300 years and a decay time of 50 years, there would be 14.3 dead trees per 100 trees. The ratio for ≥5m girth oaks in High Park is about 13 dead trees per 100, indicating a similar decay time. However, the 'theoretical grounds' were not explained so it is not possible to repeat the calculations for High Park based on this reference. A study using CO_2 release measurements as an indicator of decay rates (Liu et al. 2013) found an average exponential decomposition rate (K) of 0.09 $year^{-1}$, but with decay rates varying among pieces of wood debris by almost a factor of 18. Decomposition rates also varied greatly between six different species in Swedish subarctic forests (Freschet et al. 2012); this study did not include *Quercus robur*. Oberle et al. (2018), in a survey of standing dead trees ('snags') across the eastern United States and many species, found that, within an area with similar air temperature (climate), species wood durability and individual tree diameter were the two most important variables determining how long a snag would stand in a storm-free period. No data on species differences were presented in this paper. A workshop presentation by Parminter (2002) of the Ministry of Forests in British Columbia, Canada, gave residence times, mainly for conifer trunks on the forest floor based on wood density measurements, of up to 300 years. I found no comparable research literature on residence times for dead oaks in north-west Europe, so for now all we have are the dated stumps that are still present after 60 to 65 years. On this basis we may perhaps infer a residence time for the largest fallen oak trunks in excess of 100 or even 150 years. The question is obviously in need of research.

2.9. Ancient oaks in a wider context

Within Blenheim Park, there is another area with similar ancient and veteran oaks: Lower Park. This area is situated to the east of High Park and separated from it by the valley of the River Glyme. Two veteran oaks with girths of 6.10m and 6.80m stand between the two areas along the road from Spring Lock Lodge to New Bridge, but I have counted these with Lower Park since they are outside High Park and nearer to Lower Park. Some ancient oaks may have disappeared as a consequence of the flooding of the Glyme in the eighteenth century and the subsequent creation of gardens and tree planting, or perhaps earlier when the valley was managed for hay making (Chapter 1). Including the two oaks mentioned above, there are 71 oaks with ≥5m girth in Lower Park, the largest of which has a girth of 8.92m. Only two of these are pollards. A total of 43 oaks have girths of ≥6m. The geographical distribution of ancient and veteran oaks with a minimal girth of 5m in High Park and Lower Park is presented in Figure 2.18; their size distribution in Figure 2.1.

The oaks mapped in Figure 2.18 are the only oaks of this size category in Blenheim Park. The nearest cluster of ancient oaks in the same size range is in Cornbury Park (11 oaks ranging from 5.50m to 10.05m in girth) 9km to the west. Two other oaks with girths of 6.38m and 7.38m have been recorded within this radius, both several kilometres from Blenheim Park. Given the maximum natural dispersal range of acorns, the ancient oaks in Blenheim Park can be considered to function as a population in the limited sense that their acorns will likely remain in the area. Cornbury Park is another small population of ancient oaks. Both parks were royal deer parks, emparked within Wychwood Forest, a royal hunting forest. While the extent of Wychwood Forest at the time of Domesday Book (1086) has been established with some accuracy (Schumer, 1999), the extent of tree cover during the Middle Ages is much more difficult to reconstruct. Evidence for a more or less treeless expanse in the northern half of Blenheim (Woodstock) Park is summarized in Chapter 1. The medieval pasture woodland in Woodstock Park is unlikely to have extended into this area and probably coincided more or less with the current extent of ancient and veteran oaks (a few large cut stumps from oaks possibly in this size category were found in January 2020 in a narrow section of The Lince just outside High Park). Woodstock Park as created by Henry I in around 1110, with some later extensions, thus included both areas with oaks and other trees and more or less treeless fields that may have been emparked from farmland. Landscaping and other activities in Lower Park caused the loss of ancient and veteran oaks, but at least until the twentieth century many were retained (Figure 1.13). The removal of dead oaks from Lower Park explains their absence and has further lowered the numbers in comparison with High Park. Despite this, the random distribution in space, with gaps mostly explained by alterations in the Park landscape since the eighteenth century, is wholly congruent with that in High Park. The size distribution (not shown) is also similar, showing the same exponential decrease in numbers of the larger, and thus older, oaks. This distribution is to be expected in a natural population with a continuous

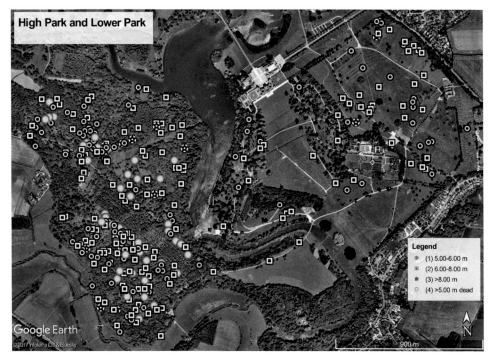

Figure 2.18. Google Earth map showing the distribution of ancient and veteran oaks with girths of ≥5m in High Park and Lower Park, divided in three size classes and all indigenous dead oaks.

rate of recruitment as well as death over a very long period. Neither in High Park nor in Lower Park is there evidence for a cohort of ancient and/or veteran oaks with a similar age. These observations agree with the idea that the ancient and veteran oaks of Blenheim Park were not planted; they are the survivors of generations of native oaks, locally indigenous and belonging to a single species, which connect back in history to at least Anglo-Saxon times. This continuity in time directly impinges on biodiversity, in particular for kinds of organisms with obligate associations with ancient oaks, as we shall see in subsequent chapters.

Acknowledgements

The author thanks Benedict Pollard for help in the field with recording the 5–6m oaks and Torsten Moller for creating the graphics in Figures 2.1 and 2.2.

CHAPTER 3

Planted and Self-Seeded Trees

Aljos Farjon and Torsten Moller

High Park is almost entirely wooded, but a distinction can be made between trees that have been planted at various times, mostly Pedunculate Oak, Sycamore, Beech, Ash and conifers, and trees that are naturally seeded. Some of the planted trees have subsequently spread naturally, while other spontaneous trees come from non-planted sources. The planted trees occur in distinct locations in some instances (e.g. conifers, earlier oak plantings) or are spread more diffusely throughout the site (e.g. later oak plantings, Ash). Examination of the planted and self-seeded trees separately in this chapter is intended to be a representative survey rather than an exhaustive account of the trees in High Park. The aim is to thereby reveal the main features of the woodland structure and associated habitats, and how conditions have changed in response to human interventions. In a concluding section the findings of the survey are compared with other known plant communities of southern England and correlated with the distribution of different soil associations in High Park.

Mapping the broad distribution of different trees and gauging their age–size relationship from girth measurements was carried out in repeated site visits using geo-referencing by hand-held GPS. Signs of competition, disease and damage by squirrels and deer were noted and assessed in relation to the growth and condition of individual trees. Further analysis was undertaken using aerial photograph sets (1945, 1961, 1981) and laser scanning imagery (LiDAR). Field observations were supplemented with archival information on tree planting campaigns implemented in the nineteenth and twentieth centuries.

3.1. Planted trees

An early sign of tree planting activity is the depiction on a Thomas Richardson map of 1771 of a nursery just inside the perimeter wall near High Lodge (Figure 3.1). However, the trees propagated here were probably all destined for Lower Park and other parts of the designed landscapes in Blenheim Park. Many trees were planted along the perimeter rides created by Capability Brown around the Great Park, but there was very little

Figure 3.1. Tree nursery in High Park. From the 1771 map by Thomas Richardson. (British Library, Add. 71602)

Aljos Farjon and Torsten Moller, 'Planted and Self-Seeded Trees' in: *The Natural History of Blenheim's High Park.* Pelagic Publishing (2024). © Aljos Farjon and Torsten Moller. DOI: 10.53061/JWXM7823

Figure 3.2.   Girth measurement records of planted oaks (coloured dots); conifer plantations (shaded). (Google Earth image)

interference with the existing woodlands in High Park by successive landscape designers in the eighteenth century.

The first oaks to be planted in High Park were probably introduced in the late nineteenth century on the initiative of the 9th Duke. Figure 3.2 shows the location and extent of various features in High Park that are described in this section.

3.1.1. The 9th Duke

During his long tenure (1892–1934), the 9th Duke conducted an ambitious tree planting campaign throughout Blenheim Park. The strength of his commitment can be appreciated from the leather-bound Tree Planting Book begun in 1892, the year of his accession, and containing records carefully entered until 1931. The various planting schemes had both aesthetic and practical aims, as shown by entries in the book stating: 'His Grace the Duke of Marlborough K.G. has planted between the years 1893 and 1919 inclusive: 459,834 Trees; 5,203 Ornamental and Park Trees; Total cost £7,736.15.0. I'm not sure what the value of the timber is at the time of writing but I fancy that in 50 years that the then owner will make a deal of money.'[1] His was also an early example of a conservationist agenda to preserve natural woodland as well as to maintain the effect of the eighteenth-century park landscape created by Capability Brown. Thus, the ornamental clumps of oak, Beech, chestnut and other trees scattered throughout Blenheim Park are a characteristic legacy of his efforts (Figure 3.3).

---

1   Entries dated 12 September 1919 and 5 October 1919 on page 151 (Spencer-Churchill, 1892–1931).

Figure 3.3.   A clump of four mature oaks in High Park. (Photograph by T.H. Moller, 13 April 2017)

The duke's basic strategy is revealed in his Tree Planting Book as one of dense planting in clumps, followed by judicious thinning a decade or so later.[2] The entries show that in High Park three different clumps were planted up in December 1902 with a total of 600 oaks 4–7ft high, aged 6–10 years and mostly grown from High Park acorns.[3] Two of the clumps were between High Lodge and Spring Lock Lodge and one clump was between High Lodge and Combe Gate (Chapter 1, Figure 1.19). Of these trees, 162 were thinned and replanted in 1910–12.

In Figure 3.4 the records are compared for different clumps of oaks in High Park of roughly equivalent ages (120–130 years). Many of the clumps are circular and have survived with relatively few gaps, but the effect of shading by nearby trees can be perceived in the different growth rates implicit in the median values of girth measurements taken in 2017–20. The blue and red clumps hemmed in by a conifer plantation and the densely planted yellow clump (74 trees, of

---

2   March 1902: '42 oaks, 6–9 feet, planted in thorn clumps in High Park with the exception of twelve planted in the holes of the limes taken to be transplanted into Low Park.' November–December 1903: '27 oaks 12/14 feet; planted near the Colonel's Tree in 1898 and now transplanted near 14-Acre Clump. Feb 1910: '18 oaks (15 feet) removed from a clump between High Lodge and Spring Lock and planted on an open space between the Lake and Clay Pits.' December 1910–February 1911: '62 oaks (10/12 feet), of which 59 taken out of a clump between Nursery and Spring Lock, and 3 from the clump near the Reservoir were planted near Queen Anne's Drive from High Lodge road to Nursery.' 1926/7: '250 oaks (10/12 feet) from the nursery planted in Love Walk' (Spencer-Churchill, 1892–1931).

3   There is evidence from crown architecture that an undetermined number of these oaks originated from other sources than the oaks in High Park; such introduced oaks have long ± straight branches spreading at an angle <90° contrasting with the shorter, bent and ± horizontal branches of 'indigenous' oaks. These two growth forms are recognized as distinct in German forestry literature (Krahl-Urban 1959).

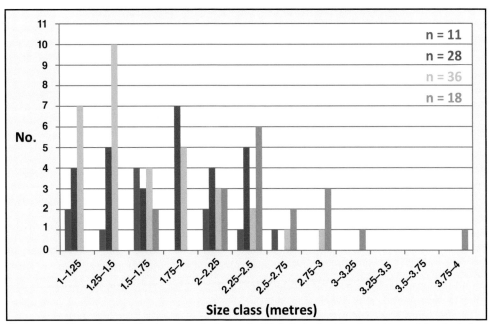

Figure 3.4.   Girth size distributions in oak clumps in High Park. Conifer plantations white-shaded.

Figure 3.5.   Regular rows of oaks in the Combe Bottom plantation, High Park. (Photograph by T.H. Moller, 13 April 2017)

which 36 were measured) are all challenged for light (average girth 1.9m), whereas the green clump (average girth 3m) is outside the plantation and has been thinned. While the median girths in Figure 3.4 differ, the compositions of each clump are comparable and approximate to a normal size distribution, implying that the trees in each clump comprise a single age cohort, probably planted as ten-year-old saplings around the turn of the twentieth century.[4]

In High Park the duke's motivation to replenish the ageing stock of oaks is set out with an entry for November to December 1898 when 148 oaks 9–10ft tall were planted: 'Many of the old oaks are dying and if the planting of young one[s] is neglected the forest will soon cease to exist, except for the existence of a few old stumps. The planting of oaks in the High Park should be continued for another 25 yrs, and then the Park will be stocked for 500 yrs to come.' These sentiments were probably based on a misconception of the apparently ailing condition of ancient oaks exhibiting hollow trunks, crown contraction ('retrenchment') and branch die-off (Farjon 2017: 246).

The area of High Park planted up and described as a 'very open spot' corresponds to the open square depicted in Chapter 1, Figure 1.14. A batch of 148 oak whips 9–10ft tall were planted in 1898 and a further 115 oaks were probably added in 1900, since it is recorded as 'a continuation of replanting carried out in this part of the Park in 1898 – N.E. side'. Together, these entries are consistent with an extensive area in the northern portion of High Park leading down to the western arm of the Great Lake (Combe Bottom) and now containing 196 surviving oaks of roughly uniform size growing in a regular grid pattern, with girths averaging 2.5m (Figure 3.5). When the girths were measured in April 2017 there were more than 43 gaps/dead oaks in the grid, implying the original planting(s) contained at least 239 trees. The grid pattern gives the impression of a commercial crop, but the trees were described

---

4   6–10m oaks were planted in three clumps in December 1902 between Springlock Lodge and Combe Lodge (Spencer-Churchill, 1892–1931).

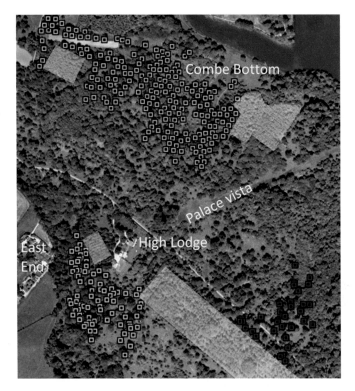

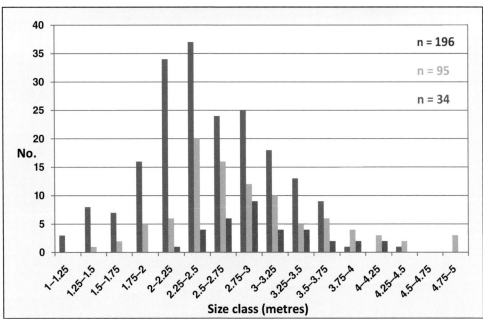

Figure 3.6. Girth size distributions in High Park oak plantations. Conifer plantations white-shaded.

by the duke as 'large fine well-grown oaks and were planted with a view of their becoming specimen trees ... to preserve the effect of the forest of the Middle Ages'.

Other parts of High Park were planted with oaks in a more irregular fashion. The girth data of the plantation oaks together with their locations are summarized in Figure 3.6. The size distributions are slightly skewed towards larger girths but broadly similar across the

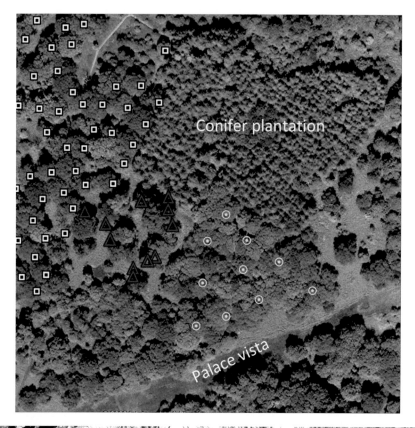

Figure 3.7.   Top: Squirrel-damaged oak plantation (yellow circles) next to an oak grove (red triangles) and the Combe Bottom oak plantation (white squares). Bottom: Squirrel-damaged oak plantation. (Photograph by T.H. Moller, 7 February 2020)

three plantations. Somewhat lower median values and hence slower growth rates apply to the denser array in Combe Bottom plantation and around High Lodge that has grown into closed-canopy woodland. In addition to the availability of light, other factors could contribute to the observed differences, such as soil composition and drainage.

Situated between the Palace Vista and a conifer plantation is a triangular assemblage of contorted oaks (Figure 3.7) that has clearly been planted since the trees are growing in parallel straight lines. The tortured stems are caused by Grey Squirrels damaging the leading shoots of the young oaks. This activity is analysed fully in Section 3.2. It is reasonable to suppose that the damage inflicted by the squirrels has stunted the growth of the oaks and that this is reflected in the size of their girths. Some of the oaks are indeed seriously challenged, but others have prevailed to form a closed canopy and grown in circumference to about 2m, revealing their likely origin during the 9th Duke's tenure, perhaps in the first decades of the twentieth century. The neighbouring oak grove (red triangles in Figure 3.7) has not been compromised by squirrel damage and in enjoying free and open growth conditions has attained an average girth of 4.3m. Next to the grove is the eastern margin of the Combe Bottom plantation (white squares, Figure 3.7) in which the average girth is 2.5m. These oaks have not been afflicted by squirrel damage either, presumably by virtue of having become established sufficiently early in relation to the squirrel infestation.

3.1.2. The 10th Duke and 11th Duke

The tenure of the 10th Duke (1934–72) coincided with the economic downturn in the post-war period and led to the formation of the first of several commercial conifer plantations in High Park (Figures 3.2, 3.8). The long plantation was established after 1961 and was followed by four smaller areas some 20 years later. Figure 3.8 from 1981 shows the nascent smaller areas cleared in preparation for planting. The management principles and methods were essentially those of the Forestry Commission, working to a crop rotation cycle that is currently at the point of harvesting. Thus, the long plantation running south-east from High Lodge has recently been clear-felled (2019), retaining most of the broadleaved trees (oak and Beech) planted there. The remaining smaller acreages are expected to follow, in recognition of the incompatibility of softwood cultivation with the priority of protecting the ecological integrity of High Park. The principal conifer species planted is European Larch (*Larix decidua*), but Corsican Pine (*Pinus nigra* subsp. *laricio*) was also used as well as some other species. Meanwhile, the imprint of the monoculture of conifers is evident in soil acidification and the scarcity or absence of an understorey in the plantations. Unlike in the long plantation (see Chapter 2) we found no stumps of cut veteran oaks in the smaller conifer plantations as evidence for their removal prior to planting. The largest of these smaller plantations contains two dead veteran oaks at its centre that were alive in 1981 and may later have succumbed to shading from closely planted conifers (Figure 3.7, top right).

The transient nature of such impact is arguably a consequence of the enlightened initiative of the 11th Duke (1926–2014) in the 1970s to appoint a management committee for the long-term restoration and preservation of park resources and the balancing of conflicting interests (Bond and Tiller 1987: 133). The current compartmented management plan for High Park forms part of a ten-year Higher Level Stewardship agreement between the Blenheim Estate and Natural England that was to run until March 2022 (Mottram and Kerans 2014). Management measures introduced to favour the oak component of the woodland include *inter alia* supplementary planting of Blenheim-sourced oaks to recreate a mosaic of woodland and grassland elements in denuded areas. Crown development of well-established oaks is promoted by selective thinning of pole-size trees and girdling of poorer specimens that are left upright to create standing deadwood for the benefit of sheltering species (the latter measure has not been implemented). Such initiatives aim to preserve the pasture woodland character and biodiversity of an ancient cultural landscape.

Figure 3.8. Composite aerial photograph of High Park from 1981 with established and nascent conifer plantations (circled). (Oxfordshire County Council imagery)

    Well after the planting of the now contorted oaks seen in Figure 3.7 there was another, more widespread effort at oak planting in the late 1980s (Mottram and Kerans 2014) as well as other broadleaf trees in some locations. Many of these oaks, now with girths averaging 1.27m (n=25), are seriously damaged by squirrels, lacking a leader and branched abruptly but irregularly above a short trunk. They occur in many locations but a concentration of these distorted planted oaks is found in an area of open grassy glades with bracken to the south of the long plantation.

3.1.3. Planted trees other than oaks and conifers

The earliest plantings in High Park of trees other than the 9th Duke's oaks were those planted on the slopes towards the Lake after this body of water had risen behind the dam created in 1763. Their apparent purpose was to create a backdrop of a wooded hill across the Great Lake as seen from the Palace or from Vanbrugh's bridge. The pasture woodland of High Park was likely more open then than it is at present. Of these trees, a few will have survived to the present, perhaps some large Beeches (*Fagus sylvatica*) and, more likely, a few Common Limes (*Tilia* x *europaea*). Planting of Beech in this peripheral zone continued in the nineteenth century and probably into the early twentieth century, now extended onto the peripheral slopes above the River Evenlode in the south and west of High Park, where it may have had the function of a shelter belt against westerly winds. More locally, trees such as Horse Chestnut (*Aesculus hippocastanum*) and Sycamore (*Acer pseudoplatanus*) were added, the latter establishing self-seeded trees in those areas.

There were probably some elms (*Ulmus procera*) among the planted trees, later killed by elm disease; a few small sucker trees survive here until also killed, but suckering again. A notable round clump of Horse Chestnut mixed with oak, Beech and Ash was planted on the south side of the Palace Vista (Figure 3.9), presumably even more visible from the garden at the west front of Blenheim Palace when High Park had a more open parkland character.

In more recent times, plantings of Wild Cherry (*Prunus avium*) and other broadleaved trees were established in a former nursery area in what was once known as Bladon Wood (Figure 1.7) at the southern end of High Park. In the vicinity of High Lodge other trees not native to High Park were planted from time to time, such as a small clump of Lombardy Poplars (*Populus nigra* 'Italica') and a larger clump of hybrid poplars (*Populus* x *canadensis*) near a pond at the top of the Palace Vista. These and other intervening trees block the view from High Lodge to the

Figure 3.9. The Palace Vista in High Park, seen from the west front of Blenheim Palace, with a planted clump of Horse Chestnut, oak, Beech and Ash on its south side. (Photograph by Aljos Farjon)

Palace.[5] Some other planted trees observed in this part of High Park are Common Whitebeam (*Sorbus aria*) and Italian Alder (*Alnus cordata*), both as specimen trees. Oak and Beech were also planted among and between the larches in the long plantation, in similar forestry plantation rows to the conifers. In the middle part of the long plantation, these broadleaved trees are of a later date than the larches and may have replaced rows of failed conifers. Most of these broadleaved trees were retained after all conifers, including a number of Western Red Cedar (*Thuja plicata*) that stood along the edge facing the road, were removed in 2019. Also removed was a nursery-style plantation of various broadleaved trees situated between the High Lodge enclosure and the long plantation of larch; it is now a clear-cut area.

Most of these tree plantings are small-scale and local and their impact on the ecology of High Park would be slight. Sycamores do spread and can 'behave' as an invasive species particularly in lowland England, but their spread in High Park is (so far) limited and can be controlled. Beech, although native and doing well on limestone substrates, seems to have been virtually absent before it was planted. It is a tree of closed-canopy high forest and does not maintain itself in pasture woodland (Vera 2000). Oak timber exploitation, such as occurred in the New Forest in early modern times, caused a shift in dominance from oak to Beech (Tubbs 1986); such exploitation did not occur here. Although self-seeded Beeches are found (there is a large old, broken tree in an old quarry), they are scarce. If Beech and Sycamore were to become abundant, they could pose a serious threat to the much more light-demanding oaks.

3.2. Self-seeded trees

Among the native oaks (all *Quercus robur*) the ancient and veteran oaks in High Park are generally considered to be self-seeded (Chapter 2). This evaluation would apply to 100% of oaks with girths of ≥5m, as their average estimated age would exceed 300 years (Table 2.2), having germinated at least 175 years prior to the earliest known planting date during the 9th Duke's tenure. This certainty diminishes statistically for oaks of smaller girth as their estimated age approaches the ages of those known to be planted. Average girths of the oaks planted for the 9th Duke range from (1.9–)2.5m to 4.3m (the latter in a grove of oaks with exceptionally large crowns). If girth alone is considered, there is clearly an overlap with the 277 veteran oaks recorded by the Estate's veteran tree survey, 2001–6, in the 2.5–4.3m size range. The term 'veteran' primarily refers to a tree's condition, not to its age in years, although there is a relation between these two parameters (Farjon 2017: 26). In theory, therefore, the 2001–6 survey may have recorded some of the planted oaks as 'veterans'. However, although we have found ailing and standing dead oaks among them, our survey of the 9th Duke's oaks did not find any with the characteristics of a veteran tree.

Other criteria can be brought to bear on the distinction between planted and self-seeded oaks now growing in High Park. These have to do with the relative situation of trees (standing in rows or at equal distance from trees of similar age/size, or the reverse, close proximity to other trees, terrain characteristics), stem and crown shape, and history of planting on the site. Oaks standing in parallel rows or at equal distance of in total more than about ten trees can be considered planted. Oaks in unlikely places, for example, the steep edge of an old quarry or very close to another tree (Figure 3.10), are self-seeded. Since acorns are actually 'planted' at random in time and space by Jays (*Garrulus glandarius*) or occasionally by other corvid birds or squirrels, cohorts of even-aged (as estimated by girth and shape) and evenly spaced trees are likely to have been planted by humans. Conversely, when we find marked differences in these parameters in a group of oaks, they are more probably self-seeded. Planted oaks (in modern times) have first been grown in a tree nursery, where they were likely

---

5   These trees were removed in 2021 in order to restore the view along the vista as well as water levels in the pond.

pruned to maintain a single, upright stem with branches retained higher up the tree. When transplanted successfully into the field, this habit is maintained at least into maturity (c. 100 years). Self-seeded oaks usually look quite different. If they started in an open situation without significant competition for light, branches will have been set much lower and be spreading. The oak can have been self-pruning under the shade of its own canopy, but scars of discarded branches remain long visible. Stems are usually not quite straight owing to this early development. Where self-seeded oaks have emerged in canopy gaps, they grow taller with less branching, but the search for light bends the trunk one way or another and often results in a lopsided crown; such trees can be distinguished from planted trees as well. In High Park, there is no history of oak coppice management (Mottram and Kerans 2014), and the oaks we found with multiple stems from near the ground probably resulted from deer browsing and are self-seeded.

Figure 3.10. A planted oak (left, damaged by squirrels) and a self-seeded oak growing in close proximity in a clay quarry area with seasonal ponds. (Photograph by Aljos Farjon)

3.2.1. Planted and self-seeded young oaks

In February 2020 five areas with young oaks, each about 1ha in extent, were surveyed to determine the proportion of self-seeded oaks. The areas were not randomly chosen since the presence of self-seeded oaks had been observed previously. The presence of self-seeded oaks among planted oaks, both categories with girths ranging from 0.3m to 0.6m in 2012–13, is mentioned in the management plan (Mottram and Kerans 2014). The results of the present survey are summarized in Table 3.1.

Although the sites were chosen for their known presence of self-seeded oaks, the results demonstrate that natural regeneration of oaks has been successful and their numbers exceed the planted oaks. With an average girth of about 0.9–1m the planted and self-seeded oaks

Table 3.1. Planted and self-seeded oaks in five areas of about 1ha (status ? = uncertain assignment to either planted or self-seeded; copp. = multi-stemmed tree), surveyed in February 2020.

Site	planted	status ?	girth (χ) m	squirrel damage	self-seeded	status ?	girth (χ) m	squirrel damage	copp.	Totals
1	2	0	0.6	50%	7	3	1.2	14%	0	**9**
2	14	0	1.4	28%	23	2	1.2	17%	0	**37**
3	3	0	1.4	100%	19	0	1.1	5%	8	**22**
4	6	3	1.2	67%	41	4	0.9	66%	2	**47**
5	2	0	0.4	100%	18	0	0.8	72%	1	**20**
Totals	27	3	1.0	44 (av.)	108	9	0.9	43 (av.)	11	**135**

are of a similar age, estimated at 30–35 years (see also Mottram and Kerans 2014) in 2020. Self-seeded oaks of significantly smaller girth are present in some locations, for example, near the track along the western periphery, but are not common. Seedlings and saplings are very rare. Potential oak regeneration in open woodlands and grasslands is more or less continuous where seed trees are present and is only suppressed under the influence of over-grazing or -browsing (Shaw 1974; Farjon and Hill 2019). It would therefore appear that there was a short period in the late 1980s of relief from this pressure, caused by a dip in the feral deer population. An apparent new arrival among oaks is the alien Turkey Oak (*Q. cerris*), of which two young self-seeded trees were discovered among native young oaks in the southern part of High Park in June 2018. The nearby presence of Knopper Galls (*Andricus quercuscalicis*) on acorns of a Pedunculate Oak is evidence that the Turkey Oaks have been flowering and may soon be spreading.

Grey Squirrels (*Sciurus carolinensis*) were first introduced in Cheshire in 1876 and were subsequently set free in many estates throughout the UK. Many new introductions took place between 1902 and 1929 (the last time) and they appeared also in Oxfordshire during this period. By the 1930s they were well established in England between Kent and Warwickshire. As far as can be ascertained, no damage caused by squirrels is evident on oaks planted under the 9th Duke's tenure prior to about 1910. It is likely that the Grey Squirrel arrived in High Park (Blenheim Park) after that date, but we do not know whether this arrival was deliberate (the 11th Duke of Bedford at Woburn Abbey had a habit of supplying other estates) or accidental. The Grey Squirrel, native in the eastern United States, probably encounters a deficiency (of calcium?) in its dietary requirements in Britain's woodlands, which it tries to compensate by stripping bark off the leader and side branches of young trees to ingest the underlying phloem (Nichols et al. 2017), in particular of Pedunculate Oak, Hornbeam (*Carpinus betulus*, not present in High Park) and birch. The oak trees are rarely killed and react by growing new branches, which are again stripped. The bark of branches of mature and veteran oaks is apparently no longer of interest to the animal, which is how the contorted oaks in the triangular plantation (Figure 3.7) eventually could grow large, if disfigured, crowns. Squirrel damage appears to vary between sites, but was found on average to have affected >40% of the oaks surveyed. At present, the estate's gamekeepers are suppressing (but not eradicating) the population by setting many squirrel traps in High Park.

Other detrimental alien 'pests' observed on oaks in High Park during our surveys are Acute Oak Decline (AOD) and Oak Powdery Mildew (OPM). AOD first appeared in Britain in the late 1980s. In winter 2013–14 eight confirmed cases had been reported in High Park (Mottram and Kerans 2014) and more were suspected. The symptom of AOD (a bacterial infection), in the form of small fissures seeping black fluid down the trunk (stem bleeds), was seen on one of the planted oaks in a clump in the long plantation (clump with orange dots in Figure 3.2) on 8 April 2019.

Another affected oak recently seen is a planted tree from the time of the 9th Duke in a glade just south of the Palace Vista; this tree shows foliage dieback as well as lesions. The affected oaks do not appear to include the ancient and veteran trees (AF, pers. obs.) and its presence and spread could be linked to tree age and provenance. It seems still to be rare in High Park, but a renewed monitoring would be necessary to establish whether its incidence has increased or decreased since 2014. Notation of provenance will be crucial in such an effort. A surge in OPM impact occurred in the summer of 2019, with many oaks of all ages (heavily) infested. This fungus (*Erysiphe alphitoides*) affects the leaves, covering them with a felty-white mycelium. It has been known in England since at least 1908, and unlike AOD it is widespread on the continent. It inhibits growth by reducing photosynthesis and has similar effects to insect-induced defoli-ation. In the years since AF surveyed the ancient oaks of High Park (2014) it has had a minor presence; outbreaks such as in 2019 may be climate/weather related as suggested by Elsa Field at Oxford University (https://www.observatree.org.uk/oak-powdery-mildew/). Interestingly, surveys of oaks by AF in August 2020 revealed relatively little mildew infestation, now seen mostly on lower, shaded branches of young trees and on lammas shoots, the latter also showing drought stress. The ancient and veteran oaks appeared almost unaffected.

3.2.2. Other self-seeded trees

After oak, Ash is the most common and widespread tree in High Park. Although there has been planting of Ash in several locations (Mottram and Kerans 2014), many Ash trees are self-seeded; Ash readily colonizes both dry and moist, base-rich soils. It grows faster than oak, but has a shorter life span and has a similar tolerance of shade (Green 2010). During recent years there has been extensive removal of young and mature Ash (with the timber harvested and 'brash' left on site) but with many scattered trees left standing.

There are few ancient and veteran Ash trees, but one specimen is exceptionally large and was recorded with a girth of 5.83m in May 2017 (Figure 3.11). Already damaged then, it has subsequently broken down

Figure 3.11. Ancient Ash. (Photograph by Aljos Farjon, 7 May 2017)

further, but parts are still standing and alive. When more intact, its girth may have exceeded 6m, but we did not see it in that stage. Ancient and veteran Ash trees are much less numerous than such oaks, with 5,849 Ashes versus 20,130 oaks (*Quercus robur* + *Q. petraea*) recorded by the Ancient Tree Inventory (ATI) in England as of February 2020.[6] Ash dieback, a fungal disease (*Hymenoscyphus fraxineus*), was first observed in eastern England on planted stock in 2012, and has subsequently spread across Britain. As at 10 October 2019 no confirmed infections had been reported to Defra from the hectad (10 x 10km) that contains Blenheim Park.[7] In AF's surveys of trees in High Park to date, possible symptoms (to be confirmed) were seen on just one group of about 30 Ash trees on 17 June 2020 at SP4324514981, which had dead branches among living ones with normally developed foliage (AF, pers. obs.). There is evidence that resistance exists in natural populations (Sollars et al. 2017) and that, while many trees will die, natural selection may eventually solve the problem in Britain's woodlands.

Centuries-long use of the land for wood-pasture will have impoverished the woodland flora of herbs as well as trees (Rackham 2003: 199) on a site such as High Park. Apart from the dominant oaks, there will have been a shift from shade-tolerant to light-demanding trees until recently, when the canopy in many places began to close again through both planting and (re)generation of self-seeded trees, mostly Ash. Two tree species that are commonly found in ancient pasture woodland, especially in deer parks, are Hawthorn (*Crataegus monogyna*) and Field Maple (*Acer campestre*). Both are present in High Park, but only Hawthorns are widespread. Among the Hawthorns are several pink-flowering trees, most likely the hybrid *Crataegus* x

---

6 https://ati.woodlandtrust.org.uk/tree-search.
7 https://chalaramap.fera.defra.gov.uk/.

Figure 3.12. Blackthorn (*Prunus spinosa*) flowering in High Park. (Photograph by Aljos Farjon, 24 March 2019)

media, which may have originated from crosses with red-flowering trees (*C. laevigata* cultivar) planted by the 9th Duke. Few of the Hawthorns are veteran or ancient; most are mature trees and there is successful regeneration, albeit browsed by deer.

Blackthorn (*Prunus spinosa*) occurs in and around old quarries and scattered in the Park south of the Bracken/False Brome line (Figure 3.12). It is therefore limited to areas with limestone near the surface. The existing shrubs and trees are large and old, with little or no suckering, indicating that they are remnants of a more open pasture woodland that managed to hang on under a closing tree canopy. Many have fallen, being top-heavy, and re-rooted ('layered') in the crown. Hawthorn and Blackthorn blossoms are an important food source for saproxylic beetles (Chapter 15). Field Maples are only common in a limited area in the west of High Park along the peripheral track. Here stands a large ancient Field Maple, likely a pollard, with a girth of 3.29m recorded on 5 June 2015. In the winter of 2019 heavy damage occurred when the largest limb broke out, exposing the hollow interior of the bolling. Smaller veterans grow nearby, as well as mature and young trees. Elsewhere in High Park Field Maples are found occasionally.

Hazel (*Corylus avellana*) occurs throughout High Park in very low density, except for one location in the northern half, near an old and very large self-coppiced tree (Figure 3.13). Although all trees found appear to be multi-stemmed, there is no record of past coppice management (Mottram and Kerans 2014) of Hazel; this species has a strong tendency to self-coppice and single-stemmed trees are rarely found. The large Hazel stool has a circumference at ground level of about 10m and some fallen stems are layering, establishing separate clonal trees when the connection to the mother tree disintegrates. The Hazel trees nearby may have originated in this way, but also from seed, which the old tree produces in abundance. The paucity of Hazel in this type of woodland (see below) in High Park could be attributed to nut predation by the Grey Squirrel, but quantitative evidence for this assumption is lacking. Hazel is a common sub-canopy tree in oak woodland, but its abundance in such woods

Figure 3.13. Large Hazel coppice (natural) in the northern part of High Park. (Photograph by Aljos Farjon)

is much enhanced by centuries of coppice management (Rackham 1990, 2003). While there is evidence of past Hazel coppice management elsewhere in Blenheim Park (Chapter 1), in High Park the presence of this tree seems more likely to be natural.

Willows (*Salix caprea, S. cinerea* ssp. *oleifolia, S. triandra*) occur in and around wet flushes and in old quarries reaching down into the Oxford Clay. They also grow in semi-permanent ponds. Of the three species found, Goat Willow (*S. caprea*) is the most common and Almond Willow (*S. triandra*) is rare, only found in the southernmost pond. Several willows have grown large, branching low, or layering in the wettest locations. The willows, but especially birches (*Betula pendula*, less commonly the hybrid *B.* x *aurata*) are rare or absent where limestone is near the surface, demarcated by the Bracken/False Brome line (Figure 3.14).

Birches were presumably more abundant in the past when much of the pasture woodland was more open; most trees seen are mature and regeneration is limited to a few locations, for example, between Combe Gate and High Lodge. The large numbers of

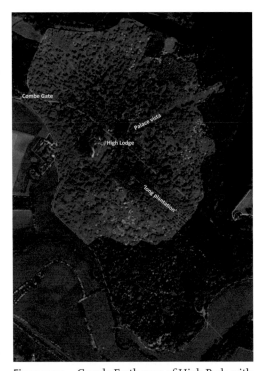

Figure 3.14. Google Earth map of High Park, with a white-shaded area in which Bracken dominates. No bracken ferns occur beyond this line, where False Brome dominates instead (park perimeter marked in red). (Annotated Google Earth image)

Figure 3.15.   Rowan (*Sorbus aucuparia*) (left) and Holly (*Ilex aquifolium*) (right) growing as 'epiphytes' high up on the trunks of veteran (dead and living) oaks. (Photographs by Aljos Farjon)

feral Fallow Deer are likely to be responsible for this, since the near clear-cutting of Ashes in the former Combe Leys area (Figure 1.21) some years previously has not resulted in colonization by birch despite the nearby presence of mature trees. Another indicator of acid soil, Rowan (*Sorbus aucuparia*) is virtually absent on the ground, but grows from a few standing dead stumps of veteran oaks in this part of High Park (Figure 3.15). Is this an edaphic avoidance situation or are the oak stumps refuges from the deer? Next to one of the 'stump rowans' another Rowan grew to a 5–6m tall tree, but then began to die back and is now dead. We have seen almost no seedlings despite the area of Northern Drift deposits over Oxford Clay (Figure 3.16, Chapter 1) providing a widespread acidic substrate suitable for these to thrive, at least to the sapling stage.

Among the shade-tolerant broadleaved trees, the first to mention is Beech, which seems to be slowly spreading from planted trees. Probably most have come from large planted trees on the peripheral slopes of High Park towards the Lake and the valley of the River Evenlode. A second source may now be the planted Beeches in the long plantation, coming to maturity. Yet self-seeded Beeches remain scarce in the more central areas of High Park. The same is true of Sycamore, a tree not native in this part of England (or all of England, depending on your view), which has been spreading from planted trees in the western periphery and a zone above the Lake. For both species, recent management (thinning, 'haloing') has reduced the numbers so that an estimate of how abundant they were very recently cannot be reliably made, unless we were to identify and count sawn stumps. Both Beech and Sycamore, if left to colonize and grow large, could become serious competitors for light to the oaks, in particular low-crowned, slow-growing ancient oaks. The long-term (unmanaged) relationship of Sycamore with native trees in High Park is unclear; there is some evidence that it would not successfully compete with Ash (Morecroft et al. 2008). Wych Elm (*Ulmus glabra*) is locally common in the northern half of High Park, more or less in a transition zone demarcated by the Bracken/False Brome line to the south of the Palace Vista. Here the Northern Drift with pebbles is still present, but the soil is free draining; there are also several large, dry old quarries. A few trees have attained mature size and one is growing through the crown of one of the 9th Duke's oaks in the manner of Beech, killing the oak by overtopping it. Wych Elm is much less susceptible to Dutch elm disease (*Ophiostoma ulmi*) and the trees seen in High Park appear unaffected. They are spread by seed and are not clonal and have developed genetically based resistance to the disease. Under ancient oaks, Elder (*Sambucus nigra*) is often present, or even abundant. Together with Nettles (*Urtica dioica*) this tall shrub or small tree thrives on the enrichment with nutrients, mainly nitrogen, of the soil from decomposing branches and litter that have accumulated over centuries under these trees. There is a small presence of Sweet Chestnut (*Castanea sativa*), both planted and self-seeded, and confined to more acidic soils. Small-leaved Lime (*Tilia cordata*) is very rare, with only one planted and two self-seeded (and fruiting) trees found. Similarly, Buckthorn (*Rhamnus cathartica*), twice found by Martin Corley, is scarce in High Park.

3.3. Woodland classification

In the British National Vegetation Classification (NVC) system the plant community to which the pasture woodland of High Park compares most closely is W8: *Fraxinus excelsior-Acer campestre-Mercurialis perennis* woodland (Rodwell ed. 1991). The 'constant' species Field Maple, Hazel, Ash, Dog's Mercury (*Mercurialis perennis*), bramble (*Rubus fruticosus* agg.) are all (abundantly) present. The moss *Kindbergia praelonga* indicating this community has also been found. These are the diagnostic species that separate W8 from other plant communities, but it does not have to be an Ash–maple–Hazel wood in the sense of Rackham (2003). Oak is the next most common tree in W8, and in many woods the three diagnostic tree species are forming an understorey (often as coppice) among standard oaks. In High Park we have the distinction that the large oaks are not (managed) standards but, apart from the 9th Duke's planted ones, mostly ancient and veteran trees that were self-seeded. Other trees commonly found in this community are Wych Elm and Sycamore and both are present in High Park, although self-seeded Sycamores likely originated from planted trees. Two of the willows, Goat Willow and Grey Willow (*Salix cinerea* ssp. *oleifolia*) are also characteristic in this type of woodland, although usually in lower frequencies than here in High Park. A common small tree in the *Fraxinus-Acer-Mercurialis* community, often next in abundance to Hazel, is Hawthorn. Here, the situation is reversed, and Hawthorns are much more common than Hazels. Less frequent in this community are Blackthorn and Elder; the latter associated with the ancient oaks and therefore more abundant than normally. Dogwood (*Cornus sanguinea*) and Spindle (*Euonymus europaeus*) are other shrubs listed as associates, but both are rare in High Park. On base-rich soils birch is uncommon and Rowan rare or absent, as is indeed the case in High Park. Both are confined to the highest section of the Park with wetter, more acidic topsoil. Holly (*Ilex aquifolium*) also mentioned as an uncommon tree in this community, is, like Rowan, only found as an 'epiphyte' on veteran oaks (Figure 3.15) and is very rare.

Soil characteristics are likely to have a significant influence on all parts of the flora in High Park and are manifested in acidity, nutrient supply, grain size, drainage, susceptibility to compaction, poaching and other potential factors. The W8 (*Fraxinus-Acer-Mercurialis*) woodland community is typical for calcareous mull soils in the lowlands of southern Britain (Rodwell ed. 1991). It occurs most commonly over sedimentary limestones, shales and clays and superficial deposits such as glacial drift, exactly the geological formations and associated soil types we find in High Park. Figure 3.16 is a block diagram indicating the distribution of soil associations in and around Blenheim Park in relation to topography and the underlying geology (Figure 1.2). The five soil associations represented in High Park divide into two groups; the Oak and Denchworth lying on impermeable clay tend to become waterlogged, while the Elmton1, Elmton3 and Aberford associations are drier and relatively free-draining by virtue of the limestone substrata.

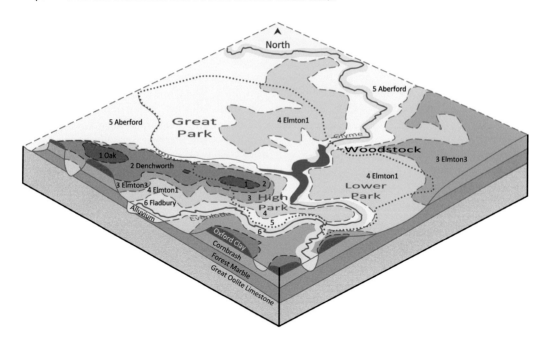

Figure 3.16.   Block diagram of soil associations in Blenheim Park. (adapted from Cranfield, 2019).

1. Oak – Northern Drift; thin cover of pebbles over less permeable and seasonally wet clay subsoils.

2. Denchworth – Oxford Clay; thin, slightly acidic, stoneless, less permeable clayey soils.

3. Elmton3 – Cornbrash; shallow, slowly permeable loams on interbedded calcareous/clay subsoil.

4. Elmton1 – Forest Marble; shallow free draining loamy soil over a brashy limestone.

5. Aberford – Great Oolite; shallow free-draining lime-rich fine loams prone to nitrate leaching.

6. Fladbury – Alluvial clayey stoneless floodplain soils with gravel outliers and a high water table.

(Graphics by T.H. Moller)

On the more acidic and wetter soils within the area with Bracken dominance a weak tendency towards W10, the *Quercus robur-Pteridium aquilinum-Rubus fruticosus* woodland (Rodwell 1991) can be observed. However, the third 'constant' species, Honeysuckle (*Lonicera periclymenum*) is only present in a very small area close to Combe Gate. The scattered mature birches can be seen as elements indicating W10, but where they occur Ash is much more common and there is no case of an oak-birch woodland (*Betulo-Quercetum roboris* on the continent). The apparently very slow spread of Sweet Chestnut and the near absence of Wood Anemone (*Anemone nemorosa*) likewise indicate the incomplete development, or indeed virtual absence, of this community.

The field and ground layers are to be discussed in more detail in Chapter 4. Suffice it here to mention that, besides the dominant Dog's Mercury, Bluebell (*Hyacinthoides non-scripta*), Enchanter's Nightshade (*Circaea lutetiana*), Wood Avens (*Geum urbanum*), Lords-and-Ladies (*Arum maculatum*) and Common Dog-violet (*Viola riviniana*), listed as indicative of the W8 community, are all present in abundance.

CHAPTER 4

Flora and Vegetation

David M. Morris, Aljos Farjon and Jacqueline Wright

4.1. Introduction

In Chapters 2 and 3 the trees of High Park are described and discussed; here we consider all other plants, from mosses and liverworts to ferns and flowering plants, including climbers and shrubs.

The two major habitats for plants within High Park are woodland and grassland. In some areas the two are sharply defined, such as the Palace Vista (Figure 1.21), but what makes the flora and vegetation of the Park interesting is the presence of a mosaic of different sized patches of woodland and grassland, and the fuzzy areas that are not quite either. This structure is called wood-pasture or pasture woodland. We carried out surveys to capture the diversity of plants across these habitats and to sample the pattern of vegetation they form across High Park. The methods and results of these surveys are described in the sections '4.2 Flora' and '4.3 Vegetation'.

Obviously, High Park has a very long history, and our survey represents but a snapshot of its composition. In the not-too-distant past, the Park would have had a much more open structure, dominated by ancient and veteran oaks (Chapter 2), but it has gradually filled in with self-seeded and planted oaks and other trees and shrubs leaving reduced areas of grassland (Chapter 3). Change continues, with widespread thinning of trees and the introduction of grazing cattle having taken place recently. Far from being a fixed backdrop to these changes, over the centuries the plants of the Park will have responded according to their individual natures and in proportion to the magnitude of the changes wrought, with species by turns dying out or colonizing the site, doing well or poorly, or just doggedly surviving. Within this chapter, we therefore describe some of the dynamic components of the flora as well as more persistent elements indicative of the ancient origin of its woodland and other habitats, and identify the similarities of the flora to other sites and conversely the plants that make High Park special.

4.2. Flora

4.2.1. Methods

The primary purpose of the survey of the flora of High Park was to compile lists of the species of bryophytes (mosses and liverworts) and vascular plants (ferns and flowering plants) occurring in the Park, and to characterize their abundance.[1] The principal methods for the compilation of both lists were unstructured walks covering the Park and its main habitats

---

1   The term 'species' is here used to refer to taxonomic species, hybrids, subspecies and other ranks.

David M. Morris, Aljos Farjon and Jacqueline Wright, 'Flora and Vegetation' in: *The Natural History of Blenheim's High Park*. Pelagic Publishing (2024). © David M. Morris, Aljos Farjon and Jacqueline Wright. DOI. 10.53061/NSNS7740

and investigating smaller-scale features where specialist species were likely to be found. The pens for rearing pheasants within the Park were not surveyed.

The bryophytes were mainly surveyed during a visit led by David Morris and Jacqueline Wright in March 2017, with a small number of records added subsequently during other surveys. Records were largely of the asexual gametophyte life stage, but the presence of the spore-producing sexual sporophyte stage was also recorded.

The vascular plants were similarly surveyed during a limited number of focused visits in 2017 and 2018 carried out by Aljos Farjon and David Morris. These visits yielded the majority of species, and the list was augmented by ad hoc records from numerous site visits by Aljos Farjon, David Morris and other surveyors. Records were also supplemented by the results of a grassland survey carried out in 2015 by the Thames Valley Environmental Records Centre (TVERC 2015).

We believe the two lists compiled accurately represent at least 95% of the flora of High Park in the second decade of the twenty-first century.

The abundances of the species recorded were scored on a semi-quantitative scale using the DAFOR notation (Dominant-Abundant-Frequent-Occasional-Rare). Species with a heterogeneous distribution were scored as 'locally dominant', 'locally abundant' or 'locally frequent'.

To describe the 'naturalness' of their occurrence in the Park, vascular plants were grouped into four categories. Spontaneously occurring species were divided into:

Native – plants that are native to the UK or became naturalized before 1500 CE (archaeophytes) and the occurrence of which in High Park is likely a result of natural colonization.

Naturalized – the presence in High Park of any plant likely to be a result of accidental or deliberate human introduction, but where populations have spread and are self-sustaining.

Species found to have been deliberately introduced to the Park but that had not spread were divided into:

Planted native – plants that are native or archaeophytic in the UK but are present in the Park only as planted populations that have not become naturalized.

Planted neophyte – plants introduced to the UK since 1500 CE (neophytes) present in the Park only as planted populations that have not become naturalized.

The UK statuses of species follows the *New Atlas of the British and Irish Flora* (Preston et al. 2002). The nomenclature for vascular plants followed in this account is that of the *New Flora of the British Isles* (Stace 2019). The nomenclature for bryophytes follows *A Checklist and Census Catalogue of British and Irish Bryophytes* (Hill et al. 2008).

4.2.2. Vascular plants

A total of 279 vascular plant species have been recorded in High Park, summarized by status and life form in Table 4.1. The occurrence of 37 species (15%) is a result of human action. Planting has been the major agent of introductions, but some species occur 'in the wild' either because they spread spontaneously from plantings within High Park, or they arrived from outside.

Trees are discussed in Chapters 2 and 3 and form the only group for which planting has influenced the numbers of species and their populations significantly. Some of the planted conifers recorded during the surveys may now have been removed. Table 4.1 presents the cumulative records of tree species since 2017. The plants from the other layers of the vegetation of the Park – the shrubs and climbers and the field layer – are described below.

Table 4.1.   Numbers of vascular plant species recorded in High Park, summarized by life form and status. 'Native' means native to the site. Under 'Planted native' we list species that do not also occur naturally; if they do, they are listed only under 'Native'.

Life form	Native	Naturalized	Planted native	Planted neophyte	Totals
Trees	17	4	3	16	**39**
Shrubs	7	-	1	2	**11**
Climbers	5	-	-	-	**5**
Field layer – Ferns	6	-	-	-	**6**
Field layer – Annuals/biennials	46	2	-	-	**48**
Field layer – Other perennials	161	7	-	2	**170**
Totals	**242**	**13**	**4**	**20**	**279**

Shrubs and climbers

Shrubs have a low presence across most of High Park, both in number of species and abundance: just seven species are native and nearly all are rare, with only Black Currant (*Ribes nigrum*) and Dog-rose (*Rosa canina* agg.) occasional, and only Elder (*Sambucus nigra*) frequent.

Elder is an indicator of nitrogen enrichment and is strongly associated with the ancient oaks in the Park, the soil under which has been enriched by centuries of bird droppings and decomposition of litter and wood. It often grows under the trees with Common Nettle (*Urtica dioica*), another well-known nitrogen-loving species. The rare shrubs in the Park are Buckthorn (*Rhamnus cathartica*), Dogwood (*Cornus sanguinea*), Gooseberry (*Ribes uva-crispa*), Red Currant (*R. rubrum*), the hybrid rose *Rosa x scabriuscula* and Spindle (*Euonymus europaeus*), of which Gooseberry, Red Currant and the rose were seen in one locality only.

Two shrubs have been introduced to the Park: Box (*Buxus sempervirens*) and Rhododendron (*Rhododendron ponticum*).

Figure 4.1.   Black Bryony (*Tamus communis*) climbing on Elder. (Photograph by Aljos Farjon)

Box is present on the steep slope in the south of High Park above the floodplain of the River Evenlode and on the slope southward into The Lince it extends more abundantly by suckering. Presumably it was introduced as evergreen cover for pheasants, though it could have arrived by bird-sown seed. As a native species, Box is Nationally Rare,[2] and the nearest populations thought to be native are in the Chiltern Hills near Watlington, 32km to the south-east of High

2   Online Atlas of the British and Irish Flora; https://www.brc.ac.uk/plantatlas/plant/buxus-sempervirens.

Park. Undoubtedly non-native is Rhododendron, which is rare and was planted in a shelter belt above the artificial waterfall at the end of the Lake. It has not spread from there into High Park, probably owing to unsuitable soil in this part of the Park.

The climbers Black Bryony (*Tamus communis*, Figure 4.1), Hedge Bindweed (*Calystegia sepium*) and Traveller's-Joy (*Clematis vitalba*) were found occasionally, and White Bryony (*Bryonia dioica*), Ivy (*Hedera helix*) and Honeysuckle (*Lonicera periclymenum*) are rare. Indeed, Ivy is restricted to a single location. The scarcity of these species is discussed further below.

The field layer

The richest component of the flora of High Park is the field layer, made up of the low-growing, mostly herbaceous plants beneath the canopy of trees and shrubs and out in the glades and grasslands. Our list of 'field layer' species includes the fern Intermediate Polypody (*Polypodium interjectum*), which was actually found not on the ground but growing epiphytically on the ancient and veteran oaks and on the inside face of the wall that bounds High Park along the valley of the River Evenlode.

The flora of High Park has a significant proportion of species often found in woodland in Britain. Kirby (2020) identified two kinds of woodland vascular plant, the 'generalists' and 'specialists', of which there are 72 and 41 in High Park, respectively. The specialists are listed in Table 4.2; all but four are field layer species, with two climbers and two shrubs. The generalists are not listed here and include a wide range of species that also occur in grassland in High Park, such as Bracken, Bugle (*Ajuga reptans*), Marsh Thistle (*Cirsium palustre*), Red Fescue (*Festuca rubra*) and Selfheal (*Prunella vulgaris*).

The distinction between specialists and generalists considers traits such as tolerance of shade, competition and stress, and dependence on disturbance (Hill et al. 2004). Woodland specialists will generally be better adapted to shade and more tolerant to stress such as low light levels and low nutrient levels, but will be poor competitors and will often react poorly to soil disturbance. Woodland specialists are generally also sensitive to frost and drought, being adapted to a more stable and humid microclimate in the shelter of trees and shrubs. Some can also occur in the shelter of hedges and are not restricted to woodland: Lords-and-Ladies (*Arum maculatum*) is an example on our list.

The two dominant species are False Brome (*Brachypodium sylvaticum*) and Dog's Mercury (*Mercurialis perennis*), while Wood-sedge (*Carex sylvatica*) and Bluebell (*Hyacinthoides non-scripta*) are abundant. Six species are found occasionally and 11 are rare (see Table 4.2). All of the rare species here may be more common or even abundant in other sites. These are Thin-spiked Wood-sedge (*Carex strigosa*), Bearded Couch (*Elymus caninus*), Broad-leaved Helleborine (*Epipactis helleborine*), Tutsan (*Hypericum androsaemum*), Yellow Archangel (*Lamiastrum galeobdolon*), Honeysuckle, Adder's Tongue (*Ophioglossum vulgatum*), Wood-sorrel (*Oxalis acetosella*), Red Currant, Sweet Violet (*Viola odorata*) and Early Dog-violet (*Viola reichenbachiana*). Meadow Saffron (*Colchicum autumnale*) is rare, at least in Oxfordshire, Snowdrop (*Galanthus nivalis*) can be locally abundant, and Tutsan (Figure 4.2) locally frequent in England; Snowdrop is naturalized in High Park and was recorded near Springlock Lodge. We found Adder's Tongue in grassland areas, not under trees.

Of the woodland specialists, some are also indicative of ancient woodland (see Chapter 1 for definition). The concept of 'ancient woodland indicators' was developed by Peterken and Rose, who compiled regional lists of vascular plants that could be used to indicate the ancient origin of woods where documentary evidence was lacking (Peterken 1974, 2000; Rose 1999). The history of High Park (Chapter 1) leaves no doubt that the site is ancient woodland, although there are open glades and its tree cover may not always have been as high as it is now. We might therefore expect an abundance of ancient woodland indicators.

Most of the woodland specialists in High Park are indeed indicators of ancient woodland, and listed in Table 4.2, but only six species are strong indicators. Of these, only Wood-sedge

Table 4.2. Woodland specialists occurring in High Park, and indication for the presence of ancient woodland (AWI). Lists are taken from Kirby (2020); AWI values from Rackham (2003).

Scientific name	Common name	Abundance	AWI
Arum maculatum	Lords-and-Ladies	frequent	-
Athyrium filix-femina	Lady-fern	locally frequent	-
Brachypodium sylvaticum	False Brome	dominant	-
Bromopsis ramosa	Hairy-brome	rare	-
Calamagrostis epigejos	Wood Small-reed	locally frequent	weak
Carex pendula	Pendulous Sedge	frequent	strong
Carex remota	Remote Sedge	locally frequent	moderate
Carex strigosa	Thin-spiked Wood-sedge	rare	strong
Carex sylvatica	Wood-sedge	abundant	strong
Colchicum autumnale	Meadow Saffron	rare	weak
Conopodium majus	Pignut	frequent	moderate
Elymus caninus	Bearded Couch	rare	moderate
Epipactis helleborine	Broad-leaved Helleborine	rare	moderate
Fragaria vesca	Wild Strawberry	occasional	-
Galanthus nivalis	Snowdrop	rare	-
Geranium robertianum	Herb-Robert	frequent	-
Holcus mollis	Creeping Soft-grass	locally abundant	-
Hyacinthoides non-scripta	Bluebell	abundant	moderate
Hypericum androsaemum	Tutsan	rare	weak
Hypericum hirsutum	Hairy St John's-wort	occasional	moderate
Hypericum tetrapterum	Square-stalked St John's-wort	locally frequent	-
Lamiastrum galeobdolon	Yellow Archangel	rare	strong
Lonicera periclymenum	Honeysuckle	rare	-
Mercurialis perennis	Dog's Mercury	dominant	weak
Moehringia trinervia	Three-nerved Sandwort	frequent	weak
Neottia ovata	Common Twayblade	locally abundant	-
Ophioglossum vulgatum	Adder's-tongue	rare	moderate
Oxalis acetosella	Wood-sorrel	rare	strong
Potentilla sterilis	Barren Strawberry	locally abundant	weak
Primula vulgaris	Primrose	occasional	weak
Ribes nigrum	Black Currant	occasional	-
Ribes rubrum	Red Currant	rare	-
Schedonorus giganteus	Giant Fescue	rare	-
Scrophularia nodosa	Common Figwort	frequent	weak
Silene dioica	Red Campion	rare	-
Stachys sylvatica	Hedge Woundwort	frequent	-
Stellaria holostea	Greater Stitchwort	occasional	-
Tamus communis	Black Bryony	occasional	-
Veronica montana	Wood Speedwell	occasional	strong
Viola odorata	Sweet Violet	rare	-
Viola reichenbachiana	Early Dog-violet	rare	weak
Viola riviniana	Common Dog-violet	frequent	-

Figure 4.2.   Tutsan (*Hypericum androsaemum*) in High Park is considered a native species and only a weak indicator of ancient woodland largely because it may also be a garden escape. (Photograph by Aljos Farjon)

Figure 4.3.   Deadly Nightshade (*Atropa belladonna*) occurs in a few places in High Park. It favours woods on chalk and limestone but can also be found on ground disturbed by cultivation. (Photograph by Aljos Farjon)

is abundant, Pendulous Sedge (*Carex pendula*) is frequent, Wood Speedwell (*Veronica montana*) is occasional and the other three species are rare. For a woodland as ancient as High Park, probably dating back to the Anglo-Saxon period, this seems rather meagre. We suggest some possible causes of this under '4.4. Controls on the flora and vegetation'.

The vascular plant flora of the Park is enriched by the diversity of open habitats within the wood-pasture mosaic. These include acid grassland, with species such as Heath Bedstraw (*Galium saxatile*) and Tormentil (*Potentilla erecta*), and calcareous grassland with species such as Common Rock-rose (*Helianthemum nummularium*) and Tor-grass (*Brachypodium rupestre*). However, these habitats tended to be less rich in species than shaded areas owing to (until very recently) lack of grazing.

Notable plants

Thirteen vascular plant species were recorded that are listed as Near Threatened or Vulnerable on national red lists of vascular plants (Cheffings et al. 2005; Stroh et al. 2014) or that are scarce in Oxfordshire (Erskine et al. 2018). These are listed in Table 4.3. Most are typical in Oxfordshire for the kinds of habitat within High Park, but Meadow Saffron (Figure 4.4) and Confused Eyebright (*Euphrasia confusa*) are notable locally, having declined significantly in the county. The latter was not assessed in Erskine et al. (2018) but is rare owing to the scarcity of acid grassland and heathland in Oxfordshire, and the record from our survey is the only recent one from the county.

There are a few other species that are not scarce in the county but were nice to find, indicative of long-established habitats of high nature conservation value. These include Adder's Tongue, Deadly Nightshade (*Atropa belladonna*), Fragrant Agrimony (*Agrimonia procera*) and Southern Marsh-orchid (*Dactylorhiza praetermissa*). Deadly Nightshade (Figure 4.3) is a characteristic species of glades and scrubby calcareous grassland, mostly found in Oxfordshire in the Chilterns and the former Wychwood Forest area. Fragrant Agrimony is a woodland plant that seems to have declined or is under-recorded in north Oxfordshire.

4.2.3. Bryophytes

The survey recorded 66 species, of which 57 were mosses and 9 were liverworts. This result indicates that High Park has relatively low species diversity for bryophytes compared with other mature woodlands that have been surveyed in Oxfordshire (for example, Wright and Wright 2018). Nevertheless, bryophyte vegetation does make an important contribution to the small-scale diversity of habitats that are important for invertebrates.

Table 4.3. Notable plants recorded in High Park.

Scientific name	Common name	Abundance	Conservation status
Fragaria vesca	Wild Strawberry	occasional	England Near Threatened
Campanula rotundifolia	Harebell	rare	England Near Threatened
Carex strigosa	Thin-spiked Wood-sedge	rare	Oxfordshire Scarce
Colchicum autumnale	Meadow Saffron	rare	Great Britain Near Threatened
Cruciata laevipes	Crosswort	abundant	England Near Threatened
Cynoglossum officinale	Hound's-tongue	rare	England Near Threatened
Euphrasia confusa	Confused Eyebright	rare	England Vulnerable
Helianthemum nummularium	Common Rock-rose	occasional	England Near Threatened
Mentha arvensis	Corn Mint	occasional	England Near Threatened
Oxalis acetosella	Wood-sorrel	rare	England Near Threatened
Potentilla erecta	Tormentil	frequent	England Near Threatened
Silene flos-cuculi	Ragged-robin	rare	England Near Threatened
Veronica officinalis	Heath Speedwell	rare	England Near Threatened

Figure 4.4. Meadow Saffron (*Colchicum autumnale*) in flower; it is known in High Park from just this one small population. (Photograph by Aljos Farjon)

Much of the site is on a hill slope, where exposure to wind and direct sun creates a dry habitat that causes desiccation of most bryophytes in drier months. Generally, humidity levels at High Park are not high enough for long enough to sustain the abundant growth of such small moisture-loving plants. Although the geology of a site often has a bearing on which bryophytes may be expected, the bryoflora at High Park shows only slight geological influence owing to the more dominating effects of the overall dryness of the habitats and predominance of vascular plants in the vegetation.

Bryophytes on trees

The ancient parkland trees at High Park (principally Pedunculate Oak) are widely spaced and exposed to wind and sun. Moreover, Pedunculate Oak has an acidic bark that generally supports a lower diversity of bryophytes than trees such as Ash with its alkaline bark. In addition, because of the length of time crustose and foliose lichens have had to gain dominance over many of the branches, they outcompete all but the strongest-growing bryophytes. Indeed, mosses and liverworts are noticeably lacking on the ancient trees at High Park. Among those that can be found are Slender Mouse-tail Moss (*Isothecium myosuroides*) and Cypress-leaved Plait-moss (*Hypnum cupressiforme*) around the lower parts of the trunks.

Figure 4.5.   Straw Bristle-moss (*Orthotrichum stramineum*) grows in small tufts on Ash trunks. (Photograph by David Morris)

Most of the mosses and liverworts occur on the younger trees in denser woodland, although not usually in great abundance as they are similarly affected by the dryness. On younger trees, bryophytes are more able to colonize before the lichens have had a change to take over. Trees with low branches in the woodland areas are not common, yet this is often the place where the greatest bryophytic diversity can be found, especially on the more horizontal branches.

A number of epiphytes are found in small numbers on the woodland trees, especially those with alkaline bark such as Ash, Elder, Goat Willow and Grey Willow. The predominant epiphytes are the cushions and tufts of the moss genera *Orthotrichum*, *Lewinskya* and *Pulvigera* (the latter two formerly in *Orthotrichum*) with six species recorded at High Park: Wood Bristle-moss (*Lewinskya affinis*); White-tipped Bristle-moss (*O. diaphanum*); Lyell's Bristle-moss (*Pulvigera lyellii*); Elegant Bristle-moss (*O. pulchellum*); Straw Bristle-moss (*O. stramineum*); Slender Bristle-moss (*O. tenellum*).

A few of the young Ash trees can be seen with round brown patches on the pale trunks. This is the liverwort Dilated Scalewort (*Frullania dilatata*). On a more mature Ash in the survey area a small shiny patch of the moss Blunt Feather-moss (*Homalia trichomanoides*) was found with its capsules showing dark against the green of the leaves. Very rarely, a few threads of the delightful and tiny liverwort Minute Pouncewort (*Myriocoleopsis minutissima*) were discovered in the deep recesses of fissured willow bark. Both Ash and willow also support small patches of Flat-brocade Moss (*Platygyrium repens*) with its distinctive, seasonally deciduous shoots looking like tiny bottle brushes. Elder stands out bright green in the High Park winter. These shrubs can be densely covered in bryophytes, but where moss cover is less thick, usually on the higher branches, the upright secondary shoots of Lateral Cryphaea (*Cryphaea heteromalla*) can be spotted contrasted against the sky on tree branches at eye level.

Bryophytes on rotting logs and fallen trees

Dead-wood habitat contributes a most important component of a diverse ecology, but the surface of most dead wood at High Park is too dry for bryophytes. Heath Star-moss (*Campylopus introflexus*), a native of the southern hemisphere, doesn't mind drier substrates, however, and another common non-native moss Cape Thread-moss (*Orthodontium lineare*) can also be found on soft, damp dead trees and stumps. Broom Fork-moss (*Dicranum scoparium*) was encountered in abundance on a fallen, shaded and long-dead ancient oak trunk (AF on 4 May 2021). Certain liverworts also favour the rotting substrate, such as the aromatic Variable-leaved Crestwort (*Lophocolea heterophylla*) and Bifid Crestwort (*L. bidentata*) which are abundant. The aroma given off by the chemical compounds in the oil bodies of the leaves can be a most pleasant olfactory experience when the plants are bruised.

Bryophytes on walls
Although superficially very different to trees as a substrate, walls have certain characteristics in common: the most important is the lack of competition from other plants. With such little encroaching vegetation, the distinctive liverwort Wall Scalewort (*Porella platyphylla*) can be seen freely colonizing the calcareous stone walls of the Park, together with the mosses Rambling Tail-moss (*Anomodon viticulosus*), Fox-tail Feather-moss (*Thamnobryum alopecurum*) and the fern-like fronds of Common Tamarisk-moss (*Thuidium tamariscinum*). Rambling Tail-moss is abundant in several places, creating extensive patches (Figure 4.6). Standing out with its glossy leaves on the top of the walls is Beech Feather-moss (*Cirriphyllum crassinervium*).

Figure 4.6.   Mosses on the inside of the wall around the south of High Park, with extensive patches of the matt green tinged with rust-red Rambling Tail-moss (*Anomodon viticulosus*); the shiny Flat Neckera (*Neckera complanata*) and Silky Wall Feather-moss (*Homalothecium sericeum*). (Photograph by David Morris)

Bryophytes in grassland
Neither mosses nor liverworts can get much of a foothold in grassland where the sward is dense, such as in High Park. The grasslands here support very few species. In the few places in which mosses can grow, typical species of grassland were present, the most robust being Neat Feather-moss (*Pseudoscleropodium purum*) and Springy Turf-moss (*Rhytidiadelphus squarrosus*).

4.3. Vegetation

4.3.1. Methods
The vegetation of the Park was investigated during the spring and summer of 2019. We recorded the vegetation within 20 randomly located 50 × 50m plots, avoiding areas of planted trees. Within these plots the tree, shrub and ground vegetation was recorded, and two to three 2 × 2m plots were chosen to sample the variation in the ground vegetation within the plot, such as across gradients of shade or management. A total of 55 smaller

plots were recorded. Within all plots, the presence and abundance of species was recorded, together with various vegetation parameters, such as canopy cover and vegetation height. The abundance of species within the different layers of vegetation was analysed to identify assemblages of species defining different types of vegetation across the Park, and compared to the plant communities described by the National Vegetation Classification (NVC; Rodwell 1991–2000). These were related to environmental factors; those we believe to be the main determinants of the vegetation patterns observed are discussed in Section 4.4: 'Controls on the flora and vegetation'.

4.3.2. Results

There are striking differences in the vegetation across small and large scales within High Park. As in most woodlands, the survey found almost all of the variation to be in the field layer, which varied fairly independently of canopy composition. Maps showing aspects of the canopy structure are provided in Figure 4.7. and the various vegetation types and the patterns these form across the Park are described below.

The canopy and shrub layers

Most of the variation in the shrub and canopy layers is in their structure rather than species composition. As described in Chapter 3, the dominant tree species across the Park is Pedunculate Oak. Only Ash approaches it in abundance, but older trees are rare, and it does not form a significant part of the canopy. Stands of scrub were generally too small to sample by our method, and include the stands of low-growing Grey Willow scrub around springs on the north side of the hill and the stands of Blackthorn and Hawthorn scrub on the south side of the hill. Stands of scrub are shown in Figure 4.7 as areas with a shorter canopy height, typically less than 5m.

Structurally, the canopy layer is much more complex, with considerable variation in the canopy cover and sizes and ages of trees. This complexity reflects past and current disturbance events, most notably the death and falling of veteran oaks, creating gaps but also long-remaining large dead-wood, and the succession of formerly open areas to scrub and woodland. Large areas of the southern part of the Park are dominated by more or less closed canopies of short young trees and scrub, forming small glades with scattered scrub, while much of the northern part of the Park has a more open character with larger glades. The latter is partly due to recent thinning of Ash trees.

The field layer

The sample recorded 15 bryophytes and 128 herbaceous species (respectively, 23% and 46% of the total diversity of bryophytes and vascular plants), and these formed several more-or-less distinct types of vegetation. Two of these types formed recognizable grassland habitats while the others formed a continuum between grassland and the ground flora under almost complete canopy cover. Most of the latter types were defined by abundance of the very gregarious species Bracken, Bluebell, Dog's Mercury and False Brome.

A striking vegetation pattern across the Park is the abrupt disappearance of Bracken from lower parts of the hill, from approximately 20m above the level of the Lake (below around 95m above Ordnance Datum). The approximate extent of Bracken is shown in Chapter 3, Figure 3.14. Bracken is present in varying abundance within most types of vegetation across the upper parts of the Park, and in many unmown areas it forms dense species-poor stands. Among the few associated plants in such stands are vernal species able to complete their life cycle before the Bracken canopy closes over in summer, such as Bluebell and Dog's-mercury, and tall plants intolerant of mowing such as Bramble and Stinging Nettle. This type of Bracken-dominated vegetation has clear affinities to the type of field layer vegetation found in the NVC woodland plant community W10 *Quercus robur-Pteridium aquilinum-Rubus fruticosus* woodland and, outside woodland, the W25 *Pteridium aquilinum-Rubus fruticosus* underscrub.

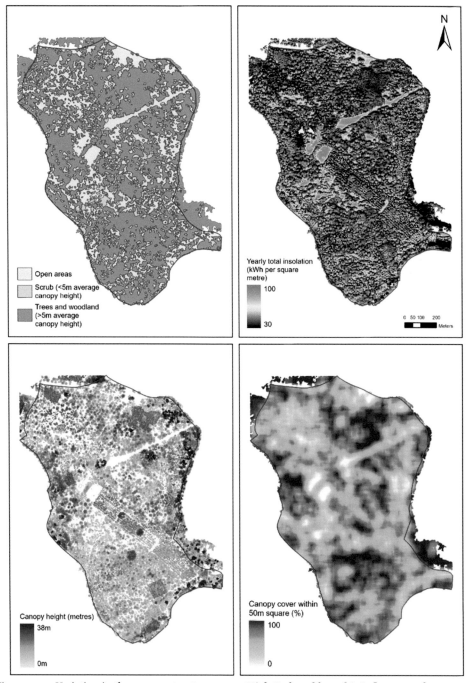

Figure 4.7. Variation in the canopy structure across High Park and how this influences solar energy (insolation) reaching the field layer (top right). Produced using National LiDAR Programme 2020 survey data obtained under Open Government Licence.

The mown Palace Vista and large glades in the northern half of the Park support stands of acid grassland, a rare habitat in Oxfordshire. The vegetation of most stands is grass-dominated, comprising the grasses Common Bent (*Agrostis capillaris*), Red Fescue (*Festuca rubra*), Sweet Vernal-grass (*Anthoxanthum odoratum*) and Yorkshire Fog (*Holcus lanatus*), and with abundant mosses such as Springy Turf-moss. The presence and abundance of other herbs varies between areas, but Common Dog-violet (*Viola canina*) and Germander Speedwell (*Veronica chamaedrys*) are constant, and Barren Strawberry (*Potentilla sterilis*), Bluebell, Creeping-Jenny (*Lysimachia nummularia*) and Cuckoo Flower (*Cardamine pratensis*) are frequent, while Bugle (*Ajuga reptans*) and Crosswort (*Cruciata laevipes*) are occasional. Barren Strawberry and Bluebell are usually woodland plants in Oxfordshire. Calcifugous species present in our sample include frequent Bracken and Lesser Stitchwort (*Stellaria graminea*), occasional Field Wood-rush (*Luzula campestris*) and rare Pignut (*Conopodium majus*) and Tormentil (*Potentilla erecta*); others such as Heath Bedstraw (*Galium saxatile*) and Wood-sorrel were locally abundant in the glades south of High Lodge but not widespread enough to be sampled. The sample from this type included one plot recorded from the quite different and species-rich grassland along the entrance from Combe Gate, with species such as Adder's-tongue, Common Spotted-orchid (*Dactylorhiza fuchsii*), Fragrant Agrimony (*Agrimonia procera*), Meadowsweet (*Filipendula ulmaria*), Meadow Saffron and Wood Small-reed (*Calamagrostis epigejos*). In terms of the NVC, some of the sampled areas conform to the acid grassland plant community U4 *Festuca ovina-Agrostis capillaris-Galium saxatile* grassland, *Holcus lanatus-Trifolium repens* subcommunity, while others seem difficult to accommodate within the NVC, particularly the area near Combe Gate. The latter vegetation is more like a wood meadow than any other in the Park, combining woodland and grassland species and tall herbs indicative of humid conditions (Peterken 2017).

In shady but still quite open areas on the plateau across the northern part of the site, the above grassland grades into grassy vegetation with abundant Bluebell and comprises the more shade tolerant grasses Rough Meadow-grass (*Poa trivialis*), Sweet Vernal-grass and Yorkshire Fog. Barren Strawberry, Bracken and Common Dog-violet are again frequent (Figure 4.8). This type of vegetation forms magnificent displays of bluebells in spring. A number of other

Figure 4.8.   Glade in the northern half of High Park with Bluebell and shade-tolerant grasses. (Photograph by David Morris)

Figure 4.9. Vegetation in the southernmost part of High Park, with rank grassland dominated by False Brome and Tor-grass. (Photograph by David Morris)

woodland species were present within the sample of this type of ground flora vegetation, including Common Twayblade (*Neottia ovata*), Three-veined Sandwort (*Moehringia trinervia*), Wood-sedge (*Carex sylvatica*) and Wood-sorrel. This is the kind of ground flora one might find in woodlands of the NVC type W10 *Quercus robur-Pteridium aquilinum-Rubus fruticosus* woodland.

Throughout the Park there is vegetation characterized by False Brome. In the northern part of the site, examples of this type of field layer have constant Bracken and Bramble at low cover, Common Dog-violet, Tufted Hair-grass (*Deschampsia caespitosa*) at low cover and the mosses Common Feather-moss (*Kindbergia praelonga*) and Common Smoothcap (*Atrichum undulatum*), and a small number of other species such as frequent Wood Sedge. This type of field layer vegetation has affinities to the NVC woodland plant community W8c *Fraxinus excelsior-Acer campestre-Mercurialis perennis* woodland, *Deschampsia caespitosa* subcommunity. Within the Park, it occupies an intermediate elevation below the previously described field layer with abundant Bluebell and above that found at lower elevations described below.

The field layer around the lower parts of the hill is generally species-poor, often overwhelmingly dominated by False Brome. Although scattered at higher elevations, it is only lower down that Dog's Mercury becomes a constant feature of the field layer. Otherwise, the field layer at lower elevations lacks distinctive features, though with affinities to that found under W8 *Fraxinus excelsior-Acer campestre-Mercurialis perennis* woodland. Similar vegetation was also found under low scrubby and often incomplete canopies in areas that had evidently succeeded from grassland relatively recently.

In the southern half of the Park and on its northern edge, glades are occupied by calcareous rather than acid grassland (Figure 4.9). This vegetation is now grazed but formerly was unmanaged, and was very rank at the time of survey, dominated by the tall coarse grasses False Oat-grass and Tor-grass. Other than Tor-grass, calcicolous species are few, but include occasional Hairy St John's-wort (*Hypericum hirsutum*), Hairy Violet (*Viola hirta*) and Spiked Sedge (*Carex spicata*) in our sample, as well as, but not represented in the sample, Common Rock-rose (*Helianthemum nummularium*) and Narrow-leaved Meadow-grass (*Poa angustifolia*). This vegetation is related to the NVC plant community CG4 *Brachypodium pinnatum* grassland, a type of semi-natural grassland typical of the oolitic limestone bedrock that underlies much of north and west Oxfordshire.

4.4. Controls on the flora and vegetation

4.4.1. Geology, soils and topography
Across High Park the main control on the vegetation is the underlying rocks and the soils derived from them (Chapter 1; Figure 3.16), which exert an influence primarily through their permeability and chemistry. The underlying Jurassic limestone sinks rainwater quickly and where this is not overlain with less permeable deposits, the soil dries out in summer. This

is the situation in the southern parts of the Park, beyond the limits of Bracken, where soils are neutral to alkaline and well-drained. The field layer is consistent with this, marked by the abundance of False Brome across the southern parts of the Park, and larger glades and rides support a rather species-poor but recognizably calcareous grassland vegetation, with strong calcicoles such as Common Rock-rose and Tor-grass confined to such areas. The same situation prevails on the steeper slopes above the Lake.

On the plateau, gradually sloping down from the highest point near High Lodge in all directions, there is a cap over the limestone of different deposits. The highest ground is covered by a layer of sand, loam and pebbles from the Northern Drift, which covers bedrock of Oxford Clay and Cornbrash. This drift layer is permeable, but retains rainwater longer than the limestone, is moderately acidic and marks the occurrence of Bracken, Bluebell and a range of calcifuge species mentioned above. Locally, historic diggings for clay and limestone have created exposures of deeper geology or created ponds where the clay was not broken through. The flora has reacted to these disturbances with species ranging from Goat Willow and Blackthorn (the latter where water drains away) to Soft-rush (*Juncus effusus*) and Common Water-starwort (*Callitriche stagnalis*). The damper conditions on the top of the hill also support a range of species of damp grassland and woodland that are absent from lower down the hill.

The topography in High Park is characterized by a more or less level plateau flanked on two sides and in the farthest south by more or less steep slopes. To the west and south they lead into the valley of the Evenlode, to the east that of the Glyme (now an artificial lake). Surrounded by a largely open landscape, this topography exposes High Park to high winds with a desiccating effect, as emphasized in the section on bryophytes. It would also affect other epiphytes: ferns do occur on trees (mainly Intermediate Polypody) but they are scarce. The slope also influences drainage, interacting with geology: the poorer drainage of the top of the hill likely exacerbated by its flat topography reducing run-off of rainwater, while the steep slopes around the south of the hill remove water from the already well-drained soils.

Shade, glades and succession

The division between wooded and open areas is complex, arbitrary and, moreover, continually changing. The contrast between the warm, sunny and grassy open areas and adjacent areas under trees is very obvious. Clearly, plants need sunshine, and how much of this is intercepted by trees and shrubs before reaching the field layer exerts a strong control; a model of how much energy from the sun reaches the field layer in different parts of the Park is shown in Figure 4.7. Openness of the canopy also controls the climate experienced by the field layer more generally: not only are smaller glades and areas under the canopy shadier, but also the air is more still and humid, and changes in temperature are less extreme than in larger open areas. Thus, the smaller glades have an abundance of both grassland and woodland plants, with Bluebells growing out in the open in the north of the Park and False Brome in smaller glades further south. Bluebell in particular likes the soil and air cool and humid, and in Oxfordshire it seldom grows outside woodland.

Much of the tree cover across High Park is of relatively recent origin (Chapter 3), and this process of succession is also evident, particularly in the southern part of the Park. Not only are there many young trees and much scrub, but the field layer is rather species-poor, made up of species able to colonize shaded places relatively quickly, such as False Brome and Wood Sedge, and plants of open situations able to persist in light shade, such as Crosswort and Tor-grass. Species of long-established woodland such as Dog's Mercury and ancient woodland indicator species are noticeably rare across the south of the Park.

As described above, in some areas the closing canopy that had developed by infill of trees has lately been opened up with the thinning of Ash and Sycamore. The recent introduction of cattle has already made formerly impenetrable thickets of willows, Bracken and Common Nettle more accessible and will open these up further. It remains to be seen whether the cows will be more effective than the mowing (if this were discontinued) in pushing back the Bracken that invades the glades. The situation has become more dynamic than it was when our surveys began – in just four years.

Grazing

It is too early to tell what influence the new herd of cattle will have on the flora and vegetation of the Park (Chapter 19). However, though more furtive beasts, deer have been grazing in High Park for centuries and their relentless nibbling and dietary preferences are apparent in the vegetation.

First, in a site as relatively lightly shaded as High Park and where much of the canopy has formed recently, one would expect an abundance of Bramble and Ivy, but Bramble patches are few and far between, while Ivy is almost completely absent. As evergreen species, both are grazed by deer in winter. In field layers with heavy deer grazing, grasses and grass-like plants such as False Brome and Wood Sedge, tend to proliferate while broadleaved herbs such as Primrose (*Primula vulgaris*), Red Campion (*Silene dioica*) and Wood-sorrel are eaten (Kirby, 2001). Honeysuckle can of course climb out of reach of deer, but the plant spreads on the ground, where it is vulnerable. Near Combe Gate there is a concentration of Bramble, Honeysuckle, Primrose and Wood Sorrel, presumably because deer avoid this area owing to human disturbance.

Evidently, there are some plants deer won't eat, such as Bracken. Bluebell is mildly toxic, though Muntjac certainly eat the flowers and leaves; it would seem that as geophytes these plants must be resilient, quickly replacing lost leaves. However, Cooke (2006) found in Monks Wood (Cambridgeshire) that long-term exposure to this animal's browsing reduced the size of re-emergent plants as well as the number of flowers, reducing seed set. It is probably all a matter of density and numbers of the plants as much as of the deer. Other toxic plants include Dog's Mercury and Lesser Celandine (*Ficaria verna*).

Missing plants

What plants occur and where is often as much a matter of chance as anything else, but it is striking that several species that are common in woodlands in Oxfordshire are absent from High Park. One of the species expected to occur, but which we didn't find, is Wood Anemone (*Anemone nemorosa*).[3] A plant of damper woodland soils, it often occurs with species such as Bluebell and Lesser Celandine, both of which are common in the Park. Like these, Wood Anemone is toxic, so presumably its virtual absence is not attributable to deer. The common grasses of ancient woodland Wood Melick (*Melica uniflora*) and Wood Millet (*Milium effusum*) are also absent, but Bearded Couch (*Elymus caninus*) is present, though rare.

There is a more forthcoming explanation for the apparent absence of Wood Spurge (*Euphorbia amygdaloides*). The seeds of Wood Spurge lie dormant in the soil for many decades, ready to take advantage of major disturbances, so that it is often most prolific in sites historically managed as rotational coppice. The historic wood-pasture system of High Park, in which most of the site would once have been grassland, perhaps precluded the development of a seed bank that could maintain a population between suitable disturbance events.

3   A very small patch of Wood Anemone was discovered at SP43321568 by Martin Corley after our botanical surveys.

CHAPTER 5

Fungi (Excluding Lichens)

A. Martyn Ainsworth, Richard Fortey,
Alona Yu. Biketova and Laura M. Suz

5.1. Introduction

Fungi are an essential part of the ecology of High Park's ancient pasture woodland, although their ubiquity and diversity are likely to be far from obvious to the general visitor. On a hot summer's day, it is possible to take a stroll between the ancient oak trees seeing virtually no evidence of fungal life at all except, perhaps, for a few bracket fungi whose tough and woody fruit bodies can survive for many years on standing or fallen trunks.

Fungi play vital roles in recycling and redistribution of the chemical constituents of woodland vegetation. Remarkably this even extends to the decomposition of those relatively durable plant structural materials cellulose and lignin. These essential ecological processes are continuously taking place above and below the surface of the soil, but they are largely invisible and therefore the environmental role of fungi is often taken for granted or simply overlooked. The woodland habitat is highly heterogeneous, and fungi are usually very particular about their microhabitats, substrates and associates. Whereas some soil-dwelling species are generalists that consume all litter components lying in their path, for example, the fairy ring-forming *Clitocybe* and *Lepista* spp., others are very fastidious feeders and are usually only seen during deliberate searches of their favoured substrates. Examples of these seen in High Park include the tiny mushrooms of Pink Oakleaf Bonnet (*Mycena smithiana*), pale pink when fresh, which are only found on old damp fallen oak leaves, and the brown goblet-shaped fruit bodies of *Ciboria batschiana*, which only emerge from old acorns buried in the soil (Figure 5.1).

Owing to the limited time allowed for our field surveys, our principal focus from the outset was to try to record those fungi that were associated with the ancient oak trees. However, unless we could see a direct connection between fruit bodies and oaks or sample directly from oak roots (see below), it became clear that it is not always possible to distinguish those fungi strictly associated with the ancient oaks from those depending on other elements of the Park's flora. The 'working part' of any fungus, known as the mycelium, moves through the environment in ways that are mostly invisible to the surveyor, and even when mycelial threads are observed spreading under leaf litter or fallen logs they cannot readily be attributed to a particular species. Stands of Bracken, bramble and Stinging Nettles surround many of the oaks, so it is likely that some of the fungi collected there are more concerned with the local ground cover than with the dominant tree species. There are several naturally damp areas of

A. Martyn Ainsworth, Richard Fortey, Alona Yu. Biketova and Laura M. Suz, 'Fungi (Excluding Lichens)' in: *The Natural History of Blenheim's High Park*. Pelagic Publishing (2024). © A. Martyn Ainsworth, Richard Fortey, Alona Yu. Biketova and Laura M. Suz. DOI: 10.53061/OZOM2670

Figure 5.1.   A cluster of brown goblet-shaped fruit bodies of *Ciboria batschiana* growing from the remains of a buried acorn. (Photograph by Aljos Farjon)

High Park in which willows (*Salix* spp.) predominate, and dead wood lying in these localities has its own suite of associated fungi. Some of these inevitably found their way into our lists. This was particularly evident during survey visits following prolonged dry weather when fungi were generally hard to find and the residual dampness of the 'willow oases' became irresistibly attractive to the survey team. Birch (*Betula*), Hazel (*Corylus avellana*), Ash (*Fraxinus excelsior*) and occasional Beech (*Fagus sylvatica*) trees are interspersed among the oaks, and these also contributed fungal species to our total inventory. Our list of visible fungi therefore inevitably embraces more than those obligatorily connected with the signature oaks, although it falls very far short of being a full fungal inventory. Although some fungi live entirely on other fungi or on animals, the fungal diversity at a site is largely a reflection of the diversity of the plant species present. Each plant species has its own suite of associated fungi. A complete fungal inventory for High Park would therefore only be achievable following many years of detailed mycological study of all the site's plant species and microhabitats, both above and below ground.

5.2. Historical fungal recording at Blenheim

High Park is a private woodland and so, prior to our surveys, hunting for fungi within its bounds was almost entirely confined to the margins of the public footpaths. Nevertheless, the Thames Valley Environmental Records Centre (TVERC) database holds the results of two limited fungal surveys of High Park, dating from 1968 and 2002, and these records were made available to the High Park Biodiversity Survey (HPBS) through a data exchange agreement. Twenty-three species were recorded, including six species not recorded in the current project: Batchelor's Buttons (*Bulgaria inquinans*), Shaggy Bracket (*Inonotus hispidus*), Beech Milkcap

(*Lactarius blennius*), Giant Polypore (*Meripilus giganteus*), Oyster Mushroom (*Pleurotus ostreatus*) and Lumpy Bracket (*Trametes gibbosa*).

The true number of fungi previously recorded in High Park is difficult to ascertain owing to the once common practice of referring records to site names such as 'Blenheim Park' or 'Blenheim' rather than to any specific subunit of the site or, more precisely, to a set of geographical coordinates. Well-annotated and georeferenced fungal recording in High Park is therefore very much in its infancy. This severely limits the value of comparisons between our fungal list and those generated from longer-term surveys at other sites. Indeed, this was one of the major reasons why fungal conservationists shifted their attention away from trying to assess the conservation importance of sites based on their seemingly endless fungal inventories and towards the ranking of sites based on the presence of habitat-characteristic species (further details of the oak-associated dead wood assemblage currently used in the selection of protected sites are given below).

Although High Park has undoubtedly evaded mycological attention thus far, this is not the case for the more accessible parts of the Blenheim Estate. For example, one of the highlights of the many Blenheim collections made by N.H. Sinnott in the 1960s was a collection of a small brown and scaly parasol mushroom. This was subsequently examined by Reid (1966), who identified it as *Lepiota hymenoderma*, a species new to science.

The British Mycological Society's (BMS) 1949 Autumn Foray was held in Oxfordshire and the day visit to Blenheim Park 'provided some of the best collecting of the week' (Ainsworth, 1950). Two species from this foray helped to put Blenheim on the mycological map. First, a station for the uncommon Devil's Bolete (*Rubroboletus satanas*) was found under a clump of Beeches overlooking the Lake. An apparently site-faithful species, it has subsequently been documented, sometimes in quantity, in the same location for over half a century (Marren 1997, 2000, 2012). The second outstanding fungus recorded on the 1949 foray was Oak Polypore (*Buglossoporus quercinus*) (=*Piptoporus quercinus*). This find represented its first record on Beech rather than on oak (Cartwright 1951), which would have been highly significant – had it been true. On re-examining the collection at Kew in 2001, Roberts (2002) redetermined it as the common Blackfoot Polypore (*Polyporus varius*) (=*Cerioporus varius*). Oak Polypore was thereby excluded from the Oxfordshire list – only to be reinstated just one year later, this time on oak and fully verified. Full details of its known Blenheim stations were recorded in the present survey.

5.3. Survey methods and limitations

The core survey team comprised members of the local fungus recording group (Fungus Survey of Oxfordshire) with a few professional and volunteer mycologists assisting on some of the visits. The route of each visit always included some previously unrecorded areas to try to maximize geographic coverage. For the most part, our visits to the site followed the format of a traditional 'fungus foray' with collectors spreading out to see what they could find in the way of visible fruit bodies. These above-ground and often ephemeral fungal structures support the development of the reproductive spore-bearing organs from where the microscopic spores are discharged and dispersed. Sampling and examination of these structures forms the basis of traditional fungal identification methods. Fruit bodies that could be identified in the field were left *in situ* to minimize habitat disturbance. However, collections of fungal materials are often required because correct identifications are increasingly dependent on microscopic examination and/or DNA sequencing and analysis. For those genera accommodating many morphologically similar species, DNA-based approaches are rapidly becoming an essential component of the identification process. Collections of the rarer species were often subsequently returned to the Park to continue to disperse their spores in suitable habitat. Those sampled for DNA analysis were dried and preserved as voucher specimens for future reference in the national collections held in the fungarium of the Royal Botanic Gardens, Kew.

We recorded and collected fungi that are visible to the naked eye in the field – that is, macrofungi. Macrofungi are capricious in appearance. A species once seen may never be discovered again. In general, fungi only appear after plentiful rainfall, and then the optimal season is usually in autumn, often in late September to October or November. Many fruit bodies are relatively evanescent, remaining in perfect condition to disperse their airborne spores for no more than a day or two before decaying to a point that defies identification. Hence visits were concentrated in the main autumnal mushroom season, but these were supplemented by some earlier surveys to pick up vernal or summer-fruiting species. During the five years of the project many species have been found only once, and doubtless if the survey had been extended for another decade the list of fungi recovered would have been considerably longer. This pattern is known in many well-studied localities: common species are sure to be discovered each year; others only appear when conditions are exactly right.

Another fungal world exists at a microscopical level: minute sporing structures that may be seen as no more than dark spots, dots or dust on twigs and leaves; fungi that inhabit the exoskeletal surfaces of insects (e.g. Laboulbeniales) or consume them from within; minute pathogens and parasites of animals, plants and even of other fungi; fungi that live unseen inside healthy plant parts or travel through animal guts before beginning to feed on deposited dung. Hardly any of these appear in our lists. Our lists are necessarily a reflection of our limited personal interests, experience, resources and the availability of personnel. It is well known that the length of a fungal inventory can always be increased by inviting more taxonomic specialists to survey and study a site at the appropriate times of year.

Fifteen collections, mostly representing different species of *Cortinarius* s. lat., *Hebeloma* and *Inocybe* s. lat., were identified by molecular methods based on analyses of their barcode (internal transcribed spacer, ITS) sequences obtained from dried fruit body samples (by AYB). DNA extraction and amplification were carried out using the Phire™ Plant Direct PCR Kit (Thermo Scientific™) following the manufacturer's protocols and parameters using primers ITS1F and ITS4B (Gardes and Bruns, 1993) and an annealing temperature of 55 °C. Sequencing was carried out by LGC Genomics GmbH (Berlin, Germany). Sequences were manually edited and assembled using Sequencher 5.4.6 software (Gene Codes Corporation, Ann Arbor, MI, USA) before their submission to the GenBank repository with accession numbers OQ745720– OQ745734. A further batch of four sequences was kindly provided by K. Liimatainen (see Acknowledgements).

Many mushroom- or toadstool-forming fungi (agarics), for example, *Amanita*, *Cortinarius*, *Hebeloma*, *Inocybe*, *Lactarius*, *Russula*, *Tricholoma* and most of the bolete genera, including the Devil's Bolete discussed earlier, exist in underground partnerships with living tree roots. Similar partnerships occur between trees and truffles or truffle-like fungi, for instance *Elaphomyces*, *Genea*, *Hymenogaster* and *Tuber*; cup fungi, for example *Humaria* and *Otidea*; and corticioid or crust-forming fungi, such as *Sebacina*, *Thelephora* and *Tomentella*. The fungus-root bridge is where they trade sugars, originally formed in the leaves of the woodland canopy, for water and nutrients such as nitrogen and phosphorus. These below-ground exchanges occur in composite plant/fungal structures resembling miniature corals or Christmas trees (Figure 5.2) known as ectomycorrhizas (literally 'outside fungus roots'), which are vital to the healthy functioning of our woodland ecosystems.

In a departure from the traditional above-ground-only survey of fungal fruit bodies, we included a small below-ground DNA-based survey (by Laura M. Suz) of mycorrhizal root-tips collected from beneath seven of the ancient oaks on two site visits. Once in the lab, the roots were rinsed with water, examined under a dissecting microscope and fungal DNA sequences were obtained from individual mycorrhizal root tips. These were analysed by comparison with publicly available reference sequences to try to identify the ectomycorrhizal species present following standard methods (Suz et al. 2014) and each of these was given a species hypothesis code (Appendix 3) derived from the UNITE database (Kõljalg et al. 2013).

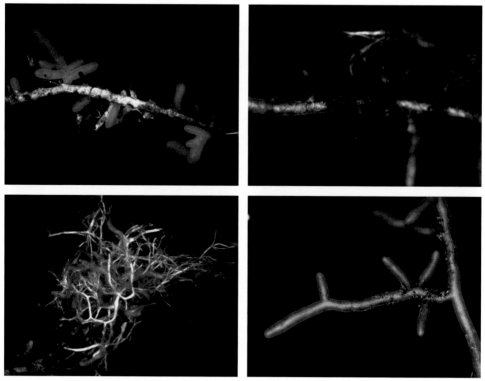

Figure 5.2.   Four mycorrhizal roots of oak showing the distinctive plant/fungal structures (ectomycorrhizas) formed by *Russula* sp. (upper left), *Cenococcum geophilum* (upper right), *Cortinarius* sp. (lower left) and *Lactarius quietus* (lower right). (Micrographs by Laura M. Suz)

These methods enabled us to detect some of the many fungi that were present but not producing fruit bodies on the day of the survey including those that have never been known to produce fruit bodies at all. Root tip analysis can also detect fungi that produce subterranean (truffle-like) or otherwise inconspicuous fruit bodies that are usually overlooked in above-ground surveys. Thus, our below-ground survey, albeit limited in scope, provided a rare glimpse into the subterranean fungal communities of High Park and yielded data on species that were invisible to the fruit body surveyors.

The application of DNA-based methods has revolutionized fungal taxonomy, and the number of species being added to the British list is currently accelerating beyond 100 species per year as a result. Many of our native fungal species are being split into smaller and more tightly defined segregate species while other species retain their boundaries but shift their allegiances between larger- and smaller-sized genera according to differing taxonomic opinions. Any taxonomic view we adopt is clearly going to represent a very limited snapshot of a highly dynamic situation and one that will soon be regarded as out of date, probably even before this book reaches the press. We have based our nomenclature and taxonomy on the online Index/Species Fungorum (2022) and Læssøe and Petersen (2019). We have cited recommended English names, where these exist, according to the list published by the BMS (2022).

Finally, it should be mentioned that the main fungal season overlaps with the pheasant shooting season, and access to the Park was not always possible when it should have been most productive. Our survey must, therefore, be regarded as preliminary, and it is certain that further surveys would repay the effort with copious additions to the inventory. To put this into context, we may quite confident about our knowledge of the flora of High Park (see Introduction and Chapter 4), but such confidence certainly does not extend to what we might term the 'funga'. Not only are mycologists unable to list the complete funga of any site in the country yet, but we have also even had to invent the word 'funga' to try to overturn the outdated, but obstinately persistent, 'flora and fauna' two-kingdom view of biodiversity. Since it is estimated that over 93% of the world's fungi are still unknown to science (Cannon et al. 2018), the present mycological survey merely represents a small step towards a better understanding of the fungal populations and communities currently living in High Park.

5.4. Above-ground survey results

Our five-year fruit body survey yielded 862 records comprising 428 fungal species with a wide range of feeding strategies (trophic modes) (Figure 5.3).

Of these, 60% were recorded on one visit only. Year one of the survey yielded 216 species. This was followed by 103 additions over the next three years (necessarily hampered by periods of national COVID-19 restrictions) and a further 109 species were recorded during the final year (2021). Our DNA-verified fruit body collections of the mycorrhizal species *Cortinarius alboadustus* and *Inocybe jucunda*, found under two oak trees in 2021, represented two new additions to the British checklist (Ainsworth and Henrici 2023).

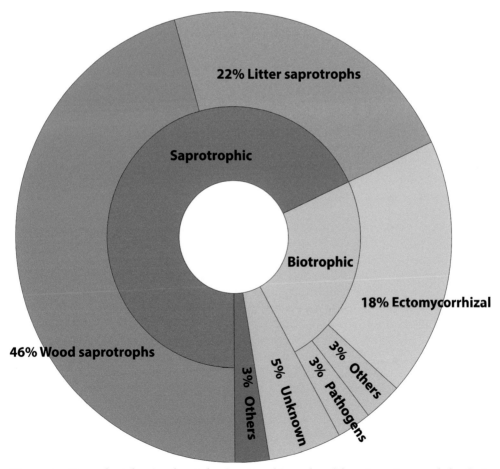

Figure 5.3.   Krona chart showing the predominant trophic modes of the 428 species recorded as fruit bodies during the five-year above-ground fungal survey. Saprotrophs feed on dead matter whereas biotrophs obtain nutrients from other living organisms.

It would be very informative to view the total of 428 species within a national context. However, few comparable datasets from fungal surveys of oak-dominated pasture woodlands are readily available. Those that exist were derived over a much longer recording period and/or a larger geographic area. Nevertheless, a compilation of site history and biodiversity records was published in 2000 for Moccas Park, Herefordshire, in a book that, in some respects, can be regarded as a prototype for the current work (see Introduction). This included an estimated 654 fungal species recorded at Moccas between 1873 and 1998 (Blackwell 2000). Blackwell's chapter also compares Moccas with the Dudmaston Estate, Shropshire, where 30 years of regular recording had yielded a list of approximately 700 species. The personal records of E.E. Green and those of one of the current authors (AMA) were compiled in 2003 to produce a list of almost 1350 non-lichenized fungal species occurring on the Crown Estate at Windsor. The Berkshire component was subsequently extracted and published in a county 'flora' for Berkshire (Crawley 2005). Although the Windsor Estate is famous for its collection of open-grown oaks, the tally of recorded species was undoubtedly boosted by its expanses of conifer plantations, bogs and heaths. Comparing High Park's funga with that of other sites therefore requires caution, because some have been surveyed for a longer period (Dudmaston and Moccas) and/or display greater habitat diversity (Windsor). It is likely that we have not yet recorded half of the fungal species that live in High Park.

5.4.1. Species of conservation interest

Saprotrophic fungi are those that feed on dead organic matter. In the High Park survey, our 'saprotrophic focus' was on plant saprotrophs. These range from wood-inhabiting (lignicolous) species found within the dead portions of standing live trees to those that decay fallen branches, twigs, leaves, flowers and fruits. Lignicolous or wood-inhabiting saprotrophic fungi (wood decayers) are the mycological counterpart of saproxylic invertebrates.

An assemblage of 16 oak-associated lignicolous saprotrophs, developed for use in the UK, was published in Table 10.1 in Ainsworth (2017). Their recorded presence at 25 sites, 23 of which are regarded as the most important in England for ancient oaks, was used to rank the sites in a league table, headed by the New Forest with 15 recorded species. This predated our High Park fungal survey as shown by its score of just three members of this oak assemblage. Following the incorporation of our survey results in an updated version of Table 10.1 (Ainsworth 2022), High Park had moved up to ninth place with a much-improved score of 10 species.

The assemblage-scoring approach was further developed in Bosanquet et al. (2018) and used to evaluate sites that were officially eligible for consideration as Sites of Special Scientific Interest (SSSI). The resulting guidelines retained the 16 species of the oak deadwood assemblage and introduced a threshold score. Sites with eight or more assemblage members can now be submitted for consideration as candidates for legal protection based on their special fungal interest. It is also useful to publicize those areas within existing SSSIs, such as High Park (within the existing Blenheim Park SSSI), which now have wood-recycling fungi as a potentially additional qualifying feature. It is hoped that such efforts will help to raise the profile of fungi within SSSI-associated documentation and/or site management plans in the future.

Notes on some members of the oak-associated lignicolous saprotrophic assemblage follow (*Buglossoporus* and *Podoscypha* are treated separately in the two subsequent sections):

Oak Mazegill (*Daedalea quercina*): A robust creamy buff bracket fungus with an unmistakeable maze-like lower surface and named after the labyrinth of Daedalus. It is a frequent oak associate that fruits on hard, decorticated and often desiccated surfaces of stumps, fallen branches and more rarely on standing trees. It was one of the species we expected to record, and its absence was a puzzling aspect of our survey, since there are appropriate substrates throughout the Park, and the brackets are durable and unlikely to be overlooked.

Figure 5.4.   Beefsteak Fungus (*Fistulina hepatica*) on an old oak stump. (Photograph by Aljos Farjon)

Beefsteak Fungus (*Fistulina hepatica*): One of the commonest bracket fungi seen fruiting at High Park (Figure 5.4) and its mycelia are expected to inhabit many of the brown-rotted (red rot) and hollowed oak trunks that are present on site. Fruiting commences early, often in July, as soft red juicy blisters, but the spongy, meat-like red mature brackets can be found throughout the autumn. The wood decomposing activities of this species, Chicken Of The Woods (*Laetiporus sulphureus*) and the rarer Oak Polypore (*Buglossoporus quercinus*) probably account for most of the brown-rotted oak wood in the Park. These fungi feed by using enzymes to soften the centres (heartwood) of the oak trunks, eventually hollowing them out and leaving suitable habitats for a great diversity of the Park's animals, ranging from tiny saproxylic insects to Tawny Owls.

Lacquered Bracket (*Ganoderma lucidum* and *G. resinaceum*): These two reddish brackets with shiny 'lacquered' upper surfaces were both absent from our list. The absence of the generally more frequent latter species, a characteristic species of parkland oaks, was as mysterious as that of Oak Mazegill (*D. quercina*) (see above).

Hen-of-the-Woods (*Grifola frondosa*): This species forms large, 'leafy', grey-tiered brackets near the base of mature or ancient oak trees. It appears repeatedly on the same tree in successive years, but it is not particularly common in the British Isles and was only recorded on two trees in High Park. Although its brackets have a distinctive mousy smell, it is a much sought-after edible fungus in continental Europe.

Spindle Toughshank (*Gymnopus fusipes*): This usually starts fruiting in July and forms clusters of mushroom-shaped fruit bodies close to the base of living trunks, often of oak. Its tough and grooved stems (stipes) can be traced below ground to white-rotted roots. It is common in Britain and numerous examples were seen in High Park.

Oak Curtain Crust (*Hymenochaete rubiginosa*): This common but inconspicuous, thin, tough, dark brown bracket is usually found on decorticated oak wood. The spore-bearing under surface appears smooth to the naked eye but is covered in dark brown 'stubble' (projecting hymenial setae) when viewed under a lens.

Figure 5.5.   Clustered Bonnet (*Mycena inclinata*) on a mossy fallen oak trunk. Photograph by Aljos Farjon.

Clustered Bonnet (*Mycena inclinata*): This forms multiple clusters of soapy-, mealy- or oily-smelling mushroom-shaped fruit bodies along fallen oak branches and trunks (Figure 5.5). A common species, it produces a white rot in the underlying wood.

Oak Bracket (*Pseudoinonotus dryadeus*): This is a large bracket fungus that usually appears quite low down on living oak trunks. When young and expanding rapidly, the fruit bodies are very pale and characteristically beaded with large droplets of liquid exudate. The brackets become drier and dark brown with age and are not very durable. Recorded on three oaks in the Park.

Oak Porecrust (*Riopa (Ceriporia) metamorphosa*): The fruit bodies of this uncommon or overlooked species consist of a white poroid surface developing on deadwood. It is usually recorded in July during Oak Polypore surveys and it was found on six trees in the Park (June to November). Microscopy is usually required to confirm its identification, but it can be named in the field if it is accompanied by its orange powdery asexually sporulating state (this is so different from the poroid sexual fruit body that the two lifecycle stages were formerly thought to be two different species and each was given its own scientific name).

5.4.1.1. Oak Polypore (*Buglossoporus quercinus*)

The top tier of fungi deemed to be of conservation concern in England comprises four non-lichenized species that have been protected by law since 1998 under Schedule Eight of the Wildlife and Countryside Act 1981. Although recording of one of these protected species, Oak Polypore (*B. quercinus*), has had a chequered history on the Blenheim Estate, as mentioned above, a good population of this distinctive golden yellow bracket fungus is now known to be extant in High Park. It is also a Section 41 species in England (a conservation priority species) and similarly prioritized in Scotland and Wales, where it is only known from a handful of sites. Furthermore, on a global scale, it is officially recognized as a threatened species with an assessment of Vulnerable (Kautmanova et al. 2021) largely owing to the disappearance of its specialized ancient and veteran oak habitat.

Figure 5.6.   Mature fruit bodies of Oak Polypore (*Buglossoporus quercinus*) on a live oak almost 10 m above the ground (left) and a fallen oak trunk (right). (Photographs by Aljos Farjon (left) and Martyn Ainsworth (right))

The Oak Polypore is a member of the oak deadwood saprotrophic assemblage (Bosanquet et al. 2018) and is now generally regarded as a conservation flagship species (Ainsworth 2022). With around 300 occupied trees in England, of which at least 100 occur in a global stronghold at Windsor, it was decided to pay special attention to this species and carry out dedicated surveys to georeference every tree in High Park supporting its spongy yellow bracket-shaped fruit bodies. Twenty different occupied oaks were recorded during our survey with a widespread distribution in OS 1km grid squares: SP4215, SP4314 and SP4315 and a site stronghold with a cluster of records in the vicinity of SP433151. Following discussions with K.N.A. Alexander, we concluded that he had recorded an additional occupied tree in 2002, which brings the site total to 21 occupied trees. A five-year survey within a comparable area of Epping Forest documented 28 occupied trees, including one that had been discovered four years prior to the survey (Ainsworth 2010).

In High Park, Oak Polypore brackets were found in many different microhabitats: on fallen and sawn trunks and branches; on rootplates of fallen trees; inside dead and living hollowed trunks, charred or not; on dead standing trunks and dead parts of standing live trees including one large branch situated almost 10m above the ground (Figure 5.6). During a visit on 26 July 2018, one exceptional oak trunk lying on the ground at SP43391512 was recorded with brackets of Oak Polypore and Chicken Of The Woods together with the rarely recorded asexual sporulating stage of Beefsteak Fungus, all of which were emerging in close proximity.

5.4.1.2. Zoned Rosette (*Podoscypha multizonata*)

This is a root-rotting species whose highly distinctive fruit bodies resemble cauliflower-or rosette-like clusters of pinkish brown fans (Figure 5.7). It is characteristic of open-grown trees and, although recorded on a range of broadleaved species, it shows a marked preference for parkland and pasture woodland oaks. Like Oak Polypore, it is a conservation priority species in England (Section 41 species) and is included in the saprotrophic assemblage listed in the SSSI guidelines for Great Britain. It is noteworthy that it is of conservation interest across Europe and most of its European populations lie in France and southern England (Fraiture and Otto 2015). The SSSI guidelines therefore allow for its stronghold sites in Britain to be considered for designation as SSSIs (Bosanquet et al. 2018).

Our survey recorded two oaks with Zoned Rosette in the Park. It might be expected that continued cattle grazing of the undergrowth in High Park would favour the fruiting of Zoned Rosette, since it is a characteristic species of pasture woodlands or areas that were formerly managed in this way. Mown parkland seems to provide a suitable substitute for grazing animals, although fruiting success then becomes dependent on the timing and frequency of the mowing events. It is possible that maintenance of such open conditions favours the

Figure 5.7.   Zoned Rosette (*Podoscypha multizonata*) fruiting at the base of living ancient oak trunks. (Photographs by Aljos Farjon)

fungus by increasing the likelihood of drought-stressed, and hence dysfunctional, roots in which its mycelia can become established. Further detailed studies are required to improve our understanding of the below-ground ecology of this species.

5.4.2. Other oak-associated saprotrophs

5.4.2.1. Saprotrophs fruiting on attached twigs and branches
The fact that decay begins not on the woodland floor but much higher up in the canopy might come as a surprise at first. However, a deliberate examination of the lower and more shaded branches of oaks, particularly those of younger trees, will often reveal at least a few that are wholly or partially dead. These are inhabited by saprotrophic fungi battling each other for territory within the wood and initiating the process of recycling the shaded and redundant limbs, a process known as natural pruning.

Among the fungal structures expected to be found during any examination of the thinnest attached oak twigs are the swarms of blackened and lip-like remnants of the fruit bodies of *Colpoma quercinum*. This species was abundant in High Park, but the pink jelly-like fruit bodies of its obligate associate, *Marchandiomyces quercinus*, were not recorded. This is another species which often requires a careful survey of its specialized habitat in order to demonstrate its fruiting presence and further survey work in suitably damp spring weather would be justified in this case. Other common pioneer saprotrophs we recorded fruiting on dead attached twigs and branches included: Oak Blackhead (*Diatrypella quercina*) resembling swarms of black warts bursting through the bark; Witches' Butter (*Exidia glandulosa*), producing black, wet, gelatinous or rubbery fruit bodies that dry to a thin skin in dry weather; Cinnamon Porecrust (*Fuscoporia ferrea*), a white rotter that reduces the branches to 'ghosts' and forms brown, velvety poroid fruit bodies; two species producing spores on downward-pointing spines, Oak Tooth Crust (*Radulomyces molaris*) (Figure 5.8) often fruiting on branches high in the canopy and then recorded after these are blown to the ground during high winds, and the bright orange *Steccherinum ochraceum*; and a series of less showy species with crust- or patch-like fruit bodies including Oak Crust (*Peniophora quercina*), Bleeding Oak Crust (*Stereum gausapatum*) and Waxy Crust (*Vuilleminia comedens*).

5.4.2.2. Saprotrophs fruiting on or inside oak trunks
Many of these saprotrophs produce enzymes that hollow out the deadwood cores of both living and dead tree trunks. Their mycelia can occupy very large volumes of wood which, owing to their elongated shape, are often called decay columns. These can be regarded as feeding territories that are occupied by the mycelia and provide them with sufficient resources to construct substantial fruit bodies. However, the presence of living water-conducting

Figure 5.8.   Oak Tooth Crust (*Radulomyces molaris*) on a fallen dead oak branch. (Photograph by Aljos Farjon)

sapwood, which surrounds the central heartwood in the living tree, represents a relatively hostile environment for saprotrophs. Their fruiting presence on the bark surface, usually in the form of shelf-like brackets, depends on their finding a conduit of dead, dysfunctional or missing sapwood that enables them to emerge from the core regions. Most of the characteristic oak-associated bracket fungi have already been considered within the oak saprotrophic assemblage (see above). However, two other species deserve special consideration, although for quite different reasons.

One of our commonest bracket-forming species, visible in the Park during the hottest and driest periods, is *Ganoderma adspersum*, a relative of Artist's and Southern Brackets. This species forms extremely hard, tough, woody brackets that persist from year to year as the fungus periodically adds a new layer of spore-producing tissue to the under surface. The poroid lower surface is white, contrasting strongly with the dark brown upper surface. However, when the fungus is actively discharging clouds of brown spores (millions per day), the upper surface and any nearby bark and surrounding vegetation become coated in a characteristic cocoa-like layer of spore powder. This species is found on several broadleaved tree species, both living and dead, and is easy to spot in the Park.

At the other extreme of the frequency scale, the bright yellow *Meruliporia* (*Serpula*) *pulverulenta*, a relative of the timber-inhabiting Dry Rot fungus, was a very poorly known British species at the start of the HPBS. The British and Irish checklist noted that it 'appears to be restricted to conifer timber in buildings' and it was classified as an alien species not found in woodland habitats (Anon., 2009). Nevertheless, it was found in High Park fruiting inside one fallen hollow

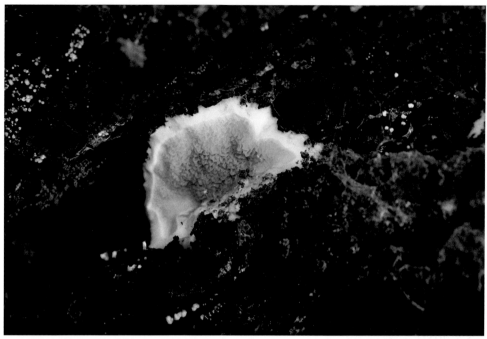

Figure 5.9. *Meruliporia (Serpula) pulverulenta* found deep inside a fallen dead hollow ancient oak. (Photograph by Aljos Farjon)

oak tree from July to November 2021 (Figure 5.9). Samples were taken for microscopy and DNA analysis and compared with previous British collections from seven oaks at Windsor, and one each at Ashton Court, Burnham Beeches and Epping Forest. All ten samples were determined as *M. pulverulenta* and further details of the taxonomic study are given in Ainsworth and Liimatainen (2022). Clearly the status of this brown-rot fungus as a characteristic species of ancient oak landscapes had been overlooked for a long time. It was deleted from the checklist of 'alien' fungi and recognized as a rare oakwood saprotroph in Britain (Ainsworth and Henrici 2023), based on research that was prompted by the current HPBS.

5.4.2.3. Saprotrophs fruiting on fallen wood
The process of natural pruning results in partly decayed twigs and branches falling to the ground. Here they encounter a very different environment where they remain damper for longer periods; a change in conditions that favours fruiting populations of species not seen while the branch was attached to the tree. The saprotrophic pioneer species, which were so successful within the attached branch, are thereby potentially exposed to a new set of competitors arriving as mycelia from the soil. These are relatively combative species capable of invading territory occupied by the pioneers and replacing their mycelia. Examples seen in High Park include Sulphur Tuft (*Hypholoma fasciculare*), which produces dense clusters of poisonous yellow fruit bodies, and *Phanerochaete velutina*, whose pale and waxy fruit bodies usually remain hidden from view beneath fallen branches in the litter. Although their fruit bodies are highly dissimilar, both species are characterized by having systems of white mycelial cords that grow through and explore the soil in search of woody food sources.

Fallen oak branches in High Park supported fruiting of troops of Bonnets (*Mycena* s. lat.) including Angel's Bonnet (*M. arcangeliana*), which has a medicinal smell; Burgundy Drop Bonnet (*M. haematopus*) whose fruit body stems (stipes) 'bleed' a dark reddish juice when cut; and Bark Bonnet (*Phloeomana speirea*) usually fruiting on mossy wood and often seen

Figure 5.10. Green Elfcup (*Chlorociboria aeruginascens*) on rotting fallen oak wood. (Photograph by Aljos Farjon)

in the winter months. Well-rotted fallen wood yielded three Shields: the brown Deer Shield (*Pluteus cervinus*), dark greenish Willow Shield (*P. salicinus*) and brown wrinkled Veined Shield (*P. thomsonii*), all of whose fruit bodies produce characteristically pinkish spores.

Examination of damp fallen wood with a hand lens often reveals swarms of minute cup fungi, which at High Park included Green Elfcup (*Chlorociboria aeruginascens*) (Figure 5.10), which has blue-green fruit bodies, accounting for its former common name Copper Nails, and whose thinly sliced, green-stained mycelial territories were historically used in marquetry; Snowy Disco (*Lachnum virgineum*) and other species with white discoid fruit bodies such as *Dasyscyphella nivea* and *Hyaloscypha daedaleae*; and *Phaeohelotium nobile*, which forms yellow fruit bodies. This microhabitat also yielded a variety of relatively inconspicuous crust or patch (corticioid) fungi: *Aphanobasidium pseudotsugae*, whose greyish filmy fruit bodies smell of phenol when scratched, and various species of *Botryobasidium*, *Hyphoderma*, *Hyphodontia*, *Peniophorella*, *Trechispora* and *Xylodon*.

5.4.2.4. Saprotrophs fruiting on the ground

Saprotrophs fruiting on the woodland floor are mainly generalist litter recycling species, specialists fruiting on specific components of the oak-derived litter, or wood inhabitants fruiting from roots, stumps or other buried woody elements. A few species, for example the aptly named Orange Peel Fungus (*Aleuria aurantia*), seem to fruit directly on bare soil and may depend on humus. Generalist litter inhabitants have mycelia that form outwardly extending patches or rings (fairy rings) and the extent of their territories is often revealed by the extent of bleached substrate and/or visible patches of aggregated mycelium observed beneath their fruit bodies. Examples recorded at High Park include Russet Toughshank (*Gymnopus dryophilus*), Funnels (*Clitocybe* spp. and *Infundibulicybe* spp.), Dapperlings (*Lepiota* spp.), Bonnets (*Mycena* spp.) and Brittlestems (*Psathyrella* spp.).

Species whose fruit bodies develop from mycelia feeding on buried wood include: Stump Puffball (*Apioperdon pyriforme*), whose puffballs have a herring-like smell and are supplied by nutrients obtained from the wood-decomposing parts of its mycelium by a network of white mycelial cords; Inkcaps (*Coprinellus* spp.), whose delicate fruit bodies rapidly collapse and disappear; Rooting Shank (*Hymenopellis radicata*), whose stem (stipe) can occasionally be carefully excavated to the point where it emerges from the dead roots on which it is feeding; and one of its lookalikes, *Xerula pudens*, which is an uncommon species with a brownish hairy (tomentose) stipe originating on a decaying root deep underground.

5.4.3. Ectomycorrhizal species

Our field records of ectomycorrhizal fungal fruit bodies include the names of the nearby tree species suspected to be the photosynthetic partner in each case. However, the tree or trees forming underground connections to a given ectomycorrhizal mushroom are not always obvious in field surveys. Even mushrooms picked well away from a given tree may be connected to its roots by fungal hyphae and the roots themselves often extend further from the trunk than the spreading branches of the tree's canopy. For the same reason, where several tree species are close neighbours, it is often impossible to determine which tree is connected to which fruit body. However, many of the High Park oaks grow in solitary splendour or surrounded by other oaks, and in those cases the suspected association is more likely to be correct. Curiously, some ectomycorrhizal fungi (e.g. *Inocybe* spp.) seem to favour fruiting along track and path sides, possibly a response triggered by the subterranean mycelium detecting the change from bare soil at the path edge to the increasingly compacted soil of the pathway itself.

Our fruit body records included a disappointingly low number of ectomycorrhizal species (79 of 428 species) for reasons that are not entirely clear, but probably include the site's rather high levels of nitrogen deposition (see below). Spooner and Roberts (2005) list 16 such species, which they regarded as 'typical fungi found with oak'. Twelve of these were recorded at Moccas and reported in Blackwell (2000). However, we only found five of these species in the five years of our survey. For instance, we expected that numerous warmth-loving (thermophilous) parkland boletes might be found fruiting in the shorter grass of the woodland rides, especially as Blenheim Park is mycologically known for its enduring population of Devil's Bolete (*Rubroboletus satanas*). However, we only recorded six bolete species fruiting near oak in High Park, of which perhaps only Bluefoot Bolete (*Xerocomellus cisalpinus*) qualifies as a characteristic species of such habitat. Moreover, we did not record any of the five boletes regarded as typical of this habitat in Spooner and Roberts (2005). Even the common Cep (*Boletus edulis*) was not seen during our survey.

Notes follow on a selection of other ectomycorrhizal fungi assumed to be associated with oak and recorded fruiting in High Park.

Amanita: This is a largely ectomycorrhizal genus that usually has a prominent fruiting presence in autumnal woods. Interestingly, that most iconic toadstool – the Fly Agaric (*Amanita muscaria*), usually associating with birch – was not among our High Park records. There was a similar absence of False Deathcap (*A. citrina*), which is, if anything, commoner than the Fly Agaric. However, the deadly poisonous Deathcap (*A. phalloides*) itself was recorded in 2017 and 2021. On the other hand, Snakeskin Grisette (*A. ceciliae*), a distinctively blackening species and generally one of the less frequently recorded members of the genus, was recorded in three of our survey years. Our DNA analyses confirmed the fruiting presence of two of its even rarer close relatives: Pale Grisette (*A. lividopallescens*) and *A. alseides* (Figure 5.11). Both species are poorly known in Britain, the latter especially so since it was only recently described as new to science (Hanss and Moreau 2020) and still lacks a recommended English name.

Figure 5.11. Two rarely recorded Amanita species: (left) Pale Grisette (*A. lividopallescens*) and (right) *A. alseides*, which were identified following DNA analysis of the photographed fruit bodies. (Photographs by Aljos Farjon)

Figure 5.12. Yellow Capped Webcap (*Cortinarius azureovelatus*) was identified from a DNA sequence derived from these fruit bodies found on the side of a track near one of the giant ancient oaks. (Photograph by Aljos Farjon)

Cortinarius, Hebeloma and Inocybe: These three brown-spored genera are particularly speciose. Although some of the species we recorded produce fruit bodies with highly distinctive odours, for example, smelling of fruit, *Pelargonium*, radish or rubber, many of them appear to be remarkably similar and are among the most difficult fungi to identify using traditional field and microscopic methods. For this reason, dried fruit bodies of these genera were targeted for our DNA analyses, and led to the identification of several poorly known species including Yellow Capped Webcap (*C. azureovelatus*) (Figure 5.12).

Lactarius and *Lactifluus*: These are genera of Milkcaps, so-called because, when damaged, their gills produce a latex-like liquid that is white and milky in many of the species. This was another group whose fruiting diversity was lower than expected in High Park. Oakbug Milkcap (*L. quietus*), so-named because the bruised fruit bodies allegedly smell of bedbugs, is ubiquitous under oak; however, other oak specialists such as Yellowdrop Milkcap (*L. chrysorrheus*), whose white milk rapidly becomes sulphur yellow, and the uncommon Whiskery Milkcap (*L. mairei*), with a fringed cap margin, were not recorded.

Russula: The Brittlegills are often a dominant component of fungal collections made in late summer and autumn. Purple Brittlegill (*Russula atropurpurea*), Scarlet Brittlegill (*R. pseudointegra*) and The Flirt (*R. vesca*) are often recorded under oaks but are not confined to them. Most of the other species we recorded are commonly found with various broadleaved trees. As with *Amanita* and *Lactarius*, there were surprising absences. The section *Compactae* was absent, including species that are usually ubiquitous in Oxfordshire woods such as Milk White Brittlegill (*R. delica*) and Blackening Brittlegill (*R. nigricans*). The strongly smelling group of species around Stinking Brittlegill (*R. foetens*) was also absent, including Sepia Brittlegill (*R. sororia*), a typical oak associate (Spooner and Roberts 2005).

5.5. Below-ground survey results

We detected 30 fungi associated with oak roots, of which 28 were ectomycorrhizal species, including the recently described *Inocybe syringae*, a new British record. The two non-mycorrhizal fungi comprised a member of the family *Hyaloscyphaceae*, probably an endophyte growing inside the oak root cells, and *Oidiodendron majus*, considered to be an ericoid mycorrhizal fungus, but frequently found in ectomycorrhizal tree roots and also likely to be living as a root endophyte.

Of the 28 ectomycorrhizal fungi, only five species were found in both our above-ground (fruit body) and below-ground (ectomycorrhizal root tip) surveys. These five were: *Cortinarius megacystidiosus*, *Inocybe jucunda*, *Russula odorata*, Scaly Earthball (*Scleroderma verrucosum*) and White Knight (*Tricholoma album*). The below-ground survey, although restricted to just seven oak trees, thus increased our total number of recorded ectomycorrhizal species from 79 to 102. Furthermore, these results were in good agreement with the small species overlap frequently detected between above- and below-ground studies of woodland fungal communities (Gardes and Bruns 1997; Spake et al. 2016).

Most of the 25 fungi found only in our DNA-based analysis of ectomycorrhizal root tips belonged to genera also represented in our fruit body survey. These included *Clavulina*, *Cortinarius*, *Inocybe*, *Russula*, *Scleroderma*, *Sebacina*, *Tomentella* and *Xerocomellus*. However, there were also new additions to the list of genera, corresponding mostly to cup fungi and their relatives, for example, *Humaria* and *Peziza*, and fungi forming subterranean fruit bodies, for example, *Genea hispidula*, or whose reproductive structures are inconspicuous in other ways. Some fungi that were recorded below-ground are thought to reproduce entirely asexually, having lifecycles that lack a fruit body stage. An example of this found in High Park is *Cenococcum geophilum* (Figure 5.2), a fungus that is dominant in the roots of over 200 tree species as revealed by DNA analysis of mycorrhizal root tips. Its dark grain-like resistant propagules (sclerotia) can remain in the soil for hundreds of years, contributing to the soil carbon stocks (Watanabe et al. 2007). The taxonomic composition (at ordinal level) of the ectomycorrhizal species found in our above- and below-ground surveys is compared in Figure 5.13.

Despite recording its fruiting presence above ground, we did not find any ectomycorrhizas of Oakbug Milkcap (*Lactarius quietus*) (Figure 5.2), which is an oak specialist and dominant in oak woodland in general and in UK ancient oak woodland in particular (Suz et al. 2014; Spake et al. 2016). Conversely, Ochre Brittlegill (*Russula ochroleuca*), a generalist fungus also dominant in old-growth woodland, was found in the roots but not as fruit bodies, despite

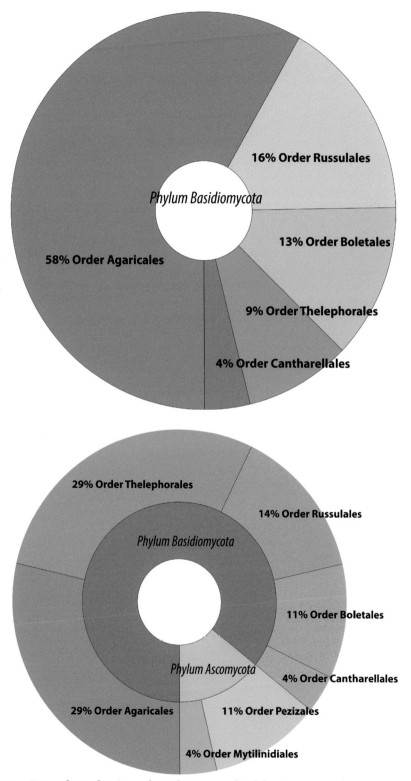

Figure 5.13. Krona charts showing orders of ectomycorrhizal fungi represented in our above-ground (top) and below-ground (below) surveys. Note that ectomycorrhizal members of the phylum Ascomycota (blue) were only detected below ground in the root tip study.

being one of the most frequently recorded British species above ground. Both species have been identified as indicators of high nitrogen and low pH (Suz et al. 2014; van der Linde et al. 2018). During 2018–20, the wooded areas of High Park received about 32kg of N/ha/yr (https://www.apis.ac.uk/app). The maximum critical load for the habitat classified as W10 *Quercus robur – Pteridium aquilinum – Rubus fruticosus* woodland, one of the habitat types for which High Park is legally protected as a SSSI, is stated to be 15–20kg of N/ha/yr (https://www.apis.ac.uk/app). However, recent work has shown that some ectomycorrhizal fungal communities can be affected if N levels rise above about 6kg/ha/yr (van der Linde et al. 2018) and N levels of between 9.5–13.5kg/ha/yr have been shown to reduce the richness and evenness of oak ectomycorrhizal communities (Suz et al. 2014). These findings might explain the absence of records of, for example, Yellowdrop Milkcap (*Lactarius chrysorrheus*), often associated with woodland sites receiving below 10kg N/ha/yr across Europe (Suz et al. 2014).

Overall, our limited above- and below-ground studies highlight the importance of complementing long-term fruit body surveys with the study of mycorrhizal fungi within roots to gain a better understanding of the fungal diversity in High Park. It should be remembered that our root sampling was restricted to small bags of soil collected under just seven of the ancient oaks. Further extensive and standardized root sampling would therefore be needed to obtain a more complete picture of the ectomycorrhizal fungal communities associated with what is aptly described in the Blenheim Park Site of Special Scientific Interest documentation as 'one of the finest areas of ancient oak-dominated pasture woodland in the country' (see also Chapter 2).

5.6. New Oxfordshire (VC 23) records

The total number of fungi recorded in High Park is 459 species, which includes the current HPBS (above- and below-ground) and historic Thames Valley Environmental Records Centre records. Thirty-two species recorded during the fruit body survey are apparently first records for Oxfordshire (VC 23). Many, but by no means all, of these records are with *Quercus* and all are listed in Table 5.1. It is important to add the caveat that some of these fungi will undoubtedly

Table 5.1.   Fungal species (fruit body surveys only) constituting new records for Oxfordshire (VC 23).

Amanita alseides	*Hyphoderma roseocremeum*
Amanita lividopallescens s.s.	*Inocybe erinaceomorpha*
Arachnocrea stipata	*Inocybe jucunda*
Basidiodendron spinosum	*Inocybe pseudodestricta*
Biscogniauxia anceps	*Jaapia ochroleuca*
Caudospora taleola	*Meruliporia pulverulenta*
Conocybe siennophylla	*Mycosphaerella aspidii*
Cortinarius alboadustus	*Peniophora reidii*
Cortinarius azureovelatus	*Peniophorella guttulifera*
Cortinarius collocandoides	*Phaeoclavulina minutispora*
Cortinarius incisior	*Riopa metamorphosa*
Cortinarius semiodoratus	*Russula odorata*
Cytospora nivea	*Trechispora dimitica*
Eutypa maura	*Trechispora stellulata*
Hyaloscypha daedaleae	*Vuilleminia cystidiata*
Hygrophoropsis rufa	*Xylodon asper*

have already been recorded in the county, but under different names. This mainly applies to those closely related and recently described species that together comprise a species complex but were formerly treated as individuals of a single, albeit variable, species. However, without re-examination and/or DNA sequencing of preserved materials (if any exist), it is often not possible to reconcile such historical records with currently accepted species names. Table 5.1 therefore includes the names of some species whose presence in the county could possibly have been inferred from reinterpretation of historic records, but which are now recorded with a greater degree of certainty and with the currently accepted name.

Acknowledgements

We would like to thank our colleague Kare Liimatainen (Royal Botanic Gardens, Kew) for generating and contributing four of the DNA sequences used in the identification of our fruit body collections.

CHAPTER 6

Lichens

Pat Wolseley, Neil Sanderson, Brian Coppins and Sandy Coppins

6.1. Introduction

High Park lies in an area of Middle England that has been occupied by people since Palaeolithic times and the forest area has been managed for hunting at least since the Normans arrived. Today, High Park contains the relics of this forest, so that when you enter from Combe Gate you are walking among giant ancient oak trees with spreading boughs characteristic of the open grown medieval forest with a sparse shrub layer and a grazed ground layer (Figure 6.1). A continuity of an ancient woodland habitat in High Park from prehistoric times onwards is not beyond the realms of possibility. The statistics of these trees are given in Chapter 2 and consist of veteran and ancient trees as well as dead trees of which some are fallen and remain lying and some are still standing. Together these trees provide a great diversity of microhabitats that, combined with long environmental continuity, provide an unusual source of ecological information from an otherwise densely populated and managed area of Middle England.

6.2. Lichen communities and their association with ancient woodland

Lichens, along with beetles and other organisms that require the specific conditions provided by veteran and ancient trees, have long been used as indicators of ecological continuity in woodlands in Britain (Coppins and Coppins, 2002; Sanderson, 2018a). Many lichens are slow-growing, and where conditions remain unchanged, they will remain on a tree for much of its lifecycle, which at High Park may mean centuries. The symbiotic nature of a fungus with one or more algal or cyanobacterial inhabitants ensures a food source through photosynthesis by the partner. Only the fungus produces spores and this means that on germination a spore must acquire an algal symbiont in order to survive. This may present difficulties in an altered environment where symbionts are not always available, so that colonization of new substrates may be infrequent and restricted to a few trees in a site. Other species respond rapidly to changing conditions by producing vegetative propagules that contain all symbionts and these are more widespread as they can be carried by wind, birds or invertebrates to a new tree or site.

A veteran tree is composed of both new and ancient surfaces, from new growth to damaged limbs and hollow trunks where the dead wood has been exposed or eaten away by fungi and insects, so that an ancient tree may support many different lichen communities characteristic of a wide range of conditions and ages (Figure 6.2). Most of the habitats for lichens are on the surface of bark or on dead and decaying wood exposed over long periods of time. The latter may support specialist communities of lichens that are now rare in disturbed and managed woodland

Pat Wolseley, Neil Sanderson, Brian Coppins and Sandy Coppins, 'Lichens' in: *The Natural History of Blenheim's High Park*. Pelagic Publishing (2024). © Pat Wolseley, Neil Sanderson, Brian Coppins and Sandy Coppins. DOI: 10.53061/RTDO9050

Figure 6.1.   Veteran oaks near Combe Gate entrance to High Park, with lichenologist for scale! (Photograph by Neil Sanderson)

environments. Lichens have been used as indicators of ecological continuity in woodlands (Rose, 1976) as well as of changes in air quality that have occurred more recently (Richardson, 1992; Welden et al. 2018). The characteristic indicator species in these communities may vary with climate and geographical location so that indicators will vary from north to south of the UK and also from the more continental east to the Atlantic west (Coppins and Coppins, 2002; Sanderson 2018a).

High Park is situated in an area of Middle England that has been affected by atmospheric pollution as well as by all types of forestry and agricultural development for centuries, so we were not expecting to find species that are sensitive to these conditions in High Park. Early records of lichens are sparse, despite Oxford being the centre for lower plant recording when Dillenius was appointed Sherardian Professor of Botany in 1734. Although he made field excursions and accumulated records of bryophytes and lichens from other people for *Historia Muscorum* (1742) there are no records from High Park (Crombie 2008). The earliest

Figure 6.2.   Brian Coppins investigating one of the High Park veteran oaks in summer. (Photograph by Neil Sanderson)

records of lichens are from Robert Paulson in 1929 when industrial development had already polluted the atmosphere such that sensitive species of macrolichens were already lost. He visited three woods around Oxford, including High Park, Tubney Wood and Wytham Woods, and recorded 67 lichens on all substrates and only 19 in High Park. There are no records of macrolichen indicators such as *Lobaria* species in any of the woods, and Paulson remarks that *Usnea* species were scarce in Bagley Wood and High Park. This is typical of a site in Middle England where many of the larger foliose and shrubby macrolichens characteristic of ecological continuity in a wood-pasture regime had long since disappeared owing to air pollution from industrial sites.

6.3. Methodology

High Park was visited from 2016 to 2019 by a number of lichenologists: Brian and Sandy Coppins, Mark Powell, Neil Sanderson, Paula Shipway, Andy Cross and Pat Wolseley. Aljos Farjon had already located the veteran and ancient trees in High Park (Chapter 2) and the lichen survey was concentrated on these. Fallen and standing veteran and ancient trees were located and recorded with photographs and a list of species made occurring on each aspect of the surveyable trunk and where possible on branches. Diagrams were made by Mark Powell of the location of rare and interesting species on the accessible sections of selected trees so that the survey could be repeated. However, owing to COVID-19 restrictions, this meant that other areas of the estate, where later plantations exist, including the 9th Duke's now 120-year-old plantation of oaks, were only briefly visited in the period of time available; likewise, isolated or boundary veteran trees in other areas of the estate were not visited during the survey. The records that are included here are from the bark and lignum of trees and shrubs in High Park and do not include records from walls or buildings.

6.4. Results of the surveys

During the surveys of High Park, we have made nearly 2,000 individual records of over 230 taxa (lichens, lichenicolous and lichen-related fungi), including many new to Oxfordshire. These records are from the bark and lignum of trees and shrubs including non-veterans. Of the 231 species identified, 196 are true lichens, of which only 42 species are macrolichens, the remaining 154 being crustose species. The foliose and fruticose macrolichens found in High Park present an interesting combination of communities that are widespread elsewhere and reflect the changes in atmospheric pollutants since the industrial development of the nineteenth and twentieth centuries. Between 1970 and 2016 levels of sulphur dioxide across the UK have fallen by 97% and nitrogen oxides by 72%, while ammonia has hardly decreased since 1980 (Defra report, 2018), and in some agricultural areas, where fertilizers are used extensively on arable crops, it has increased. Although today's atmospheric conditions are dominated by nitrogen, lichens characteristic of acid bark are still present on the veteran trees in High Park. These include species of *Hypogymnia*, *Hypocenomyce*, *Imshaugia* and *Parmeliopsis* that were formerly more common on acid barked trees in the northern uplands but had increased on trees in the south in the nineteenth and twentieth centuries owing to the influence of acid rain. In the current century some of these are now in decline in lowland England. Meanwhile, nitrogen-loving species such as foliose species of *Xanthoria*, *Candelaria*, *Physcia* and *Melanelixia* are also present on trees in High Park, but these tend to occur on younger bark of twigs and branches around the margins of the wood. Particulates were also abundant in the industrial atmosphere, and we found examples of macrolichens that are characteristic of dust impregnated bark surfaces typical of urban trees such as *Phaeophyscia orbicularis* and *Hyperphyscia adglutinata*.

Macrolichens form a large component of ancient woodland indicator species, so that the absence of macrolichen indicator species that are sensitive to pollutants in High Park makes a comparison with existing indices of ecological continuity difficult (Table 6.1). The only macrolichen found in High Park in the Southern Oceanic Woodland Index (SOWI) is *Cladonia parasitica*, while 16 crustose lichens from the same index are found on ancient trees in High Park. These include such dry bark species as *Lecanographa lyncea*, a greyish crust with elliptical fruiting bodies that grows on the rough bark of ancient oaks in open parkland and pasture woodland (Figure 6.7a). The national threshold for SOWI sites that demonstrate ecological continuity is 20 species. Although 17 is an unusually low number for

Table 6.1. Habitat quality indices species recorded in High Park from the Southern Oceanic Woodland Index (SOWI), characteristic of old growth woodland in southern and western England and in Wales, and the Pinhead Lichen Index (PLI), characteristic of dry bark and lignum on veteran and dead trees (Sanderson et al. 2018).

Species	SOWI	PLI	Abundance
Agonimia allobata	1	-	Rare
Agonimia flabelliformis	1	-	Occasional
Calicium abietinum	-	1	Rare
Calicium glaucellum	-	1	Frequent
Calicium salicinum	-	1	Rare
Calicium viride	-	1	Rare
Chaenotheca brachypoda	1	1	Rare
Chaenotheca brunneola	1	1	Occasional
Chaenotheca chrysocephala	1	1	Frequent
Chaenotheca ferruginea	-	1	Abundant
Chaenotheca furfuracea	-	1	Rare
Chaenotheca hispidula	1	1	Rare
Chaenotheca stemonea	1	1	Occasional
Chaenotheca trichialis	1	1	Abundant
Chaenothecopsis nigra	-	1	Occasional
Chaenothecopsis pusilla	-	1	Rare
Chaenothecopsis savonica	-	1	Occasional
Cladonia parasitica	1	-	Frequent
Cresponea premnea	1	-	Abundant
Lecanographa lyncea	1	-	Frequent
Lecanora quercicola	1	-	Rare
Lecanora sublivescens	1	-	Occasional
Microcalicium ahlneri	-	1	Occasional
Microcalicium disseminatum	-	1	Occasional
Mycocalicium subtile	-	1	Rare
Pachyphiale carneola	1	-	Occasional
Syncesia myrticola m. sorediata	1	-	Occasional
Thelopsis corticola	1	-	Frequent
Thelopsis rubella	1	-	Occasional
Total	**17**	**18**	
SSSI Quality Threshold	**20**	**10**	

the SOWI, it is easily the highest in the Oxfordshire region. In contrast, the proportion of species from the Pinhead Lichen Index (PLI) of crustose species associated with dry bark and lignum is high, including species of *Calicium*, *Chaenotheca* and *Chaenothecopsis* all of which have stalked fruiting bodies on a crustose thallus (Figure 6.3). Out of 51 possible species in the PLI, 18 are present on trees in High Park, where a score of 10 or more indicates a site of national interest. Coincidence maps show the clustering of species in these indices in the older parts of High Park (Figure 6.4). High Park represents the climax of this community on the dry bark and lignum of ancient trees and is the highest scoring site known from England and Wales, only exceeded by native pinewood sites in the Central and Eastern Highlands of Scotland.

Environmental conditions within High Park vary across the site so that trees have grown up in a range of conditions that vary with aspect and position across the site. With age, oak bark becomes more rugged, cracks and crevices develop, huge bosses form and where bark is lost lignum is exposed. Where rotting wood is lost trunks and limbs become hollow, creating other habitats for lichens and fungi as well as for invertebrates and other animals. As trees slowly die, they become stag-headed and lose parts of their canopy and smaller branches, so that the remaining surface becomes more exposed with less rain reaching the trunk. As the bark dries out it becomes increasingly hard and dry, and a typical dry bark community of crustose lichens develops before the dead bark falls off, exposing the lignum that some years later can support a specialist assemblage of lichens and microfungi. Aspect and weather help so that where branches fall wound tracks develop below the fall and the bark becomes increasingly base-rich, supporting a lichen community that is very different to the dry bark community that develops on exposed areas of the trunk that receive little rain. Over time, the loss of limbs and bark creates a landscape of microhabitats on these veteran trees that varies with aspect and exposure and drainage tracks, which then are associated with specialist lichens and fungi from several biogeographic elements. Each veteran tree has a different environmental history of growth and loss of parts to its neighbour, resulting in a rich diversity of microhabitats each with its specialist community of lichens.

Figure 6.3.   A pinhead lichen, *Chaenotheca chrysocephala*, with bright yellow thallus and stalked pinheads that are covered in brown spores. (Photograph by Eric Peterson)

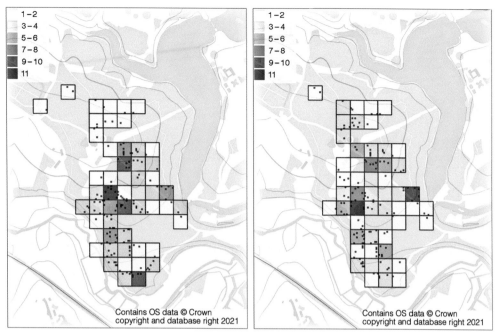

Figure 6.4.   Coincidence maps of species in High Park; **a.** in the SOWI index and **b.** in the PLI index. Species used as listed in Table 6.1.

Veteran oak tree 0428 provides a good example of this diversity (Figure 6.5).[1] The succession of lichens on this tree began when it was a sapling in a medieval forest and continues today on the bark and lignum of this ancient tree. Thirty lichen species were recorded on the trunk and accessible branches. Lichens present include common species of younger bark and pollution tolerant species – both acidophytes such as *Lepraria incana* and *Hypocenomyce scalaris* and nitrophytes such as *Physcia tenella* and *Amandinea punctata*. Species on the ancient bark and lignum include indicators of ecological continuity in the Southern Oceanic Woodland Index; both *Cresponea premnea* typical of a dry bark community and a base-rich community that includes the species *Thelopsis rubella* and *Cladonia parasitica*. The list on this tree includes 7 of the 49 Nationally Scarce (NS) or threatened species in High Park (Table 6.5). Although several specialist communities may be found on different aspects of the same tree, the diversity in High Park is due to a combination of ages and conditions across the site. Species diversity of corticolous lichens in these indicator groups is highest in the sheltered parts of High Park where conditions have remained similar over long periods of time, and lowest on the margins of the wood and in the newer plantations (Figure 6.4a).

Oak trees have a rough bark with plenty of nooks and crannies (niches) that develop over time. Generally, bark surfaces on oak have a relatively low pH of 3–4 and where the canopy is open and well-lit the bark dries out so that species characteristic of dry bark thrive (Table 6.2). These are mainly slow growing crustose species that are able to physiologically 'shut down' when conditions become too dry and continue growth during wet periods. While these conditions are widespread on oaks in High Park, the ability of the lichens inhabiting veteran trees to colonize new areas varies. This is illustrated in the coincidence map of dry bark species (Figure 6.6), where only a few lichens characteristic of veteran trees with dry bark, such as *Cresponea premnea*, have colonized the 120-year-old plantation of oaks in the north of the site,

---

1   Tree numbers used in this chapter refer to the metal tag numbers found on the trees; they refer to the Blenheim Estate's veteran tree survey carried out in 2001–6 (see Chapter 3).

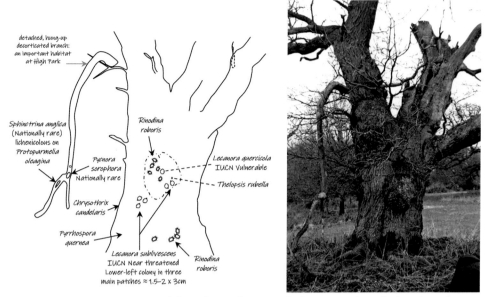

Figure 6.5.   Veteran tree 0428 viewed from the south-west with diagram to show the location of associated lichens on bark and on lignum. (Photograph by Sandy Coppins; drawing by Mark Powell)

Table 6.2.   Species associated with dry bark in High Park.

Andreiomyces obtusaticus
Chaenotheca brachypoda
Chaenotheca chrysocephala
Chaenotheca hispidula
Chaenotheca stemonea
Chaenotheca trichialis
Chaenothecopsis savonica
Cresponea premnea
Lecanographa lyncea
Microcalicium disseminatum
Milospium graphideorum
Sporodophoron cretaceum
Syncesia myrticola m. *sorediata*

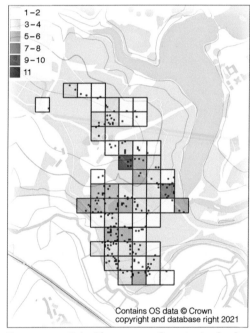

Figure 6.6.   Coincidence map of species in the dry bark community.

while the NS species *Lecanographa lyncea* and *Syncesia myrticola* are restricted to veterans in the High Park pasture woodland (Figure 6.7). This is supported by evidence from the New Forest, where trees in a 200-year-old oak plantation were just beginning to acquire a few lichens of ecological continuity from an adjacent plot (Wolseley et al. 2017).

As oak trees age, their girth increases and their bark becomes more base-rich, with an accompanying increase in the pH of their bark to around 5 or 6. This supports a specialist

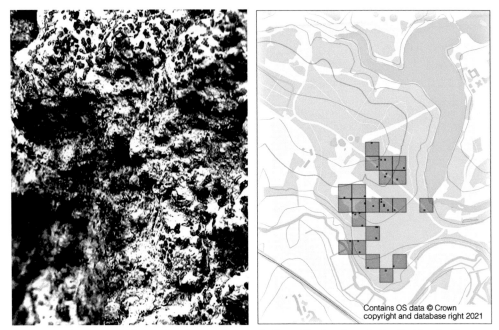

Figure 6.7.   *Lecanographa lyncea* showing a pale grey thallus with abundant darker fruiting bodies (left) and distribution of *L. lyncea* on trees in High Park (right). (Photograph by Jenny Seawright)

Table 6.3.   Species associated with base-rich bark in High Park.

Agonimia allobata
Agonimia flabelliformis
Caloplaca lucifuga
Coenogonium tavaresianum
Gyalecta flotowii
Lecanora quercicola
Lecanora sublivescens
Pachyphiale carneola
Rinodina exigua
Rinodina roboris var. *roboris*
Strigula phaea
Thelopsis corticola
Thelopsis rubella

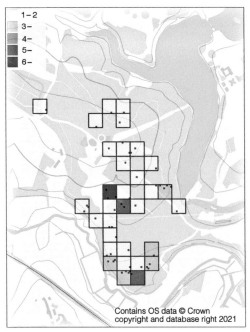

Figure 6.8.   Coincidence map of species at High Park typical of base-rich bark habitats shown in Table 6.3.

relict community of lichens that is found on veteran trees with base-rich bark in areas of High Park that have been sheltered from acid rain (Table 6.3 and Figure 6.8). This community is characteristic of oceanic woodlands in the south and west of Britain, as shown by the distribution of *Thelopsis corticola* across the British Isles (Figure 6.9). This community contains

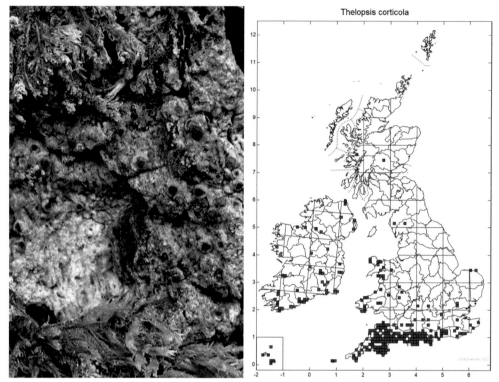

Figure. 6.9. *Thelopsis corticola* ancient oak bark (left), showing distribution of T .corticola in the UK (right), a pale grey thallus with darker pot-shaped fruiting bodies. (British Lichen Society; photograph by Neil Sanderson)

several rarities, including one of the remarkable finds on this survey: a new record for Britain. When a specimen was first encountered at High Park in June 2017, its characters did not fit any known British species. By coincidence, Neil Sanderson had collected the same taxon earlier in the year from another quality site and had not been able to identify it. *Dimerella tavaresiana* was published as new to the British Isles in *Bull. Brit. Lichen Soc.* 121: 113–114 (Winter 2017 edition) with Neil Sanderson's Hampshire record cited as the first (May 2017) and the High Park discovery as the second British record. Recent investigation has shown it to belong in *Coenogonium*, a mainly tropical genus, with a yellow trentepohlioid alga as its photobiont. Two other species that are new to the British Isles have been recorded at High Park: *Lecanora hypoptoides* and *Rinodina furfuracea*.

High Park has always been grazed both by deer and by agricultural stock, when tenants were allowed to keep stock in the Park. Open sunlit glades have been maintained by grazing animals over long periods of time. Trees on the edge of these glades benefit from the increased nutrients from the droppings. These conditions enable a characteristic association of species of enriched bark to thrive that include *Physcia tribacia* together with *Diploicia canescens*, *Pertusaria flavida* and *P. coccodes*. Elsewhere these species are mostly associated with trees in open grazed parklands in Britain.

Veteran trees may take decades to die and lose their bark, and when the lignum of the wood is exposed, they may last several decades longer. If these trees remain upright the lignum becomes hard and dry, providing a habitat for a specialist group of lichens that are now rare across lowland Britain as the management practice has been to remove dead wood. Their presence on ancient trees in High Park has hugely contributed to the diversity and

Table 6.4.   Lichen species associated with lignum.

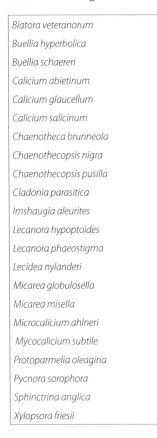

Biatora veteranorum

Buellia hyperbolica

Buellia schaereri

Calicium abietinum

Calicium glaucellum

Calicium salicinum

Chaenotheca brunneola

Chaenothecopsis nigra

Chaenothecopsis pusilla

Cladonia parasitica

Imshaugia aleurites

Lecanora hypoptoides

Lecanora phaeostigma

Lecidea nylanderi

Micarea globulosella

Micarea misella

Microcalicium ahlneri

Mycocalicium subtile

Protoparmelia oleagina

Pycnora sorophora

Sphinctrina anglica

Xylopsora friesii

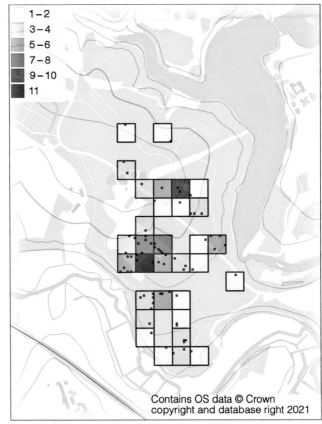

1–2
3–4
5–6
7–8
9–10
11

Contains OS data © Crown copyright and database right 2021

Figure 6.10.   Coincidence map of species of interest that are typical of lignum habitats, species used as listed in Table 6.4.

conservation value of this site (Table 6.4). Of the 17 species from this community found at High Park, eight are NS or rare and yet some such as *Buellia hyperbolica* are found on several trees at High Park (Table 6.4, Figures 6.10, 6.11, 6.12). The community is found on standing dead wood and lignum exposed on live trees, and of large fallen dead trees. These include pinhead species in the genera *Calicium, Chaenotheca, Chaenothecopsis* and *Microcalicium.* Species requiring slightly damper lignum favour the fallen tree trunks and branches such as *Lecanora phaeostigma* and *Protoparmelia oleagina.* When fallen timber gets cut and moved, many of these species are lost. One of the features of High Park is the large number of upright and fallen dead trunks and branches that support a great diversity of rare lichens on their lignum (Figure 6.13). At many locations, especially those with a general public access, such niche availability is denied for health and safety considerations.

The survey of lichens and associated fungi at High Park has revealed many threatened and notable species that are excluded from the indices as they are too rare (Table 6.5). Of the 49 species that are of conservation significance, four are on the IUCN Vulnerable list: *Biatora veteranorum, Buellia hyperbolica, Caloplaca lucifuga* and *Lecanora quercicola.* These are all restricted to veteran trees and our survey has shown that they are found on several oak veterans in the oldest part of the High Park site. None of these species are found on oaks in the plantations. Of the 231 species recorded at High Park, 58 have turned out to be new records for Oxfordshire (VC 23).

Figure 6.11. Distribution of *Buellia hyperbolica* on lignum in High Park.

Figure 6.12. Crustose *Buellia hyperbolica* (grey thallus with blackish apothecia) on lignum with foliose *Imshaugia aleurites* (above). (Photograph by Neil Sanderson)

Figure 6.13. A view of upright veterans and fallen dead trees in High Park. (Photograph by Neil Sanderson)

Table 6.5.   Threatened and Notable lichens and associated fungi recorded from High Park.
Bold = species that qualify for SSSI site selection (Sanderson et al. 2018). Conservation status:
VU = vulnerable, 4 spp.; NT = Near Threatened, 5 spp.; Nb = Notable, 36 spp.; IR = International
Responsibility, 11 spp.; S41= Section 41 spp., 5 spp.; NR = Nationally Rare, 12 spp.; NS = Nationally
Scarce, 46 spp, NE = Not Evaluated, 4 spp.

Species	Conservation Status	Abundance
Agonimia allobata	Nb (NS)	Rare
Agonimia flabelliformis	Nb (NS)	Occasional
Andreiomyces obtusaticus	NE	Occasional
Bacidia friesiana	Nb (NS)	Rare
Bactrospora corticola	Nb (NS)	Rare
Biatora veteranorum	VU (NR/IR)	Occasional
Buellia hyperbolica	VU (NR/IR/S41)	Occasional
Caloplaca lucifuga	VU (NR/IR/S41)	Rare
Chaenotheca hispidula	Nb (NS)	Rare
Chaenotheca stemonea	Nb (NS)	Occasional
Chaenothecopsis nigra	Nb (NS)	Occasional
Chaenothecopsis pusilla	Nb (NS)	Rare
Chaenothecopsis savonica	NT (NR)	Occasional
Cladonia cyathomorpha	Nb (NS)	Occasional
Coenogonium tavaresianum	Nb (NR)	Occasional
Cresponea premnea	Nb (IR)	Abundant
Cyphelium sessile	Nb (NS)	Rare
Gyalecta flotowii	NT (NS)	Rare
Lecanographa lyncea	Nb (IR)	Frequent
Lecanora albellula var. albellula	Nb (NS)	Rare
Lecanora hypoptoides	NE	Rare
Lecanora phaeostigma	Nb (NS)	Rare
Lecanora quercicola	VU (NS/IR/S41)	Rare
Lecanora sublivescens	NT (NS/IR/S41)	Occasional
Lecidea nylanderi	Nb (NS)	Occasional
Leptorhaphis atomaria	Nb (NS)	Rare
Micarea globulosella	Nb (NR/DD)	Rare
Micarea misella	Nb (NS)	Occasional
Microcalicium ahlneri	Nb (NS)	Occasional
Microcalicium disseminatum	Nb (NR)	Occasional
Milospium graphideorum	Nb (NS)	Frequent
Mycocalicium subtile	Nb (NS)	Rare
Ochrolechia arborea	NT (NR)	Rare
Porina borreri	Nb (NS)	Rare
Porina byssophila	Nb (NS/DD)	Occasional
Protoparmelia oleagina	Nb (NS)	Occasional
Pycnora sorophora	Nb (NS)	Rare
Rinodina exigua	NE	Rare

Table 6.5.   Threatened and Notable lichens and associated fungi recorded from High Park. *(Continued)*

Species	Conservation Status	Abundance
Rinodina furfuracea	NE	Rare
Rinodina roboris var. *roboris*	Nb (IR)	Occasional
Sphinctrina anglica	Nb (NS/DD)	Rare
Sphinctrina turbinata	Nb (NS)	Rare
Sporodophoron cretaceum	Nb (IR)	Occasional
Strigula jamesii	Nb (NS)	Occasional
Strigula phaea	Nb (NS)	Rare
Strigula taylorii	Nb (NS/IR)	Occasional
Syncesia myrticola m. sorediata	NT (NS/IR/S41)	Occasional
Thelopsis corticola	Nb (NS)	Frequent
Xylopsora friesii	Nb (NS)	Rare

Rarities provide a snapshot of lichen biogeography and High Park, situated in Middle England, includes several elements from different regions. Species that have a northern origin with the core or their distribution in boreal areas include *Lecanora phaeostigma*, *Lecidea nylanderi*, *Microcalicium disseminatum* and *Pycnora sorophora*. Other rarities belong in the southern oceanic group with the core of their distribution being in warmer climates than Britain, where they are on the edge of their range. These include *Agonimia allobata*, *A. flabelliformis*, *Coenogonium tavaresianum*, *Strigula phaea*, *Thelopsis corticola* and *T. rubella*, and species characteristic of the dry bark of veteran oaks such as *Cresponea premnea*, *Lecanographa lyncea*, *Sporodophoron cretaceum* and *Syncesia myrticola*. Many of these species have a yellow trentepohlioid photobiont that is sensitive to cold and dry conditions, and these lichens are restricted to veteran trees within the more humid sheltered areas of High Park. Ten of the taxa recorded in High Park are rare throughout the rest of their range and are species for which Britain is considered to have International Responsibility: *Biatora veteranorum*, *Cresponea premnea*, *Lecanographa lyncea*, *Lecanora quercicola*, *L. sublivescens*, *Thelopsis corticola*, *Rinodina roboris*, *Sporodophoron cretaceum*, *Strigula taylorii* and *Syncesia myrticola* ('*Enterographa sorediata*'). All except *Strigula taylorii* are restricted to the veteran trees.

In terms of the Guidelines for the Selection of Biological SSSIs (Sanderson et al. 2018), the standard guide to the national significance of lichen assemblages, we have ample evidence to support the importance of this site from the lichens that have been recorded in High Park during this survey. The Vulnerable species (*Biatora veteranorum*, *Buellia hyperbolica*, *Caloplaca lucifuga* and *Lecanora quercicola*) and Near Threatened species that are also International Responsibility species (*Lecanora sublivescens* and *Syncesia myrticola*, sorediate morph) are all selectable species in their own right as features of national importance. To be nationally important, the populations of these species need to have a sustainable population in the local area of selection, which here is the Upper Thames Clay Vales (national Character Area 108). All of these species pass this criterion, and it is clear from the distribution of these species on trees in High Park that many of the populations are significant nationally. In addition, two indices pass their thresholds; the Pinhead Lichen Index scores 18, with a threshold of national interest of 10, and the ecologically coherent assemblage Old Trees of Open Places scores 26, with a threshold of national interest of 16. These results indicate that the lichen assemblage of the Park is of national and probably international significance.

This survey of corticolous lichens and of lichenicolous fungi in High Park has established the importance of this site for lichen communities of ancient trees in an area of pasture woodland that has been forested since before the Normans came, allowing species that are now rare across Britain to survive. Although lowland sites in Middle England have been subject to air pollution and to encroachment from agriculture and forestry for centuries, this site is an example of ecological continuity of lichen communities over a millennium. A comparison with other sites of long ecological continuity and high conservation value for lichens in lowland England and Wales (Table 6.6) highlights the importance of this site for a special group of lichens associated with lignum and ancient bark. The PLI of 18 is higher than any other site in lowland Britain, while the SOWI remains low until we reach Dinefwr, a site in west Wales. That this is due to the loss of lichens that are sensitive to air quality rather than the loss of their habitat is illustrated by the distribution records for the classic macrolichen indicator of ancient woodlands, the Lungwort – *Lobaria pulmonaria* – where we can see that this lichen existed in many sites in central and eastern parts of England before 1959 and is now restricted to western sites across Britain (Figure 6.14).

Table 6.6.   Lowland sites in England and Wales with total lichens, associated macrolichens, crusts, other lichens and SOWI and PLI indices.

Site	County	Total	Macro	Crust	Lichen	SOWI	PLI
High Park	Oxfordshire	231	40	170	21	17	18
Windsor Great Park	Berkshire	126	41	85	3	11	15
Chatsworth	Derbyshire	165	55	108	4	6	11
Moccas Park	Herefordshire	235	64	161	10	16	8
Dinefwr	Carmarthenshire	341	78	253	10	40	11

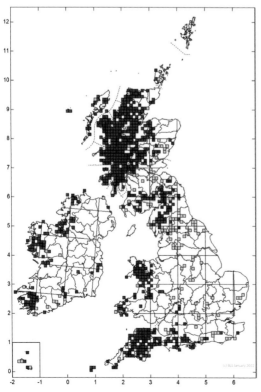

Figure 6.14.   Distribution of *Lobaria pulmonaria* in 10 km squares in the UK: Green – sites lost before 1959; Blue – recorded 1960–99; Red – recorded since 2000. (British Lichen Society)

Our records from Paulson show that by the 1920s species of *Usnea* were scarce in the Oxford area and that there are no records of highly sensitive cyanolichens in his list. Yet where continuity of ancient trees both standing and fallen exists, as in High Park, Windsor Great Park and Chatsworth, we retain a wealth of lichen diversity in crustose species that has been present from primeval times. To walk among trees such as those in High Park gives us the experience that Tolkien created with his vision of the Ents – derived from an Old English word for giant. Tolkien lived in Oxford and we pondered whether he visited High Park. There are many unanswered questions, and High Park has shown us that there is more to find out about ancient woodlands in Middle England.

Acknowledgements

The authors thank Mark Powell for contributing many of the field records as well as diagrams of individual trees. Thanks also to Zdněck Palice (Prague) for helping with identification of *Lecanora hypoptoides*.

CHAPTER 7

Snails, Slugs and Bivalves (Mollusca)

Rosemary Hill, Peter Topley, Thomas Walker and Rosemary Winnall

7.1. Introduction

There are about 234 species of land and freshwater molluscs in Great Britain, excluding alien species found only in hothouses. This figure is slowly creeping up as molluscs are adept at dispersal and reach British shores, often aided by humans and their vehicles (including forestry equipment) and the import of plants into garden centres. Despite restricted access, new molluscs make their way into woodlands, adding to those that arrived after the last Ice Age and before the marine flooding of the land bridge with Europe. This was not a complete barrier to colonization as small molluscs coated in mucus and their eggs might still be transported by birds or humans in boats.

An important feature of the shelled molluscs is that their shells may remain *in situ* in sediments and when these are datable, the timing of the colonization of species may be determined. For instance, there are no records of the Roman Snail (*Helix pomatia*, Figure 7.1), that predate the arrival of the Romans in Britain. Since this snail is edible, later and repeated introductions are also likely and this may account for the presence of this species in High Park, although introduction by the Romans cannot be ruled out. This species is mostly found in the Cotswolds and other areas of southern England with the friable limestone soils that it needs for egg-laying. Introductions further north in Britain have invariably died out.

A total of seventy mollusc species was found in High Park, only nine of which are freshwater species, compared with 37 species held by the Thames Valley Environmental Records Centre prior to 2017. All the species found earlier are still present. It is difficult to compare High Park with other ancient forests and pasture woodlands. Some of these locations will be highly acid, for instance Windsor Great Park, which has a very limited mollusc fauna. However, Windsor Great Park has species that are regarded as indicators of ancient woodland in the lowlands of Britain, such as the Ash-Black Slug (*Limax cinereoniger*) and Lemon Slug (*Malacolimax tenellus*). Forty-eight species have been found in Moccas Park, Herefordshire (140ha) including *L. cinereoniger* (Harding and Wall 2000).

Some ancient forests include a wider mosaic of habitats or cover a greater area. The lists are out of date for some locations and the amount of recording effort varies among sites. Comparing with two ancient woodlands where the recording is currently up to date, the Wyre Forest ten-km grid square (1,000ha) with predominantly Upper Coal Measures with sandstones, siltstones and clays with bands of *Spirorbis* limestone and tufa outcrops has 99 species (including *L. cinereoniger* and *M. tenellus*) (Westwood et al. 2015); whereas Shrawley Wood, Worcestershire (141ha) has 51 species (also including *L. cinereoniger* and *M. tenellus*) but

Rosemary Hill, Peter Topley, Thomas Walker and Rosemary Winnall, 'Snails, Slugs and Bivalves (Mollusca)' in: *The Natural History of Blenheim's High Park*. Pelagic Publishing (2024). © Rosemary Hill, Peter Topley, Thomas Walker and Rosemary Winnall. DOI: 10.53061/ZAFH4557

Figure 7.1.   Roman Snail (*Helix pomatia*). (Photograph by Aljos Farjon)

is mainly on Keuper sandstone with some Pleistocene glacial drift, and while the fertility of the soil increases in the lower parts of this wood, there are no calcareous areas (Green et al. 2022). In all these locations, some mollusc species are restricted to very small areas and so finding this many species represents substantial recording effort.

7.2. Survey methods

After the initial general searches to gain an idea of High Park, a search protocol was put in place to include sites in the vicinity of the ancient (mainly oak) trees, delimited by the extent of the surviving branches on the trees and the region in which their dead wood had fallen (Figure 7.2). Any identifying tags found on the tree were noted, and the predominant species in the ground flora, the presence or absence of fallen wood and notable features of the ground/soil were recorded. Each tree location was photographed to ensure a site could be returned to if required, a grid reference

Figure 7.2.   Mollusc survey in progress. (Photograph by Rosemary Hill)

taken, a sample of leaf or grass litter was sieved and the smaller species of molluscs found, collected and identified under the microscope. This was balanced by the selection of more open sites, including grassland, scrub, ponds, wet areas, quarries and limestone walling. Some vacuuming of the grassland and of tree trunks was undertaken using a small hand-held vacuum cleaner, but this did not add to the species obtained. The small debris samples taken from within the limestone walls were more fruitful.

A total of 113 locations within High Park were recorded yielding 1,214 records over a total of 15 visits. Efforts were made to visit all parts of High Park. Sufficient records were obtained for the more frequent species for distribution maps to be produced (e.g. Figure 7.3). On the

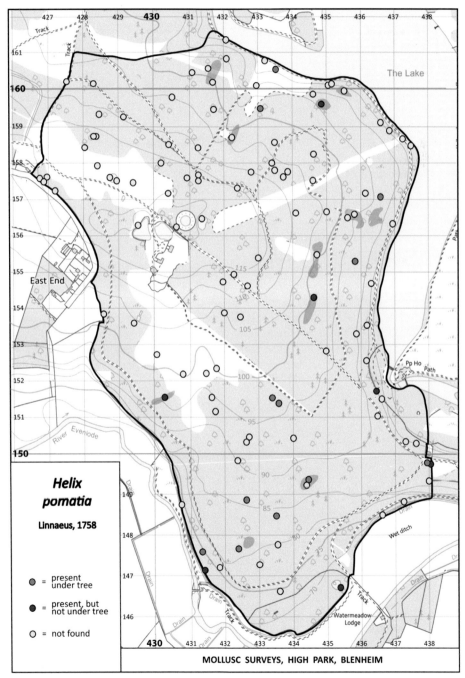

Figure 7.3.   Roman Snail (*Helix pomatia*) distribution map.

maps, woodland is shown as green, grassland as white and quarries as grey, and all surveyed locations are indicated. Freshwater molluscs were sampled by dipping in the three ponds in High Park, two ditches and some persistently wet areas, but excluding the Great Lake, which was outside the remit of the survey.

7.3. Accounts of the terrestrial molluscs found

Prickly Snail (*Acanthinula aculeata*) – A very small species, which has spines and is disguised by particles attached to these spines. Finds in High Park were sporadic, in calcareous areas and at the bottom of slopes that remain more damp, next to the Lake. Waxy Glass Snail (*Aegopinella nitidula*, Figure 7.4) – a common species throughout High Park.

Clear Glass Snail (*Aegopinella pura*) – Recorded sporadically in deciduous areas, with a preference for damper areas at the bottom of slopes. Copse Snail (*Arianta arbustorum*) – Generally but patchily distributed in areas that remain damp throughout the year either because of moisture-retentive substrates or deeper accumulations of leaf litter, including bottoms of small former quarries. No clear association with veteran trees; not in conifer plantations. Large Black Slug (*Arion ater*) – Occasionally observed, often under veteran trees but mobile at night and in rainy conditions and otherwise at rest in shady places during the day. Most records are in the northern half of the site. Brown Soil Slug (*Arion distinctus*) – Sporadically present across High Park, with a preference for damper areas. Green-soled Slug (*Arion flagellus*) – Sporadically present in High Park. Blue-black Soil Slug (*Arion hortensis*) – Sporadically present, not in conifers, with most records at bottoms of slopes on east side of High Park. Hedgehog Slug (*Arion intermedius*) – Occasionally present in the litter layer, tolerant of acid woods. Dusky Slug (*Arion subfuscus*) – Frequently found throughout High Park. Herald Snail (*Carychium minimum*) – Occasionally present, only in permanently wet areas. Not found in coniferous areas. Slender Herald Snail (*Carychium tridentatum*) – Occasional and generally in more calcareous areas. Not seen in the higher area of High Park. White-lipped Snail (*Cepaea hortensis*) – Present throughout High Park except in conifer plantations. Brown-lipped Snail (*Cepaea nemoralis*) – Throughout High Park except in conifer plantations but twice as commonly found as *C. hortensis*. Common Door Snail (*Clausilia bidentata*) – Found in leaf litter and under deciduous trees, but not under conifers. A climbing species also observed on walls but avoiding the highest part of High Park. Slippery Moss Snail (*Cochlicopa lubrica*) – Widely

Figure 7.4. Waxy Glass Snails (*Aegopinella nitidula*) mating. (Photograph by Peter Topley)

distributed but less common in higher areas of High Park. More tolerant of acid areas than *C. lubricella*. Least Slippery Snail (*Cochlicopa lubricella*) – Less frequently seen than *C. lubrica*, it avoids acid areas and the higher part of High Park and prefers the bottom of slopes (damper areas). Plaited Door Snail (*Cochlodina laminata*) – Prefers base-rich areas and the presence of deciduous trees. If seen away from trees, then in a quarry or next to a wall with accumulated leaf litter and so able to climb (Figure 7.5).

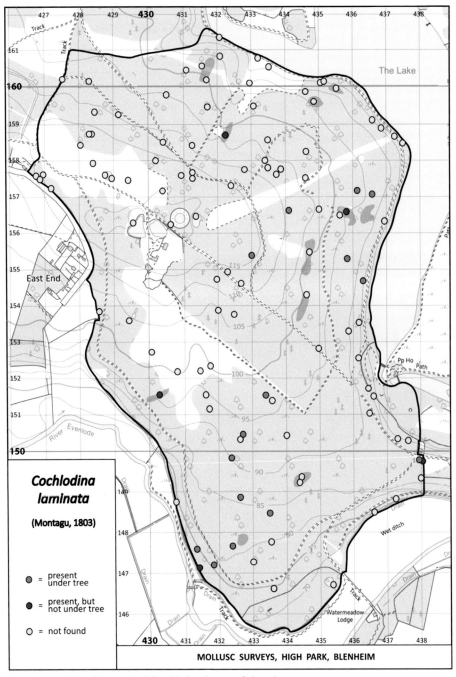

Figure 7.5   Plaited Door Snail (*Cochlodina laminata*) distribution map.

Garden Snail (*Cornu aspersum*) – Of the five records of this synanthropic species, three were near the gates into High Park and the remaining two were near to tracks. It was not seen in open areas or woodland away from tracks. Tramp Slug (*Deroceras invadens*) – Five observations were made of this rapidly spreading species, which reached Britain in 1931, two near a gateway, one near the main tarmac track through the site and all in base-rich areas. Netted Field Slug (*Deroceras reticulatum*) – Widely distributed in wooded and grassy areas of High Park and found throughout the UK. Rounded Snail (*Discus rotundatus*) – This was the most commonly seen of all species, being present in 66 of 113 locations; it is a widely distributed woodland species but not present in open areas or wetter woodland areas (Northern Drift). Tawny Glass Snail (*Euconulus fulvus*, Figure 7.6) – Distributed across High Park but more frequently in the more acid areas.

Roman Snail (*Helix pomatia*, Figure 7.1) – Restricted to calcareous areas and often but not necessarily observed under trees. It needs to be able to burrow into friable soil, which is more likely where leaf litter accumulates regularly. Not in the more acid areas in the highest area of High Park (Figure 7.3). Common Chrysalis Snail (*Lauria cylindracea*) – Restricted to calcareous areas and found most abundantly in crevices in limestone walls. Tree Slug (*Lehmannia marginata*) – Associated with trees and quarries and regularly climbs in wet weather. Usually absent from conifer plantations. Leopard Slug (*Limax maximus*) – Located throughout High Park and showing no association with trees. Lesser Bulin (*Merdigera obscura*) – Sporadically present, mostly in calcareous areas. Not in conifer plantations or the highest area of High Park (Figure 7.7).

Kentish Snail (*Monacha cantiana*) – A few records were in calcareous areas, except one. Rayed Glass Snail (*Nesovitrea hammonis*) – Sporadically distributed but more common in the acid

Figure 7.6.   Tawny Glass Snail (*Euconulus fulvus*). (Photograph by Peter Topley)

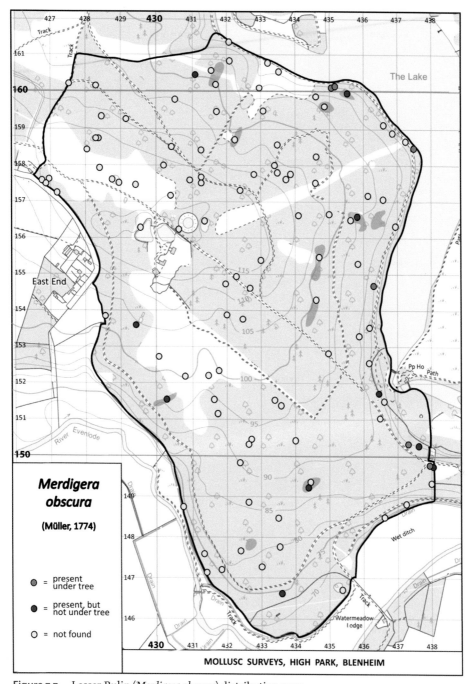

Figure 7.7.   Lesser Bulin (*Merdigera obscura*) distribution map.

non-coniferous areas. Garlic Snail (*Oxychilus alliarius*) – Common across the site and tolerant of acid areas (Figure 7.8). It shows some association with trees and is found throughout Britain. Cellar Snail (*Oxychilus cellarius*, Figure 7.9) – Common across High Park.

Glossy Glass Snail (*Oxychilus navarricus helveticus*) – Sporadically distributed, often under trees, avoiding the higher/more acid ground. Often found in disturbed places with nettles. This species is now widespread in lowland Britain. Dwarf Snail (*Punctum pygmaeum*) – Present sporadically,

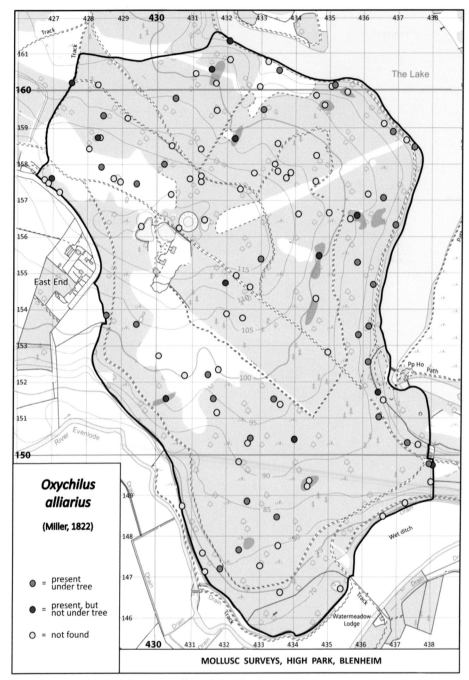

The legend within the map reads:

Oxychilus alliarius

(Miller, 1822)

- ◑ = present under tree
- ● = present, but not under tree
- ○ = not found

MOLLUSC SURVEYS, HIGH PARK, BLENHEIM

Figure 7.8. Garlic Snail (*Oxychilus alliarius*) distribution map.

mostly in wooded areas but not specifically associated with veteran trees, it prefers base-rich soils at the bottoms of slopes next to the Lake at the north-east edge of the site (Figure 7.10).

Rock Snail (*Pyramidula pusilla*) – Observed at four sites on the perimeter wall; this is a limestone rock crevice species restricted to suitable areas and walls with lime mortar in Britain. Budapest Keeled Slug (*Tandonia budapestensis*) – Sporadically present and a species of disturbed areas. Not found in the higher parts of High Park. Hairy Snail (*Trochulus hispidus*) – Present sporadically, often in quarries, but not in conifer plantations. Strawberry Snail (*Trochulus striolatus*) – Seen

Figure 7.9.   White form of the Cellar Snail (*Oxychilus cellarius*). (Photograph by Rosemary Winnall)

in most areas but avoiding the wetter (Northern Drift) area and conifer plantations. A white form was seen. Ribbed Grass Snail (*Vallonia costata*) – Present in open grassy areas, not under trees, in calcareous parts of the site (Figure 7.11). This species is mainly found in base-rich parts of southern and eastern England. Eccentric Grass Snail (*Vallonia excentrica*) – Found in open areas, not under trees, in calcareous parts of the site. It tolerates less calcareous soils than the Ribbed Grass Snail (*V. costata*). Milky Crystal Snail (*Vitrea contracta*, Figure 7.12) – Sporadically present in damp base-rich areas, including some quarries, and in the perimeter wall.

Crystal Snail (*Vitrea crystallina*) – Found throughout High Park more frequently than *V. contracta*. Pellucid Glass Snail (*Vitrina pellucida*) – Present mostly in coniferous and deciduous woods in acid and calcareous areas. Often the records were shell only, as this species is most active in the autumn/winter and is short-lived. Brown Snail (*Zenobiellina subrufescens*) – Recorded sporadically in damper deciduous woodland; it is rare in Oxfordshire.

7.4. Species with three or fewer records

The following terrestrial species were not mapped: Rusty False-keeled Slug (*Arion fasciatus*) is a species less widespread in Southern England. Tawny Soil Slug (*Arion owenii*) was present on one area of tussock grassland. Tree Snail (*Balea perversa*) was recorded on the perimeter wall and is known to feed on lichens so it is restricted to areas with low atmospheric pollution, as in High Park (Chapter 6). Worm Slug (*Boettgerilla pallens*) is an introduced species that has spread widely in Britain and is usually subterranean. The Toothless Chrysalis Snail (*Columella edentula*) is a very small species, found in moist base-rich places. Marsh Slug (*Deroceras laeve*) and Shiny Hive Snail (*Euconulus alderi*) are only found in permanently marshy areas. Lapidary Snail (*Helicigona lapicida*, Figure 7.13) is associated with the limestone perimeter wall where there are deep crevices in which it can hide. It is generally only observed active after prolonged wet weather and its distribution in Britain is declining. The Girdled Snail (*Hygromia cinctella*) is an introduced species, rapidly spreading northwards in the UK. The Green Cellar Slug (*Limacus maculatus*) is also an introduced species, originally seen only in old damp brickwork but now often found under logs as the winters have become milder. It is spreading rapidly in Britain. Draparnaud's Glass Snail (*Oxychilus draparnaudi*) is an introduced species of disturbed areas. Round-mouthed Snail (*Pomatias elegans*) is present only in calcareous places such as limestone quarries or friable limestone soil as it burrows to hibernate. This is a species that is declining in eastern England. Large Amber Snail (*Succinea putris*) lives in waterside vegetation. Sowerby's Keeled Snail (*Tandonia sowerbyi*) lives below the soil surface and frequents lowland habitats. Ear Shelled Slug (*Testacella haliotidea*, Figure 7.14) was seen twice, in both cases under trees just above the lakeshore. This uncommon subterranean species only comes to the surface of the soil in very wet weather, typically in May. Smooth Grass Snail (*Vallonia pulchella*) was found once in an improbable habitat and may have been displaced by a bird. Common Whorl Snail (*Vertigo pygmaea*) was found once in damp calcareous grassland. Striated Whorl Snail (*Vertigo substriata*) is considered rare in the Midlands but was located in one sample of damp leaf litter.

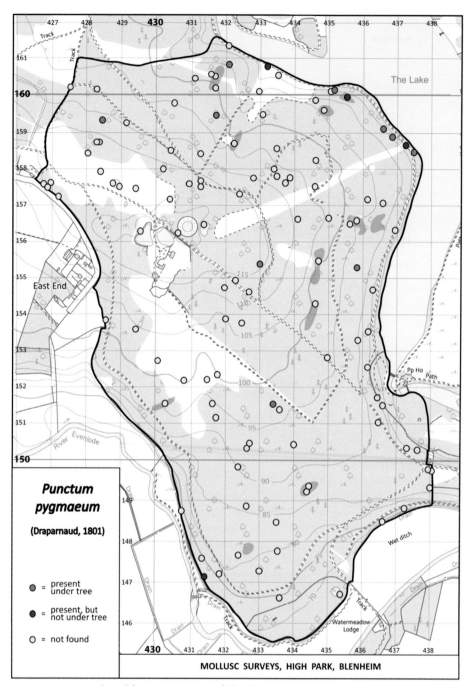

Figure 7.10.   Dwarf Snail (*Punctum pygmaeum*) distribution map.

7.5. Freshwater molluscs

The limited freshwater ponds on the site would explain the small number of records for White-lipped Ram's-horn (*Anisus leucostoma*), Common Bithynia (*Bithynia tentaculata*), Dwarf Pond Snail (*Galba truncatula*), Nautilus Ramshorn (*Gyraulus crista*), Great Pond Snail (*Lymnaea stagnalis*), Porous Pea Mussel (*Euglesa obtusalis*) and Wandering Snail (*Ampullaceana balthica*). *Anisus leucostoma*, *Gyraulus crista*, *Galba truncatula* and Red-crusted Pea Mussel (*Euglesa personata*)

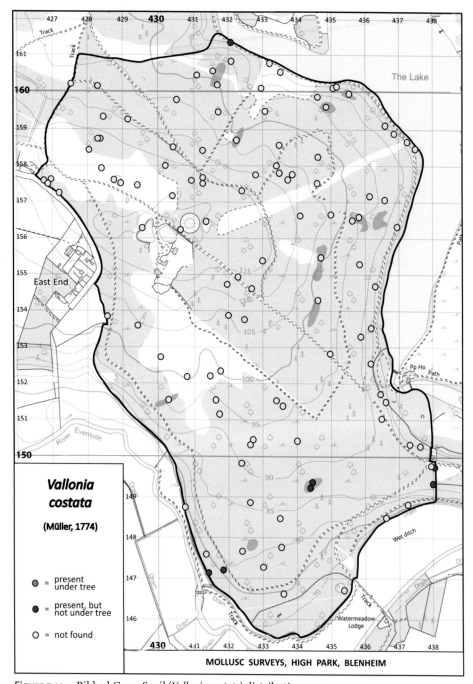

Figure 7.11.   Ribbed Grass Snail (*Vallonia costata*) distribution map.

are found in poor waterbodies that dry out. Jenkin's Spire Snail (*Potamopyrgus antipodarum*) was present at the edge of two ponds. This now ubiquitous species was probably introduced from New Zealand. *Euglesa personata* was recorded in some numbers in a small ditch. *Bithynia tentaculata* was present in a small pond surrounded by tree leaf litter. *Galba truncatula* was seen in two ponds, one of which was dried up, and also in two wet sites with tree leaf litter. Eleven Nautilus Ram's-horn (*Gyraulus crista*) were present at the margin of one pond. *Lymnaea staynalis*

Figure 7.12.   Milky Crystal Snails (*Vitrea contracta*) mating. (Photograph by Rosemary Hill)

Figure 7.13.   Lapidary Snail (*Helicigona lapicida*). (Photograph by Rosemary Hill)

Figure 7.14.   Ear Shelled Slug (*Testacella haliotidea*). (Photograph by Rosemary Winnall)

was found only in a small ditch. *Euglesa obtusalis* was located at the margin of a pond and in the ditch. *Ampullaceana balthica* was present in two ponds. None of these species is uncommon.

7.6. Species abundance

While the number of individuals of a species was noted, the accuracy of these data would be affected by several factors as individual molluscs are easier to find on survey days with recent wet weather, or may aggregate in response to food sources, shelter and so on. Most records were of a single individual at a site, but there were some obvious exceptions where conditions were favourable to that species, including *Aegopinella nitidula*, *Carychium tridentatum*, *Cepaea nemoralis*, *Cochlicopa lubrica*, *Cochlodina laminata*, *Discus rotundatus* (which was most often found in caches of higher numbers), *Euconulus fulvus*, *Lauria cylindracea* (where a small debris sample from one section of the perimeter wall contained a record 337 individuals), *Oxychilus alliarius*, *Trochulus striolatus*, *T. hispidus*, and the aquatic species *Gyraulus crista* and *Euglesa personata*.

7.7. Number of mollusc species per location

The abundance mapped in Figure 7.15 ranged from 0 species in two sample locations up to 33 species in one location. One location with three distinct habitats (open ground, ancient oak and limestone wall, all adjacent) gave 33 species, but for further analysis was treated as three locations with 11 species each. The High Park site was then divided into those locations at or above the 100m contour line (51 locations) and all those at lower altitudes (62 locations: 3 aquatic locations are excluded). This provided an unbiased way of separating locations. The average number of mollusc species per location at or above 100m was 6.2 and below 100m was 11.0. This is not an effect of altitude as such; the area above 100m contains the bulk of

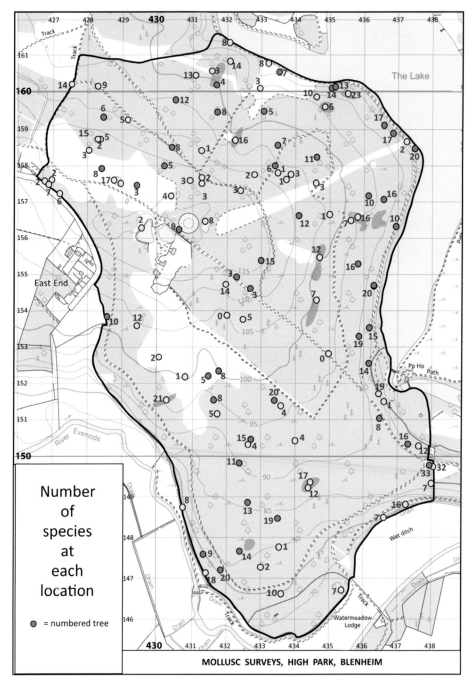

Figure 7.15. Map with numbers of species per recorded location. 'Numbered tree' refers to an ancient oak with a metal tag number from the Estate's veteran tree survey, 2001–6, and was used in this survey of molluscs as an identifier of the site.

the more acid sites (except where quarrying had exposed the Cornbrash) and is also more likely to be dry during drought in the summer, whereas the locations in the area below 100m were more likely to be calcareous and more likely to remain moist over the summer. Acid locations contain fewer species and numbers of molluscs as many species require calcium to make their shells. So calcareous locations are always more species rich. While some molluscs are well adapted to drought, more species and numbers of molluscs are generally found in damp and wet areas. The limited occurrence of species found in persistently wet areas in the more acidic soils in the Oak – Northern Drift and Denchworth – Oxford Clay suggested that while species tolerant of acid substrates such as *Deroceras laeve* and *Euconulus alderi* were present, species requiring calcareous conditions and permanent wetness such as *Carychium minimum* were not.

It is noticeable that all but one of the sites with 20 or more species of mollusc were at the periphery of High Park, coinciding with the Aberford Great Oolite deposit and benefiting from runoff from rain falling on higher areas.

7.8. Influence of ancient trees

The numbers of species per location for the ancient tree sites (Figure 7.15) were compared with those for the other survey sites (with or without trees). This is clearly not intended to be an absolute comparison. Locations with ancient trees had on average 11.83 mollusc species whereas the general sites had 7.45. Looking at the notes made on the ancient tree locations with high numbers of species (19 or 20), the ground flora associated with them was not predictive of this. All the sites with high species counts did not have bracken in the ground flora, indicating less acidic sites. The presence of fallen wood left *in situ* contributed to an increase in mollusc species and numbers found under the ancient trees, and forms an important aspect of the management of the site.

It is notable that two species, Ash-Black Slug (*Limax cinereoniger*) and Lemon Slug (*Malacolimax tenellus*), normally associated with ancient woodlands (see Section 7.1), were not found at High Park. This may be related to historic ground disturbance probably from high grazing pressure or human activity such as quarrying or hunting. *Malacolimax tenellus* has been recorded in woodland about 4.8km from High Park. The nearest records for *L. cinereoniger* are in Bagley Wood, to the south of Oxford, about 19km away. Both of these species are rare in north Oxfordshire.

7.9. Molluscs found in quarries

There is evidence of a lot of hand quarrying across High Park in soils where Cornbrash or other limestone is exposed. Some of these quarries are very small but are a part of the general disturbance of the ground that has taken place in High Park. The quarries dug did not contain ancient oak trees – either because hand quarrying around large trees would have been impractical or because quarrying that damaged trees would not have been permitted. The 11 locations sampled that lie in the larger quarries showed an average of 12.8 mollusc species per location, slightly higher than the average for locations with ancient trees. If quarries are excluded from the general sites, the average number of species per general site is 6.38. It is probable that base-rich conditions are the foremost factor determining the presence of molluscs that require calcium.

The perimeter limestone wall added to the species count by providing an artificial cliff with crevices suitable for *Helicigona lapicida*, *Balea perversa* and *Pyramidula pusilla*, which were found nowhere else. While *Lauria cylindracea* is occasionally found elsewhere, it was regularly found from small debris samples taken from the wall.

Acknowledgements

The authors thank Thomas Walker for producing the distribution maps.

CHAPTER 8

Spiders and Relatives (Arachnida)

Aljos Farjon

8.1. Introduction

Arachnids (Class Arachnida) are arthropods including spiders, scorpions, false-scorpions (Pseudoscorpiones), harvestmen and mites, among lesser-known subgroups. The vast majority of arachnids recorded in High Park are spiders and the only three other groups are harvestmen (four species), false-scorpions (one species) and mites (one species). Most of the records made during the High Park Biodiversity Survey (HPBS) are by the arachnologist Bill Parker, a leading member of the British Arachnological Society. Records prior to 2017 were made available by the Thames Valley Environmental Records Centre (TVERC). Not all records made from 2017 to 2021 were available at the time of writing, in particular not those specimens found in autumn 2021. These too will be submitted to TVERC when identifications are completed, but unfortunately will have come too late to be included here. Despite this, there is sufficient cover of the diversity of arachnids present in High Park to merit an account.

This chapter should of course ideally have been written by an arachnologist, but time constraints prevented those associated with the HPBS to complete the task. Arachnids are an important group that merit inclusion as a chapter and therefore I decided, as editor of this book, to report on them using information presented in the literature or available on the internet, including suitable images. It would be inappropriate to describe the survey methods Bill Parker used, although indications are given in the Excel spreadsheet of the HPBS records. Similarly, I cannot comment on the significance of High Park for arachnids other than to pick out notable or interesting species from the checklist in Appendix 6 and to give a brief account of each, with some illustrations. The chapter ends with some concluding remarks.

8.2. Harvestmen

Harvestmen or harvest spiders (order Opiliones), popularly known as 'daddy-long-legs' (among some other similar taxa) differ from true spiders in having a single body part rather than two, suspended between eight often extremely long and thin legs, and two simple eyes instead of six to eight in spiders. Of the four species found in High Park, the Fork-palped Harvestman (*Dicranopalpus ramosus*) is of note as a species originating from the Mediterranean Region and spreading northwards. It was first recorded in England on the south coast in 1957 and reached Scotland in 2000 (BAS 2010–23). A female specimen was swept from a small yew tree at the edge of High Park near the Springlock Gate by Bill Parker. The other three species found in High Park are all common in Britain.

Aljos Farjon, 'Spiders and Relatives (Arachnida)' in: *The Natural History of Blenheim's High Park*. Pelagic Publishing (2024). © Aljos Farjon. DOI: 10.53061/KRTC5727

8.3. Mites

Mites are small arachnids belonging to two separate orders. The single species recorded in High Park, *Eriophyes rubicolens*, is a member of the order Acariformes. It causes galls on the leaves of brambles (*Rubus fruticosus* agg.). The record consists of the gall and was made by Ivan Wright on 19 August 2021. Another species in this genus causes galls on the leaves of lime trees (*Tilia*), but this one was not seen.

8.4. False-scorpions

These are very small arachnids not at all related to scorpions (order Scorpiones) but belonging to the order Pseudoscorpiones. The resemblance is, for a layperson like me, at first striking owing to the two frontal appendices (pedipalps) having scorpion-like pincers (chelae) which they seem to wield in the same manner as scorpions. But closer inspection – they are tiny (2–8mm long) – reveals a rounded, segmented abdomen without a tapering 'tail' terminating in a sting (Figure 8.1). The species found, *Dendrochernes cyrneus* (Chernetidae), measures 3.5–4.2mm in body length and is red-brown or grey-brown with almost black 'hands'. It is the largest of the British species and very rare; it is also an indicator of ancient woodland. It is phoretic, that is, it disperses from tree to tree by grabbing hold of the leg of an ichneumon (parasitic wasp) or a beetle and hitching a lift. Three adult animals were found by R. Fleetwood inside the hollow trunk of a recently fallen veteran oak and identified by Keith Alexander during the excursion of the Saproxylic Group of the British Entomological and Natural History Society (BENHS) on 15 June 2019. The false-scorpions were living in the brown decaying wood now exposed by the fall. Another specimen was found by Steve Gregory under oak bark during the same excursion.

Figure 8.1.   The pseudoscorpion *Dendrochernes cyrneus* in High Park. (Photograph by Keith Alexander)

8.5. Spiders

The vast majority of recorded arachnids are spiders (order Araneae). In Britain there are approximately 670 species (Bee et al. 2017). There are 100 species (including five 'indets'), divided over 19 families, in the available records for High Park (Appendix 6). This compares well with the results in Moccas Park: 121 species (see Introduction) if we roughly estimate that the records for 2021 could have added another 20 species. Below, I shall describe interesting or notable species in order of the families listed in the Checklist of Arachnids given in Appendix 6.

Agelenidae

The funnelweb spiders are represented in Britain by 14 species. The four species (one 'indet') found in High Park that belong in this family are all common in Britain. *Tegenaria silvestris* is however rare to the north of central England and Wales and is commonly associated with woodland habitats. This species is much smaller than other members of the genus, with a body length of 5–7mm. An adult specimen was found inside the largest ancient oak (10.37m girth) by Lawrence Bee and another appeared from under loose oak bark. The Large House Spider (*Eratigena (Tegenaria) gigantea*) is best known – and feared, but not dangerous – from its indoor occurrence, but is also found in sheds or other outbuildings and of course in the hollowing ancient oaks of High Park, a natural habitat. There is a record of a male of *Coelotes atropos* (Figure 8.2) held by TVERC and dating from 1990. Although widespread and common

Figure 8.2.   The spider *Coelotes atropos* – female guarding egg sac. (Photograph by A.J. Cann)

in the west, it is very rare in south-eastern England (BAS 2010–23), with only 12 records for Oxfordshire (VC 23) since 1980. It seems to be declining in the east, with more recent records increasingly concentrated in the western uplands; the decline is perhaps associated with more frequent spells of summer drought. The female makes tube web tunnels in tree cavities or under logs.

Amaurobiidae
There are three species of laceweb spiders in Britain and the two common *Amaurobius* species were recorded in High Park. These spiders are active at night and hide during the day in a lace-like web spun around holes. One is most common on buildings (*A. similis*) and the other (*A. fenestralis*) is more often found under the bark of trees. Both spiders are very similar in appearance, but *A. similis* is the larger of the two.

Anyphaenidae
Of this family just three species in the genus *Anyphaena* are known in Britain and only one species, the Buzzing Spider (*Anyphaena accentuata*), has been found in High Park, present in both the HPBS and earlier TVERC records. During courtship the male is known to emit a high-pitched buzzing sound by vibrating its abdomen on a leaf.

Araneideae
This family comprises most of the British orb-weavers and includes the familiar Garden Spider or Cross Spider (*Araneus diadematus*), which appears in late summer to early autumn sitting in the centre of its symmetrical web suspended vertically between all kinds of erect plants or other supports. It has been found in High Park, so it is not confined to gardens. Much less common is *A. triguttatus*, a species fairly common in south-east England, but with only a single record in Oxfordshire: found in High Park on 26 April 1984 (TVERC). It makes its webs especially on oaks in the lower canopy (BAS 2010–23). Also commonly found on oaks is *Araniella cucurbitina*, which has a cucumber-green or apple-green colour, especially conspicuous in the swollen abdomen of the females. A rare species, *Leviellus (Stroemiellus) stroemi* (Figure 8.3), has been recorded in High Park in 1985–6 (TVERC) as one of only two confirmed records in Oxfordshire (VC 23). It seems to be one of the few British spiders exclusively associated with ancient trees (Bee et al. 2017: 201). This species is listed as Nationally Rare (NR) in Britain and as Near Threatened (NT) on the IUCN Red List (Bee et al. 2017:430; Harvey et al. 2017).

The Missing Sector Orbweb Spider (*Zygiella x-notata*) is so named because it spins a web like the other spiders in this family, but usually with one sector missing. It is similar in appearance to the previous species, but with a silvery-grey rather than brown folium on top of the abdomen. Like the Garden Spider, it is most frequently found near human habitation (typically in its web spun in the corners of door and window frames), but can, as found here, also occur in open woodland.

Clubionidae
The sac spiders are represented in Britain by two genera and 24 species. In High Park only the genus *Clubiona* was found, with five species recorded. All are common in Great Britain, but *C. corticalis* is rare in the northern half of the island. It is most commonly found beneath the loose bark of dead and dying trees. These roving spiders build sac-like silken nests in low vegetation or under loose bark. *Clubiona brevipes* is most often found on oaks.

Dictynidae
The meshweb spiders count 17 species in Britain. The single representative of this family recorded in High Park is *Lathys humilis*. It is common in the south of Britain but rare in the north and west. The legs of this small (body length 1.7–2.5mm) woodland spider are

Figure 8.3.   The rare spider *Leviellus (Stroemiellus) stroemi*. (Photograph by Ondřej Machač)

conspicuously marked with dark annulations. Its habitat is usually the dense foliage of evergreens such as Holly, Yew, pine and gorse, but these are absent or rare in High Park (Chapters 3, 4) so the single record from TVERC is likely to have come from oak, the other tree it has been recorded on (Bee et al. 2017: 267).

Dysderidae
The woodlouse spiders have only four species in Britain. *Harpactea hombergi* is the only species of this family found in High Park. It is widespread in Britain and common in the south. It is a fast-moving nocturnal hunter that hides during the day in a silken cell, often under loose bark. Its common prey are woodlice (order Isopoda), which are caught with large, pincer-like chelicerae.

Gnaphosidae
The ground spiders, with 33 species in Britain, is another family in High Park represented by a single species, *Zelotes latreillei*. Widespread but local in southern England and scattered further north, this almost entirely black-brown spider is often found under stones or logs.

Linyphiidae
The money spiders, with around 280 species, are the largest family in Britain. The majority are very small and need the use of a high magnification microscope for certain identification, but some are larger. Most of the species recorded for High Park were found prior to the HPBS and come from TVERC. Many species in the Checklist for High Park (Appendix 6) are reported to be declining in Britain (Harvey et al. 2017). There are 31 species recorded in High Park. Of these, *Collinsia inerrans* is widespread but uncommon in Britain. There are nine records in

Figure 8.4.   A female of *Walckenaeria acuminata*. (Photograph by Arno Grabolle)

Oxfordshire (VC 23), but only three were made from 1992 onwards (BAS 2010–23). The High Park record from TVERC dates from 1986. This little spider is a frequent aeronaut, dispersing by air on its strand of silk that provides upward lift, known as 'ballooning'. Much rarer is *Sintula corniger*, considered Nationally Scarce and flagged as 'amber' and recorded on the Amber List with an estimated decline of >40% (Bee et al. 2017: 447; Harvey et al. 2017). A male specimen was caught in a pitfall trap in High Park in 1990 (Oxfordshire Biological Record Centre, OBRC = TVERC); there are five other records in Oxfordshire (VC 23). This species generally occurs in wet swampy areas but is also sometimes found in litter of woodlands, the more likely habitat for the High Park record. This very small spider (body length 1.5–2.5mm) with its shining black abdomen is easily missed and may be under-recorded (BAS 2010–23).

Two species in the genus *Walckenaeria* were found in High Park, *Walckenaeria acuminata* (Figure 8.4) and *W. atrotibialis*. The first species is widespread and common throughout Great Britain, the second is also widespread but more locally distributed and uncommon in most places. According to Harvey et al. (2017), both are in decline in the UK. These spiders are very small, only up to 4mm long, with a black head and abdomen in *W. acuminata* and a black abdomen only in *A. atrotibialis*. They are ground dwelling spiders, mostly in grassland but sometimes also in open woodland.

Lycosidae

The wolf spiders are ground-running hunters most active in warm sunny weather. The female carries an egg-sac on the abdomen attached to the spinnerets (from which the silk threads emerge) and the newly hatched young cling to its abdomen. There are 38 species in Britain; eight species were recorded in High Park. Of these, *Pardosa lugubris* is a rare species associated with ancient woodland. The record for High Park comes from three male specimens caught in a pitfall trap by John Campbell in 1990 (OBRC/TVERC). This is almost certainly an erroneous identification; it is given a ? in the checklist in Appendix 6. Most verified records are from Scotland and just one from north-west Yorkshire. *Pardosa saltans* was described by Töpfer-Hofmann (in 2000) as a species distinct from *P. lugubris* and, so far, nearly all British specimens examined have proven to be *P. saltans*. *Pardosa saltans* is common and 'largely restricted to over-mature and ancient woodland sites' (Bee et al. 2017: 216). This latter species was recorded several times by Bill Parker in 2018–19. The other three species in the genus *Pardosa* recorded in High Park are all common.

Philodromidae
The running crab spiders or Philodromidae are represented in High Park by two of the three British genera, *Philodromus* (two species) and *Tibellus* (one species). *Philodromus albidus* is confined to southern England and is often found on the foliage of old oak trees. It is one of the smallest of the running crab spiders at 3–5mm body length. The other two species (Appendix 6) are widespread and common.

Pisauridae
This small family has three species in Britain. One species, the Nurseryweb Spider (*Pisaura mirabilis*), of this family with the same English name was found in High Park. It is a distinctive long-legged spider and the female carries a large white egg sac held by its pedipalps underneath the body. It is abundant and common in much of England and Wales.

Salticidae
This family comprises the jumping spiders, usually small and with a compact body and short legs. They have eight eyes in two rows, with the central pair in the front row much larger than the others; this enables them to judge accurately the jumping distance to catch prey. There are 17 genera with 38 species known in Britain. Four genera, each represented by one species, were recorded in High Park. *Neon reticulatus* is widespread and fairly common, but reported as declining by >10% (Harvey et al. 2017). In woodland habitat it occurs in dry leaf litter. Much less common is *Pseudeuophrys erratica* (Figure 8.5), which occurs mainly in northern and western Britain. High Park is the single location where it has been recorded in Oxfordshire (VC 23), and a female specimen was found under oak bark by Steve Gregory during the BENHS Saproxylic Group excursion on 15 June 2019. This species is also reported as declining.

Segestriidae
The tubeweb spiders are represented in High Park's records by the species *Segestria senoculata*. There are in Britain just three species and this one is the most widespread and common. They construct a tubular web, the opening of which is often seen in hollows of the ancient oaks in High Park. Delicate lines of silk radiate from this entrance and the spider lurks behind in the tunnel, its front legs touching the radiating threads. When an insect touches this network, the spider dashes out to seize its prey. Nature is a dangerous world!

Tetragnathidae
These are orbweb spiders like the Araneidae (see above), but their webs differ and are often associated with damp or wet areas (see Bee et al. 2017: 78–9 for images). Britain has 14 species. All six species identified in High Park are widespread and very common.

Theridiidae
This is a large family with more than 3,000 species worldwide; 58 species are known in Britain, of which 10 were recorded in High Park. *Parasteatoda (Achaearanea) lunata* is widespread only in southern England and generally uncommon. *Parasteatoda (Achaearanea) simulans* is also

Figure 8.5.   The jumping spider *Pseudeuophrys erratica* is just 3–4mm long. (Photograph by Luc Gizart)

restricted to southern England, but with an even more scattered distribution. Only six records are mapped in BAS (2010–23) from Oxfordshire (VC 23), but these appear not to have included an OBRC/TVERC record from High Park made in 1986.

Thomisidae
This family is known in English as crab spiders and is treated as separate from Philodromidae, the Running Crab spiders (see above). The first two pairs of legs are much longer than the two following pairs, which gives them a superficial resemblance to crabs. *Diaea dorsata* is a small crab spider with a bright green cephalothorax and legs and a brown abdomen with yellowish-green sides. This species is widespread but local in the southern half of England. It was found on small yew trees near the Springlock Gate at the edge of High Park by Bill Parker. These are planted evergreens; yews are not present in the wild in High Park. *Xysticus lanio* (Figure 8.6) is a brown crab spider found on shrubs and young trees, particularly oak – Bill Parker found it on epicormic shoots of *Quercus robur*. It is widespread but scarce in southern England, and according to BAS (2010–23) there is some evidence of decline in the last 30 years.

8.6. Concluding remarks

The arachnids mentioned in this chapter and in Appendix 6 do not present an entirely complete account of the result of surveys in High Park, as explained in Section 8.1. Even if these missing records were included, it is evident that in the relatively limited time of surveying (TVERC 1985–7, 1989–90; HPBS 2018–20) not all species present will have been found; this is as true for spiders as it is for flies (Chapter 9), moths (Chapter 12) and beetles (Chapter 13). Conclusions concerning the biodiversity of arachnids in High Park are therefore at best provisional and in any case beyond the competence of this author. However, some general observations may be made.

Figure 8.6.   The crab spider *Xysticus lanio*. (Photograph by Zdeněk Hyan)

The majority of arachnids (mainly spiders) found are common in Great Britain, but a number of these are restricted or nearly restricted to the southern half of England. Six arachnids (four spiders, one mite and one false-scorpion) are rare, with two spiders officially noted as NS, amber (*Parasteatoda simulans*) and NR, Near Threatened (NT) (*Parasteatoda lunata*). The latter is the only one listed on the IUCN Red List as NT in Britain and found in High Park. Five spiders are local or uncommon. A rather large number of species of spider (22) have been assessed as declining by >10% nationally, most of these in the family Linyphiidae. Among these are species still common and widespread. If they are rare, I have used that status preferentially in Appendix 6.

I thought it could be interesting to compare our records of arachnids with those of Moccas Park, Herefordshire (see also Introduction). Of the 106 species of arachnids recorded in High Park, 56 were also found in Moccas Park (marked in Appendix 6). This leaves 50 species only found in High Park. Conversely, 62 species were found in Moccas Park that are not in the records for High Park. Perhaps the fairest conclusion to be drawn from this comparison is that the inventory of arachnids in both these otherwise similar medieval deer parks remains incomplete.

Acknowledgements

The author thanks Lawrence Bee for his comments and corrections, which improved the accuracy of the chapter.

CHAPTER 9

Two-Winged Flies (Diptera)

Peter Chandler

9.1. Introduction

There are more than 7,000 species of two-winged flies (Diptera) now known from the British Isles (Chandler 2023), and it is likely that well over 1,000 of these are present in High Park. Flies have many different lifestyles, including predation, parasitism of other insects, development in all parts of higher plants, in fungi and in all types of decaying plant or animal material. Many of them have saproxylic associations with ancient trees and other dead wood habitats. In contrast to Coleoptera and with some insects in other orders, where both adult and larval stages may have the same food requirements, with Diptera it is only the larval stage that has the saproxylic association, while the adults feed elsewhere, if at all – an exception being those that develop in sap runs, which are also attractive as a food source for the adult flies. Adults of many flies in general visit flowers to feed on nectar or pollen, and so are important as pollinators. The presence of flowers at a site is therefore critical to the survival of many saproxylic Diptera, and flowering plants are therefore an important part of saproxylic habitats.

Alexander (2002) compiled a list of British saproxylic invertebrates, in which he listed around 740 species of Diptera, including some which were unconfirmed but considered likely to be saproxylic. Chandler (2010) provided a partial update for the mycophagous species with a list of Diptera and their associated fungus hosts, and further species have since been confirmed to develop in saproxylic fungi. Fungus feeders are considered as saproxylic if at least one of their known hosts grows on wood. Probably at least 1,000 species of British Diptera have some saproxylic associations. Chandler (2021) described the range of such associations and discussed their habitat requirements.

High Park, Blenheim is expected to be a rich site for saproxylic Diptera, though insufficient fieldwork has yet been completed to demonstrate this conclusively. The total Diptera list currently stands at 679 species, of which at least 188 can be regarded as saproxylic. The saproxylic species form a greater proportion of the total Diptera species so far recorded in High Park than for some better-worked sites.

Knowledge of the Diptera fauna in High Park is still incomplete. Prior to the present survey there were records of 88 species obtained over the years from 1982 to 2002, and 36 of these species have not been recorded more recently. Thirty species of Anthomyiidae were recorded by Michael Ackland from 1984 to 1996, 13 of which have not been recorded again. Otherwise, the most productive visit was on 20 May 1992, by Roger Morris, Mark Parsons and Stuart Ball, when 24 species of Diptera were recorded, of which 16 were hoverflies (Syrphidae); there are no more recent records of 12 of these 23 species, and 8 of these 12 were hoverflies. I first visited

Peter Chandler, 'Two-Winged Flies (Diptera)' in: *The Natural History of Blenheim's High Park*. Pelagic Publishing (2024).
© Peter Chandler. DOI: 10.53061/PEVB9384

on 1 June 2017 and have since made only four further visits (14 June 2017, in the company of several mycologists; 12 October 2017; 15 June 2019, a saproxylic field meeting of the British Entomological and Natural History Society; 30 October 2019, coinciding with a visit by several conchologists). On the autumn visits access was restricted to the vicinity of the main public path. From my visits only 230 Diptera species have been identified, of which 65 were fungus gnats. Some further Diptera were also identified by Ivan Wright, and by Graham Collins and Jovita Kaunang, while they were recording other insects. However, the majority of Diptera records came as a by-catch from trapping: by Ivan Wright in 2017 to 2019, by Benedict Pollard in 2020 and 2021, and by Jonathan Cooter in 2021. I am grateful to these recorders for sorting Diptera from their catches and passing them to me for identification.

By way of comparison, a recent account of the Diptera fauna of Windsor Forest and Great Park (Chandler 2021) included records of 1,817 Diptera species, of which 455 could be considered saproxylic; this resulted from a considerable amount of fieldwork over many years and covered a large estate with diverse habitats. That article included a comparative discussion of the available data for Windsor and for some other sites for which substantial species lists existed. These included two sites that are comparable with Windsor in the abundance of old and decaying trees – Burnham Beeches, Buckinghamshire, and Epping Forest, Essex, for both of which manuscript lists compiled by John Ismay were consulted. Burnham Beeches, dominated by ancient pollarded Beeches, also includes deep and shaded stream valleys, ponds and bog habitats. Epping Forest also has many ancient oaks and Beeches, with extensive woodland as well as heath and grassland habitats and some streams.

The other sites discussed in the 2021 Windsor article covered a range of habitats, although all included some old trees and were all shown to support a substantial number of saproxylic species. The most comprehensive Diptera list for a single site was that for Wicken Fen, Cambridgeshire (Perry and Langton 2000), principally a fenland habitat but including some carr woodland. The other published lists considered were for Bushy Park, Middlesex (Chandler 2015), which includes areas of woodland as well as some ancient oaks in a parkland setting, and the records for a well-developed hedge on an extensively managed Devon farm (Wolton et al. 2014). The latter site includes species-rich wet woodland, and subsequent recording in that habitat greatly increased the total of Diptera species and the number of saproxylics (Rob Wolton, pers. comm.), as indicated separately in Table 9.1. Records from Dinton Pastures

Table 9.1.   Comparison of overall total of Diptera species, and the saproxylic and mycophagous subsets of these totals, recorded in High Park, Blenheim, and seven other well-recorded sites.

Site	Total Diptera species	Saproxylic (including in fungi)	Saproxylic in fungi	Other fungus-associated	Total fungus gnats	Total spp. also at Windsor	Saproxylic also at Windsor
Windsor Park	1,817	455	242	70	300		
High Park	679	188	115	18	147	584	179
Epping Forest	1,628	303	150	34	204	1,107	279
Burnham Beeches	1,572	351	165	33	273	1,079	297
Dinton Pastures	1,237	239	142	31	154	843	220
Bushy Park	1,037	240	136	30	168	813	161
Locks Park Farm (2014 hedge list)	830	163	98	23	140	638	152
Locks Park Farm (all records to 2020)	1,391	297	227	51	272	1,001	272
Wicken Fen	1,849	199	88	24	133	984	174

Country Park, Berkshire, were also included, based on a manuscript list, compiled by the author in 2002, which resulted from fieldwork by members of the British Entomological and Natural History Society after its headquarters were established there in 1992. A recent account has also been published (Wright and Wright 2018) of the fauna and flora of the Shotover area in Oxfordshire, which includes many ancient trees; the Diptera list was reported to include more than 1,300 species, of which 196 species were considered saproxylic – a full list was not provided.

The comparative table from Chandler (2021) is reproduced here (Table 9.1), with the addition of the numbers for High Park. Shotover was excluded as precise totals for each category were not available. It was noted in the original article that the results for the sites covered were not strictly comparable because of the different emphasis given to particular families by individual recorders.

9.2. The Diptera fauna in High Park

Out of the 108 fly families recognized in the latest British Isles checklist (Chandler 2023), representatives of 69 families have been recorded in High Park compared with 88 at Windsor; most of the families missing at both localities contain few species. Here comment is provided for all families that include saproxylic species, with some mention of the more interesting or significant non-saproxylic species. The families are dealt with in taxonomic order.

9.2.1. Tipulidae, Pediciidae and Limoniidae (craneflies)
The large brightly coloured comb-horn craneflies (Tipulidae) are among the more spectacular of the saproxylic Diptera. The common name refers to the pectinate antennae of the males. Of the five species of comb-horns recorded at Windsor only the two most widespread – the Orange-sided Comb-horn (*Ctenophora pectinicornis*) and the Twin-mark Comb-horn (*Dictenidia bimaculata*) – have been recorded in High Park, mostly by trapping. *Ctenophora pectinicornis* has also been observed flying over a pile of Ash logs on 14 June 2017 and over large Beech rounds in the timber yard on 8 July 2021.

Two species of *Ula* (Pediciidae) develop in saproxylic as well as terrestrial fungi. Those Limoniidae found in High Park that develop in decaying wood (*Austrolimnophila ochracea*, *Epiphragma ocellare*, *Neolimonia dumetorum*) are generally common in woodland.

9.2.2. Bibionidae (St Mark's flies)
Most members of this family are conspicuous spring-flying insects, and half of the 18 British species have been recorded in High Park. Of most interest is *Dilophus bispinosus*, a rarely seen but widespread species, of which one female was trapped in oak woodland in High Park in July 2021. Until recently only females (which are brownish yellow while males are black with yellow legs) had been found in Britain, but males have recently been discovered at other localities and new observations were made on its habitat associations (Alexander and Chandler 2021).

9.2.3. Bolitophilidae, Diadocidiidae, Ditomyiidae, Keroplatidae and Mycetophilidae (fungus gnats)
There are records for High Park, Blenheim of 147 species in these families, less than half the Windsor total but comparing well proportionally with other sites (see Table 9.1), and these include ten species that have not been recorded at Windsor. Among the latter the most significant record is of *Neoempheria striata*, of which one male was trapped by Ivan Wright in June 2018. It is a distinctive species, yellow with brown markings on the body and wings, which has been accorded Vulnerable conservation status. There are only two previous British records; both of these were from carr woodland (Cothill, Berkshire V.C., 13 July 1985; Osier Lake, Huntingdonshire, July 1998) so its occurrence in High Park was unexpected. However,

it is more frequent elsewhere in Europe, where there are records of larval webs on saproxylic fungi, and it does have a wide range of fungus hosts.

The fungus gnats found in High Park and associated with saproxylic fungi include some of the larger and more distinctive species such as *Cerotelion striatum, Keroplatus testaceus* and *Leptomorphus walkeri*, as well as some Nationally Scarce (NS) species such as *Grzegorzekia collaris* and *Mycomya insignis. Sciophila baltica* is an uncommon species that develops in the terrestrial tooth fungus *Hydnum repandum*. Two recent additions to the British list have also been found in High Park: *Mycetophila stylatiformis*, of unknown biology, was first recorded in Britain at Windsor in 2014 but is now known from 23 hectads across southern England, while *Epicypta fumigata* was first found in Devon in 2013, with only a few subsequent records reaching Gloucestershire, Cambridgeshire and a site south of Oxford; it has been reared from a wood encrusting fungus in Finland.

9.2.4. Sciaridae (black fungus gnats)
This family has been poorly recorded in High Park with records of only 16 species; of these six species known to develop in rotten wood are all common nationally.

9.2.5. Anisopodidae (window gnats)
The common species *Sylvicola cinctus* develops in sap runs as well as in rotten wood and decaying fungi.

9.2.6. Culicidae (mosquitoes)
Only *Dahliana geniculata*, which develops in water-filled rot holes in decaying trees, has been recorded.

9.2.7. Xylomyidae (wood-soldierflies)
The Drab Wood-soldierfly (*Solva marginata*) (Figure 9.1), the most widespread of the three species comprising this family in Britain and the only one found in High Park, develops under loose bark, with a preference for poplar, but has also been recorded from a wide range of broad-leaved trees. One male was trapped in oak woodland in High Park in July 2021.

9.2.8. Xylophagidae (awl-flies)
Xylophagus ater is a frequent species that develops under loose bark of rotten trunks and logs. There is one record of an adult at High Park in May 2016.

9.2.9. Stratiomyidae (soldierflies)
Of the 11 species recorded in High Park, five are at least in part saproxylic, all common except *Eupachygaster tarsalis*, which develops in rot holes in standing trees and is more restricted to ancient woodlands, with most records from Beech. One female was trapped in oak woodland in High Park in July 2021. The Twin-spot Centurion (*Sargus bipunctatus*), once found dead under pine bark in High Park, is a large distinctive species that may develop in decaying fungi as well as in dung.

Figure 9.1.   *Solva marginata* ♂. (Photograph by Steven Falk)

9.2.10. Asilidae (robberflies)

Six species of these large predaceous flies have been recorded in High Park. One of these, the Golden-haired Robberfly (*Choerades marginatus*) (Figure 9.2), develops in rotten wood, where its larvae feed on the larvae of wood-boring beetles.

9.2.11. Atelestidae, Hybotidae, Brachystomatidae and Empididae (dance flies)

This group is well represented here, with records of 58 species. Many are predaceous on other insects as adults and probably as larvae and many species are flower visitors. Only a small number

Figure 9.2.   *Choerades marginatus* ♂. (Photograph by Steven Falk)

are saproxylic, developing in rotten wood, including *Leptopeza flavipes*, *Oedalea* and *Tachypeza* species (Hybotidae), and *Hilara lurida* (Empididae). Two NS species of Hybotidae, *Tachypeza fuscipennis* and *Platypalpus mikii*, have been recorded. *Tachypeza fuscipennis* develops in rotten wood; it is widespread but much more local in occurrence than the very common *T. nubila* – adults of both species run about seeking prey on trunks and logs. *Platypalpus mikii* may also be saproxylic as it has been found in an emergence trap over rotten Beech wood in Germany; like the other 11 (not saproxylic) species of this genus recorded in High Park, adults search for prey on the leaves of trees and shrubs.

9.2.12. Dolichopodidae (long-legged flies)

Only 34 species of this family, which occur mainly in wet habitats, have been recorded. These include some common saproxylic species in the genera *Medetera* and *Neurigona*, and *Sciapus platypterus*; the adults of these species are usually found on tree trunks, and larvae develop under bark. An NS species *Hercostomus nigrilamellatus* has been trapped in both oak and Beech woodland in High Park; it is usually to be found in ancient woodland and may have saproxylic associations. Some dolichopodids (*Australachalcus* and *Systenus*) are characteristic members of the rot-hole fauna in ancient woodlands but have yet to be recorded in High Park.

9.2.13. Opetiidae and Platypezidae (flat-footed flies)

The one common species of Opetiidae, *Opetia nigra*, develops in rotten wood, while most Platypezidae develop in saproxylic fungi. Seven species of Platypezidae have been recorded in High Park, compared to 24 of the 35 British species at Windsor. These include *Paraplatypeza bicincta*, found on 12 October 2017, which was first recorded in Britain in Surrey in 2001 but is now widespread in the south. While the male is all black, the female has striking silvery markings, so it was certainly a recent arrival in this country that is unlikely to have been overlooked previously; it develops in the fungus Deer Shield (*Pluteus cervinus*). *Bolopus furcatus* is uniformly dark grey in both sexes; it was found on 14 June 2017, ovipositing on its host Dryad's Saddle (*Cerioporus squamosus*), some brackets of which were found on a Horse Chestnut log. Other species recorded in High Park that develop in polypores are *Agathomyia unicolor* in Smoky Bracket (*Bjerkandera adusta*) and *Polyporivora picta* in Turkeytail (*Trametes versicolor*), while *Callomyia amoena* larvae feed on wood-encrusting fungi, and the two species of *Protoclythia* recorded here develop in Honey Fungus (*Armillaria mellea*).

9.2.14. Syrphidae (hoverflies)

The saproxylic species of Syrphidae found in England are listed in Table 9.2, reproduced from Chandler (2021), with the addition of a column for High Park and omission of sites with fewer than ten species of saproxylic Syrphidae recorded.

Table 9.2.   The 36 species of saproxylic Syrphidae known to occur in England, with occurrence in High Park, Blenheim and four other sites with more than ten of these species.

Species	High Park	Windsor Park	Epping Forest	Burnham Beeches	Locks Park Farm	Post-1980 hectads (2021 total)
Brachyopa bicolor	X	X	X		X	50
Brachyopa insensilis	X	X	X			158
Brachyopa pilosa		X	X	X	X	80
Brachyopa scutellaris		X			X	305
Brachypalpoides lentus	X	X	X	X	X	410
Brachypalpus laphriformis	X	X	X	X	X	140
Caliprobola speciosa		X				6
Callicera aurata		X	X	X		104
Callicera rufa						45
Callicera spinolae						16
Chalcosyrphus eunotus						37
Chalcosyrphus nemorum	X	X	X	X	X	553
Criorhina asilica		X				194
Criorhina berberina	X	X	X	X	X	659
Criorhina floccosa		X	X	X	X	491
Criorhina ranunculi		X	X		X	357
Ferdinandea cuprea		X	X	X	X	897
Ferdinandea ruficornis		X	X		X	63
Mallota cimbiciformis		X	X	X		62
Myathropa florea	X	X	X	X	X	1,660
Myolepta dubia		X	X			97
Myolepta potens						4
Pocota personata	X	X	X	X		43
Psilota anthracina		X				36
Sphegina clunipes		X	X	X	X	907
Sphegina elegans		X	X	X	X	443
Sphegina sibirica						271
Sphegina verecunda		X			X	266
Volucella inflata	X	X	X	X	X	390
Xylota abiens		X	X	X		80
Xylota florum		X			X	176
Xylota jakutorum						352
Xylota segnis	X	X	X	X	X	1,871
Xylota sylvarum	X	X	X	X	X	971
Xylota tarda		X				95
Xylota xanthocnema		X	X		X	99
Total species	11	30	23	17	20	
Total all Syrphidae (283 species in Britain)	**63**	**160**	**134**	**91**	**97**	

Figure 9.3.   *Brachyopa bicolor* ♂. (Photograph by Steven Falk)

The Windsor total of 30 exactly matched that for the New Forest, Hampshire, for which a comprehensive Diptera list is not yet available. It was also significantly greater than the totals for the other sites, although the number of presently known hectads (2021 totals provided by Stuart Ball, pers. comm.) showed that most of these species are now known to be more widespread than formerly thought, thanks to the increase in records submitted to the Hoverfly Recording Scheme in recent years. The 15 species with fewer than 100 post-1980 hectads were retained as having conservation status by Ball and Morris (2014). Only three common species were recorded from all the sites and they are now also known from High Park; these include *Myathropa florea* and *Xylota segnis*, which are generalists rather than specialist saproxylics.

The overall total of 63 hoverfly species recorded for High Park is still low, and eight of them were only recorded on the visit by Roger Morris and Stuart Ball in 1992, including three of the 11 saproxylic species recorded in Table 9.2. The latter included *Pocota personata*, a striking bumble-bee mimic that is scarce though widespread nationally; this record was from a decaying Beech, though it is also known to develop in rot holes in other trees. *Brachyopa bicolor* (Figure 9.3), once found by trapping, is the only other relatively scarce saproxylic syrphid recorded in High Park – it has recently been increasing its range nationally. Most of the saproxylic syrphids recorded in High Park can be found visiting tree blossoms in the spring and early summer. A mating pair of *Volucella inflata* (Figure 9.4), which develops in sap runs, was swept from low vegetation in High Park on 15 June 2019, the only species of syrphid recorded on a rather dull and wet day.

Figure 9.4. *Volucella inflata* ♂. (Photograph by Steven Falk)

The aphidophagous hoverflies also produced two records of uncommon species, *Platycheirus sticticus* and *Syrphus nitidifrons*; the latter, recorded by Jovita Kaunang and identified by Graham Collins, is probably a recent arrival in Britain, first recorded in Dorset (Parker 2011), and which is still known only from a few scattered localities in southern England.

9.2.15. Lonchaeidae (lance flies)
Many species of this family develop under bark or in rotten wood, but only two species, *Lonchaea tarsata*, of uncertain biology, and *L. patens*, known to develop in wet decay under bark of Ash and poplar, have yet been recorded in High Park.

9.2.16. Pallopteridae (trembling-wing flies)
Of the three species recorded, *Palloptera ustulata* is one of several species in this family that develop under bark.

9.2.17. Conopidae (thick-headed flies)
The larvae of these flies are parasitoids of adult bees and wasps. One of the three species recorded in High Park, *Myopa pellucida* (Figure 9.5), was formerly considered scarce but is becoming more frequent in southern England; the solitary bee *Andrena nitida* has recently been confirmed to be a host.

9.2.18. Lauxaniidae (lauxaniid flies)
Most of the 13 species recorded are considered likely to develop in decaying vegetation; only the two species of *Pseudolyciella* recorded (*P. pallidiventris* and *P. stylata*) are associated with decaying wood, in which they are presumed to develop.

Figure 9.5. *Myopa pellucida* ♀. (Photograph by Steven Falk)

9.2.19. Sciomyzidae (snail-killing flies)
Most members of this family are associated with wet habitats and only six common species have been recorded in High Park; most of these have larvae that are predators of aquatic or emergent snails, so are probably strays from the Lake margin. However, the most frequently recorded species *Pherbellia dubia* is a predator of woodland snails.

9.2.20. Sepsidae (ensign flies)
These small flies are usually seen wing-waving on low vegetation. The three species recorded in High Park develop in dung or decaying vegetation.

9.2.21. Clusiidae (druid flies)
All the ten British species develop in rotten wood. The three species recorded in High Park, *Clusia flava* (Figure 9.6), *Clusiodes gentilis* and *C. verticalis*, are common generally in woodland.

9.2.22. Chloropidae (grass flies)
This large, mainly phytophagous family has been poorly recorded in High Park, with records of only four species. One of these, *Tricimba cincta* develops in decaying fungi.

9.2.23. Heleomyzidae (heleomyzid flies)
Only eight species have been recorded in High Park. Some common species of *Suillia* (*S. atricornis*, *S. bicolor*, *S. variegata*) and *Tephrochlamys flavipes* are polyphagous in soft fungi, so are partially saproxylic. Some other *Suillia* species are confined to terrestrial fungi, including truffles.

Figure 9.6.   *Clusia flava* ♀. (Photograph by Trevor and Dilys Pendleton)

9.2.24. Sphaeroceridae (lesser dung-flies)

This is another large family that has been poorly recorded in High Park, with only seven species identified. Among these *Crumomyia fimetaria* is one of the many species that develop in decaying fungi as well as other decaying materials. With the low level of recording, it is not possible to say whether the intro- duction of cattle has had any impact on the numbers or diversity of the dung-feeding species. However, there is a good amount of data on dung associated flies in other families, which suggests that they already had a strong representation owing to the population of deer present in High Park (see also comments under families Stratiomyidae, Sepsidae, Scathophagidae, Anthomyiidae and Muscidae).

9.2.25. Drosophilidae (fruit flies)

Only 11 species of this family have been recorded in High Park, compared with 34 of the 65 British species at Windsor; these include several species that develop in saproxylic fungi, as well as a record of *Chymomyza fuscimana*, which is usually observed wing waving on cut

Figure 9.7.   *Phortica variegata*. (Photograph by Steven Falk)

ends of logs, as happened on a pile near the entrance to High Park (Combe Gate) on 15 June 2019. *Phortica variegata* (Figure 9.7), was formerly known in Britain only from the New Forest, where it is associated with sap runs on oak, especially those resulting from activity by larvae of the Goat Moth *Cossus cossus*. However, in recent years it has proved to be more widespread in southern England, and it was particularly noticeable at Windsor as males are attracted to eye secretions and consequently fly around the observer's head. The only High Park record is from one of Ivan Wright's traps in 2019. Chandler (2021) discussed the possibility that the recent records result from new invasions from the continental population. A certain new arrival in this country is *Drosophila suzukii*, first recorded in Kent in 2012, which is a pest of fruit cultivation that also develops in wild blackberries and is now widespread throughout Britain. It was numerous in High Park on both of my October visits, and like some other *Drosophila* species may be found visiting decayed Honey Fungus.

9.2.26. Hippoboscidae (flat flies)
One member of this family, all of which are parasites of birds and mammals, has been recorded – the Deer Ked *Lipoptena cervi*, which is probably as numerous as at other woodlands with deer present. It was observed in the timber yard when it flew into the recorder's eye, remaining under the lid for around ten seconds before it was extricated; it then spent a while rearranging its wings. After alighting on a suitable host, the winged adult usually tears off its wings as they have then become an encumbrance; on this occasion that stage wasn't quite reached.

9.2.27. Scathophagidae (dung-flies)
Four common species of *Scathophaga* recorded in High Park are dung feeders as larvae. Some other members of this family are plant feeders, an example being *Norellisoma spinimanum*, which develops in stems of Dock (*Rumex* species).

9.2.28. Anthomyiidae (root-maggot flies)
There are High Park records for 48 species of this family, of which most members either develop in higher plants or in decaying materials, or are fungus feeders. However, the four British species of *Eustalomyia* are kleptoparasites of solitary wasps that nest in decaying wood. Three species of this genus have been recorded in High Park, including the widespread *E. festiva* (Figure 9.8), and two less frequent species, *E. hilaris* and *E. vittipes*, both designated as provisionally NS by Falk and Pont (2017). They are widespread in the south, with *E. hilaris* having more recent records and *E. vittipes* seemingly scarcer. The common species *Anthomyia procellaris* develops in saproxylic fungi as well as in bird nests, while species of several genera recorded in High Park (*Hylemya, Hylemyza, Puregle* and *Pegoplata*) develop in dung.

9.2.29. Fanniidae (the lesser house-fly family)
Ten of the 61 British species have been recorded in High Park. Most members of this family develop in decaying materials, with some also occurring in rotten wood or decaying fungi; these include

Figure 9.8. *Eustalomyia festiva* ♀. (Photograph by Trevor and Dilys Pendleton)

the lesser house-fly *Fannia canicularis*. Another species recorded in High Park, *F. aequilineata*, is more restricted to decayed wood and rot holes; although it is proving to be widespread from trapping at ancient trees at other sites, it is accorded provisional NS status.

9.2.30. Muscidae (the house-fly family)

This family is moderately well recorded in High Park, with 37 species confirmed. Most species have predaceous larvae in various substrates, including dung (in High Park *Mesembrina meridiana*, *Graphomya maculata*, *Musca autumnalis*, *Hydrotaea irritans* and some species of *Helina*), fungi (*Muscina levida* and several

Figure 9.9.   *Macronychia dolini* ♂. (Photograph by Trevor and Dilys Pendleton)

Phaonia species) and decaying wood (*Helina abdominalis*, *Phaonia palpata*). Some *Phaonia* species that are attracted to sap runs, in which their larvae develop, have not yet been found in High Park.

9.2.31. Calliphoridae (blow-flies)

Apart from the several genera of blow-flies that develop in carrion or are snail associated, members of the subfamily Rhinophorinae are specialist parasitoids of woodlice, and four of the eight British species have been recorded in High Park.

Figure 9.10.   *Gonia picea* ♀. (Photograph by Steven Falk)

9.2.32. Sarcophagidae (flesh flies)

Most of the 16 species of this family that have been recorded are carrion feeders or have larvae that are predators of earthworms, while most of those that are kleptoparasites of ground-nesting bees and wasps have not been found at this site. Only two species of *Macronychia* in that group have been recorded here; *M. dolini* (Figure 9.9) is associated with the wood-nesting solitary wasp *Ectemnius cavifrons*, while *M. polyodon* is known to associate with both wood- and ground-nesting solitary wasps.

9.2.33. Tachinidae (parasitic flies)

This large family, which includes parasitoids of a wide range of invertebrates, has been poorly recorded in High Park. There are records of only 11 species, most of which parasitize larvae of Lepidoptera; *Gonia picea* (Figure 9.10) is an example of these. Exceptions are *Ocytata pallipes* which parasitizes earwigs, and *Phasia obesa* which parasitizes Heteroptera. *Redtenbacheria insignis* also belongs to subfamily Phasiinae, but its hosts have not been confirmed. It is a large, distinctive but scarce species, which has also been recorded at Windsor; one female was trapped in High Park in oak woodland.

Acknowledgements

I thank Steven Falk and Trevor and Dilys Pendleton for kindly enabling their photographs to be included. I also thank Graham Collins for comments on a draft.

CHAPTER 10

Sawflies, Wasps, Bees and Ants (Hymenoptera)

Ivan Wright

10.1. Introduction

The Hymenoptera are a very wide-ranging and species-rich order of insects. There are about 6,500 species in Britain alone, the vast majority of which go quite unnoticed, and yet within the order are some of the best known of insects: the honey bee, the bumblebees, the Common Wasp and the ants. Other groups in the order demonstrate the great variety of insects that the Hymenoptera encompass, including the gall wasps, 'solitary' mining bees and digger wasps, sawflies (which are not flies but primitive ancestors of the wasps) and the Parasitica. The bees, wasps and ants are collectively known as the aculeates by virtue of the females, in the majority of species, having a defensive sting at the tip of their abdomen. The Parasitica are a very large group of wasps that are mostly parasites of other insects and in terms of the British fauna outnumber the total of all other Hymenopteran species; they are also particularly difficult to identify and are only infrequently studied or surveyed.

10.2. Recording history

Before systematic recording for the High Park Biodiversity Survey (HPBS) began in February 2017, there had been trapping of beetles and other insects including Hymenoptera commissioned by Natural England in 2016. Before this, there were incidental records of Hymenoptera dated 1982–2004. The results were made available to this project by the Thames Valley Environmental Records Centre.

10.3. Survey visits and methods

In the survey period 2017–21 about 20 specifically targeted visits were made to High Park, during which most of the Hymenopteran species of the project were recorded. Numerous other surveyors contributed to the total of records by visual observation and through the by-catch of invertebrate traps set primarily for other taxa. Owing to an initial restriction on late summer visiting, the data from 2017–18 were somewhat biased towards species that are active only in the earlier months of the year; however, notwithstanding the additional limitations on visits imposed by the COVID-19 pandemic, summer visits in the latter years probably compensated quite adequately to achieve a reasonable five-year sample for the site.

Ivan Wright, 'Sawflies, Wasps, Bees and Ants (Hymenoptera)' in: *The Natural History of Blenheim's High Park*. Pelagic Publishing (2024). © Ivan Wright. DOI: 10.53061/NZVF3441

Figure 10.1. A male of the rare Long-horned Nomad Bee (*Nomada hirtipes*), a 'cuckoo' species of the Big-headed Mining Bee (*Andrena bucephala*). (Photograph by Steven Falk)

The majority of specimens (60%) were caught in yellow pans of water. This is a common sampling method that is substantially selective for Hymenoptera; the colourful pans are mistakenly seen by insects as a flower while foraging for pollen or nectar. The pans used in this study were 20cm in diameter and enhanced on the inside with fluorescent paint. Typically, 10–15 pans would be deployed around the survey site, in selected warm and floristic locations, for a single sunny day. Each pan is half filled with water, with a little detergent added to break the surface tension.

A further 20% of specimens were caught in the various forms of flight interception trap that have been deployed both during and before the current project. These traps sample flying insects at various heights above the ground, from the tent-like Malaise traps for insects near the ground, to 'vane' traps hung high in tree canopies (see Chapter 13 for details). The remaining 20% of records are from the hand-netting of specimens, direct observations and other unspecified methods of field collecting and identification.

10.4. Overall results

The total species richness including all Hymenopteran groups was 233 species. In more detail this includes: 83 bees of which 13 are 'social' species (such as the bumblebees) and 22 are cleptoparasites of other bee species; 62 wasps of which six are 'social' species (such as the Common Wasp *Vespula vulgaris*) and eight are parasites of other aculeate species; seven species of ant; 49 sawflies; and 34 species of Parasitica of which 11 were identified from the plant galls that they form.

10.5. High Park as a site for Hymenoptera

The ancient pasture woodland of High Park is a habitat that is well suited for some groups of Hymenoptera but, perhaps surprisingly, less attractive to others, even though they are relatively common in the wider countryside. The ancient trees and shaded woodland will suit Hymenoptera that nest in dead wood, holes made by other insects and hollow stemmed plants, such as bramble (*Rubus fruticosus* agg.). The woodland shade, leaf litter and grassy woodland glades are particularly suitable for various ant species, and the abundance of moths and small flies will host a wide range of Parasitica.

However, a significant limiting factor for bees, wasps and sawflies is the restricted diversity of vascular plants. Although the total species richness of plants is fairly impressive at High Park, beyond a relatively small number of common plants that appear in great abundance, most of the remaining species appear in very small numbers indeed (Chapter 4). Some more common species are often browsed by the abundant deer, preventing them from flowering. This presents a significant limitation in the diversity of supply of pollen, nectar and foodplants to support a wider variety of Hymenoptera, especially those that require a reasonable abundance of a particular plant type for their lifecycle. A further consideration is the type of soil; High Park is mostly sited on damp heavy clay (Chapter 1), which is not well suited for most soil-nesting bees and wasps, that is, those that require a lighter dry soil in which to burrow. Wherever these two limiting factors combine, such as at High Park, a comparatively reduced fauna, particularly for solitary bees, would be expected and a bias towards aerial-nesting species would be evident.

Within the aculeates, the 83 species of bee and 60 wasps can be compared with totals from other sites surveyed in Oxfordshire in recent years (author's data, Bees Wasps & Ants Recording Society – BWARS 2000–11). A total of 143 recorded species of bee and wasp would seem to be quite high and is a very satisfactory result. However, this total has been accumulated over five years of surveying and represents an unusually thorough sample, with about a third of species represented by only one or two specimens in the whole of the five-year survey. In comparison, for example, many Oxfordshire sites are only sampled for a single year and yet return a sample total of over 100 species. There are also prime sites in the county that have been surveyed over many years, and these can accumulate recorded totals of bees and wasps in excess of 200 species.

10.6. Solitary Bees

Of the 53 species of nest-building (non-cleptoparasite) solitary bees recorded at High Park since 1982, 17 nest above ground, which at 32% is a high proportion compared with other sites in England (Wright and Gregory 2006) as well as sites in Oxfordshire recently surveyed by the author (BWARS data 2000–2011). This high proportion might reasonably be expected in an ancient woodland on a heavy damp soil and nesting opportunities biased towards aerial nesting. Many of the aerial-nesting species use pre-existing holes in dead wood, some use dry hollow stems of plants such as bramble, and the rest are mostly generalists and will use any suitable cavity including dead wood and hollow stems. Two species at High Park, *Osmia bicolor* and *O. spinulosa*, nest exclusively in empty snail shells. For the species of bee in the survey that are cleptoparasitic in the nests of other bees, all of the required host species, or species groups, were also recorded.

Among the particularly noteworthy bees found at High Park is *Nomada facilis*. This species was added to the British list in 2017 (Notton and Norman 2017); however, re-examination of older specimens of *Nomada integra* (with which *N. facilis* was previously confused) showed that the species was found to be present in Britain since before 1900. As there are only seven British specimens found with location data, it is assumed to be a rare native species of southern England. The host species is not known, but deduction by Notton and Norman (2017) suggests it is quite likely to be *Andrena fulvago*. At High Park both *N. facilis* and *A. fulvago* were recorded on the same day and in the same (yellow pan) water trap.

The host and cleptoparasite pair of bees *Andrena bucephala* and *Nomada hirtipes* were recorded together at various locations in May, which is when Common Hawthorn (*Crataegus monogyna*) was in flower. *Nomada hirtipes*, or Long-horned Nomad Bee, is a rare 'cuckoo' bee (Figure 10.1), and is thought to be exclusively dependent upon its Nationally Scarce host, *A. bucephala*, Big-headed Mining Bee (Figure 10.2). The host bee, in turn, requires the pollen of trees and shrubs, particularly hawthorn, to supply its brood (Else and Edwards 2018). Both species of bee are restricted to southern Britain.

Heriades truncorum is a small stem-nesting species which has historically been restricted to sites in south-east England, and there appears to be no record for the species in Oxfordshire until 2018 (Else and Edwards 2018). Since that date *H. truncorum* has been seen at several sites in the county in

Figure 10.2.   A male Big-headed Mining Bee (*Andrena bucephala*). (Photograph by Steven Falk)

most years, and the records from High Park in 2019 and 2021 are among the first to indicate a clear westward expansion of is range into Oxfordshire.

10.7. Social Bees

10.7.1. Bumblebees

Twelve species of bumblebee were recorded, all of which are common and widespread in southern England. Seven of these are colony-founding species and five (of the six British species) are nest inquilines – commonly known as 'cuckoos' – that parasitize the colony of one or more host species. With the exception of *Bombus hypnorum*, all colony-founding species have an associated 'cuckoo'.

The recent arrival, *B. hypnorum*, was first recorded in Britain in 2001 and first appeared in Oxfordshire in 2008; to date no 'cuckoo' species has been confirmed for it in Britain (Else and Edwards 2018). High Park is particularly suited for *B. hypnorum*, also known as the Tree Bumblebee, as it has a strong preference for nesting above ground, including in hollow trees.

All bumblebee species were seen in abundance at High Park except for the colony-builder *Bombus ruderarius* and four of the 'cuckoo' species of which fewer than four specimens of each were recorded: these were: *B. barbutellus*, *B. campestris*, *B. rupestris* and *B. sylvestris*.

10.7.2. Honey bees

Throughout the survey foraging domestic European Honey Bees (*Apis mellifera*) were a common sight, collecting pollen and nectar, and these bees would be coming from hives or feral colonies, both near and far, and not necessarily from within the area of High Park. Feral colonies that have swarmed from domestic hives and become established in natural hollows are well known in the countryside, and are commonly seen in hollow oak trees. However, with at least 50 such colonies having been counted in the oaks at High Park (Filipe Salbany, pers. comm. 2020) and the unusually large area of ideal habitat at High Park, there are a number of interesting possibilities with this species of bee. For instance, the proportion of honey bee colonies that are founded by swarms originating from within High Park is not known. If the proportion is high, there will be a component of inbreeding and the gene pool will derive colonies that are variously weak or strong depending upon the traits that are passed on and persist.

Considering the high density of tree-nesting honey bee colonies at High Park, a number of factors point to the possibility of a genetically isolated race of honey bee, and in particular – as suggested by Salbany – the Dark European Honey Bee (*Apis mellifera mellifera*), which is a subspecies native to Britain commonly known as the 'British Black Bee'. For example, many of the honey bees at High Park are very dark and some display the wing-vein character associated with *A. m. mellifera* (Filipe Salbany, pers. comm. 2020). A further contributing factor is that the circumstances for mating by *A. m. mellifera* do not always coincide with that of other races of honey bee (Denwood 2017), supporting the case for a robust semi-isolated race among High Park's tree-nesting colonies. If there is a resident race of honey bee at High Park, there would be a strong case for conservation and further research, but this would require DNA analysis to establish its type and purity.

DNA analysis has established that *A. m. mellifera* is a native subspecies of British honey bee, derived from a race that arrived naturally in Britain not long before the land-bridge to France became the English Channel (Pritchard 2008) and was probably well established in Britain 4,000 years BP and still survives today in colonies in various parts of the UK (Carreck 2008). However, the native British honey bee is currently a scarce and threatened subspecies and there is some doubt whether any of the existing colonies are pure *A. m. mellifera* because of the unavoidable hybridization with domestic and imported bees (Muñoz et al. 2017).

Whether or not the frequency of honey bee colonies is currently stable at High Park cannot be readily assumed, particularly when considering the rapid increase in Hornets (*Vespa crabro*) over the past decade (see below). Honey bee colonies in the countryside are open to attack from various parasites and predators, including woodpeckers and especially Hornets, but strong, healthy and well-located colonies can continue for at least seven years (Wright and Wright 2018).

10.8. Solitary and Social wasps

Of the 43 species of nest-building (non-parasitic) solitary wasp recorded at High Park during the survey, 31 nest above ground, which, as with the solitary bees, is a high proportion (72%) compared with other sites in England (Wright and Gregory 2006) as well as sites in Oxfordshire recently surveyed by the author (BWARS data 2000–11) but this is to be expected at High Park. The aerial-nesting species recorded at High Park will typically use any suitable cavity including pre-existing holes in dead wood and dry hollow stems of plants.

All wasps and their larvae are carnivorous and the solitary wasps mostly feed on the adults or larvae of other invertebrates. While foraging, solitary wasps are very adept at finding their preferred invertebrate prey type, which is killed or immobilized and then removed to the nesting cavity where the eggs are laid and where the hatched larvae can feed and develop. The social wasps, such as the Common Wasp predate a wide range of invertebrates but unlike the solitary wasps the prey is masticated to a pulp before being fed to the larvae.

10.8.1. Crabonidae

The majority of solitary wasps at High Park are in the Crabonidae, which is a very large family of Hymenoptera, nationally and globally, and individuals range widely in size and coloration. At High Park the 23 species recorded prey mostly on aphids and small flies, although some Hemiptera, Coleoptera and Lepidoptera are targeted as prey, particularly while in their larval stage.

Most of the Crabonids recorded are common to southern England, the exception being a single specimen of *Crossocerus walkeri* caught in a flight interception trap in July 2021 by Jon Cooter. Although *C. walkeri* is not considered to be particularly rare, it is scarce and perhaps elusive; the few British records in recent decades are from locations that are very thinly scattered throughout the country. There may, however, be some association with the habitats at High Park as it is typically found in wetter places and is a predator of mayflies (Ephemeroptera).

10.8.2. Pompilidae
The spider-hunting wasps (Pompilidae) (Figure 10.3) are well represented at High Park, with 14 of the 44 British species recorded. The structural diversity of habitats at High Park, from grassy glades to a mix of woodland types, is well suited to an abundance of spiders (Chapter 8) and the population of Pompilids reflects this. The spider-hunting wasps are not strongly dependent on a rich floral diversity and their nest requirements are relatively simple. Most species are nominally ground nesting but often use any suitable cavity at ground level, such as under stones and under fallen tree trunks and branches. All species recorded at High Park are fairly common in Oxfordshire and only one, *Evagetes crassicornis*, is a cleptoparasite of other Pompilids – typically in the Pompilinae sub-family.

10.8.3. Vespidae
The Vespidae all share some common structural characters but comprise two quite distinct sub-groups depending upon whether they nest solitarily or socially. There are the 23 species of solitary Potter Wasp in Britain, with life cycles not unlike the Crabonid wasps (see Section 10.8.1), and 14 species of social wasp, most of which are the well-known yellow and black creatures with their large colonies of queens and workers in an impressive 'paper' nest.

Four species of solitary Vespid wasps were recorded at High Park; all are fairly common species and nest in the dry stalks of hollow-stemmed plants. However, they differ quite widely in their prey preferences, which include small beetles and moth larvae.

Six species of social wasp were recorded during the survey. The common wasps *Vespula vulgaris* and *V. germanica*, and the Hornet (*Vespa crabro*) were all a fairly common sight in High Park during the survey and numerous observers recorded their observations. A small number of specimens of the larger social wasps of the genus *Dolichovespula* were also caught by Jon Cooter and Benedict Pollard as by-catch in interception traps on trees for beetles. The Red Wasp (*Vespula rufa*) was last recorded in 1990 by John Campbell.

During the twentieth century the Hornet was not a common species throughout much of south-east England and was an uncommon sight in Oxfordshire; it also would seem to have been declining in the latter half of the century. However, with High Park being such an ideal habitat for the species it is possible

Figure 10.3.   The spider-hunting wasp *Priocnemis perturbator*. (Photograph by Steven Falk)

Figure 10.4.   The 'cuckoo'-wasp *Chrysis illigeri*. (Photograph by Steven Falk)

that its continuance there was reasonably secure and atypical for the county overall, as was Richmond Park in Surrey (Baldock 2010) for instance, where Hornets were considered resident in 1985. Then from about 2010, for reasons not fully understood, there was a rapid increase in abundance and an eastward extension to its range, such that it is now a frequent sight in Oxfordshire.

10.8.4. Other solitary wasp families

A few specimens of common cleptoparasitic solitary wasps from small taxonomic families were found during the survey, namely Chrysididae, Dryinidae, Mutillidae and Tiphiidae. Four species of the highly colourful 'cuckoo' wasps (Chrysididae) were recorded, and included *Chrysis illigeri* (Figure 10.4) and *C. terminata*, a species added to the British list in 2016 (Wood and Baldock 2016) but which has probably been in Britain for many years having now been split from the *C. ignita* s. lat. species aggregate. *Chrysis terminata* is shiny metallic green and blue with a red abdomen. There are so few confirmed British records of this species that there are currently no details of its distribution or habitat requirements in this country.

10.9. Sawflies

Sawflies (suborder Symphyta: Hymenoptera) are a primitive form of wasp (Figure 10.5), appearing in fossil records that date from the early Jurassic period, and long before bees appear in such records. Consequently, as an order of insects that originated before flowering plants, some sawflies continue to be associated with similarly primitive plants, such as horsetails and ferns. The name 'sawfly' derives from the ovipositor of the female, which is adapted to act like a saw and so has the combined function of sawing a slot in the larval foodplant as well as enabling subsurface placement of her eggs. Most species are associated with a limited group of plants and the 'saw' is not just adapted to cut into the foodplant; the shape of the 'saw' and its teeth are often unique to the species and can be used for identification.

Figure 10.5.   The common sawfly *Selandria serva*. (Photograph by Steven Falk)

Even though many sawflies are common, they tend to receive less attention from entomologists compared with other groups of insects and there are far fewer records. Therefore, it is only in recent years that national statuses have been attempted, and for some species it is only possible to infer scarcity by their few records compared with other species of sawfly. Most of the sawflies recorded at High Park over the past 30 years are considered to be common; however, the three sawflies that are specifically associated with oak trees (*Periclista albida*, *Apethymus filiformis* and *A. serotinus*) are all 'local' with few records on the national database (National Biodiversity Network Atlas Partnership 2017). *Athalia scutellariae* is locally common in Britain and restricted to damper habitats where its foodplant grows: Skullcap (*Scutellaria galericulata*).

Since 1992, 42 species have been trapped or netted at High Park, 38 of which are from the current survey project (HPBS). There is no correlation between the method or location of capture and the foodplant preference of the species. Nine species are associated with tree species, nine with grasses, seven with dicotyledonous herbs and four with Bracken and other ferns; the remaining 19 species are more general in their foodplant requirement. Apart from the sawflies on oaks mentioned above, other tree associations are with Ash (*Fraxinus excelsior*), Aspen (*Populus tremula*), Hawthorn (*Crataegus monogyna*) Rowan (*Sorbus aucuparia*) and conifers (Pinaceae). The one species associated with conifers is the Giant Wood Wasp (*Euceras gigas*) which was recorded at the northern edge of the Park in 1992 by Simon Grove (see also Section 10.11).

10.10. Ants

Ants are closely related to primitive wasps, and have developed a social biology that is quite different from their solitary ancestors. Although ants, bumblebees, honeybees and the social wasps all form social colonies that are based around an egg-laying queen, they have independently evolved social strategies that seem similar but differ in many respects. For ants, most importantly, they are much more integrated and interactive with the ecology of their surroundings. Some species build conspicuous ant hills, while others live in smaller colonies in hollow trees or within the nests of other species of ant, but their contribution en masse is of vital importance to the ecology of all but the wettest habitats. Despite their small size, the total volume of ants can be a significant fraction of the living biomass of a habitat, and their great abundance and ceaseless activity make indispensable contributions to many species of flora, fauna and fungi.

At a basic level they serve as food for birds: Green Woodpeckers feed almost entirely on ants in the winter months. The ant hills of the Yellow Meadow Ant (*Lasius flavus*) develop a fine and friable soil structure that is conducive for the germination of seeds. The ant hills also have a particular ecology and microclimate that can be exploited, for example by basking insects and other small animals. At a more specific level ants' nests have many complex associations with other species, including the mutually beneficial hosting and exploitation of 'guest' invertebrates. This includes the overwintering larvae of 'blue' and hairstreaks butterflies, and aphids that the ants manage and 'milk' for their sugary honeydew.

Perhaps one of the most important and intriguing contributions of ants for a diverse flower-rich habitat is the movement of plant seeds by the process of myrmecochory. The seeds of some plants have a gland that exudes a substance to which the ants are strongly attracted. Ants collect the seeds and move them nearer to their nests, thereby transporting the seeds some distance from the plant; the seeds are moved to where they are less likely to be eaten and, in some cases, to an environment that is more suitable for dormancy or germination. Some common plants such as violets (*Viola* sp.) produce seeds with an eliaosome for the exploitation by ants. Four species of violet were recorded in High Park, of which one, Common Dog-violet (*Viola riviniana*), is widespread (Chapter 4).

Of the 54 species of ant found in Britain, seven species were recorded at High Park, of which four are widespread in Britain and three are much less frequently recorded and more restricted to the woodlands of southern England. *Lasius brunneus* is very common in High Park; typically, it nests in old living oaks as well as dead stumps, and chews a labyrinth of galleries within the tree. Evidence for *L. brunneus* nesting in a tree is the large volume of brown dusty frass at the ground in and around the tree and caught in spiders' webs in a hollow trunk (Figure 10.6).

Stenamma debile and *Temnothorax nylanderi* are seldom seen or recorded as these species have much smaller nests, with colonies of less than 200 individuals and only a single queen.

Figure 10.6.   Fine dust or frass from brown-rotten wood on an ancient oak, excavated by ants (*Lasius brunneus*) from the hollowing interior. (Photograph by Benedict Pollard)

Temnothorax nylanderi was observed on oak trees by Rosemary Winnall and Steve Gregory in different years and locations, and Rosemary Winnall found a colony of *Stenamma debile* under a stone in the south-eastern corner of the Park.

10.11. Parasitica

Relative to the number of species of British Parasitica and owing to the difficulty of identification it has been possible to identify only a very small number to species level in this survey. The list is dominated by the Ichneumonidae family of parasitoid wasps and the Cynipidae family of gall-wasps.

10.11.1. Ichneumonid wasps

Ichneumonid wasps are a large and diverse family of parasitoid Hymenoptera and include many that are conspicuously longer and thinner than most other wasps. The female ichneumonids differ from the solitary and social wasps in retaining their tube-like 'sting' structure as an ovipositor for placing eggs as well as delivering liquid poisons that can modify the behaviour of the host species.

Notable records at High Park are for two particularly large Ichneumonid wasps *Pseudorhyssa alpestris* and *Rhyssa persuasoria*, with both male and female specimens of the latter being recorded near the timber yard at the southern end of High Park. The females of these species can have a total body length of more than 3cm, not including a 3cm ovipositor. There are very few records of *Pseudorhyssa alpestris* in Britain; it is therefore probably quite rare and little is known of its biology in this country. *Rhyssa persuasoria* is widely distributed in Britain but seldom recorded and is a parasite of the larvae of Longhorn beetles (Cerambycidae) and Wood Wasp sawflies (Symphyta). At High Park, numerous species of Longhorn beetle have been recorded in recent years, and also the Giant Wood Wasp (*Euceras gigas*) in 1992.

10.11.2. Gall-wasps and other gall-causers.

A gall can be defined as an abnormal growth produced by a plant under the influence of another organism that, once formed, provides both protection and nutrition for the gall causer. At High Park the majority of galls observed and recorded were found on oak trees. In Britain, over 70 different galls are known to form on oaks and of these 90 per cent are caused by wasps of the Cynipidae family. This accounts for this section on galls being placed here under 'Hymenoptera', even though there are a smaller number of other galls observed on other plant hosts and most of these are not caused by Hymenopteran species (for a review of oak galls see Tyler 2008: 180–8).

Of the 18 different galls observed at High Park since 1982, 10 are by wasps (Cynipidae: Hymenoptera), four by flies (Cecidomyiidae: Diptera), two by psyllid bugs (Triozidae: Hemiptera) and two by mites (Eriophyidae: Arachnida); and of the host plants, 10 galls were on oaks, four on other trees (Beech, Buckthorn, Hawthorn and willow) and three on herbs (nettle, rose and bramble).

Figure 10.7.   Oak Apple Gall, caused by the wasp *Biorhiza pallida*. (Photograph by Aljos Farjon)

All of the galls found at High Park are common and many are a familiar sight in the countryside, including the spongy spheres of the Oak Apple Gall, caused by the wasp *Biorhiza pallida* (Figure 10.7), and the Knopper Gall that harbours the wasp *Andricus quercuscalicis* in a sticky and distorted acorn. This gall wasp requires the non-native Turkey Oak (*Quercus cerris*) for the sexual generation and Pedunculate Oak (*Q. robur*) for the asexual generation. The Marble Gall appears as a woody brown 'marble' and remains on the twig long after the adult wasp, *A. kollari*, has emerged via the clearly visible exit hole. *Andricus kollari* was introduced into Britain for dyeing and tanning, as the tannin content of the gall is unusually high. The galls were also ground into powder to make a long-lasting ink for writing legal documents.

Some gall wasps have a two-year lifecycle with contrasting sexual and asexual generations every other year. At High Park, besides *A. quercuscalicis*, the other species is *A. lignicola*; the latter inducing the formation of Cola-nut Galls. In the case of *Neuroterus quercusbaccarum* a different gall is produced for each generation of the same species of wasp: flat Common Spangle Galls on the underside of leaves from the asexual generation and spherical Currant Galls from the sexual generation the following year.

A common sight on Dog-rose (*Rosa canina*) and Field-rose (*Rosa arvensis*) is the distinctive Robin's Pincushion, or Bedeguar Gall, with its tangle of long wiry reddish-green hairs and caused by the Cynipid wasp *Diplolepis rosae*. This gall can host a very complex micro-community of insects, including herbivores that feed directly on the gall tissue, as well as insect eating parasitoids that prey on the herbivores, which in turn may themselves be preyed upon by secondary parasitoids.

Acknowledgements

The images of Hymenoptera kindly provided by Steven Falk are available from him at https://www.flickr.com/photos/63075200@N07/collections/72157629294459686/.

CHAPTER 11

Butterflies (Lepidoptera)

Phillip Cribb and Caroline Steel

A total of 22 species of butterfly were recorded during the survey period, only half of the 44 species recorded from the County of Oxfordshire (https://www.upperthames-butterflies .org.uk). It is likely that this is an underestimate of the number of species as at least two common species were not recorded. Of the species recorded, the majority are generalists, species of the wider countryside (grassland, hedgerows and woodland margins), and only three are habitat specialists.

11.1. Methods

Daytime walks along access roads, paths and glades were undertaken periodically throughout the spring and early summer, limited from late July onwards by lack of permission to leave the tarmac-surfaced road. A total of 36 visits were made during the survey period 2017–21 and records were received from 16 recorders. The main recorders were Caroline Steel, Margaret Price, Graham Collins, Phillip Cribb and Anthony Cheke. Butterfly names follow those provided by Aggasiz et al. (2022). Plant names are taken from Stace (2019). The Thames Valley Environmental Record Centre provided some historical records for the period 1980–2015.

11.2. Woodland

The Purple Emperor (*Apatura iris*, Figure 11.1), one of the habitat specialists, is a notable new record for Blenheim High Park. The adults spend much of their life in the canopy of the tallest Pedunculate Oaks (*Quercus robur*) and the females lay their eggs on Goat Willow (*Salix caprea*). Two were seen in July 2019 by Caroline Steel and Margaret Price flying high around tall oak trees in the middle of an area of pasture woodland that also contained Goat Willows. Penny Cullington recorded one male flying above an oak tree towards the Lake in 2020. In 2021 there were three sightings – a male recorded near High Lodge by Bob Cowley and Linda Losito and two males by Paulo Salbany when he was climbing tall oak trees in search of honey bee nests. Its range within England currently appears to be expanding. It is found to the south in the Wytham Woods area and to the north-west in Foxholes Wood.

The Purple Hairstreak (*Favonius quercus*, Figure 11.2) is also associated with the oak canopy and seldom comes down to ground level. Its larvae feed on oak leaves and the adults spend much of their time near the tops of oak and Ash trees feeding on honeydew. In 2018 the butterfly was common, with over 30 individuals recorded on oaks along the main track (tarmac-surfaced road).

Phillip Cribb and Caroline Steel, 'Butterflies (Lepidoptera)' in: *The Natural History of Blenheim's High Park*. Pelagic Publishing (2024). © Phillip Cribb and Caroline Steel. DOI: 10.53061/REZP4047

Figure 11.1.   Male Purple Emperor (*Apatura iris*). (Photograph by Peter Gasson)

Figure 11.2.   Male Purple Hairstreak (*Favonius quercus*) on leaves of an oak. (Photograph by Jim Asher)

Figure 11.3.   Silver-washed Fritillary (*Argynnis paphia*). (Photograph by Phillip Cribb)

The third species of note here and the second habitat specialist is the Silver-washed Fritillary (*Argynnis paphia*, Figure 11.3), which spends more time in the glades among the oaks. The butterflies have a strong sweeping flight and are frequent visitors to bramble and thistle blossoms. They lay their eggs on the leaves of violet species, notably Early Dog-violet (*Viola reichenbachiana*) and the Common Dog-violet (*Viola riviniana*) which is the commonest violet in High Park. The eggs are laid on the trunks of oak trees where clumps of violets are growing near the bases of these trees. There are no historical records for High Park for the period 1980–99, but since 2005 the Silver-washed Fritillary has expanded its range dramatically, and by the end of 2014 it had been recorded from many of the wooded areas in Oxfordshire. (Asher et al. 2016). The butterfly was especially common in High Park in 2018, with over 30 individuals recorded along the main track on 2 July.

11.3. Butterflies of the countryside preferring woodland, rides, glades and margins

Butterflies that overwinter as adults are the first to appear in early spring. In High Park, these include the Comma (*Polygonia c-album*), Peacock (*Aglais io*), Red Admiral (*Vanessa atalanta*) and Brimstone (*Gonopteryx rhamni*). The first three lay their eggs on Common Nettle (*Urtica dioica*) and occasionally on other plants such as English Elm (*Ulmus procera*). Overwintering adults can be seen on the wing into early June. The eggs of the Brimstone are laid on Buckthorn (*Rhamnus cathartica*); their adult offspring begin to emerge in late July and move into hibernation in the autumn. Buckthorn is rare in High Park, but the butterfly is a strong flier and often seen well away from buckthorn bushes.

The Speckled Wood (*Pararge aegeria*) is a common butterfly in High Park and may be seen flying in dappled sun and shade in woodland glades and rides as early as March, as the butterfly can overwinter either as a caterpillar or pupa. It is also one of the latest, its second or third brood often surviving into early November. Its larva feeds on a variety of grasses, including

False Brome (*Brachypodium sylvaticum*), Cock's-foot (*Dactylis glomerata*), Yorkshire-fog (*Holcus lanatus*) and Common Couch (*Elymus repens*). The adults feed predominantly on honeydew, the sticky secretion of aphids, particularly in the early months of the year, while the summer and autumn broods can be seen on flowers and blackberries.

The Orange-tip (*Anthocaris cardamines*) is also one of the first butterflies on the wing to emerge from an overwintering pupa. Its preferred foodplants are Cuckooflower (*Cardamine pratensis*) and Garlic Mustard (*Alliaria petiolata*) – the second not found in High Park – but it can also lay its eggs on other cresses. At almost the same time, the Green-veined White (*Pieris napi*) and Small White (*Pieris rapae*) can also be seen on the wing, the Green-veined White preferring damp grassland and woodland rides. Their larvae feed on various *Brassica* species and other cresses, such as Hedge Mustard (*Sisymbrium officinale*), Garlic Mustard and Hoary Cress (*Lepidium draba*), and also on Wild Mignonette (*Reseda lutea*). None of these foodplants have been recorded in High Park.

In June, the Ringlet (*Aphantopus hyperantus*) is the first grass-feeder to emerge, followed a few weeks later by the Gatekeeper (*Pyronia tithonus*). The adults feed on brambles (*Rubus fruticosus* agg.) and various thistles (*Cirsium spp.*). In most years, they occur in abundance along the woodland margins and glades.

11.4. Grassland and meadows

There are two main areas of open grassland suitable for grassland butterflies in High Park – the first in the southern half, which is largely acidic grassland, and a strip of mown grass in the northern half known as the Palace Vista, which downslopes towards the Lake and has plants indicative of calcareous soils. Other patches of grassland are in pasture woodland with ancient oaks, and some of these areas have self-seeded with other trees and shrubs or have been invaded by Bracken (*Pteridium aquilinum*), reducing their suitability for grassland species of butterfly. The Small Copper (*Lycaena phlaeas*, Figures 11.4, 11.5) is usually the earliest butterfly

Figures 11.4 and 11.5.   Small Copper (*Lycaena phlaeas*). (Photographs by Phillip Cribb)

to be encountered in the grassy areas of High Park, its first brood appearing in May. This butterfly lays its eggs on Common Sorrel (*Rumex acetosa*), Sheep's Sorrel (*Rumex acetosella*) and docks (*Rumex* spp.). A second brood appears in summer and sometimes a third one later in the year. Only one individual was recorded during the survey period.

The first brood of the Common Blue (*Polyommatus icarus*) can be seen in May. Its usual foodplant here is Bird's-foot Trefoil (*Lotus corniculatus*). A second brood begins to emerge in late July. Only two individuals were recorded during the survey period. The smaller Brown Argus (*Aricia agrestis*) resembles a small female Common Blue but lacks the bluish hairs on its thorax and abdomen and also has a series of rich orange spots along the margins of both the upper fore and hind wings. Its larvae feed predominantly on Rockrose (*Helianthemum nummularium*), various Geranium

Figure 11.6.   Small Skipper (*Thymelicus sylvestris*) (top) and Essex Skipper (*Thymelicus lineola*) on Common Ragwort (*Jacobaea vulgaris*). (Photograph by Phillip Cribb)

species (*Geranium* spp.) and Common Stork's-bill (*Erodium cicutarium*). Six individuals were recorded during the survey period, and interestingly four of these were recorded along a shaded ride in the south of High Park.

Several grass feeders are on the wing by late June and early July. These include three species of skipper, first the Large Skipper (*Ochlodes sylvanus*), followed shortly by the similar Small Skipper (*Thymelicus sylvestris*) and Essex Skipper (*Thymelicus lineola*). The last two are most readily distinguished by examination of the tips of their antennae, the former having brown tips, the latter black ones (Figure 11.6). The Large Skipper larvae feed on Cock's-foot, Tor-grass (*Brachypodium rupestre*) and False Brome and was the commonest of the three skippers. The Small Skipper usually lays its eggs on Yorkshire-fog but can feed on other common grasses, while the larvae of Essex skipper usually feed on creeping Soft-grass (*Holcus mollis*), Timothy (*Phleum pratense*) and Meadow Foxtail (*Alopecurus pratensis*). Only one Essex Skipper was recorded, but it is likely to be under recorded because of the challenge of distinguishing it from the Small Skipper.

By late June, the first male Meadow Browns (*Maniola jurtina*) emerge, followed a few days later by the larger more brightly coloured females. Almost simultaneously, the Marbled White (*Melanargia galathea*), the third habitat specialist, can be seen on the wing. Both these butterflies can be seen in numbers, often on the heads of Common Knapweed (*Centaurea nigra*) – not found in High Park – and other grassland flowers. The Meadow Brown lays its eggs on Cock's-foot, False Brome, bent (*Agrostis* spp.) and fescue (*Festuca* spp.) while Marbled Whites lay their eggs on Yorkshire-fog and fescues. The summer months can also bring migrants, such as the Large White (*Pieris brassicae*) and Red Admiral (*Vanessa atalanta*) occasionally in large numbers. The Large White will lay its eggs on a range of *Brassica* species and the Red Admiral on Common Nettle (*Urtica dioica*) or Small Nettle (*Urtica urens*).

11.5. Missing or overlooked species

The low butterfly species count from the survey period requires some explanation. Almost half of Oxfordshire's butterflies, particularly habitat specialists such as the chalk-grassland species, are unlikely to be found in High Park. Others that might be expected here may have been overlooked or absent because the Park lacks suitable habitat or their foodplant. Of the grassland butterflies, the most surprising absentee is the Small Heath (*Coenonympha pamphilus*). Its foodplants (meadow-grasses, fescues and bents) are common here. We have searched for it but without success. It is a nationally declining species and we cannot rule out that it may have disappeared from High Park in recent years. Surprisingly, no examples of the Small Tortoiseshell (*Aglais urticae*) or Painted Lady (*Vanessa cardui*) were seen during our survey. The former might be expected in good migration years. There are historical records for Small Tortoiseshell in the decade 1982–92. Most individuals of the Small Tortoiseshell are resident, but Thomas and Lewington (2020) state that migratory flights have been recorded in both directions. Its numbers in any one year are affected by the parasitic fly *Sturmia bella*, which can also affect the Peacock butterfly. According to Asher et al. (2016), the population of the Small Tortoiseshell dipped between 2000 and 2012 but has made a significant recovery since 2013. The latter was recorded in both 1990 and 2015, and might well reappear in the occasional good migration years.

Among those butterflies for which habitat might be absent or rare, we can list Brown Hairstreak (*Thecla betulae*) and Black Hairstreak (*Satyrium pruni*), both Nationally Rare species but occurring in woodlands in the vicinity of Oxford. Their foodplant is Blackthorn (*Prunus spinosa*) which has a restricted range within the Park and is mostly made up of old shrubs. The Green Hairstreak (*Callophrys rubi*) is readily overlooked; however, its principal foodplant is Gorse (*Ulex europaeus*), which is absent here, although its larvae will also feed on brambles (*Rubus* spp.). The White-letter Hairstreak (*Satyrium w-album*) occurs locally nearby while its main foodplant, Wych Elm (*Ulmus glabra*), is frequent in some parts of High Park. It has probably been overlooked. The White Admiral (*Limenitis camilla*) also occurs locally, but its foodplant, Honeysuckle (*Lonicera periclymenum*), is rare and overgrazed in the Park. Given that it has been expanding its territory in England, enough foodplant might persist to sustain a small population. Similarly, the Holly Blue (*Celastrina argiolus*), a ubiquitous species in the region, needs Ivy (*Hedera helix*) for its overwintering brood, but that is very rare in the Park. The Holly Blue was recorded from High Park in 1990 and 1992.

There are historical records for the Grizzled Skipper (*Pyrgus malvae*) in 1983 and 1990 and Dingy Skipper (*Erynnis tages*) in 1983, and certainly these two could still be present, especially the Dingy Skipper, which lays its eggs on the Common Bird's-foot Trefoil (*Lotus corniculatus*). The Grizzled Skipper needs its foodplants, Wild Strawberry (*Fragaria vesca*), Agrimony (*Agrimonia eupatoria*) and Creeping Cinquefoil (*Potentilla reptans*), where the vegetation is sparse and where there are good nectar sources in early spring. All three foodplants are present in the Park. The Wall (*Lasiommata megera*) was recorded in 1982, but since 2005 there has been a drastic decline in the Upper Thames Region. The last record in Oxfordshire was at Sandford-on-Thames in 1992. The female lays its eggs on a range of grasses and prefers dry areas with short grass.

11.6. Concluding remarks

In conclusion, Blenheim High Park has a good butterfly population (Table 11.1), but is rather species-poor in comparison with some nearby localities. Wytham Woods had 30 species recorded between 2020 and 2022, but is a larger area of woodland (404ha) with more diverse habitats (Charlie Hackforth, pers. comm.). Foxholes Wood, which was once part of the ancient Wychwood Forest, has a historical total of 28 species, but only 19 species were recorded on the butterfly transect in 2021 (Colin Williams, pers. comm.). A number of species also have

Table 11.1.   Butterflies recorded in High Park, Blenheim (2017–21).

Species	Breeding	Migrant	Number recorded	Years recorded
Aphantopus hyperanthus (Ringlet)	√		150+	2018, 19, 20, 21
Maniola jurtina (Meadow Brown)	√		120+	2017, 18, 19, 20, 21
Ochlodes sylvanus (Large Skipper)	√		70+	2018, 19, 21
Pyronia Tithonus (Gatekeeper)	√		50+	2018, 19, 21
Pararge aegeria (Speckled Wood)	√		50+	2017, 18, 20, 21
Melanargia galathea (Marbled White)	√		40+	2018, 19, 20, 21
Argynnis paphia (Silver-washed Fritillary)	√		50+	2018, 19, 20, 21
Favonius quercus (Purple Hairstreak)	√		40+	2018, 19, 21
Pieris napi (Green-veined White)	√		16	2017, 18, 20, 21
Pieris rapae (Small White)	√	√	14+	2017, 18, 19, 21
Pieris brassicae (Large White)	√	√	14	2018, 19, 21
Gonopteryx rhamni (Brimstone)	?		11	2017, 18, 21
Anthocharis cardamines (Orange Tip)	√		10+	2018, 20, 21
Polygonia c-album (Comma)	√		9	2017, 18, 20, 21
Thymelicus sylvestris (Small Skipper)	√		8+	2018, 20, 21
Vanessa atalanta (Red Admiral)	?	√	8	2017, 20, 21
Apatura iris (Purple Emperor)	√		8	2019, 20, 21
Aricia agestis (Brown Argus)	√		6	2018
Aglais io (Peacock)	?		5	2018, 21
Polyommatus icarus (Common Blue)	√		2	2018
Thymelicus sylvestris (Essex Skipper)	√		2	2021
Lycaena phlaeus (Small Copper)	√		1	2018

Notes: Limited recording took place in 2017 and 2020. Larger numbers (e.g. 50+) are estimates only.

very small populations, such as the Small Copper, Common Blue, Small and Essex Skippers, but without standardized counts such as the Butterfly Monitoring Transect Method, it is not possible to compare annual fluctuations in abundance with nearby sites.

Deer grazing and predation by pheasants in High Park may be factors limiting species abundance and diversity. The pheasant population is supplemented annually by reared young birds. Climate change may have impacted some species more widely, but its effects on butterfly abundance in High Park are difficult to estimate in the absence of more detailed historical records. It is possible that the recently introduced cattle grazing, particularly of the calcareous grassland, will open some opportunities for a greater diversity of habitat for butterflies in the future.

Acknowledgements

The authors thank Martin Corley for comments on two earlier drafts of this chapter.

CHAPTER 12

Moths (Lepidoptera)

Martin Corley

A total of 684 moth species was recorded during the course of the survey. This is 27% of the 2,476 total of British and Irish species (Agassiz et al. 2022). It is undoubtedly incomplete, as there are historic records of some species that were not found in the survey. Additional species continued to be added right through 2021, the final year of the survey, and there are a few common species that have not been recorded but must surely be present. The main factors limiting the survey were the available time of surveyors, COVID-19-related restrictions in 2020 and the seasonal restrictions regarding access to High Park imposed by the Estate.

12.1. Surveyors

The lead surveyor was Martin Corley (MC), author of this chapter. Others who brought equipment (generators and light traps) at one time or another for night work were Martin Townsend, Peter Hall and Marc Botham. Several others accompanied the main surveyors on one or more visits, including Pedro Pires, who took valuable photographs, Dinis Pires, Benedict Pollard, Steve Nash, Mary Elford and Julian Howe. MC was involved on all nights except one April night session by MT and MB. We also received records from daytime work by Graham Collins and Jovita Kaunang. Benedict Pollard's beetle traps produced abundant by-catch, which he sorted into insect groups. Sifting through the moths obtained this way added further species. All the above are thanked for their help.

12.2. Methods

The great majority of records were made at night using lights of various types. MC's preferred method was a 125 watt mercury vapour light bulb over a white sheet, with egg boxes on the sheet for moths to shelter in. Similar setups with a 160 watt blended bulb and a 20 watt actinic bulb and a Robinson type trap with a 125 watt MV bulb, sitting on a white sheet were also used regularly and on a few occasions a Heath trap with a 6 watt actinic light. MT, MB and PH all ran Robinson traps with 125 watt MV bulbs. A few night sessions in high summer continued through to dawn, but the majority of sessions ended during the night when moth activity had fallen to a low level. We carried out surveys on 33 nights during the period 2017–21, two in February, one in March, four in April, two in May, six in June, five in July, five in August, four in September, three in October and one in November.

Although light is the single most efficient way of sampling moths, there are many species that rarely if ever come to light and there are species that are exclusively day-flying, therefore other methods of sampling were also used. MC and MT used wine ropes on most visits (Figure 12.1). These are pieces of cord (old-fashioned hemp washing line is ideal), soaked in

Martin Corley, 'Moths (Lepidoptera)' in: *The Natural History of Blenheim's High Park*. Pelagic Publishing (2024). © Martin Corley. DOI: 10.53061/WZHD1567

Figure 12.1. Dorsal view (left) and ventral view (right) of Clifden Nonpareil (*Catocala fraxini*) feeding on wine rope. (Photographs by Pedro Pires)

a strong solution of sugar in red wine and hung over twigs and low branches. At times they can be very productive, but results are highly unpredictable. In High Park they were never covered in moths, but they did add species that were not seen at light.

Pheromones are increasingly used for sampling insects. These are synthetic approximations to the pheromones that female moths produce to attract males. They are most effective when used with a suitable trap. They are particularly effective in the case of Clearwing moths (Sesiidae), which are day-flying and mimic Hymenoptera, mainly ichneumon wasps. They are rarely seen without the aid of pheromones. However, in High Park, pheromone traps were employed on two occasions in 2021 and produced a single non-target species (*Nemapogon koenigi*) that had already been recorded at light.[1] There are other species for which there are commercially available pheromones, such as Emperor Moth (*Saturnia pavonia*), but we did not use these.

Particularly in the spring months, many species, especially smaller species of moth, fly by day, mainly on sunny days. The main reason is that the nights are still cold, often right through May into early June, so there is little opportunity for smaller moths to fly at night. Probably another reason is that many species of plant are in flower at this time, so the moths can take advantage of this for nectar or pollen and often for egg-laying. Moths can be netted while flying, or found on flowers, or sometimes can be obtained by sweeping vegetation with a net. Daytime visits were made in May, adding a number of species.

The end of the afternoon and early evening up until dark can be a good time to catch moths, some of which have their main flight period at this time of day, when they can be netted. This simple technique, known to lepidopterists as 'dusking', can be effective, but for reasons that are not clear this was never profitable in High Park.

In addition to the above methods, which target adult moths, the early stages, particularly the larvae can be found, and identified either as larvae or after rearing through to the adult stage. One technique is beating branches of trees and bushes over a tray and identifying the dislodged caterpillars. This was not much used as most larvae obtained this way are Macrolepidoptera,[2] which can usually be found by other means. Larvae obtained by beating cannot be easily returned to their tree, so MC prefers not to use this method.

1   Nomenclature of moths follows Agassiz et al. (2013) and subsequent updates.
2   The separation of moths into Macrolepidoptera and Microlepidoptera that is used in this chapter has little basis in science, but reflects the amount of interest that each category receives. The availability of identification aids for Microlepidoptera has always lagged far behind. There are nearly twice as many Microlepidoptera species and many are very small. Macrolepidoptera include those moth families in which most species are larger than is the case with Microlepidoptera, but actually there is a wide overlap. English names for Microlepidoptera have no vernacular basis and are not used in this chapter.

Larvae of Microlepidoptera, which are often concealed in leaves spun together or into tubes with silk, can be searched for on their hostplants. Searching is particularly appropriate for leaf miners. These are larvae of some of the smallest Microlepidoptera, which live inside leaves. The combination of chosen hostplant and the detailed form of the leaf mine allows easy identification of many species. However, some species that share hostplant and mine characters cannot be separated, and therefore require rearing of the adults, which is easy in some cases but more difficult in others. The best season for searching for leaf mines is autumn, at which time the signs of earlier vacated mines remain on the leaves, while other larvae are still present in their mines and may be possible to keep over winter for identification when they emerge the following year. Because of the regular curtailment of access to High Park in the autumn, it was only in the final autumn of the survey in 2021 that there was any real opportunity to work on leaf mines.

The pupal stage is little used in surveying, but in the survey empty pupal cases of the Hornet Moth (*Sesia apiformis*) were found protruding from tunnels at the base of poplar trees. This species has larvae that feed in the wood of poplar trunks. The adult moth was not seen. It is remarkable for its mimicry of a Hornet, not only in appearance but when flying it sounds and moves like a Hornet (Figure 12.2). This is an extremely effective protection against predators.

There are other methods of capturing and recording moths. The Sallow-feeding micro-lepidopteran *Batrachedra praeangusta* was seen just once at light, but over 100 were captured in interception traps set for catching beetles. This technique, which is clearly effective for some moth species, is not favoured by lepidopterists because the samples obtained are immersed in alcohol, making visual identification from wing markings difficult or impossible.

Figure 12.2.   The Hornet Moth (*Sesia apiformis*), a perfect example of Batesian mimicry. (Photograph by Iain H. Leach, by permission from Butterfly Conservation)

12.3. All about moths

12.3.1. Seasons

Moths can be found flying in all months of the year, particularly in woodland. Activity is restricted by the coldest winter weather, but as soon as the weather turns milder there are moths flying. Some are winter moths with their only flight time during the winter, a strategy that avoids predation by bats. In many cases the females of these species are flightless, making them less at risk from strong winds. There are also moths that emerge from the pupa in autumn and lay eggs the following spring. Some of these hibernating species will fly in the winter months during mild spells.

As spring arrives, the number of species flying rises, reaching a peak from the end of June through to early August, after which numbers fall off quite rapidly. There is a smaller peak in numbers in March and early April when species of genus *Orthosia* are flying, which coincides with the flowering of sallow (*Salix* species).[3] There is a minor peak in number of species in late September and October, although the numbers of moths at this time in High Park were lower than expected. Possible reasons for this are discussed later in this chapter (Section 12.5.4).

While many species of moth have a single generation each year, there are also many others that have two or sometimes even three generations per year.

12.3.2. Longevity

The adult life of a moth varies from just a day or two in the case of Water Veneer (*Acentria ephemerella*) to as much as eight months in the case of Oak Nycteoline (*Nycteola revayana*) for example, which may emerge about September, hibernates over winter and lays eggs in May and early June. Probably the great majority of species that do not overwinter as adults live for around one to three weeks, but there are some long-lived species in summer too. The Large Yellow Underwing (*Noctua pronuba*) hatches in June, but soon goes into a period of aestivation, often not reappearing before late August, with some individuals still flying as late as early November. Moths that feed in the adult stage are likely to be longer lived than those that are unable to feed. Some of the larger moths that come freely to wine ropes such as Old Lady (*Mormo maura*) and Red Underwing (*Catocala nupta*) have a potential adult life of two months or more.

12.3.3. Flight distance

Individual moths may hardly travel any distance in their adult life, or they may travel moderate or great distances. Some small micros with larvae on abundant hostplants have little need to travel. These are unlikely to come to light and are best found by sweeping. Examples are *Cauchas fibulella* found on flowering Germander Speedwell (*Veronica chamaedrys*) and *Elachista gangabella*, which is a leaf miner of False Brome (*Brachypodium sylvaticum*). Moths with larvae on dominant trees in woodland also have little need to travel far as the hostplant is all around. Species with larvae on herbaceous plants that are of local and transitory occurrence must travel widely to find their hostplants. Mullein Moth (*Cucullia verbasci*) is a species of this type, with larvae on mullein (*Verbascum*) and figwort (*Scrophularia*) (Figure 12.3).

On hot nights in summer, unexpected species can appear at light, such as Gold Spot (*Plusia festucae*) and Marsh Oblique-barred (*Hypenodes humidalis*), which are species associated with marshy habitats, but both came to light in woodland in the middle of the southern half of High Park, well away from the Lake margin. Marshy habitats change with time and hostplants may be lost. Females that disperse may find new egg-laying sites, so the species benefits from a proportion of females dispersing away from their larval home.

---

3   Nomenclature of plants follows Stace (2019).

Some moths, mainly larger species but also including a few Microlepidoptera, are known to be long-distance migrants. Every year millions of Silver Y moths (*Autographa gamma*) come to northern Europe from southern Europe and North Africa. They arrive in spring, usually in May, and breed here. Later in the year, the second or third generation – or at least part of it – returns southwards, where further generations will be raised in warmer climates. The Silver Y is not frost tolerant so cannot normally survive the British winter. Many other species show migratory tendencies, but the migration is often a more random process, generally partly wind assisted. Migration may be necessary to escape an area where the hostplants have dried up, or it may simply offer the opportunity of potential colonization of new territory. In High Park there was a single record of the well-known migrant Scarce Bordered Straw (*Helicoverpa armigera*).

While the great majority of moth species recorded in High Park will have also been in High Park as larvae, there will be a small proportion of species from further afield, maybe from as near as the Great Lake or the

Figure 12.3.   Caterpillar of Mullein Moth (*Cucullia verbasci*) on figwort. (Photograph by Benedict Pollard)

Evenlode valley, but potentially from considerably further. The species list resulting from the survey (Appendix 10) does not attempt to distinguish the few species that may not have their origin in High Park from the remainder that almost certainly do.

12.4. Moths and plants

With very few exceptions, moth larvae are vegetarians. Adult moths, if they feed, also feed on the products of plants. Most moths have a proboscis, a hollow tube that can be coiled at rest or extended when in use through which liquids can be imbibed. In some genera or whole families this is rudimentary. Such species cannot feed as adults and have to rely on stored fat laid down in the larval stage for their energy requirements in adult life. These species, including the Swift moths (Hepialidae), the Poplar Hawkmoth (*Laothoe populi*) and Drinker Moth (*Euthrix potatoria*) (Figure 12.4) (the caterpillar drinks dew) tend not to have a long adult life.

Of those that do feed, some Erebidae and Noctuidae in particular are strongly attracted to wine ropes. In nature these feed on nectar, sap runs, honeydew, damaged and over-ripe fruit and so on. The primitive moth family Micropterigidae, with two species of *Micropterix* in High Park, have jaws in place of a proboscis and feed on pollen of flowers such as buttercups (*Ranunculus*) and sedges (*Carex*).

The growing stage, the larva, feeds on living plants or dead plant material. The few non-vegetarian exceptions are considered at the end of this section. Larvae of various species can be found feeding on all parts of living plants, and most species of plant have some Lepidoptera species that can feed on them. The majority of High Park moth species have larvae that feed on flowering plants, including trees. Some larvae feed on a wide range of

Figure 12.4.   Drinker Moth male (left) and female (*Euthrix potatoria*). (Photograph by Pedro Pires)

unrelated plant species. These moth species, including many Noctuidae and Tortricidae, are said to be polyphagous. Certain polyphagous species such as some *Acronicta* species and many of the larger Geometridae only feed on trees, but on a range of trees from different plant families. Other larvae are oligophagous, feeding on several plant species from a single family, or sometimes on just a few plant species that are not particularly closely related. The Poplar Hawkmoth is oligophagous on willows (*Salix*) and poplars (*Populus*), which belong to the same plant family. The most specialized larvae are monophagous, feeding on a single plant species, or at most a single genus of plants. An example is the Chimney Sweeper (*Odezia atrata*), restricted to Pignut (*Conopodium majus*).

In Table 12.1 different types of association of moth larvae with their larval food are analysed, with numbers of High Park moth species given for each type of association.[4]

Moth larvae can be found feeding externally or internally on all parts of flowering plants. Common Swift (*Korscheltellus lupulina*) larvae feed in and among the roots of herbaceous plants; Mere Wainscot (*Photedes fluxa*) feeds in stems of Wood Small-reed (*Calamagrostis epigejos*); Leopard Moth (*Zeuzera pyrina*) (Figure 12.5) feeds in the living wood of several species of tree; *Argyresthia glaucinella* feeds in the bark of oak trees; Yellow-legged Clearwing (*Synanthedon vespiformis*) feeds under bark of oak trees where this is freshly exposed by branches breaking off or by felling; *Spuleria flavicaput* feeds inside young woody shoots of hawthorn (*Crataegus*); buds are eaten by some species as they begin to open in spring; innumerable species feed on leaves, sometimes within shelters made by spinning leaves together with silk or rolling leaves or leaf margins into tubes; many species live inside leaves as leaf miners; flowers are eaten by several species of Pug Moths (*Eupithecia*); developing seeds of bulrush (*Typha*) are eaten by *Limnaecia phragmitella*; fleshy fruits such as those of hawthorn are eaten by *Grapholita janthinana*; *Cydia splendana* lives in acorns. Spongy galls on oak provide nourishment for

---

4   Information for Tables 12.1, 12.2, 12.3 and 12.5 is extracted from Langmaid et al. (2018) for Microlepidoptera and Henwood and Sterling (2020) for Macrolepidoptera.

Table 12.1.   Associations of moth larvae in High Park with their larval food.

Feeding habit	Food sources	Total spp.
Polyphagous	Trees, shrubs and herbs	40
Polyphagous	Trees and shrubs	75
Polyphagous	Herbs	44
Polyphagous	Grasses	49
Monophagous	Single tree genus	180
Monophagous	Single herb genus	105
Monophagous	Single shrub genus	16
Oligophagous	Two genera of shrubs	5
Oligophagous	Two genera of trees	79
Oligophagous	Three genera of trees	14
Oligophagous	Three genera of herbs	7
Other	Plant detritus	14
Other	Animal detritus	8
Other	Dead wood	5
Other	Fungi	6
Other	Lichens	17
Other	Mosses	11
Unknown	Unknown	3

Figure 12.5.   Leopard Moth (Zeuzera pyrina). (Photograph by Steve Nash)

Table 12.2.   Associations of moths in High Park with herbs.

Species	Common name	Total spp.
Urtica dioica	Common Nettle	5
Galium	Bedstraws	4
Hypericum	St John's-worts	4
Juncus	Rushes	3
Jacobaea	Ragworts	3
Pteridium aquilinum	Bracken	3
Cirsium	Thistles	2
Deschampsia cespitosa	Tufted Hair-grass	2
Luzula campestris	Field Woodrush	2
Rhinanthus minor	Yellow Rattle	2
Stachys sylvatica	Hedge Woundwort	2
Veronica chamaedrys	Germander Speedwell	2
Vicia	Vetches	1
Trifolium	Clovers	1
Centaurea nigra	Common Knapweed	1
Achillea millefolium	Yarrow	1
Arctium	Burdocks	1
Carex	Sedges	1
Epilobium	Willowherbs	1
Phragmites australis	Common Reed	1
Pulicaria dysenterica	Common Fleabane	1
Brachypodium sylvaticum	False Brome	1

Pammene albuginana, but it is not the gall causer. A few moth species do cause galls on plants, but no example was found in High Park.

Dead plant material also provides food for some moth species. Gold Triangle (*Hypsopygia costalis*) larvae feed on a variety of dead plant material often in dry situations. *Agnoea josephinae* makes a portable case from which it feeds on dead leaves on the ground.

In Table 12.2 the number of species of moth larvae that are monophagous on High Park herbaceous plants is given for each genus of herb. Not all these plants were found in High Park during the botanical surveys.

When the survey began in 2017 there was a long plantation in High Park with numerous larch trees (*Larix decidua*), but these and some other coniferous trees were removed before the survey ended. *Cydia illutana* which has larvae in the developing cones of larch is probably no longer present. Some conifer plantations including a few larch trees remain. Several species have larvae that feed on leaves of various conifers, including Pine Hawkmoth (*Sphinx pinastri*) and Spruce Carpet (*Thera britannica*).

Rather few species of moth feed on ferns, but Brown Silver-line (*Petrophora chlorosata*), which feeds on Bracken (*Pteridium aquilinum*), is extremely common in High Park.

Mosses are eaten by larvae of *Eudonia mercurella* and *Bryotropha senectella*. *Aplota palpellus* was refound in Britain in Savernake Forest in 1986 (Sterling, 1987) after a gap of over 80 years. Larvae were found in High Park in 1987 feeding on *Hypnum cupressiforme* on oak trunks. This Nationally Rare (NR) species was seen at light on three occasions during the survey (Figure 12.6).

Figure 12.6.   *Aplota palpellus.* (Photograph by João Nunes)

Table 12.3.   Saproxylic moth species recorded in High Park.

Species	Substrate
Morophaga choragella	Bracket fungi
Nemapogon cloacella	Various fungi (*Trametes spp.*)
Nemapogon koenigi	Fungi including *Annulohypoxylon* and *Fomitopsis*
Nemapogon variatella	Fungi including *Fistulina* and *Fomitopsis*
Triaxomera parasitella	Fungi especially *Trametes versicolor*
Crassa tinctella	In dead wood
Batia lunaris	On lichen and dead wood
Crassa unitella	On fungus on and under dead bark
Esperia sulphurella	On dead wood and under bark
Parascotia fuliginaria (Waved Black)	Bracket fungi on rotten wood, particularly birch

Lichens provide food for a number of moth species, in particular a group of Erebidae known collectively as Footmen (including the genera *Eilema, Lithosia, Miltochrista, Thumatha, Nudaria* and *Atolmis*). Some lichen-feeders such as the wingless parthenogenetic case-bearing *Luffia lapidella* feed also on algae such as *Pleurococcus* on tree trunks.

In contrast to beetles and flies, there are rather few saproxylic moths, but there are a few species with larvae that feed in bracket fungi on dead wood and others that feed under dead bark or in rotten wood, where they may obtain nutrition from fungal hyphae rather than the decaying wood. The saproxylic species recorded in High Park are listed in Table 12.3.

Some of the saproxylic species belong to the Clothes Moth family Tineidae, in which nearly all species have larvae that do not feed on living green plants. Besides dry fungi, some tineid larvae may feed on material of animal origin, including wool, fur, feathers, owl pellets and dung of carnivores. Some species live in birds' nests. Table 12.4 includes all the Tineidae recorded in High Park.

Table 12.4.   Moths of the Tineidae family recorded in High Park.

Species	Substrate
Morophaga choragella	Bracket fungi
Nemapogon cloacella	Various fungi (*Trametes spp.*)
Nemapogon koenigi	Fungi including *Annulohypoxylon* and *Fomitopsis*
Nemapogon variatella	Fungi including *Fistulina* and *Fomitopsis*
Triaxomera parasitella	Fungi especially *Trametes versicolor*
Monopis laevigella	Refuse of animal origin, in birds' nests, owl pellets etc.
Monopis weaverella	Fox dung, dead animals and probably other material of animal origin
Monopis obviella	Refuse of plant and animal origin, birds' nests
Monopis crocicapitella	Refuse of plant and animal origin, birds' nests
Niditinea striolella	In birds' nests in holes and nest boxes
Tinea semifulvella	In birds' nests, preferring those in the open
Tinea trinotella	In birds' nests

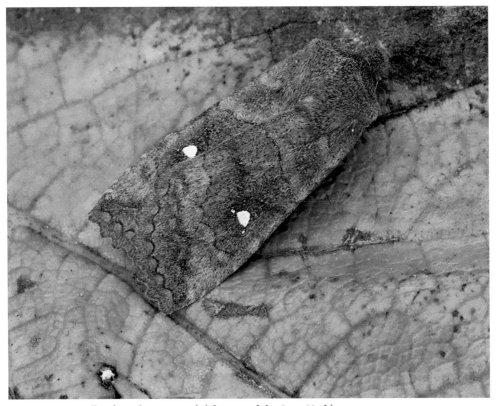

Figure 12.7.   Satellite (*Eupsilia transversa*). (Photograph by Steve Nash)

The Bee Moth (*Aphomia sociella*) has gregarious larvae that live in nests of bumblebees and wasps, where they feed on detritus, wax and often also on the bee or wasp grubs.

A very small number of species have carnivorous larvae. High Park species include Dun-bar (*Cosmia trapezina*), Satellite (*Eupsilia transversa*) (Figure 12.7) and Grey Shoulder-knot (*Lithophane ornitopus*), of which the larger larvae are predatory on other caterpillars.

12.4.1. Hostplants of larvae of High Park moths
In Table 12.5 the number of species of moth larvae that are monophagous on High Park trees, shrubs and climbers is given for each genus or species of woody plant.

12.4.2. Moths and oak trees
There are more species of Lepidoptera that feed on oak trees than on any other genus of plants. There is no evidence that the foliage of ancient oaks is more or less suitable as larval food than that of younger oaks, but there is a very small number of species that prefer young plants. *Zimmermannia atrifrontella* makes a gallery mine just under the surface of young bark of trunks and branches under about 12cm in diameter, which are much more likely to be available on faster growing young trees.

Table 12.5.   Trees, shrubs and climbers in High Park with associated moth species.

Species	Common name	Total spp.
Quercus robur	Pedunculate Oak	58
Betula	Birches	22
Crataegus	Hawthorns	16
Salix	Sallows and Willows	13
Prunus spinosa	Blackthorn	11
Populus	Poplars and Aspen	10
Fraxinus excelsior	Ash	9
Fagus sylvatica	Beech	7
Acer pseudoplatanus	Sycamore	6
Ulmus glabra	Wych Elm	6
Acer campestre	Field Maple	5
Corylus avellana	Hazel	5
Rubus	Brambles	4
Rosa	Roses	4
Clematis vitalba	Traveller's Joy	4
Rhamnus cathartica	Common Buckthorn	4
Larix decidua	European Larch	3
Picea abies	Norway spruce	3
Euonymus eiropaeus	Spindle	3
Pinus	Pines	2
Carpinus betulus	Hornbeam	2
Tilia	Limes	2
Solanum dulcamara	Woody Nightshade	2
Ligustrum vulgare	Privet	1
Malus	Apples	1
Aesculus hippocastanum	Horse Chestnut	1
Hedera helix	Ivy	1
Lonicera periclymenum	Honeysuckle	1
Prunus avium	Wild Cherry	1
Prunus padus	Bird Cherry	1

Dystebenna stephensi is a NR species that was found new for Oxfordshire in High Park. It feeds in the bark of large mature oaks, and the reddish-brown frass (caterpillar excrement) is visible in the bark crevices. Of the two High Park specimens, one was taken in flight and the other in a pheromone trap for beetles, so it is not known if the affected trees were mature, veteran or ancient. *Argyresthia glaucinella* is a somewhat more common species that feeds in the same way, also requiring large trees with deeply fissured bark.

Pammene albuginana larvae feed on the tissue of spongy galls on oak as inquilines, feeding on the gall tissue but not causing the gall. Other *Pammene* species with similar habits were not recorded but are likely to be present.

Aplota palpellus, which feeds on moss on mature oaks (already mentioned in Section 12.4), is a rare species. There is sometimes very little moss on the ancient oaks (Chapter 4).

The ancient oaks are certainly important for moths in two ways. They provide the decaying wood and fungi that the few saproxylous moth species require. The deep bark crevices and hollow trunks offer sheltered hiding places for resting moths, which must be particularly important for those species that survive the winter as adults. In many landscapes, ivy (*Hedera helix*) is very valuable in this respect, but as it is scarce in High Park, the hollow trees can provide the same service.

A curious and unexplained phenomenon was observed on the night of 23 July 2019. One ancient oak beside the road through High Park was observed to have numerous Svensson's Copper Underwings (*Amphipyra berbera*) (Figure 12.8) resting on the trunk. Over 40 were counted on the one tree, while other nearby trees had no more than one or two. The two species of Copper Underwing are very common visitors to wine ropes, yet these moths were not feeding and they were not aggregated in any way but scattered all around the trunk. There were no visible exudations from the trunk for them to feed on and no evidence that there was any social purpose. The moths did not appear to be all newly emerged. MC did not have the opportunity to discover if this was a regular nightly event or if it was repeated in other years. The explanation remains baffling.

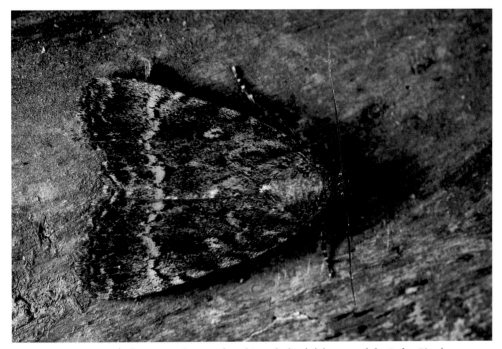

Figure 12.8. Svensson's Copper Underwing (*Amphipyra berbera*). (Photograph by Pedro Pires)

12.5. Changes in the moth fauna

The moth fauna of any area, large or small, will always gradually change over time. Species become scarce or disappear while others become more common or establish themselves in new areas. In the present author's entomological lifetime there were notable periods of change in the hot summers of 1975 and 1976. The poor summers of 1985–7 caused some losses. The hot summers of 1989 and 1990 also produced many changes, both gains and losses, and appear to mark the beginning of the current climatic period widely referred to as climate change (which is nothing new), characterized in Oxfordshire most obviously by earlier springs, at least compared with the period 1962–88.

On top of climatic changes, moth populations are greatly affected by changes in land use. In some ways High Park is a very stable environment, but there have been significant changes during the five years of the survey that must surely have had effects on some moth species. The long plantation was cleared, removing all larch trees, a group of tall hybrid poplars opposite High Lodge was felled, and there has also been substantial thinning of 'infill' Ash trees and in some areas Sycamore. It is probable that the introduction of cattle will eventually have a significant effect on the vegetation.

There is very strong evidence that moth populations have generally declined over recent years and several possible causes have been suggested, such as agricultural intensification, nitrogen pollution and light pollution. Less often mentioned are vehicle headlights. A car travelling at night attracts moths from the roadside habitat into the path of the car, which kills them. This occurs much less frequently now than formerly, presumably because of the drain in moth numbers in roadside habitats over many years. Of these possible causative factors, nitrogen pollution, which is everywhere, is the only one likely to have much effect in High Park. Together with climate change, these factors are sure to have affected High Park's moth fauna.

12.5.1. Declining species

There are some historic records of moths from High Park from the years 1987 to 1990, with a total of 167 species.[5] Of these, 14 were not recorded during the survey (see Table 12.6). This is obviously too small a sample to provide an earlier baseline for the High Park moth fauna and the survey period of five years is too short to detect significant declines or disappearance of species from High Park although three of the species in Table 12.6 are known to have declined nationally in the last 30 years. Indeed, the survey period was even shorter for some parts of the year. There was no night-time moth session in August before 2019. A few species that are now rarer than formerly at least in western parts of Oxfordshire, such as Tawny Pinion (*Lithophane semibrunnea*) and August Thorn (*Ennomos quercinaria*) were each recorded once during the survey, perhaps suggesting that these species have survived longer in High Park than in some other localities. However, September Thorn (*Ennomos erosaria*) (Figure 12.9), which is considered to have declined significantly nationally in recent years, was common in High Park.

12.5.2. Increasing species

In the twenty-first century a number of moth species have apparently benefited from climate change and either become resident in England or increased their distribution area from limited beginnings. White-point (*Mythimna albipuncta*), once an irregular migrant, then established as a resident species on the south coast, had spread into Oxfordshire as a resident species before the survey began. Least Carpet (*Idaea rusticata*) (Figure 12.10) and Kent Black Arches (*Meganola albula*) were southern species that had spread into Oxfordshire just a few years before the survey began.

---

5   This number and the contents of Table 12.6 are extracted from data supplied by Thames Valley Environmental Record Centre (TVERC). Declining species after Randle et al. (2019).

Table 12.6.   Moths only recorded in High Park prior to the 2017–21 survey.

Species	Common name	Year	National status	Change (%)
Infurcitinea argentimaculella		1987		
Phyllonorycter hilarella		1988		
Ochsenheimeria urella		1988		
Aethes dilucidana		1989		
Pammene aurana		1987		
Hellinsia carphodactyla		1989		
Malacosoma neustria	Lackey	1989	Declining	91%
Idaea subsericeata	Satin Wave	1989	Declining	51%
Xanthorhoe ferrugata	Dark-barred Twin-spot Carpet	1989	Declining	89%
Eupithecia satyrata	Satyr Pug	1988	Declining	75%
Eupithecia icterata	Tawny Speckled Pug	1989	Declining	81%
Diaphora mendica	Muslin Moth	1989	Increasing	58%
Abrostola triplasia	Dark Spectacle	1990	Increasing	75%
Sideridis rivularis	Campion	1989	Increasing	3%

Figure 12.9.   September Thorn (*Ennomos erosaria*). (Photograph by Steve Nash)

Figure 12.10.   Least Carpet (*Idaea rusticata*). (Photograph by Pedro Pires)

Figure 12.11.   Dark Crimson Underwing (*Catocala sponsa*). (Photograph by Steve Nash)

Other species that had extended their range into southern or south-eastern England in recent years reached parts of Oxfordshire shortly before the survey started and apparently arrived in High Park during the course of the survey: Clifden Nonpareil (*Catocala fraxini*) (Figure 12.1) was present in 2019, but several were seen so it very probably had arrived in the previous year and bred on the poplar trees; Dark Crimson Underwing (*Catocala sponsa*) (Figure 12.11) and Tree-lichen Beauty (*Cryphia algae*) were recorded for the first time in High Park in 2021.

12.5.3. Is High Park a rich site for moths?

This is a hard question to answer. Comparisons are only available for Macrolepidoptera, with a total 312 species in the survey. There are several large woods in Oxfordshire that have been surveyed for moths at one time or another. Bernwood Forest straddles the Buckinghamshire/ Oxfordshire border. The all-time list of Macrolepidoptera species for that site stands at 458, of which 399 have been seen since 2000 (Waring, 2022). The figures for Wytham Woods in Oxfordshire are 487 all-time species recorded and 335 recorded between 2019 and 2021.[6]

MC has surveyed two other West Oxfordshire woods of comparable size. Wychwood Forest was visited a number of times in the 1980s and early 1990s. This was not an official survey and there were fewer visits than to High Park, but a total of 246 Macrolepidoptera species was recorded. In 2002–3, a survey was conducted in Bould Wood, Bruerne, for BBOWT. This produced 236 Macrolepidoptera species.

Unlike so much old woodland in Britain, High Park was relatively little affected by timber felling in the twentieth century, nor was there as much conifer planting as in many woods. This continuity of habitat would allow for the possibility of species surviving in High Park that have been lost elsewhere in the county. There is no clear example of this among the Macrolepidoptera, but it may be true for the Microlepidoptera *Aplota palpellus* and *Dystebenna stephensi*.

12.5.4. Are there species that are absent or poorly represented in High Park?

We did not record the generally common Muslin Moth (*Diaphora mendica*). It has been suggested that this species only flies very late at night, so is readily trapped in a garden moth trap left overnight, but was probably not picked up in the survey because the surveyors left before its flight time. Garden Tiger (*Arctia caja*) certainly only flies late at night, but some surveys in its flight period continued until dawn. This species has shown a striking decline in recent years and woodland is not its primary habitat. This cannot explain the absence of Beaded Chestnut (*Agrochola lychnidis*), a common autumn species, but there may be another explanation, see below.

Some species that were common in Wychwood Forest or Bould Wood were not recorded in High Park, but it is not known if these are still common in those woods. *Olindia schumacherana*, which has larvae feeding mainly on Dog's Mercury (*Mercurialis perennis*) and Lesser Celandine (*Ficaria verna*), was seen in Wychwood and Bould Wood, but not in High Park despite the abundance of its hostplants. Likewise, the birch-feeding Scarce Prominent (*Odontosia carmelita*) was present in those woods but not found in High Park.

Some common woodland species were only seen in very small numbers, but there are possible explanations. Early Grey (*Xylocampa areola*) was rare, but its sole hostplant, Honeysuckle (*Lonicera periclymenum*) has a limited distribution in the north-west of High Park. There are other species which were not seen because their hostplants are absent or very scarce, including Campion Moth (*Sideridis rivularis*) which feeds on seeds of Red Campion (*Silene dioica*). Reasons for some plants being rare or absent are discussed in Chapter 4. Only single individuals of two oak-feeding species that can be abundant, Great Prominent (*Peridea anceps*) in 2017 and Oak Lutestring (*Cymatophorina diluta*) in 2021, were seen. Although both are said to have declined nationally (Randle et al. 2019), their apparent scarcity is greater than

---

6   To put these figures into perspective, MC runs moth traps in various sites on his small Oxfordshire farm two or three times a week in reasonable weather throughout the year. The all-time Macrolepidoptera total (going back to 1956) is 475 species, but annual totals over the last 15 years vary between 233 and 297 species. This suggests the possibility that a significant proportion of the species of an all-time list are not necessarily present during shorter time-spans, but include locally or nationally extinct or severely declining species together with those that were transitory, vagrant or migrant.

expected. Birch-feeding species were not as well represented as would have been expected. As well as Scarce Prominent, Grey Birch (*Aethalura punctulata*) and Scalloped Hook-tip (*Falcaria lacertinaria*) were not seen. Leaf miners on birch were remarkably few with only a single mine of one species of Nepticulidae seen, even though a number of species were possible which can be found elsewhere in Oxfordshire. Wild Rose is common in High Park, but no leaf mines were found on it.

A number of species of Noctuidae that fly in autumn were unexpectedly rare, including Red-lined Quaker (*Agrochola lota*) and Yellow-lined Quakers (*A. macilenta*) (Figure 12.12), Brick (*Agrochola circellaris*), Sallow (*Cirrhia icteritia*) (Figure 12.13) and Pink-barred Sallow (*Xanthia togata*), Green-brindled Crescent (*Allophyes oxyacanthae*) and several others, including Beaded Chestnut, mentioned above, which was not seen. The low numbers were not attributable to poor weather, as some surveys in autumn were on near-perfect moth nights with warm conditions, little wind and no moon. Many of these species are brightly coloured yellow or orange-brown and often spend the day hiding among fallen autumn leaves of similar coloration.

A possible explanation is that they are subject to excessive predation. By September, High Park is full of released pheasants, which spend their time scratching around the woodland floor, where any moths hiding among fallen leaves will be very much at risk. Another predator with peak numbers in September is the Hornet (*Vespa crabro*). High Park's ancient oaks provide wonderful nesting sites for hornets and they are everywhere in late summer and autumn. They are active at night, at least if a moth light is in operation, as we found when one trap was left too close to an undetected nest. We must have had to remove over 100 worker hornets from the trap and its surrounds before we could go home. On that occasion no-one was stung and no hornets were killed, but the moths in the trap had been chopped to pieces.

Not all autumn-flying moths were so uncommon. Merveille du Jour (*Griposia aprilina*) (Figure 12.14), which has a green, black and white pattern and is quite cryptically coloured on some lichens, spends its days on tree trunks and branches, and is therefore not at the same risk from pheasants. The same is true for Brindled Green (*Dryobotodes eremita*), Chestnuts (*Conistra vaccinii*) and Satellites, which hibernate, presumably in hollow trees. If they also roost in such sites during their main autumn and spring flight periods they would be at less risk.

Figure 12.12.   Yellow-line Quaker (*Agrochola macilenta*). (Photograph by Pedro Pires)

Figure 12.13.   Sallow (*Cirrhia icteritia*). (Photograph by Steve Nash)

12.5.5. The future of moths in High Park

Further species that have reached some parts of Oxfordshire already can be expected to appear in High Park over the next few years, including Light Crimson Underwing (*Catocala promissa*), Gypsy Moth (*Lymantria dispar*), Hoary Footman (*Eilema caniola*) and Toadflax Brocade (*Calophasia lunula*). On the other hand, some species that have declined regionally or nationally may disappear. Changes in the management of High Park will have knock-on effects on the moth fauna. The effect of cattle grazing is unpredictable and will largely depend on how this affects the flora. Further removal of conifers may eliminate a few species, although these are unlikely to be rare or scarce.

Figure 12.14.   Merveille du Jour (*Griposia aprilina*). (Photograph by Steve Nash)

Acknowledgements

I am most grateful to Aljos Farjon and Benedict Pollard for organizing the biodiversity survey and for encouragement to continue through difficult times; to the Estate staff at Blenheim for allowing night time access, often at short notice; to all those mentioned at the beginning of this chapter who accompanied me on night sessions, and especially Martin Townsend and Marc Botham for a vitally important early April session when I was not free. Special thanks also to Steve Nash, Pedro Pires, Benedict Pollard, João Nunes and Iain Leach who have provided the photos.

CHAPTER 13

Beetles (Coleoptera)

Benedict John Pollard

Beetles are, for the most part, flying animals – invertebrates that are generally distinguishable by their two pairs of wings, of which the outer pair are usually hardened and known as elytra (sing. elytron), a protective covering of the delicate inner pair. There are, of course, exceptions to this broad overview. They are largely terrestrial and eyed, but include groups that can, for example, be flightless, two-winged, aquatic, blind or soft-bodied. The Coleoptera are arguably the most diverse group of living things, with over 400,000 beetle species described and named globally (Marshall 2018: 13), of which a little over 1% (c. 4,100 spp.) are found in the UK. It is likely that many more species remain to be formally described by taxonomists. Occasionally species new to science are reported from the UK (e.g. Allen et al. 2021).

Beetles can be found in almost every terrestrial and freshwater habitat, occupying a myriad of ecological niches, displaying a diverse accompanying range of morphologies and behavioural expressions. A particularly important group of species are called 'saproxylic', including some of the most threatened species in Britain (Harding and Alexander 1993). Fowles et al. (1999) defined them as being 'dependent upon microhabitats associated largely with the processes of damage and decay in the bark and wood of trees and larger woody shrubs and climbers. This includes sap runs, fungal hyphae or fruiting bodies, rot holes, etc.' Alexander (2008: 9–13) further offered a definition: 'Saproxylic organisms are species which are involved in or dependent on the process of fungal decay of wood, or the products of that decay, and which are associated with living as well as dead trees.'

Farjon (2022) revealed the international significance of High Park's ancient oak tree population; Chapter 2 expands on these discoveries. By inference, therefore, the author assumed there would be a diverse and speciose saproxylic beetle assemblage known from High Park. Upon searching the saproxylic rankings webpage (Fowles 2023, continuously updated, first accessed in 2016), it was therefore a surprise to the author in finding that neither High Park nor Blenheim were mentioned at all. 'Blenheim Park' was ranked eighteenth by Harding and Alexander (1994) as one of the 45 most important national sites for saproxylic Coleoptera. The site list included some significant species by 1994, for example *Ampedus cardinalis* (Elateridae), *Lymexylon navale* (Lymexylidae), *Grammoptera abdominalis* (Cerambycidae), *Rhizophagus oblongicollis* (Rhizophagidae) and *Plectophloeus nitidus* (Staphylinidae). It was already seen as a 'classic site' by 2003, despite only 470 beetle records having been made up to that point. It seems most likely that the relatively low site ranking pointed more to a historical lack of biological recording as opposed to a low level of saproxylic beetle diversity, given the extraordinary variety and volume of suitable habitat present.

Benedict John Pollard, 'Beetles (Coleoptera)' in: *The Natural History of Blenheim's High Park*. Pelagic Publishing (2024).
© Benedict John Pollard. DOI: 10.53061/ESXO4392

13.1. History of beetle recording in High Park (adapted from Pollard et al. 2022)

Despite its unrivalled ancient and veteran oak tree population, High Park has had limited historic recording for beetles. Commander J.J. Walker produced the *'Preliminary list of Coleoptera observed in the neighbourhood of Oxford from 1819 to 1907'* and subsequent supplements and interim reports (Walker 1907–37) based to a large degree on his own fieldwork and that of his Oxford contemporaries (e.g. J.J. Collins and W. Holland), and material housed in the Hope Entomological Collections (Oxford University Museum of Natural History). Walker (1907: 52) stated: 'For the purposes of this list, I have practically restricted myself to a radius of 11km from Carfax, the centre of the city of Oxford, this being the limit at which one can effectively work a district by the aid of one's legs unassisted by a bicycle.' High Park sits just on the border of that radius, inferred here to be too far to walk for Walker's fieldwork. In his second supplement (Walker 1910: 61) he states: 'It is only fair to say, however, that one or two localities of old repute near Oxford, which are now closed to Naturalists in general, might well be expected to yield many rare and interesting additional species of *Coleoptera* as well as other Orders of Insects were it possible to obtain the requisite facilities for working in them.' There are several old records from 'Woodstock' that cannot be geographically attributed with any more certainty, but it is plausible these were from Blenheim, as 'Woodstock Park' became part of the Blenheim Estate, so perhaps there is an argument that it was one of the sites of 'old repute'. Aubrook (1939) does contain a single Walker record from 'Blenheim Park', *Tetratoma fungorum* (Tetratomidae), which demonstrates that he visited, but the recording of just a single species suggests access was difficult. Either way, High Park remained neglected until a single record of *Anotylus sculpturatus* (Staphylinidae) from 21 March 1933, as part of the 'Wytham Survey'. This was followed by just a handful of records (<20) from the 1940s, 1950s and 1960s. A visit in the 1950s by the doyen of British Coleoptera, Mr A.A. Allen, resulted in just two records. However, the precise location for these 'Blenheim' records is unclear, so that High Park likely remained 'undiscovered' by coleopterists.

Between 1978 and 1993, John Campbell produced 283 records, and Keith Alexander (KNAA) contributed 94 records in 2002. In 2016, Ryan Clark, as part of a Natural England survey, generated 270 records, mostly saproxylic beetles, adding many important records to the site list, but it remained in 18th place nationally (Clark 2017), probably as other sites also received additional recording effort.

Extensive survey work since 2017, under the remit of the High Park Biodiversity Survey (HPBS), has greatly increased our knowledge of the site's Coleoptera. The site dataset now stands at over 4,750 records, the bulk of which (over 80%) were contributed by the HPBS, with the years 2019 (c. 570 records) 2020 (c. 815) and 2021 (c. 2,200) being the most productive. The total number of known beetle species from High Park increased from 313 in 2017 to 932 as of May 2023.

Clark (2017) collated all previous known records from the Park, based on the dataset shared by Jonty Denton, the county recorder for Oxfordshire (VC 23), which included many non-saproxylics such as leaf beetles (Chrysomelidae), ground beetles (Carabidae) and predominantly aquatic families (e.g. Hydrophilidae, Dytiscidae). This appears to have been the first in-depth list of historic records from High Park, providing a baseline dataset.

It was decided early on, given the constraints of time, resources and restricted seasonal access, to focus mainly on the species of beetles likely to be associated with the old oaks, these being the main habitat of interest.

13.2. Records, determinations, contributions (adapted from Pollard et al. 2022)

The HPBS contributed around 4,000 records, of which more than 96% were fully identified to species (Figure 13.1). A number of voucher specimens were too degraded to permit satisfactory

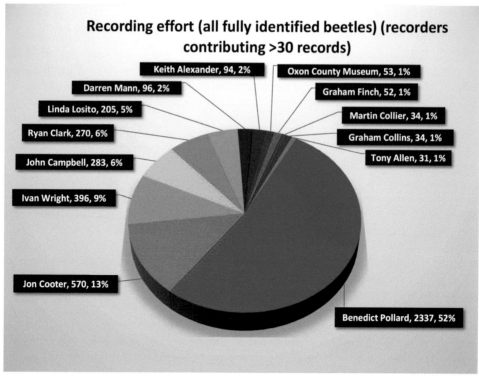

Recording effort (all fully identified beetles) (recorders contributing >30 records)

Keith Alexander, 94, 2%
Oxon County Museum, 53, 1%
Darren Mann, 96, 2%
Graham Finch, 52, 1%
Linda Losito, 205, 5%
Martin Collier, 34, 1%
Ryan Clark, 270, 6%
Graham Collins, 34, 1%
John Campbell, 283, 6%
Tony Allen, 31, 1%
Ivan Wright, 396, 9%
Jon Cooter, 570, 13%
Benedict Pollard, 2337, 52%

Figure 13.1.   Details of records made in High Park, Blenheim.

identification, and some species require male specimens for a confident identification. This is usually because diagnostic characters rely on specific features of the male genitalia, only visible through microscopic dissection. For instance, a female specimen of what was thought by Steve Lane to be the scarce saproxylic mould beetle *Corticaria alleni* (Latridiidae) was found under bark in the timber yard but cannot be verified with any certainty, and so is excluded from the species totals. Benedict Pollard (BJP) generated about 2,400 records in total, of which he determined about 20%, with 1,760 of the remainder being identified by SAL, including all the dissections. Darren Mann (DJM), Jon Cooter (JC), Jonty Denton and Ivan Wright (IW) identified about 160 of BJP's specimens. JC contributed about 600 records, of which SAL identified about 30%, including some of the trickier Staphylinidae: Aleocharinae, among other groups. Roger Booth identified some of JC's Aleocharinae, Pselaphinae and Scydmaeninae (Staphylinidae). IW contributed almost 400 records, all of which he determined, including over 70% of the records from the first two years of the survey (2017–18). Significant numbers of determined records were added by Linda Losito (205) and DJM (96). A British Entomological and Natural History Society field day was held on 15 June 2019, which resulted in 171 records by ten recorders.

13.2.1. Recording methodologies – Active sampling
Sieving
Sieving is a broad approach that can be used on a range of substrates to dislodge beetles from their 'perch' and collect them in a tray, whereby beetle movement or glint of colour can be more easily spotted, and specimens pootered up for later examination.

A variety of materials can be placed in a sieve and shaken vigorously to guide beetles into a catching tray below. These include moss, straw, grass tussocks (Figure 13.2), fungi, bark, leaf litter, twigs, bird, vole or squirrel nests/dreys, red rotten decaying wood, frass, bones and carcasses (Figure 13.3).

'Tussocking' between December and March revealed a wide variety of beetles that overwinter low down among the tightly packed culms of grasses or sedges. A large clump of ungrazed grass or sedge is removed from the ground just below ground level, using a pruning saw. This is

Figure 13.2.   Winter tussock sieving of Tufted Hairgrass (*Deschampsia caespitosa*). (Photograph by Benedict Pollard)

then teased apart into smaller clumps or groups of culms, and sieved. This is a productive method during a time of year when beetles are hibernating or generally hard to find, yielding different groups of species from what may otherwise be recorded. Some areas of grassland close to the Lake at Blenheim have extensive areas of tussocks, which host a suite of species that tend to inhabit the water edge or strandline during the warmer months, but retreat 'inland' once the cold weather arrives. For instance, in just a single visit on 20 February 2020 a mere five tussocks yielded 18 species, of which 12 had not been found in High Park before, including four species of the rove beetle genus *Stenus* (Staphylinidae).

High-water levels during flood events bring flotsam downstream along the River Glyme, which can carry a high diversity of 'clinging' beetles, teased out of their usual waterside dwelling-places. Over long timespans this will have likely increased the species diversity of the Park, as immigrating species will inevitably colonize the Lake edge and/or establish breeding populations inland. Easton (1947) reports on collecting 272 species of beetle from just a single bucket load of flood refuse on the banks of the River Thames (Isis) at Shillingford, Oxfordshire, on 1 December 1946. The species and population dynamics of the High Park lake edge habitat merit further study, especially sampling of flood refuse, which was a technique we did not apply during our surveys, despite evidence of flood events, notably the very high Lake levels during the 2020/1 winter. A single visit and sampling of vegetation along the Lake edge and by the broken jetty on 21 February 2021 yielded 26 species, of which 12 were new to the site list, suggesting this is a habitat that remains under-sampled.

Animal dung, carcasses and pellets

A variety of species feed on tissues and bones of animal cadavers. Chief among these are the Carrion Beetles (Silphidae) and many of the larger Rove Beetles (Staphylinidae), which usually predate fly larvae. Some other specialist species occur in numbers in decaying flesh including, for example, *Dermestes murinus* (Dermestidae), known from High Park from 11 individuals recorded by Linda Losito on 22 May 2019 from a dead pheasant carcass, and four specimens from 8 April 2020 on an earlier visit to the same Muntjac carcass shown in Figure 13.3. Other beetles can be found on old bones and drying animal cadavers, including some Clown beetles (Histeridae), such as *Saprinus semistriatus*, and three brightly coloured metallic blue or bluish-black species of *Necrobia* (Cleridae), predacious on the invertebrate fauna found on animal carcasses. Although these clerids were not observed in High Park, they could reasonably be expected to occur.

Figure 13.3.   Muntjac cadaver (8 Apr 2020) (left), DJM sieving the same carcass (19 July 2020) (right). (Photographs by Benedict Pollard)

In Figure 13.3, note the shrunken nature of the deer's skin, showing signs of dehydration, and the absence of flesh on the skull, suggesting an advanced state of decay. Sieving dead animals over a longer time will reveal different species of beetle, as the decomposition process advances, with some favouring very dry bones that have been in situ for many months. An additional three species new to the site were recorded from just this single animal. A variety of methods may be used with cadavers, such as pitfall traps, to thoroughly sample the chrono-ecological progression of beetles. That was beyond the scope of our study but would likely add additional species to the site list, if utilized in future.

Extensive efforts were made by DJM and LL, and to a lesser extent BJP, to sample animal dung. This included the excrement (crottie, scat, dung) of Roe and Muntjac Deer, pheasant, Badger and more extensively the introduced English White Cattle. The site species lists for the Scarabaeidae, Histeridae and Hydrophilidae in particular were greatly increased as a result. Bird pellets were also inspected by LL, adding the pill beetle *Byrrhus pilula* (Byrrhidae) to the site list.

Sweeping, beating, tapping
A sweep net is usually used to sample long grass or taller herbaceous vegetation and flowers. Repeated sweeping actions, to and fro in a fluid action, result in a range of insects being captured inside a cloth net for inspection. Beating usually involves a strong stick or wooden implement being used to beat or 'tap' (a gentler effort) branches or other plant parts, whereby the sudden jolt of energy dislodges insects down onto a white cloth beating 'tray' below. Beetles can then be sampled by way of a pooter before they take flight, especially requiring agility to do so before the more 'flighty' species take off. Many saproxylic beetles feed on flowers as adults, thus trees and shrubs such as Hawthorn, Blackthorn, Rowan and Elder are a particu-larly valuable source of nectar and pollen in High Park, where the overall floral diversity is relatively low (Chapter 4). Beating these species in flower was particularly productive, adding a range of beetles to the site inventory.

Under bark
A wide variety of beetles will live part of their lifecycle in dead wood and can often be found underneath bark. Tell-tale signs of colonization include exit holes (e.g. Figure 13.4) and the accumulation of frass and wood mould ejected from burrows. Sampling under loose bark

can be as simple as lifting up the bark with a claw hammer and inspecting the substrate below. Many species are often diminutive, sometimes only around 1mm in length, and can easily hide in small cracks and crevices, so breaking the bark up into smaller pieces and sieving it can tease out some of these harder to spot species. A small timber pile to the east of the Palace Vista yielded several species, including the first Oxfordshire record of *Uleiota planatus* (Silvanidae), which has since proved itself to be very well established in High Park, often found under the bark of various tree species.

Figure 13.4.   An 'exit hole' whereby an adult beetle has emerged from its pupa and eaten its way out from under the bark, leaving the tell-tale hole and 'sawdust' below. (Photograph by Benedict Pollard)

Rot-holes and tree fissures/gaps

Old trees can display a range of different types of cavities, scars, fissures or rot-holes, usually above ground (Figure 13.5), but sometimes even at ground level among the roots. Leaf litter or bird nest material tends to accumulate in these cavities over time, and a range of beetles will inhabit them, depending on the type of hole and the substrates present. The larvae of the only saproxylic member of the Scirtidae (Marsh Beetles), *Prionocyphon serricornis*, for example, often develop in rot holes at ground level that retain water during spring. A number of scarce or rare species were found in aerial tree fissures, including *Nemadus colonoides* (Leiodidae), *Batrisodes venustus* and *Stenichnus godarti* (Staphylinidae).

Oak frass in hollow trees

Linda Losito made a concerted effort to sample the frass and debris from inside hollow trees, often oak, but also ash and lime, with over 70% of her records being made this way. This would often yield dead

Figure 13.5.   DJM sieving red-rot from the inside of an ancient oak. (Photograph by Benedict Pollard)

beetles, sometimes as complete specimens but often only as partial remains, such as single elytra. Occasionally live beetles were recorded, including the only site record of *Leptinus testaceus* (Leiodidae) found in oak frass around a bee nest. This technique added some very important saproxylic records to the species checklist, including the only records of the Nationally Rare (NR) RDB1 species *Eucnemis capucinus* (Eucnemidae), the Nationally Scarce (NS) false darkling beetle *Anisoxya fuscula* (Melandryidae) and the rarely seen little anobiid beetle, *Anitys rubens* (Ptinidae).

Recording after dark

Artificial light at night emphasizes colours, especially reflective surfaces such as pronota or elytra, in a way that daylight tends not to do. As such, BJP conducted many night-time searches, notably in the timber yard, where species that dwell inside the logs or under the bark often emerge. Other methods included visiting sap runs, where beetles are more active at night. Spiders' cobwebs were also a fruitful repository of beetle fragments, including the first site record of *Colydium elongatum* (Zopheridae) (Figure 13.6), a remarkable looking predator of *Platypus cylindrus* (Platypodidae), and the only record of the longhorn beetle *Pyrrhidium sanguineum* (Cerambycidae). At night, a variety of species become active, especially in warmer temperatures. These include flying species, which can be attracted to

Figure 13.6.   *Colydium elongatum* (Zopheridae). (Photograph by Benedict Pollard)

light. A total of 30 species were recorded as bycatch during moth-trapping sessions, of which ten were not found by any other method, for which Martin Corley is thanked. Other species will tend to emerge from their daytime locations and can be found by inspecting tree trunks with a head torch. Species regularly seen include the colourful saproxylic Clerids *Thanasimus formicarius* and *Opilo mollis*, and the longhorn beetle *Phymatodes testaceus*. A highlight of the night-time surveying was the first record of the NR lined flat bark beetle, *Laemophloeus monilis* (Laemophloeidae) which was pootered off the end of a cut log in the timber yard.

13.2.2. Recording methodologies – Passive sampling

Passive traps

A variety of passive traps were installed by IRW, BJP and JC, being serviced on a regular basis, typically weekly, fortnightly or monthly. These include five different types of aerial traps: bottle traps, vane traps, slam traps, pheromone lures and flight interception traps. BJP also installed a single subterranean trap, in 2019 and 2021, and several trunk pitfall traps. DJM introduced dung-baited pitfall traps (Figure 13.13), which BJP also employed. IRW also utilized water traps, which typically consisted of bright yellow plastic bowls, primarily used as attractants for his survey of the Hymenoptera, but to which some beetles were also drawn.

Bottle traps

IRW installed a series of bottle traps, made of four upside-down clear plastic bottles attached to, and suspended, from a plastic tray. Each bottle has a cut-away section that insects fly into before dropping down into the killing and preserving fluid (Figure 13.7), which can be emptied for sampling by unscrewing the bottle cap and sieving the contents. These yielded many significant saproxylic records, including *Ampedus cardinalis*, *A. elongatulus* and *Procraerus*

Figure 13.7.   One of IRW's bottle traps installed inside a cut fallen oak trunk. (Photograph by Benedict Pollard)

tibialis (Elateridae), *Malthodes crassicornis* (Cantharidae), *Globicornis nigripes* and *Trinodes hirtus* (Dermestidae).

Vane traps

IRW installed two vane traps at a time, high up in the canopies of ancient oaks, positioned in different locations in different years (Figure 13.8). In late May 2021 BJP installed 11 Perspex cross-shaped vane traps (modified slightly from that described by Piper and Allen 2020). These traps were baited with synthesized pheromones of either *Elater ferrugineus* (Elateridae) or *Gnorimus nobilis* (Scarabaeidae), supplied by Dr Deborah Harvey of Royal Holloway College, University of London. Harvey et al. (2017, 2018) presented results of their use elsewhere in the UK. These traps proved to be the most productive method of sampling, in terms of numbers of specimens added. Of the 3,815 records for which a detailed method was recorded, almost 21% (796) were from vane traps (Figure 13.21 under 13.4.2 Species totals).

In September 2021 one of these traps was moved and installed on an ancient oak with a cavity close to ground level, to sample species that are known to keep low to the ground (Figure 13.9), such as *Trichonyx sulcicollis* (Staphylinidae) (KNAA, pers. comm. September 2021). It was unproductive, given the lateness of the season, but the approach would merit future use, particularly during late spring through summer.

Slam traps

BJP installed three slam traps (Figure 13.10); a taut material is most effective for animal capture. Slam traps contain two collecting bottles, one below and one above, to account for species that 'play dead' upon hitting the material and drop down, and those that cling on and climb up, respectively. These contributed many important records, including for instance the only specimen of *Scydmoraphes helvolus* (Staphylinidae).

Figure 13.8.   A vane trap low down on an ancient Ash pollard. (Photograph by Benedict Pollard)

Figure 13.9.   Vane trap installed by hollow close to the ground on ancient oak. (Photograph by Benedict Pollard)

Figure 13.10.   Slam trap, suspended on dead *Quercus robur*, partly covering exit hole, in the southern part of High Park. (Photograph by Benedict Pollard)

Pheromone lures

BJP installed 13 pheromone lures (Figure 13.11) in 2019, using pheromones of either *Gnorimus nobilis* or *Elater ferrugineus*, supplied by Dr Deborah Harvey. Neither of the target species was

Figure 13.11.   Emily Hobson, Saira Gregory and Anthony Cheke at ancient oak No. 136563 during the mini 'bioblitz' in July 2021. A pheromone lure is suspended below the vane trap. (Photograph by Benedict Pollard)

recorded with these lures, but a range of other scarce species were recorded, some for the first time, including *Microrhagus pygmaeus* (Eucnemidae) and *Tomoxia bucephala* (Mordellidae). Both were later recorded elsewhere by different methods. It is not known whether the pheromones were acting as attractants or if the beetles were merely interrupted randomly in their flight.

Flight Interception Traps
JC installed two flight interception traps (Figure 13.12) in May 2021, moving them to new locations in July 2021. These were very productive, leading to a large number of specimens and many species newly recorded for the Park. Owen (1992) gives a good general description of these traps.

Dung-baited pitfall traps
DJM installed a series of dung-baited pitfall traps, which BJP also adopted. These consisted of a plastic container dug down into the ground with the rim at ground level, and a coarse grade plastic mesh covering it, supporting a ball of damp cow dung, about 5cm in diameter. A plastic cover was inserted on three pins to keep rain out (Figure 13.13). A preservative of monopropylene glycol was added to the container. These traps added a significant number of species to the site list, predominantly Scarabaeidae, Staphylinidae and Carabidae. An occasional saproxylic species was also recorded in this manner.

Trunk Pitfall Traps
BJP installed a few trunk pitfall traps (Figure 13.14) broadly following the methodology of Luff and Towns (2018). They were mostly installed on a cluster of veteran oaks centred around a

Figure 13.12. Flight interception trap – original location, near Sycamore and elm, and a notice in the 'timber graveyard' in north High Park. A slam trap can be seen in the distance on an old dead oak. (Photographs by Benedict Pollard)

Figure 13.14. A simple aerial trunk pitfall trap, consisting of a small yoghurt pot, lodged in a natural gap on an ancient oak, filled with a vinegar, sherry and alcohol preservative mix. (Photograph by Benedict Pollard)

Figure 13.13. A dung-baited pitfall trap, set with moist cow dung collected nearby and rolled into a ball. (Photograph by Benedict Pollard)

Figure 13.15.   A subterranean trap, with its assembly shown clockwise from top left. (Photographs by Benedict Pollard)

single very large ancient oak in southern High Park. This method yielded some rare and scarce species, including single specimens of *Ampedus cardinalis* (Elateridae) and *Ptinus subpilosus* (Ptinidae).

Subterranean pitfall traps
Subterranean pitfall traps (Figure 13.15) are designed to catch invertebrates below ground, up to about a foot in depth, near to decaying roots of veteran or ancient trees, where a rarely observed beetle fauna exists (Owen 2000). The single trap installed in High Park was based on a design modified by Telfer from that of Owen (1995). Some species were added to the site list solely from this single trap, for example three species of *Cryptophagus*: *C. distinguendus*, *C. punctipennis* and *C. scutellatus* (Cryptophagidae).

13.3. Habitat descriptions in High Park

Dead and decaying wood
The most significant habitat in High Park for the scarcer beetles is the oak trees, especially the dead and decaying wood, both within standing living and dead trees, or fallen dead whole trees or parts thereof. There are a good number of very large Beech trees, mostly forming part of the shelter belt along the south-western and southern slope of High Park. These were planted in the late eighteenth and the nineteenth centuries (Chapter 3) and are a very significant habitat for a range of Beech-specific saproxylic beetle species. A number of dead Beech stumps occur, including one (Figure 13.16) that was sampled using a vane trap lowered

Figure 13.16.   Standing deadwood stump of a veteran Beech tree beside the Palace Vista. (Photograph by Benedict Pollard)

down within the central hollow. It yielded many specimens of the Lesser Stag Beetle (*Dorcus paralellepipedus*) (Figure 13.22a), one of only four specimens found of the NR *Stenichnus godarti* and the second of two specimens of the NS staphylinid *Scydmaenus rufus*, the first having been found under Beech bark on a log in the timber yard.

Occasionally beetles were also found under the bark of other tree species such as fallen willow, Ash, Aspen and Sycamore, most notably in the timber yard (see below). A smaller subset of species is specifically conifer-associated, notably with pine, of which there are a good number of Corsican Pine and Scots Pine in plantations in High Park.

Fungi-associated species
A small number of beetle families are known to associate more closely with fungi, which include the silken fungus beetles (Cryptophagidae), hairy fungus beetles (Mycetophagidae), pleasing fungus beetles (Erotylidae), polypore fungus beetles (Tetratomidae), and ciid beetles (Ciidae). In addition, the Sphindidae (cryptic slime-mould beetles) are obligately dependent on

Figure 13.17.   Anthribid weevil, *Platyrhinus resinosus* on *Daldinia concentrica* fruiting body. (Photograph by Benedict Pollard)

myxomycetes (slime moulds). Many aleocharine rove beetles and several histerids can also be found in fungal fruiting bodies, being predacious on invertebrate mycophages.

Some species from other families, such as the scarce weevil *Platyrhinus resinosus* (Anthribidae), are also mycophagous (Figure 13.17). Eggs are deposited in summer onto a suitable fungus, usually *Daldinia concentrica* (Ascomycota: Xylariaceae), known colloquially as King Alfred's Cakes. Several larvae develop in the fruiting body, bore into the wood and after as much as two years pupate, the adults emerging in spring or summer.

From time to time, mature fungal fruiting bodies were inspected visually or hand-picked from the ground or from logs, tree branches or trunks, then broken into pieces and sieved or extracted using a Berlese Funnel. In the genus *Cis* (Ciidae), by way of example, 59 individuals were recorded, representing nine species. Of these, *C. bilamellatus* and *C. boleti* were relatively common, but the other six species were found at most on three occasions: *C. castaneus* (3), *C. fagi* (1), *C. festivus* (1), *C. micans* (1), *C. pygmaeus* (2), *C. vestitus* (1) and *C. villosulus* (2). Four of these were found by inspecting fungal fruiting bodies, and two species (*C. festivus*, *C. micans*) were only found this way. Given that this method was only used on an ad hoc basis and given the restricted site access during more than one autumn, it is reasonable to suggest that further targeted sampling of fungi for beetles would yield additional species records for the site.

Timber Yard

In the southernmost part of the Park, adjacent to the perimeter wall, lies an area of about 0.2ha that is used as a timber yard (Figure 13.18). This yard has been in use since at least 1980 (Nick Baimbridge, pers. comm.) and receives wood from all parts of the Estate. Logs are left *in situ* for up to seven years but are generally used within two or three years, being cut into firewood logs in winter as and when required for use on the Blenheim Estate. Brash is piled

Figure 13.18.   The author in the timber yard. *Laemophloeus monilis* (Laemophloeidae) was recorded under bark in the stacked logs in the top left of the picture (Pollard 2021). (Photograph by Darren Mann)

Figure 13.19.   *Platypus cylindrus* burrowing into a log in the timber yard (left). Female *Lymexylon navale* ovipositing in crack (c. 0.4mm diameter) in oak log (right). (Photographs by Benedict Pollard)

up for bonfires, which are lit intermittently. The logs range from about 20cm to over 100cm in diameter, comprising a multitude of species, including Ash, Beech, birch, pine, poplar, Sycamore and Pedunculate Oak, of which Beech was particularly common in the larger sizes.

Beetle sampling of the timber yard was conducted in 2020 and 2021. It is debatable whether the yard species data should be included in the overall High Park totals or not, as there is an argument that logs brought from other parts of the estate could be importing beetle species not truly found in High Park. The freshly cut logs, however, were observed to generally display intact bark, minimal signs of decay and rarely hollowing, and little evidence of prior habitation by beetles and fungi. From a beetle's perspective, this large volume of dead wood of varying age represents an ideal habitat for colonization and reproduction. Indeed, saproxylic species such as *Platypus cylindrus* (Platypodidae) were observed alighting on newly deposited timber to burrow into the wood before egg-laying. One of Britain's rare saproxylic species, *Lymexylon navale* (Lymexylidae), was noted ovipositing directly into minute cracks of freshly introduced oak timber (Figure 13.19).

Other habitat associations
While the main focus of the beetle surveys was oak-related, many other habitats and ecological niches are present in High Park. The floral diversity is relatively depauperate (Chapter 4), but a range of beetles associated with flowers have been recorded, through active and passive sampling. The ground level fauna was also sampled using pitfall traps, notably by IW, which added a good number of species to the site list. Leaf litter also harbours a wide range of species, and was sampled on occasion, by sieving and extraction using a Berlese Funnel.

Specialist species associated with woody climbers (Hedera *and* Clematis)
Ochina ptinoides (Ptinidae) is a small anobiid beetle known exclusively on dead branches of Ivy (*Hedera helix*). During the many visits made by the author to High Park, Ivy was only seen at two locations, one on the tree next to the Combe Gate and one on a small tree next to High Lodge; both plants were inhabited by *O. ptinoides*. Likewise, there are two species obligate on the climber Old Man's Beard (*Clematis vitalba*), one of which, *Xylocleptes bispinus* (Curculionidae) was recorded in February 2021. Unfortunately, it was observed that much of the *Clematis* and one of the ivies had been cut or removed by the end of 2021.

Ponds and puddles
There are a few seasonal ponds in High Park that were sampled on occasion, revealing a range of species not otherwise recorded. These included a variety of ground beetles, and members of the Helophoridae and Heteroceridae, which were also found in long-lasting puddles. The ponds were sampled by hand, looking for movement of beetles along the damp muddy edges of the wet zone, and under twigs, branches, stones and clumps of drying mud.

Puddles were typically sampled by tapping the surface of the water, causing ripples but without stirring up sediment, thus retaining good visibility through the water. This tends to encourage movement of otherwise static beetles that can be caught by hand or in a pot or vial.

Lake edge
Although the lakeshore could arguably not form part of High Park, there is little doubt that it has relevance. Occasional sampling of the lakeside vegetation yielded the first and only site records of a range of species such as the Spotted Marsh Ladybird, *Coccidula scutellata*, the Water Ladybird, *Anisosticta novemdecimpunctata* (Coccinellidae) and a host of rove beetles (see Section 13.2.1). Easton (1947) showed how important flood events can also be on rivers in terms of the dynamic influx and efflux of beetle species from a given site, riding on floating vegetation. The Great Lake bordering High Park can act as a repository for these sorts of flood events.

Ant-associated species
A suite of unusual beetles is linked ecologically to ants (Donisthorpe 1927). Targeted sampling of ants' nests or trees showing obvious signs of ant inhabitation was not implemented. However, some species associated with the nests of ants in decaying trees were recorded, such as *Batrisodes venustus* (Staphylinidae), which is known to occur on occasion with *Lasius brunneus* (Formicidae). At High Park it was recorded once in 2018 by sieving ground leaf litter and in 2021 from leaf litter in a vertical slit about 4m above ground on an old oak. There is therefore an opportunity for further specific research on ant-associated beetles, especially as a number of trees showed signs of ant inhabitation by *L. brunneus* (Figure 10.7). Jet Ant, *Lasius fuliginosus* has a rich associated beetle ant fauna, particularly among the aleocharine rove beetles such as the genera *Pella*, *Thiasophila*, *Haploglossa* and *Oxypoda*. Targeted sampling of ants' nests, both arboreal and terrestrial, merits future recording effort, and would likely reveal species of beetle not yet encountered during biological recording in High Park.

Nest-associated species
The nests of animals usually attract a range of beetles, which vary according to the type of nest, be they birds, voles, mice, moles, rabbits, Badgers, squirrels, for example. An old squirrel drey in a fallen Beech tree was sampled by DJM, yielding 34 specimens of *Dendrophilus punctatus* (Histeridae), a species not otherwise found during the survey. This type of specific sampling shows that some species have very narrow preferences, and to find them requires targeted recording methods. On a few occasions, birds' nests were sampled and yielded species that were not otherwise found. More extensive searching of mole and rabbit burrows and Badger setts would likely also result in leiodids (Leiodidae) such as *Catops*, and a range of aleocharine and other rove beetle species not previously found in High Park.

13.4. Results and analysis

13.4.1. Analysis of recording localities in High Park
It is curious that in entomological recording in the UK there appears to be a historical trend of creating site lists of species without any further detail on intra-site location. Many museum records and antiquarian collections only have reference to the nearest town, Woodstock, for example. The author, being originally a tropical botanist and conservationist by training, utilized a GPS for most of his records, usually with a ten-digit grid reference, to provide the most granular location possible (usually to within about 6m). This method was also followed by JC. Other recorders generally used only four- or six-digit grid references, and some contributors prior to the HPBS only offered 'Blenheim Park'.

   A spatial analysis of 2,735 of BJP's and JC's dataset records (almost 60% of the site list) gives us some idea of the geographical coverage of the surveying (Figure 13.20, top), even if it does not

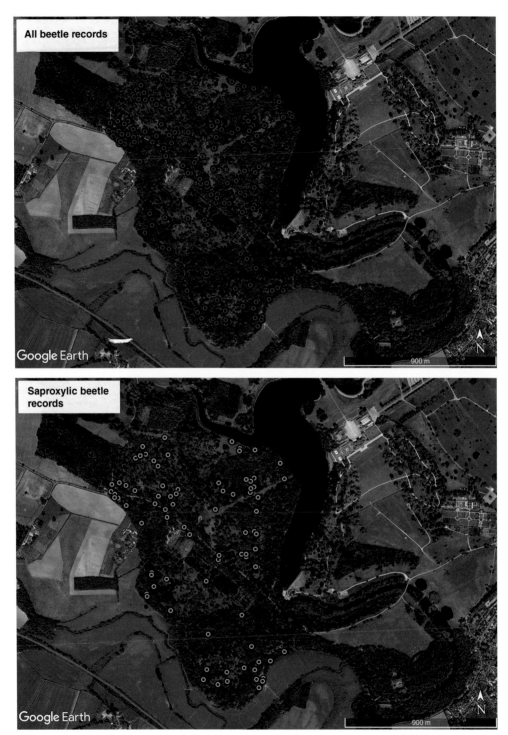

Figure 13.20.   Google maps of High Park showing sample locations of 2,735 records of beetles (top, in red) and a subset of these: records of saproxylic species (bottom, in yellow). (Graphics by T.H. Moller)

give the complete picture. It reveals a relatively even and thorough coverage of the Park, with noticeable gaps in the more modern plantations, with only a few incursions into the conifer areas. One such effort did, however, yield a record of the NR *Globicornis nigripes* (formerly *G. rufitarsis*) (Dermestidae) found by beating hawthorn blossom in the understorey of a larch plantation. There appears to also be a very noticeable gap in the coverage of the eastern section of woodland along the Lake edge and an area to the west of High Lodge. The second map in Figure 13.20 shows a subset of the same dataset, showing only the locations of saproxylic species records.

13.4.2. Species totals

The total number of species recorded prior to the HPBS was 313, largely based on collections made by John Campbell, KNAA and D. Copestake, as well as a modest number of records derived from the Oxfordshire County Museum collections. The species total now stands at 932, representing an increase of about 200%. Figure 13.21 shows the breakdown of species discoveries by methodology and/or locus. Increased recorder effort, especially in 2021, by BJP and JC, led to the discovery of many more species than in previous years. This was mainly due to the implementation of three previously untried methods: pheromone-baited vane traps (Harvey et al. 2017, 2018), slam traps and flight interception traps.

   Table 13.1 gives a breakdown of species numbers per family, and Table 13.2 lists the most significant genera in terms of species diversity. Unsurprisingly, the largest numbers of species recorded in High Park for a single family (241, or 26%, of our total) are the Rove Beetles (Staphylinidae), which is almost precisely in keeping with the percentage of the UK fauna represented by that family (c. 25%, see Duff 2018). The presence of specialists in the fieldwork team including JC (Leiodidae) and DJM and Linda Losito (Scarabaeidae) increased the numbers of species in those families considerably. Representatives of 70 beetle families have to date been found in High Park.

13.4.3. National context

Although 932 species might be considered as a high number of species for a site, representing almost 25% of the UK beetle fauna, other sites have reported much higher figures, largely on account of their higher levels of recording effort over many decades. One example

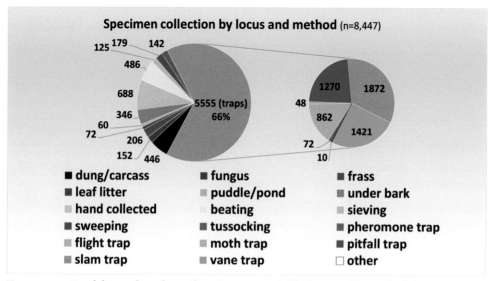

Figure 13.21. Breakdown of numbers of specimens recorded by locus and/or methodology, presented using slightly different criteria. (Graphic by T.H. Moller)

Table 13.1.   Numbers of species (Spp.) recorded in High Park, per family.

Family	Spp.	Family	Spp.	Family	Spp.
Staphylinidae	241	Buprestidae	7	Endomychidae	2
Carabidae	80	Tenebrionidae	7	Geotrupidae	2
Curculionidae	78	Dermestidae	6	Hydraenidae	2
Chrysomelidae	62	Helophoridae	6	Kateretidae	2
Scarabaeidae	27	Scirtidae	6	Lucanidae	2
Cryptophagidae	25	Throscidae	6	Pyrochroidae	2
Nitidulidae	25	Eucnemidae	5	Sphindidae	2
Elateridae	24	Mordellidae	5	Tetratomidae	2
Cantharidae	23	Oedemeridae	5	Aderidae	1
Leiodidae	23	Zopheridae	5	Byrrhidae	1
Cerambycidae	20	Corylophidae	4	Cucujidae	1
Coccinellidae	18	Dytiscidae	4	Dasytidae	1
Hydrophilidae	18	Erotylidae	4	Haliplidae	1
Latridiidae	18	Laemophloeidae	4	Heteroceridae	1
Ptinidae	17	Phalacridae	4	Lampyridae	1
Ciidae	13	Silvanidae	4	Lycidae	1
Histeridae	13	Anthribidae	3	Lymexylidae	1
Melandryidae	12	Cleridae	3	Melyridae	1
Scraptiidae	12	Salpingidae	3	Noteridae	1
Mycetophagidae	11	Anthicidae	2	Rhynchitidae	1
Ptiliidae	11	Biphyllidae	2	Trogidae	1
Apionidae	9	Byturidae	2	Trogossitidae	1
Monotomidae	8	Clambidae	2		
Silphidae	8	Cerylonidae	2		

is Bookham Common, in Surrey, where the species list stood at over 1,600 in 2013 (Natural History Museum 2013) and is likely to be already higher than this. Windsor Great Park and Forest report a list of almost 2,000 species (Windsor Great Park website 2023), and 2,600 were recorded in the New Forest (Brock 2011), albeit the latter having a land area more than 400 times larger than High Park. Moccas Park is more similar to High Park, with 960 species listed, despite a longer lasting recording effort there (see Introduction).

13.4.4. The Oxfordshire context

Species counts for individual Oxfordshire sites (in Watsonian Vice-County 23) are few and far between (Jonty Denton, pers. comm. March 2023). Wright and Wright (2017: 128) state that Shotover Hill harbours 1,081 beetle species. Without doubt a high number of species are also present at Wytham Woods, the well-studied scientific site owned by Oxford University lying just to the west of Oxford, in modern Oxfordshire, although for the purposes of biological recording it remains in the Watsonian Vice County of Berkshire (Vice-County 22). High Park is likely to already be among the top three sites in Oxfordshire in terms of overall numbers of species, despite having a much lower level of historic recorder effort than Wytham and Shotover. Recording was also initiated in 2022 in other nearby pasture woodland locations,

Table 13.2.   The most species-rich genera recorded in High Park.

Genus	Spp.	Genus	Spp.
Atheta sens. lat. (Staphylinidae)	25	*Rhizophagus* (Monotomidae)	7
Quedius (Staphylinidae)	16	*Anotylus* (Staphylinidae)	7
Stenus (Staphylinidae)	15	*Agrilus* (Buprestidae)	6
Bembidion (Carabidae)	14	*Helophorus* (Helophoridae)	6
Longitarsus (Chrysomelidae)	12	*Trixagus* (Throscidae)	6
Anaspis (Scraptiidae)	11	*Euplectus* (Staphylinidae)	6
Philonthus (Staphylinidae)	10	*Malthinus* (Cantharidae)	5
Cis (Ciidae)	9	*Malthodes* (Cantharidae)	5
Cercyon (Hydrophilidae)	9	*Amara* (Carabidae)	5
Atomaria (Cryptophagidae)	9	*Ceutorhynchus* (Curculionidae)	5
Meligethes (Nitidulidae)	9	*Phyllobius* (Curculionidae)	5
Aleochara (Staphylinidae)	9	*Agriotes* (Elateridae)	5
Cryptophagus (Cryptophagidae)	9	*Mycetophagus* (Mycetophagidae)	5
Pterostichus (Carabidae)	8	*Carpelimus* (Staphylinidae)	5
Epuraea (Nitidulidae)	8	*Gyrophaena* (Staphylinidae)	5
Tachyporus (Staphylinidae)	8	*Oxypoda* (Staphylinidae)	5
Cantharis (Cantharidae)	7	*Tachinus* (Staphylinidae)	5
Phyllotreta (Chrysomelidae)	7		

such as Kirtlington Park and Eynsham Park, and it is hoped that Cornbury Park can also be accessed to ascertain the species linkages in these ancient habitats.

Species newly recorded for Oxfordshire
Clark (2017) reported two species found for the first time in the Watsonian vice-county of Oxfordshire (VC 23) from Blenheim: *Dropephylla gracilicornis* (Staphylinidae) and *Ischnodes sanguinicollis* (Elateridae) (Figure 13.26). Pollard (2021) later added two more species to the county list: *Elater ferrugineus* (Elateridae) and *Laemophloeus monilis* (Laemophloeidae), from High Park. Pollard et al. (2022) added a further 45 species to the VC 23 list, all from High Park. Of the 45 HPBS 'vice-county firsts', 32 were also first records for the administrative county of Oxfordshire, which includes part of VC 22 (Berkshire), notably Wytham Woods. Many VC 23 'first modern' records were also listed, having not been seen since the completion of the Victoria County History of Oxfordshire (Aubrook 1939) as well as other records notable in the county context. In Appendix 11, the new VC 23 records are marked with an asterisk, new VC 23 + modern Oxfordshire records with a double asterisk and the 'first modern' records are marked with a triple asterisk.

13.4.5. Analysis of records
The 'original' site list, comprising records prior to John Campbell's work in 1978, stood at approximately 70 species, gleaned from the Oxford County Museum records, the JNCC, English Nature, the 'Wytham Survey' and some records by R.B. Angus, D. Copestake and R.J. Barnett.

Table 13.3 shows the species added to the site list, organized by collector, with a minimum number of 50 records and 10 species. In brackets it shows the number and percentage of

Table 13.3. Analysis of specimen and species totals for recorders with over 50 records. Recorder abbreviations: BJP (Benedict Pollard); JCo (Jon Cooter); IRW (Ivan Wright); JCa (John Campbell); RC (Ryan Clark); LL (Linda Losito); KNAA (Keith Alexander); DJM (Darren Mann); OCM (Oxford County Museum).

Recorder (see codes below)	BJP	JCo	IRW	JCa	RC	LL	KNAA	DJM	OCM
Number of records	2,337	570	396	283	270	205	94	104	53
Number of species recorded	654	273	205	136	108	102	83	56	53
Number and percent (%) of species new to site list	348 (53.2%)	96 (35.2%)	81 (39.5%)	132 (97.1%)	59 (54.6%)	17 (16.7%)	40 (48.2%)	26 (46.4%)	43 (81.1%)
Number & percent of recorder's total species count only ever found once during HPBS	154 (23.5%)	50 (18.3%)	19 (9.3%)	13 (9.6%)	7 (6.5%)	8 (7.8%)	3 (3.6%)	8 (25.0%)	n/a
Percent of recorder's 'new to site list' species only ever found once during HPBS	44.3%	52.1%	23.5%	9.8%	11.9%	47.1%	7.5%	30.8%	n/a

those species that were found only once, and therefore not later recorded by others or by the original recorder.

Earlier recorders inevitably have a greater chance of adding species to the list, and a lesser chance that the species they find will remain known only from a single record. Later recorders have a lower chance of adding species to the list. We can also use these datasets to infer what the likely actual total of species in High Park might be by looking at species accumulation curves over time, and at what point there might be a plateau in the numbers of species added.

The site species list stood at 313 prior to the HPBS, which then added 618 species – representing, more or less, a 200% increase. In total the HPBS found 873 species, so this gives us a figure of 71% of the HPBS species that were new to the site list. There were 56 of the original pre-HPBS species that were not rediscovered, marked in Appendix 11 with a X. When including these in the calculations, we find that roughly two out of every three species recorded by us were newly recorded for High Park. This is a remarkable figure as it suggests that we are a long way from knowing the Park's full beetle diversity, and we suggest that future survey work is merited to further our understanding of the real species tally.

These results are more or less a reflection of time spent in the field, and the number of passive traps that were installed by each recorder, giving a measure of 'recorder effort'.

13.4.6. The saproxylic beetle fauna

Overview

The main focus of the beetle study was to record the Coleoptera associated with veteran and ancient oak trees, and to a lesser degree the beetle fauna associated with other tree species, including Beech, Ash and lime. Deadwood (saproxylic) species are indicators of woodland quality and condition (Chapter 15), which can serve as a useful proxy for the overall health of the habitat.

There are over 55 beetle families in the UK that have one or more saproxylic species. These are a range of different sizes, colours and habits, from the well-known Stag Beetle (*Lucanus cervus*), Britain's largest beetle, the males of which can reach up to 75mm in length, to the tiniest Feather-wing beetles (Ptiliidae) which are often well under just 1mm in length.

To illustrate some of the diversity of forms, a selection of saproxylic beetles known from High Park is presented in Figures 13–22a, b, in taxonomic order, following Duff (2018). Subfamilies are noted where they have been subsumed, having previously been recognized at family rank.

Rare and scarce species – saproxylics

A number of Red Data Book saproxylic species have been recorded, which are listed below in taxonomic order (following Duff 2018). A selection of these is illustrated and discussed in more detail. The species names are followed by codes that detail their National Red Data Book Status following Hyman (1992, 1994): RDB1, RDB2, RDB3 or RDBK, RDBI, followed by two numerical codes: the Saproxylic Quality Score (SQS: 32, 24 or 16) and the Index of Ecological Continuity score (IEC: 3, 2 or 1 (or 0)), its IUCN status (if evaluated) and its British Nature Conservation Status. For instance, *Aeletes atomarius* is a Red Data Book 3 species (RDB3), with an SQS of 16 and an IEC score of 3 and an IUCN status of Least Concern (LC). The analysis of saproxylic beetle scores in High Park is presented and discussed in Chapter 15.

Histeridae

Aeletes atomarius RDB3 16 3 LC – this species was christened 'The Invisible Clown' (Lane et al. 2020) on account of it being the smallest member of the UK's Clown Beetles (Histeridae). Known from High Park by just a single record in the 'English Nature' dataset, 1 May 1964, BJP then recorded it on three further occasions in vane traps, one each in June, July and August 2021, interestingly on Ash, Beech and Pedunculate Oak respectively.

Staphylinidae

Stenichnus godarti RDB3 24 2 – Four records of this species were made, from May to July 2021. The first was from inside a vertical slit on an ancient oak, around 4m above ground level, accessed by ladder. A large bagful of leaf litter was harvested and then extracted using a Berlese Funnel. The second specimen resulted from the author laying a white sheet inside the largest girthed oak in High Park and using an extendable modern 'feather' duster to brush off the sides of the internally decaying wood. The other two specimens were found in a flight interception trap and a vane trap inside the old dead Beech stump of Figure 13.16.

Scydmaenus rufus RDB2 24 1 (proposed to be reduced to 0) – Denton (1999) posed the question as to whether *S. rufus* should still be considered to be scarce. There are some species formerly considered to be rare or scarce that appear to be extending their geographical range. *S. rufus* is often found in manure heaps and other decaying vegetation, and its range has certainly expanded in the last decade or two (Steve Lane, pers. comm. 2023). Alexander (in prep.) proposes to downgrade its IEC score from 1 to 0.

Euplectus tholini RDB3 24 3 – There has been some taxonomic confusion between this species and *Euplectus punctatus*. KNAA (pers. comm. April 2023) indicated that the latter species is now thought to be restricted to the Caledonian pine forests. *Euplectus tholini* is one of the

Abraeus perpusillus (Histeridae) (TDP)

Scaphidium quadrimaculatum (Staphylinidae: Scaphidiinae) (TDP)

Scydmaenus rufus (Staphylinidae: Scydmaeninae) (TDP)

Dorcus paralellepipedus (Lucanidae) on log (BJP)

Sinodendron cylindricum (Lucanidae) on old oak (BJP)

Agrilus sinuatus (Buprestidae) (TDP)

Platycis minutus (Lycidae) on log (BJP)

Ptinomorphus imperialis (Ptinidae) (TDP)

Thanasimus formicarius (Cleridae) (TDP)

Dasytes aeratus (Melyridae: Dasytinae) (TDP)

Triplax aenea (Erotylidae) (TDP)

Rhizophagus dispar (Monotomidae) (TDP)

Biphyllus lunatus (Biphyllidae) (TDP)

Pediacus dermestoides (Cucujidae) under bark (BJP)

Epuraea unicolor (Nitidulidae) (TDP)

Figure 13.22a.   A selection of saproxylic beetles from High Park. (Photographs by Benedict Pollard (BJP) and Trevor and Dilys Pendleton (TDP))

Enicmus testaceus (Latridiidae) (TDP)

A group of *Mycetophagus piceus* (Mycetophagidae) under bark of fallen dried willow branch (BJP)

Phloiotrya vaudoueri (Melandryidae) (TDP)

Mordellochroa abdominalis (Mordellidae) (TDP)

Bitoma crenata (Zopheridae) (TDP)

Prionychus ater (Tenebrionidae) on old oak (BJP)

Ischnomera sanguinicollis (Oedemeridae) (TDP)

Pyrochroa coccinea (Pyrochroidae) (AF)

Poecilium alni (Cerambycidae) (TDP)

Leptura quadrifasciata (Cerambycidae) on bracken (BJP)

Platystomos albinus (Anthribidae) under bark, on small log in timber stack (BJP)

Magdalis memnonia (Curculionidae) (TDP)

A group of *Dryocoetes villosus* (Curculionidae: Scolytinae) creating galleries, under bark in log in timber stack (BJP)

Scolytus intricatus in cop. (Curculionidae: Scolytinae) (TDP)

Anisandrus dispar (Curculionidae: Scolytinae) (TDP)

Figure 13.22b.   More saproxylic beetles from High Park. (Photographs by Benedict Pollard (BJP), Trevor and Dilys Pendleton (TDP) and Aljos Farjon (AF))

Carabus violaceus (Carabidae)
(BJP)

Elaphrus riparius (Carabidae)
(TDP)

Badister bullatus (Carabidae)
(TDP)

Cercyon haemorrhoidalis
(Hydrophilidae) (TDP)

Leptinus testaceus (Leiodidae)
(TDP)

Necrodes littoralis (Silphidae)
(TDP)

Aleochara bipustulata
(Staphylinidae) (TDP)

Stenus juno (Staphylinidae)
(TDP)

Aphodius pedellus (Scarabaeidae)
(TDP)

Lampyris noctiluca (Lampyridae)
(TDP)

Cantharis rustica (Cantharidae)
(TDP)

Dermestes murinus
(Dermestidae) (TDP)

Coccidula scutellata
(Coccinellidae) (TDP)

Gastrophysa viridula
(Chrysomelidae) (TDP)

Cionus hortulanus
(Curculionidae) (TDP)

Figure 13.23.   Some of the non-saproxylic beetles encountered. (Photographs by Benedict Pollard (BJP) and Trevor and Dilys Pendleton (TDP))

few scarce or rare species in our dataset that has still only been found in the timber yard, where three specimens were encountered on 27 February 2021, under the bark of a Beech log after dark with a head torch.

Amarochara bonnairei RDBI 24, *Stichoglossa semirufa* RDBI 24 – These were both recorded by JC in his flight interception traps. Cooter (2022a) provides extensive notes on the historical context of these species. They are both considered to be rare in the UK.

Plectophloeus nitidus RDB2 32 3 – A rare short-winged mould beetle, only 1.3–1.5mm long (Figure 13.24),

Figure 13.24.   *Plectophloeus nitidus.* (Photographs by Christoph Benisch (left) and Lech Borowiec (right))

it favours red-rotten oak wood. It was first recorded on 26 July 1978, with two further records, by IRW on 19 June 2019 in a water trap and by BJP between 8 and 23 July 2021 in a vane trap on a large oak at the transition zone between the timber yard and pasture woodland to the west.

Elateridae

Ampedus cardinalis RDB1 32 3 NT – There is clearly a healthy population of this species at High Park, it having been recorded on 18 occasions. Species of this genus are generally considered hard to identify, and often require very close analysis of the shape of the antennomeres (segments of the antennae) and the punctation on the pronotum (Figure 13.25). A record of what was purported to be *A. quercicola* (Pollard et al. 2022) was redetermined as *A. cardinalis* by Howard Mendel in April 2023, albeit it being a very small specimen indeed and outside the usual size range for this species.

Ischnodes sanguinicollis Notable A 16 2 VU – This species is considered to be threatened with extinction in a global context, being classified as Internationally Vulnerable (VU) (Nieto and Alexander 2010), following the criteria set out by IUCN. Its stronghold is in the UK, and in High Park it has been found on a single occasion, by Ryan Clark, on 4 August 2016 (Figure 13.26).

Figure 13.25.   *Ampedus cardinalis.* (Photograph by Trevor and Dilys Pendleton, Eakring Birds)

Figure 13.26.   *Ischnodes sanguinicollis* (Elateridae). (Photograph by Pavel Krásenský)

Elater ferrugineus RDB1 32 3 NT (Figure 13.27) – A singleton was recorded in the vane trap shown in Figure 13.11, hanging just above the pheromone lure (Pollard 2021). The vane trap was baited with pheromones of *E. ferrugineus* (Harvey 2017). This was the first and only known record for Oxfordshire.

Procraerus tibialis RDB3 16 3 – KNAA recorded a singleton on 1 June 2002, with seven further specimens resulting from aerial traps; four by Ryan Clark in 2016, two by IRW in 2018 and one in 2019, then a singleton was beaten off a fallen dead oak by BJP in 2021. A specimen mistakenly identified as *Ampedus nigrinus* (Pollard et al. 2022) was reidentified by Howard Mendel in April

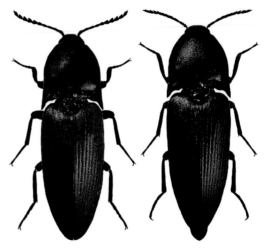

Figure 13.27. *Elater ferrugineus.* (Photograph by Lech Borowiec)

2023 as *P. tibialis*, and a number of specimens known only from fragments found inside rot holes and general litter inside old oaks recorded by LL were also identified by him as *P. tibialis*.

Eucnemidae

Eucnemis capucinus RDB1 32 1 – Hyman (1992) lists this species as Endangered. LL regularly sampled red rotten leaf litter and frass inside old oaks, and extracted fragments of beetles (elytra, heads, pronota) with a view to identifying these species where possible. Her collection of Elateroidea samples were sent to Howard Mendel in April 2023, wherein a single fragment of *E. capucina* was identified, making this a new county record (Pollard, Mendel and Losito 2023) and the only known specimen of this species from High Park.

Hylis cariniceps RDB1 32 3 – Hyman (1992) lists this species as Endangered, then known only from two locations, in Hampshire and Dorset. It has since also been recorded in Somerset (Parsons 2008) and a second Hampshire record (Harrison 2012). Pollard et al. (2022) list High Park as being the fifth known UK location. However, Fowles (2023) also lists Petworth Park, West Sussex, the record being derived from Telfer (2020), so High Park could more accurately be said to represent the sixth known locality.

Hylis olexai RDB3 24 – Considered by Hyman (1992) as Rare. See Pollard et al. (2022) for further notes on this species.

Cantharidae

Malthodes crassicornis RDB3 24 3 NT NR – According to Alexander (2014), this diminutive solider beetle 'develops in the red-rotting heartwood of old, open-grown oaks in relict, old lowland forest and ancient wood-pastures. The site list reads like a Who's Who of classic sites including Windsor and Epping Forests and the ancient parks of Moccas, Duncombe, Blenheim and Staverton. In June 2021, it was also found by Steve Lane at Blo' Norton Fen in Norfolk, unusually a site with no veteran trees. Adults may be found from mid-May until late June.' This is a NR species, assessed as Near Threatened (NT) with extinction (Lane et al. 2019), on the basis of it being known from 12 sites, which narrowly misses the threshold of ten sites to qualify as VU following IUCN criteria (Alexander 2014). IRW first recorded this in a bottle trap, in May 2019, and BJP subsequently found three individuals, two in a slam trap and one in a vane trap, in May and June 2021.

Dermestidae
Globicornis nigripes RDB1 32 3 VU NR – A rather unremarkable-looking beetle, this is actually NR, known from only 15 records in the UK, and assessed as being VU to extinction (Lane et al. 2019) following IUCN criteria. It was first recorded in High Park by KNAA in 2002 (Alexander 2003). Subsequent specimens were found by IRW and BJP.

Trinodes hirtus RDB3 24 3 NT NR – This unusual looking beetle is unmistakable, as the specific epithet suggests, by its covering of spiky long hairs over the elytra and pronotum. It has been assessed as NT in a European conservation context but was found relatively frequently in High Park, with 15 individuals on 12 occasions from 9 different localities being recorded.

Trogossitidae
Nemozoma elongatum RDB3 24 VU NR – This is a rather striking beetle with a delightful, banded patterning on the elytra (Figure 13.28). It is considered to be NR (Alexander 2014) and was listed as VU by Lane et al. (2019) in a British context, following the IUCN Red List categories and criteria. Some apparently unpublished records from Oxfordshire were made known in Pollard et al. (2022). Just a single example was recorded, by JC, in one of his flight interception traps on a south-west facing slope in Beech-dominant woodland, in the first week of September 2021.

Lymexylidae
Lymexylon navale RDB1 32 2 LC NS – See Figure 13.19 for a photograph of this rather unusual-looking beetle. In High Park, the author observed several individuals alighting on freshly sawn oak timber, whereby the females appear drawn to oviposit inside the tiniest of cracks. It seems that there is a healthy population present at Blenheim, with 27 specimens seen in 12 localities, mostly between late June and late July. It has been seen at both ends of High Park, suggesting it is well-distributed in the landscape.

Monotomidae
Rhizophagus oblongicollis RDB1 24 3 – 'Rhizophagus' literally means root-eater, and indeed many of the members of this genus are subterranean, found in the vicinity of dying or dead roots. True to form, the only modern records were by the author, in his single subterranean pitfall trap, set adjacent to a very large oak that had died in 2019. The only previous Oxfordshire record was by A.A. Allen at High Park in 1954.

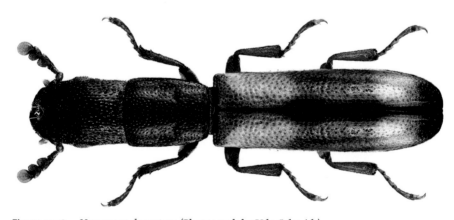

Figure 13.28. *Nemozoma elongatum.* (Photograph by Udo Schmidt)

Laemophloeidae

Laemophloeus monilis RDB1 32 3 – This lined flat-bark beetle was first found by the author on 11 October 2020, after dark, using a head torch, where it formed part of a mass jumble of different apparently nondescript species pootered off the ends of smallish cut logs in the timber yard. Upon microscopic inspection back home later that night, it became clear this was a very special find, a Red Data Book 1 species, only known from eight UK locations (Pollard 2021), mostly in southern England. A repeat visit to the yard on 15 November 2020 revealed three further specimens, under the bark of a non-fungoid Beech log, around 50cm to 60cm in diameter. Additional adults were recorded from the timber yard on 5 January, 30 May, 13 June and 26 June 2021. Two other records were made about 200m away, on a veteran lime (*Tilia* sp.), one in a cobweb on 30 May and one in a vane trap that was set for the period 1 to 8 July 2021. These latter two records confirm that the population is present in the Beech-dominant woodland beyond the timber yard.

Tenebrionidae

Platydema violaceum RDB1 32 LC NR – See Pollard (2022) for additional notes about this rare beetle. A specific hunt for this species was undertaken by the author, under the guidance of Wil Heeney, who knew to look for it in old, dried Jelly Ear Fungus on old Elder bushes. Indeed, that was where we found it (Figures 13.29, 13.30), after spending around two hours looking in an area of about 0.4ha in which the dominant understorey woody species is Elder. Some researchers consider it to be spreading rapidly across the country, whereas others are convinced it is being found more often because we now have a better understanding of its specific habitat requirements. It was first reported from Oxfordshire at Shotover Hill (Wright 2015).

13.4.7. Rare and scarce species – non-saproxylics

One NR species was recorded during our surveys that does not exhibit a saproxylic life strategy:

Leiodidae

Leiodes macropus – JC reported a singleton from a flight interception trap between 8 and 27 July 2021. Hyman and Parsons (1994) considered this species to be Red Data Book 'Insufficiently Known', but only known from eight vice-counties, mostly in the south of England. The Global Biodiversity Information Facility (GBIF 2023) suggests it is also rare in a European context.

A number of NS species were also recorded: the ground beetles (Carabidae) *Acupalpus exiguus* by sieving lakeshore material, *Amara montivaga* found as roadkill on a tarmac road near Combe Gate and *Bembidion octomaculatum* on the edge of a pond by a cattle grid in the timber

Figure 13.29.   *Platydema violaceum* on *Sambucus nigra*. (Photograph by Benedict Pollard)

Figure 13.30.   Stringy frass of Jelly Ear Fungus, on Elder, resulting from larval feeding. (Photograph by Benedict Pollard)

yard area. These were only found as singletons. A few leaf beetle species (Chrysomelidae) were found: *Epitrix atropae* in 1949 and 1992; three examples of *Longitarsus lycopi*, one by IRW in a water trap in June 2018 and two by JC in a flight interception trap in May and June 2021; *Psylliodes luteola* found by JMC in 1985 and 1990, and a singleton by BJP in the top bottle of a slam trap, 31 July to 3 August 2021. *Helophorus dorsalis* (Helophoridae) was recorded by BJP in July 2020 in a puddle. *Orsodacne humeralis* (Orsodacnidae) was swept on two occasions: in June 2019 by Graham Finch, and then in May 2020 by BJP. The dung beetle *Melinopterus consputus* (Scarabaeidae) was recorded in a pheromone lure in late 2019 by BJP. *Zyras haworthi* (Staphylinidae), an ant-generalist, is rarely encountered, but JC reported two individuals from a flight interception trap in June 2021. It was formerly considered to be RDB 3 (Rare) (Shirt 1987), but was downgraded in Hyman (1994) to Notable A. See Pollard et al. (2022) for more information on this latter species.

13.5. Conclusion

The beetle fauna of High Park is clearly exceptionally rich in species and includes many NR and NS species, especially in relation to its saproxylic fauna, which is of international significance. Our increased knowledge of the wealth of Coleoptera resulting from the HPBS is closely correlated with recorder effort and utilization of novel methods of sampling. Without doubt, continued targeted surveying would lead to the discovery of many more species not yet encountered. The saproxylic fauna of High Park ranks third in the UK in terms of its ecological continuity (Fowles 2023), and in terms of species density it is arguably the most important British site for saproxylics (see Chapter 15 for further discussions of the significance of the saproxylic fauna). The historic and ongoing management of the Park, especially in relation to sympathetic deadwood habitat management is likely to have played a significant part in the continuity of such a remarkably rich beetle fauna and is to be encouraged for the future. We can consider High Park to be what one might call a 'conservation ark' of beetle species diversity at local, regional, national and international levels.

Acknowledgements

I am grateful to the following: the editor and reviewer (Steve Lane) for helpful comments on earlier versions of the manuscript, Tosh Moller for data analysis, producing graphics and document formatting assistance; Trevor and Dilys Pendleton of Eakring Birds, Lech Borowiec, Udo Schmidt and Pavel Krásenský for permission to use their photographs.

CHAPTER 14

Bugs (Hemiptera) and other Insect Orders

Graham A. Collins and Jovita F. Kaunang

Introduction

The order Hemiptera consists partly of insects known as 'bugs'. This is a precise entomo-logical definition and not to be confused with the American use of 'bug' to indicate any insect, something that sadly seems to be coming into popular usage in this country.

Traditionally, Hemiptera have been divided into two suborders – Heteroptera and Homoptera – but a modern treatment splits them into four suborders. One of these is restricted to the southern hemisphere and so we have three to consider here:

> Heteroptera – e.g. shieldbugs, groundbugs, capsid bugs.
> Auchenorrhyncha – e.g. froghoppers, planthoppers.
> Sternorrhyncha – e.g. aphids, whiteflies, jumping plant lice, scale insects.

The characteristic feature of the order is the modification of the mouthparts into a sucking rostrum. The majority of species feed on vascular plants, mostly extracting the sap but with a few species utilizing fruits or ripening seeds. Some, exclusively within the Heteroptera, are predacious, and a small number of these even suck the blood of vertebrates.

The Hemiptera are hemimetabolous insects, that is, they develop through a series of nymphal stages resembling the adult but lacking wings or functional genitalia. There is no pupal stage.

In Britain we have around 1,700 species, so although we may consider them a 'large' order they are very much less speciose than, for example, Diptera or Coleoptera.

14.1. The Hemiptera fauna (true bugs) in High Park

Only 86 species of Hemiptera in 15 families were recorded, a considerably smaller proportion of the British fauna than was recorded, for instance, for flies (Diptera, Chapter 9). One reason for this is that a considerable proportion of Hemiptera (c. 40%) lies within the Sternorrhyncha, which includes some 500 species of aphid alone in Britain and constitutes an obscure group not often studied except by specialists.

A number of constraints caused the total number of site visits to be quite low. More specialized recording techniques, for example suction sampling, which might have revealed many more ground-dwelling Auchenorrhyncha, were not employed. However, a number of additional species were recorded as by-catch from the trapping employed by the coleopterists.

Graham A. Collins and Jovita F. Kaunang, 'Bugs (Hemiptera) and other Insect Orders' in: *The Natural History of Blenheim's High Park*. Pelagic Publishing (2024). © Graham A. Collins and Jovita F. Kaunang. DOI: 10.53061/NQKI9199

Hemiptera, because of their method of feeding, particularly their lack of biting jaws, have very few species that might be considered saproxylic. Alexander (2002) identified only 15 species that were associated with living and decaying timber. These comprised: Aradidae (flatbugs), which feed on fungal mycelia in decaying wood; Reduviidae (assassin bugs), one species of which can be found in old hollow trees; Anthocoridae (as Cimicidae), a few predatory bugs found beneath bark; and Microphysidae, also predatory around epiphytes and wood-decay fungi.

Of the species on Alexander's list, only one, *Aneurus avenius* (Aradidae) was recorded during the High Park Biodiversity Survey (HPBS). Three other species, *Aneurus laevis* (Aradidae), *Xylecoris cursitans* (Anthocoridae) and *Loricula elegantula* (Microphysidae), appear in the Thames Valley Environmental Records Centre (TVERC) data.

14.1.1. Heteroptera

Aradidae (flatbugs and barkbugs)

Aradids have extremely flattened bodies, appearing almost two-dimensional, and live beneath the bark of recently dead wood where they feed on fungal hyphae, thus being some of the very small number of bugs that justify the term saproxylic. One species, *Aneurus avenius*, was recorded during the HPBS and the very similar *A. laevis* is listed in the TVERC data.

Acanthosomatidae (shieldbugs)

This family, together with the next and other related families not yet found at High Park, are collectively known as shieldbugs. These are perhaps the most widely recognized of the bugs owing to their relatively large size and colourful appearance. The three species recorded on the HPBS are all common and widespread in Britain (Figure 14.1).

Figure 14.1.   Hawthorn Shieldbug (*Acanthosoma haemorrhoidale*). (Photograph by Graham Collins)

Pentatomidae (shieldbugs)

This is the largest family of the shieldbugs, although only four species were recorded. *Pentatoma rufipes*, colloquially known as the Forest Bug, is a common species associated with mature trees, particularly oak. *Dolycoris baccarum* and *Palomena prasina* are both common species of the herb/shrub layer, while *Aelia acuminata* is a grassland species feeding on the ripening seeds of various grasses.

Lygaeidae (groundbugs)

Lygaeids are small to medium-sized, hard-bodied bugs mostly found on the ground or among leaf-litter. A few species occur on trees, including *Gastrodes grossipes*, a flattened bug that feeds on the seeds of pine by squeezing between the scales of partially opened cones. The four other species found on the HPBS together with three previously recorded ones from TVERC are all common insects.

Coreidae (squashbugs)

British coreids are brown, tanned-looking bugs, giving rise to their alternative name of leatherbugs. The commonest British species, *Coreus marginatus*, is abundant on dock (*Rumex* spp.) everywhere, but has apparently not been recorded from High Park. Of the two species that were recorded, *Coriomeris denticulatus* is fairly common in southern Britain and feeds on a range of leguminous plants (Fabaceae). The other species, *Gonocerus acuteangulatus* (Figure 14.2), was once very rare and, in Britain, confined to a single site in Surrey. It existed here and here alone until 1990 when there was evidence of local spread (Hawkins 2003). This expansion has continued apace and the species is found across Oxfordshire and considerably further northwards.

Figure 14.2.   The coreid bug *Gonocerus acuteangulatus*. (Photograph by Graham Collins)

Rhopalidae

The Rhopalidae is a fairly small family with no English name. Many species have wings with window-like patches of clear membrane, and a translation of their German name, as glass-winged bugs, is apt for some but not all. One species was found on the HPBS, *Rhopalus subrufus*, which is perhaps the commonest member of the family. Some authors claim an association with St John's-wort (*Hypericum* spp.), but others (e.g. Hawkins 2003) suggest that crane's-bill and stork's-bill (*Geranium* spp. and *Erodium* spp.) are preferred. A further species, *Myrmus miriformis*, a grass feeder, is recorded in the TVERC data.

Tingidae (lacebugs)

Lacebugs are small, delicate bugs in which the wing has a lace-like reticulation. Two species associated with thistles were recorded, *Tingis cardui* on the HPBS, and *T. ampliata* from the TVERC data. A third species, *Physatocheila dumetorum*, occurs among foliose lichens on old Hawthorns.

Miridae (capsid bugs)

The Miridae is one of the larger families of the Hemiptera with over 200 species in Britain. Many of them are quite large and colourful (Figure 14.3), and several species can be pests on cultivated plants. There were 25 species recorded, all quite common. Perhaps the most notable was *Deraeocoris flavilinea*, a fairly recent arrival, first discovered in Britain in 1996 but now widely established and common across south and central Britain. A further 34 species were recorded on the TVERC data, including several additional species of *Psallus* associated with oak.

Microphysidae

A family of very small bugs, with only six species in Britain. These were not found during the HPBS, but one species, *Loricula elegantula*, occurs in the TVERC data. This bug lives among lichens growing on old trees and predates other small insects living there.

Figure 14.3. *Deraeocoris ruber*, a capsid bug. (Photograph by Jovita F. Kaunang)

Nabidae (damselbugs)

The Nabidae is a family of active, predacious bugs that hold their prey in their strong front legs. Four species of *Nabis* (two on the present HPBS and another two in the TVERC data) were recorded. They live in the ground-layer vegetation. A further species, *Himacerus apterus*, is a tree-dwelling insect.

Anthocoridae (flowerbugs)

Flowerbugs are rather small and inconspicuous bugs, although characteristic once found. They too are predacious, but not as aggressive as nabids; they do not attempt to hold their prey, which is often immobile and defenceless. Three species were recorded on the HPBS, with a further four in the TVERC data. They are all common, but *Xylocoris cursitans* appears to have a particular association with ancient woodland and parkland sites.

Saldidae (shorebugs)

These bugs are predators that hunt rapidly across bare ground; almost all species are associated with wet mud around ponds or on saltmarsh. None were encountered on the HPBS, but *Saldula saltatoria*, a common and widely distributed species, appears on the TVERC list.

14.1.2. Auchenorrhyncha

Cixiidae

Cixiidae are rather delicate bugs with net-like wings. The two species recorded, *Cixius nervosus* and *Tachycixius pilosus*, are both common species occurring on trees.

Delphacidae (planthoppers)

Delphacidae is the second largest family in the Auchenorrhyncha. Most are small and often brachypterous (with reduced wings), living on or close to the ground. All but three British species feed on grasses, sedges or rushes. Two common species were recorded on the HPBS. *Ditropis pteridis* is atypical for the family as it feeds on bracken. Five further species are listed on the TVERC data. All are common, at least in the south of Britain.

Cercopidae (froghoppers)

The Cercopidae contains a single British species, the striking red and black *Cercopis vulnerata* (Figure 14.4). Adults are found near the ground or on the shrub layer, but the nymphs are subterranean. Other froghoppers, formerly in this family, are now given a family of their own (see below).

Aphrophoridae (froghoppers)

Froghoppers are rather robust, jumping insects. The nymphs feed from the xylem of grasses, herbs or shrubs, producing copious amounts of watery excreta that is mixed with air and proteins to form a protective foam known as 'cuckoo-spit'. Two common species were recorded on the HPBS: *Aphrophora alni*, which occurs on trees and bushes, and *Philaenus spumarius*, on the shrub and ground layers. The grass-dwelling *Neophilaenus lineatus* was recorded in the TVERC data.

Cicadellidae (leafhoppers)

Cicadellids constitute one of the largest of the British hemipteran families. They occur in a range of habitats; those on trees are usually fully winged, while those that live on or near the ground are frequently brachypterous. There were 27 species recorded on the HPBS. Most of these are common, but two are worthy of note. *Ledra aurita*, a large, flattened insect

Figure 14.4.   The froghopper *Cercopis vulnerata*. (Photograph by Jovita F. Kaunang)

with ear-like lobes on the pronotum and found mainly on oaks, is considered by some as a scarce species; however, this is probably a result of its cryptic habits. The record, made on 7 September 2021, was of exuviae (remains of the cast nymphal skin). *Acericerus heydenii* is a recent arrival from the continent, first found in 2010 and now quite widespread in southern England. It occurs on Sycamore. A further 15 species are listed in the TVERC data. These are also common species.

Membracidae (treehoppers)

Only two species of membracid occur in Britain, and neither of these occur on trees. *Centrotus cornutus* is a quite large species in which the pronotum extends backwards over the wings. It is found on a range of herbaceous plants and woody saplings. It eluded us on this HPBS but was recorded in the TVERC data.

14.1.3. Sternorrhyncha

Triozidae (jumping plant lice)

As mentioned in the Introduction, the Sternorrhyncha, while comprising a considerable proportion of British Hemiptera, are mostly poorly studied. Only three species from the family Triozidae were noted. Bugs of this family are small insects with prominent, membranous wings. *Trioza remota* is associated with oaks. The nymph causes a gall on a leaf in a form of a small blister in which it lives. *Trioza galii*, a black species with white markings, does the same on bedstraws (*Galium* spp.). *Trichochermes walkeri* occurs on Buckthorn (*Rhamnus cathartica*) and the nymph makes a leaf-edge gall. The adult is a very distinctive species with patterned and pointed wings.

14.2. Other insect orders found in High Park

Introduction

Some of the larger insect orders have been covered in their own chapters (Chapters 9 to 13). There arc a number of other insect orders, some containing only a few British species, and these are discussed here.

Collecting methods included recording by the authors, records supplied directly by other recorders and the by-catch of trapping, particularly the interception traps employed to sample saproxylic invertebrates but also light traps used by the lepidopterists.

The orders are considered alphabetically.

14.2.1. Dermaptera

This order comprises the earwigs – insects familiar to most gardeners. There are four native species, and further species that exist as occasional introductions, sometimes temporarily established. The only species recorded from High Park is the Common Earwig *Forficula auricularia*. Of the other British natives, the Lesser Earwig *Labia minor* has been recorded through much of England, Wales and southern Scotland (including older records for Oxfordshire) and might be expected to occur, although it is difficult to find. Lesne's Earwig *Forficula lesnei* is restricted to south-east England, just reaching as far as Oxfordshire, and this species might occur too.

14.2.2. Ephemeroptera

These are the mayflies, well known to anglers who have their own names for some of the species, but less popular with entomologists than some other orders. The larvae are aquatic and the adults are unique among winged insects in having two adult stages – the subimago and the imago. Their name is derived from the Greek *ephēmeros* (lasting for a day), and this is the popular conception. However, adults live for only a few hours in some species but up to a couple of weeks in others. There are just under 50 British species, but only one, *Ephemera danica*, has been recorded from High Park by casual collecting. The Lake is likely to provide good breeding ground, and some of the ponds within the Park may be suitable too. More targeted recording is certain to add to the species list.

14.2.3. Mecoptera

The Mecoptera is a very small order with only four British species. Three are the scorpionflies, named after the male in which the enlarged genital capsule is held forward over the back in the manner of a scorpion's sting. The fourth is the so-called Snow Flea, a small, wingless and rarely recorded (but apparently widely distributed) insect. Two species have been recorded from High Park. *Panorpa germanica* was recorded on many occasions and seemed to be the only species present until *Panorpa communis* was identified from a female caught in a bottle trap. Both these species are widespread and

Figure 14.5.   The alderfly *Sialis lutaria*. (Photograph by Graham Collins)

common in England. The third species of scorpionfly, *Panorpa cognata*, is a much scarcer insect. There is an apparent association with calcareous soils and nearby records suggest it could possibly occur at High Park. The three species can only be separated by examination of the male genitalia or dissection of females.

14.2.4. Megaloptera
The Megaloptera contains the alderflies and is represented by only three British species, all in the genus *Sialis*. They have aquatic larvae and winged adults; all three species are very similar and require detailed examination, such as under a microscope, to separate them. The only species recorded from High Park is *Sialis lutaria* (Figure 14.5), the most common and widespread of our species. The other two are much scarcer and associated with flowing water in rivers and streams and so are unlikely to occur within the Park.

14.2.5. Neuroptera
The Neuroptera includes the lacewings. Green lacewings, of the family Chrysopidae, are the most familiar. But there are also brown lacewings, the Hemerobiidae, which have rounder and more opaque wings, together with 'white' lacewings, Coniopterygidae, which are very small and have wings and bodies covered with a white waxy substance and could easily be confused with whiteflies (Hemiptera: Aleyrodidae). They have varied life histories; some have active larvae, predacious on aphids and other small insects, while others have aquatic larvae (one family even being parasitic on freshwater sponges). There are over 60 British species, of which only seven were recorded during the HPBS.

The three species of green lacewing together with the three species of *Hemerobius* are all associated with deciduous trees, while *Micromus variegatus* tends to be associated with ground-level vegetation. Many species are primarily nocturnal, and further light trapping would undoubtedly add to the list.

Two further 'species' were listed in the TVERC dataset: *Chrysopidia ciliata*, a common green lacewing, and *Chrysoperla* 'carnea'. The latter taxon was once thought to be a single Holarctic species, but is now known to be a complex of many cryptic species of which (probably) three occur in Britain. They cannot be separated morphologically but produce different substrate-borne 'songs' that keep them reproductively isolated (Henry et al. 2002).

14.2.6. Odonata
The Odonata comprises the familiar dragonflies (Anisoptera) and damselflies (Zygoptera). The larvae are aquatic, occurring in rivers, lakes and ponds, with the scarcer species often having quite specific requirements in terms of water quality and pH. Adults are winged; the smaller damselflies tend to keep quite close to their breeding water, while the larger and stronger-flying dragonflies may move into woods and gardens some distance from where they have emerged. There are just under 50 British species, including some immigrants and others that have established in recent years.

Eleven species were recorded during the HPBS. Three large dragonflies, called 'hawkers' in popular terminology, were Southern Hawker (*Aeshna cyanea*), Brown Hawker (*A. grandis*) and Emperor (*Anax imperator*). These are all common and widespread in southern Britain. Other dragonflies were Black-tailed Skimmer (*Orthetrum cancellatum*) and two species of 'darter', Common Darter (*Sympetrum striolatum*, Figure 14.6) and Ruddy Darter (*S. sanguineum*). The latter was once considered Notable, but is now increasingly common.

Five species of damselfly were noted. Blue-tailed Damselfly (*Ischnura elegans*) and Red-eyed Damselfly (*Erythromma najas*) were each recorded on single occasions. The 'demoiselles' Banded Demoiselle (*Calopteryx splendens*) and Beautiful Demoiselle (*C. virgo*) both occurred. Their larvae breed in slow-flowing rivers and streams and they probably originated from the Rivers Glyme or Evenlode. The common species, seen regularly and in numbers, was

Figure 14.6.   The dragonfly Common Darter (*Sympetrum striolatum*). (Photograph by Graham Collins)

Common Blue Damselfly (*Enallagma cyathigerum*), no doubt breeding in ponds within the Park. Strangely, the other very common blue species Azure Damselfly (*Coenagrion puella*), which might have been expected, was not seen.

Additional species, from the TVERC dataset, were Broad-bodied Chaser (*Libellula depressa*) and White-legged Damselfly (*Platycnemis pennipes*): the former common, the latter a scarcer species sharing the demoiselles' habitat of clean, slow-flowing rivers.

14.2.7. Orthoptera

This order comprises the grasshoppers, bush-crickets and ground-hoppers. They develop through several nymphal stages, resembling small adults but always without wings, and adults of most groups occur in summer through to the autumn. The ground-hoppers differ in that they pass the winter as larger nymphs or adults and are mature in the spring. Grasshoppers and bush-crickets have songs, produced by stridulation, which are often diagnostic to species. The order is quite small in Britain with 11 grasshoppers (one confined to the Isle of Man), around 12 bush-crickets and 3 ground-hoppers.

Three grasshoppers were recorded: Field Grasshopper (*Chorthippus brunneus*), Meadow Grasshopper (*C. parallelus*) and Common Green Grasshopper (*Omocestus viridulus*), all of which are common and widespread in Britain. No other species are listed in the TVERC dataset, but Lesser Marsh Grasshopper (*C. albomarginatus*) and Mottled Grasshopper (*Myrmeleotettix maculatus*) could potentially be found.

Two of the three British ground-hoppers occur: Common Ground-hopper (*Tetrix undulata*) in drier areas, and Slender Ground-hopper (*T. subulata*) in the wetter areas around the ponds. The third species is coastal and would not be expected.

Four species of bush-cricket were recorded. Oak Bush-cricket (*Meconema thalassinum*) is a common species of woodland, the females laying their eggs in crevices in bark. Speckled Bush-cricket (*Leptophyes punctatissima*) also occurs on trees and bushes. The other two species, Long-winged Cone-head (*Conocephalus discolor*) and Roesel's Bush-cricket (*Roeseliana roeselii*), are species of grassland. Both have experienced a remarkable range expansion in the last few decades, being formerly confined to the south and east coasts respectively. A further species

Figure 14.7.   Meadow Grasshopper (*Chorthippus parallelus*). (Photograph by Jovita F. Kaunang)

Figure 14.8.   Oak Bush-cricket (*Meconema thalassinum*) (left) and Roesel's Bush-cricket (*Roeseliana roeselii*) (right). (Photographs by Graham Collins)

that might occur is Dark Bush-cricket (*Pholidoptera griseoaptera*), although it was not recorded in the HPBS and is also absent from the earlier TVERC data.

14.2.8. Psocoptera

Members of the order Psocoptera have in recent years been termed 'barkflies', a sensible term for the outdoor species that tend to be winged and to live on tree trunks. A more familiar term might be 'booklice', which is more appropriate for indoor, or synanthropic, species, which are usually wingless. All species are small insects, rarely more than a few millimetres long. There are approximately 100 British species. Adults can be beaten from tree branches, but were also well represented in trap samples from High Park. Fifteen species from eight different families were recorded.

Valenzuela flavidus, Ectopsocus briggsi, Ectopsocus petersi, Elipsocus hyalinus, Philotarsus parviceps, Graphopsocus cruciatus and *Stenopsocus immaculatus* are described as common in Britain.

Caecilius fuscopterus, Epicaecilius pilipennis, Mesopsocus immunis, Mesopsocus unipunctatus, Loensia fasciata and *Loensia variegata* are uncommon.

Peripsocus milleri is scarce and *Loensia pearmani* is rare.

14.2.9. Rhapidioptera

The Raphidioptera comprises insects generally known as snakeflies. They resemble lacewings but have an elongated pronotum ('neck') that raises the head so that they look like a snake about to strike. There are four British species and, although several are not rare, they are rarely seen, probably because they spend much of their life in the tree canopy.

One species, *Phaeostigma notata*, was recorded from a trap, as were a couple of larvae that could not be assigned to species. This is a common species associated with oaks.

14.2.10. Siphonaptera

The Siphonaptera are the fleas that, although apparently atypical, are in fact insects. Their form, smooth and strongly laterally compressed, is dictated by their lifestyle, an adaptation for moving through dense fur or feathers. Recording them is a specialized task involving examination of their hosts or hosts' nests, but one that was taken in a subterranean pitfall trap was identified as *Hystrichopsylla talpae*. This is the so-called Mole Flea. It will use the mole as a host but can equally be found on voles, mice and shrews.

14.2.11. Trichoptera

The order Trichoptera contains the caddisflies. There are just under 200 species in Britain and they range from very small to quite large in size. They closely resemble moths but have hairs rather than scales on their wings; in some species this produces a pattern sufficient to identify them. The larvae are aquatic and different species utilize a range of habitats from fast flowing streams to lakes and ponds, and even temporary waterbodies. Caddisflies can be collected from the margins of waterbodies, either swept from the vegetation or netted if they form mating swarms, but are also often taken at the lepidopterist's light traps.

Only five species were recorded: *Mystacides longicornis* (Leptoceridae); *Glyphotaelius pellucidus* and *Limnephilus lunatus* (Limnephilidae); and *Phryganea grandis* and *P. bipunctata* (Phryganeidae). All are common species and all but one were taken at the light traps. The trapping methods used in the HPBS, together with the emphasis on the pasture woodland of High Park rather than investigating the waterbodies, goes some way to account for this small species list. Many more species are likely to be found given the presence of a range of types of waterbodies within and around the Park. Indeed, there are a further six species listed in the TVERC data.

CHAPTER 15

Assessing the Importance of High Park for Saproxylic Beetles

Benedict John Pollard and Keith N.A. Alexander

The total number of saproxylic beetles – those dependent on decaying wood (Alexander 2002) – now known from High Park has reached 300 species. Although dominated by long-term natives, this list includes accidental introductions as well as species known to be expanding their ranges in response to climatic variability and including recent colonists from the near continent.

Two systems have been devised in Britain for the relative assessment of site quality for nature conservation using saproxylic beetles: the Index of Ecological Continuity (Alexander 1998, revised 2004) and the Saproxylic Quality Index (Fowles et al. 1999). These metrics are useful not only for ranking and comparing sites of high known diversity, but also to highlight gaps in our knowledge or clues about microhabitats that have not been much researched.

15.1. Index of Ecological Continuity

The Index of Ecological Continuity (IEC) has been used to identify Britain's most important sites for the saproxylic beetles of ancient trees and pasture woodland and parkland type habitats – Britain's old growth (Alexander et al. 2003) – and a hierarchical site table has been developed. The Index calculation is based on the presence or absence of a select list of beetle species (Harding and Rose 1986; Alexander 1988; Harding and Alexander 1994; Alexander 2004). The species are graded according to their degree of association with Britain's remaining areas of old growth – mainly the ancient pasture woodlands and historic parklands – and these grades are used as the basis for a scoring system (from 1 to 3). The total of these scores provides the Index. The IEC is targeted at temperate broad-leaved tree cover across Britain and so does not apply to the Scottish pine forest zone, which is more boreal in character.

The species in the qualifying list include many that are difficult to find on demand, and so the Index is best built up over a number of years. Records from earlier recording therefore contribute to the Index. A control on old records is imposed, however, with only post-1950 records being used in the calculation. The cumulative nature of the IEC means that the figure at any one time is a minimum figure; the Index can only increase as previously overlooked species are revealed. 'Missing' species can therefore be targeted for new recording effort.

The current list of IEC species is in need of revision and some species included in the 2004 revision now need removing, while others should be added as our knowledge of their ecology improves. Some species have already been recommended for removal

Benedict John Pollard and Keith N.A. Alexander, 'Assessing the Importance of High Park for Saproxylic Beetles' in: *The Natural History of Blenheim's High Park*. Pelagic Publishing (2024). © Benedict John Pollard and Keith N.A. Alexander. DOI: 10.53061/NGWD5034

Table 15.1. The richest British sites for the specialist saproxylic beetle fauna of broad-leaved trees assessed using the Index of Ecological Continuity (Fowles 2023, accessed 16 July 2023, except for High Park, which had not yet been uploaded by that date).

Site name	Calculated IEC
Windsor Forest and Great Park	251
New Forest	207
High Park, Blenheim	161
Richmond Park	153
Bushy and Home Parks	152
Hatfield Forest	147
Bredon Hill	143
Moccas Park	137
Epping Forest	128
Ashridge Estate	119
Langley Park	115
Croome Park	109

(Alexander, 2009, 2016, 2019) analysis, but this has not yet been implemented on the main saproxylic rankings website (Fowles, 2023).

Experience has suggested that sites of national (GB) importance have an IEC in the range of 25–80 while IEC values of 15–24 are of regional importance (Alexander 2004). Sites in excess of 80 are considered to be of European significance.

The IEC value of High Park has now reached 161, among the very top British sites for the saproxylic beetles of broad-leaved trees, and clearly also of European importance. Table 15.1 lists all British sites with an IEC of 100 or greater, as reported by Fowles (2023) (www.khepri.uk). It is worth noting that additional records from various published and unpublished sources remain to be uploaded to the website – the scores are therefore somewhat outdated. Sites such as Burnham Beeches, Yardley Chase and Ashridge Commons and Woods merit future inclusion. The scores of Windsor Forest and Great Park, New Forest, Richmond Park, Langley Park and Croome Park will also increase once more recent data are uploaded. It is possible that Richmond Park may move above High Park on that basis.

It is still very likely that further species remain undiscovered in High Park (Chapter 13). These could potentially fall in the category of long overlooked species where the probability of detection on demand remains small. Continued survey effort is required in order to move towards a more complete species inventory. Good examples include *Ptenidium gressneri*, *P. turgidum*, *Tetratoma desmaresti*, *Enedrytes sepicola* and *Stereocorynes truncorum*.

15.2. Saproxylic Quality Index

The Saproxylic Quality Index (Fowles et al. 1999) is a more recent development designed to take the whole saproxylic beetle fauna into account and to include some control of recording effort. The species are scored according to the level of their national status and on a geometric scale – from 1 point for common species through to 32 points for the rarest. The total of these scores is termed the Saproxylic Quality Score, and the Saproxylic Quality Index is calculated by dividing this score by the number of qualifying saproxylic species recorded and then multiplying the result by one hundred.

$$SQI = \frac{total\ SQS}{number\ of\ saproxylic\ species} \times 100$$

The SQI calculation has certain provisos:

- a threshold of 40 qualifying species have been recorded from the site.
- the list should be complete, that is, include all qualifying species recorded during surveys.
- the same attention should have been applied to recording common species as rare ones.

The 1999 list of SQI species with their scores is now out of date and care is needed in its interpretation. The list only includes species believed to be long-term British natives and so does not include introductions and recent arrivals. Some additional species were not recognized as saproxylic at that time and were inadvertently omitted. To date, 268 qualifying beetle species have been recorded within High Park (Appendix 11b), and 32 further saproxylic species have been recorded but do not merit an SQS score. The calculated SQS is 1,891 and this produces a SQI of 722.7. Fowles et al. (1999) suggest that an SQI of 500 is probably an appropriate threshold for assessing a site as being of national importance. High Park therefore substantially exceeds this provisional threshold for national importance. No threshold for European significance has been proposed for use with SQI.

The SQI approach (Table 15.2) is of particular use in comparing a series of datasets from a single site – the IEC is less useful for this as the list of qualifying species needs to be built up with continuing recording effort. Repeated surveying is recommended so that the SQI statistic can be used as a simple monitoring tool (Alexander 2014a).

The more comprehensive a survey is, the more accurate the IEC and SQI will be. There can be recording bias – for instance, some methodologies (e.g. the various types of flight interception traps) seem to record proportionally more of the scarce/rare species than common ones and an exaggerated Index may result. Perivale Wood in Greater London, for example, appears to be such a site. Consulting the UK rankings (Fowles 2023), it sits in 12th place based on the SQI rankings, and it transpires that flight interception traps were used. Apparently, the site has very few veteran trees (Farjon, pers. comm. 2021). For a species list to be as comprehensive as possible, it is preferable to employ a wide variety of recording methods, both active

Table 15.2.   Top British sites for saproxylic beetles based on SQI values (data from Fowles 2023)

Site name	No. of spp.	SQS	SQI
New Forest	326	2,794	857.1
Windsor Forest and Great Park	364	3,094	850.0
Langley Park	152	1,189	782.2
Bredon Hill (1970–2021)	218	1,600	733.9
High Park, Blenheim	268	1,891	722.7
Richmond Park	253	1,801	711.9
Bushy and Home Parks	255	1,804	707.5
Croome Estate	175	1,221	697.7
Hatfield Forest	246	1,709	694.7
Silwood Park	159	1,090	685.5
Moccas Park	239	1,518	635.1

and passive, within the time constraints of any given survey. This should yield more useful IEC and SQI values.

Windsor and the New Forest are clearly the premier sites in Britain for saproxylic beetles, with High Park sitting among a secondary cluster of sites. Some of these other historic sites have witnessed beetle recording for many decades, and in the case of Windsor, for well over 100 years; it being an easy commute from London and the Natural History Museum. High Park, on the other hand, has really only seen around six years of focused sampling of beetles, of which only four were particularly focused on saproxylics, so one might infer the true number of species present to be considerably higher than what we currently know.

A number of species included in these scores, such as *Pyrrhidium sanguineum* (Cerambycidae), *Scydmaenus rufus* (Staphylinidae), *Silvanus bidentatus* and *Uleiota planatus* (Silvanidae) are now considered to be more widespread than previously thought and in some cases no longer thought to be long-term natives; these merit a downgrade of both their SQS and IEC scores. In Table 15.6 they are marked with an asterisk. Some revision of individual IEC species has already taken place (Alexander 2009, 2016, 2019), but these have not yet been adopted by Fowles (2023). Alexander (*in prep.*) has drawn up revised IEC scores, but as this has not yet been published and there are currently no agreed or published amendments to the SQS scores, we have calculated the High Park metrics with the scores currently accepted by Fowles (2023). This ensures comparability across the assessed UK sites.

15.3. Saproxylics – the county context (Oxfordshire)

Of the 235 sites listed on the Saproxylic Quality Index rankings at www.khepri.uk (Fowles 2023), Oxfordshire is represented by three other sites: Shotover Hill, Drayton Copse and Hutchin's Copse. High Park is ranked considerably above these others in all metrics, but clearly Shotover Hill is also of national significance (Wright and Wright 2017). It is worth adding that surveys of Kirtlington Park were instigated by the first author in 2022. Provisional scores are included in Table 15.4, which suggest it is another important saproxylic site.

15.4. The European Red List of Saproxylic Beetles

Six species from High Park qualify for a European Red List category of threat, or near-threat. Two more were assessed but found to be Data Deficient (DD). Alexander (2011) provided the IUCN conservation status of species occurring in Britain and Ireland in a European context. Eight species on the IUCN Red List (Europe) found in High Park are listed in Table 15.4.

It is interesting to note that *Pediacus dermestoides*, *Pseudotriphyllus suturalis* and *Dacne rufifrons* have no conservation profile in the UK, but are more numerous and widespread in Britain than anywhere else in Europe.

Table 15.3.   Saproxylic beetle scores in five sites in Oxfordshire.

Site	No. of spp.	SQS	SQI	IEC
High Park, Blenheim	268 (3rd)	1891 (3rd)	722.7 (5th)	161 (3rd)
Shotover Hill	155 (34th)	804 (31st)	518.7 (47th)	71 (31st)
Kirtlington Park	46	300	652.17	34
Drayton Copse	76 (120th)	301 (123rd)	396.1 (112th)	19 (138th)
Hutchin's Copse	78 (117th)	288 (128th)	369.2 (138th)	18 (144th)

Table 15.4.   IUCN Red List (Europe) species recorded in High Park.

Family	Scientific name	IUCN European Status
Cucujidae	*Pediacus dermestoides*	DD
Erotylidae	*Dacne rufifrons*	DD
Eucnemidae	*Epiphanis cornutus*	NT
Elateridae	*Ampedus cardinalis*	NT
Elateridae	*Ampedus elongatulus*	NT
Elateridae	*Elater ferrugineus*	NT
Elateridae	*Ischnodes sanguinicollis*	VU B2ab(iii,iv)
Mycetophagidae	*Pseudotriphyllus suturalis*	NT

(DD = Data Deficient, NT = Near Threatened, VU = Vulnerable)

15.5. Factors involved in species-richness at site level

The species richness of a particular area or site depends on the combination of four key factors:

- the total number of trees present
- the age structure of the tree population
- the density of those trees, and
- continuity of habitat over time.

Each individual species will have habitat requirements that are met by a particular cohort of trees, whether a single species of tree or a group of species with similar characteristics. In some cases, the host trees may need to be ancient (Figure 15.1); to be available in sufficient numbers; and located within a certain distance of each other. Often the larvae may require decaying wood of a particular type, but the adult may require blossom for pollen and/or nectar.

Figure 15.1.   Dead wood of an ancient oak with beetle exit holes. (Photograph by Aljos Farjon)

The blossom may need to be in full sunshine and sheltered from the wind, to favour flight activity. Where the larval habitat is confined to ancient trees, the younger generations of that tree species will need to be constantly developing in order to maintain habitat viability in the long term. Thus, the four factors outlined above interact in complex ways.

It follows that species richness of any area or site is dependent, not only on the extent or area of available habitat, but also the total number of trees (and shrubs) present and the overall site dynamic over time. Site history is especially crucial as the existence of multiple overlapping generations of the host trees demands minimal site disturbance over centuries. Places such as High Park, with its recorded history going back to the enclosure of Wychwood Forest almost 1,000 years ago (see Chapter 1), with its earlier origins undocumented, are notably rare in modern Britain. This is why High Park has been found to be so exceptionally rich in saproxylic beetles.

15.6. Conclusions

High Park is now confirmed as a site of European significance for its saproxylic beetles. Given that two out of three beetle species recorded during the High Park Biodiversity Survey were new to the site, we can reasonably infer that a large number of species remain to be recorded there. Thirty-two saproxylic species reported from High Park do not qualify for an SQI score, and these have not been discussed in our work. Almost 30% of the current species checklist (Appendix 11a) are scoring saproxylics (268 out of 932), and this system is helpful in interpreting the diversity and rarity of species present. If future fieldwork were to be permitted, the cumulative metrics would certainly continue to rise. The fact that High Park is less than one four-hundredth of the land area of the New Forest, but contains over 82% of its scoring saproxylic species, tells us that this really is a jewel of British biodiversity, and we wish for it to be cherished and managed as such for the centuries to come.

CHAPTER 16

Amphibians and Reptiles

Aljos Farjon and Angela Julian

16.1. Introduction

Great Britain (excluding Northern Ireland) has 13 native species of amphibians and reptiles: six indigenous reptiles and seven species of amphibian. This relatively low number of species may be because of the relatively cool temperate climate but has likely been exacerbated by the final opening of the English Channel when sea levels rose at the end of the last glacial period some 9,000 years ago. Several mainland European species such as the Fire Salamander (*Salamandra salamandra*), Yellow-bellied Toad (*Bombina variegata*), Common Spadefoot Toad (*Pelobates fuscus*), European Tree Frog (*Hyla arborea*), Moor Frog (*Rana arvalis*) and the Common Wall Lizard (*Podarcis muralis*) are well distributed across northern mainland Europe, and were possibly prevented from reaching Britain by the untimely marine flooding of the land bridge. However, differences in climate or other environmental factors may also have deterred them. The Northern Pool Frog (*Pelophylax lessonae*) became extinct in England at the end of the twentieth century but was reintroduced in Norfolk from a genetically comparable population in Sweden. The Common Wall Lizard is native to the Channel Islands (close to France) and appears to be thriving in a few locations where it has been unofficially reintroduced in Southern England. However, little is currently known about the impact on our native reptiles, or what the long-term future for these populations is.

Our indigenous amphibian and reptile species can be further subdivided into 'rare' species with a restricted range and generally low population numbers, and 'widespread' species that are widely distributed and, excepting the Adder, believed to be relatively abundant (Table 16.1).

Table 16.1.   Abundance of indigenous amphibian and reptile species.

Widespread native amphibians	Widespread native reptiles
Common Frog (*Rana temporaria*)* Common Toad (*Bufo bufo*)* Great Crested Newt (*Triturus cristatus*)* Palmate Newt (*Lissotriton helveticus*) Smooth Newt (*Lissotriton vulgaris*)*	Adder (*Vipera berus*) Common Lizard (*Zootoca vivipara*)* Grass Snake (*Natrix helvetica*)* Slow-worm (*Anguis fragilis*)*
Rare native amphibians	**Rare native reptiles**
Natterjack Toad (*Epidalea calamita*) Northern Pool Frog (*Pelophylax lessonae*)	Sand Lizard (*Lacerta agilis*) Smooth Snake (*Coronella austriaca*)

*Recorded from Blenheim High Park during the survey period (2016–21)

Aljos Farjon and Angela Julian, 'Amphibians and Reptiles' in: *The Natural History of Blenheim's High Park*. Pelagic Publishing (2024). © Aljos Farjon and Angela Julian. DOI: 10.53061/PKXT2424

With the exception of the Adder, which is now considered to be locally extinct, all of the widespread species are known to be present in Oxfordshire, and most were recorded at Blenheim High Park during the survey period. The rare species are all range limited, and do not occur naturally in Oxfordshire.

16.2. Survey methods

Observations of amphibians and reptiles were occasionally opportunistic, made by recorders active with other groups of organisms, but the vast majority resulted from effort-based, systematic approaches. Because many of these species are cryptic, to improve detectability in terrestrial habitats in addition to visual searches, two types of 'reptile mats' or refugia, comprising a small quadrangular approximately 40 x 40cm flat sheet of flexible roofing felt and a larger rectangular corrugated onduline sheet of about 50 x 100cm were laid on sunny, south-facing locations in habitat considered suitable for reptiles throughout High Park. In total, there were about 20 small and 30 larger refugia, which were marked with numbered bamboo canes. The locations were also GPS referenced. Over the course of the five-year survey, some refugia had to be relocated owing to the spread of Bracken, while others were damaged by mowing machines or the canes broken by deer.

In addition to terrestrial habitats, permanent or semi-permanent ponds were investigated for amphibians using three recognized search methods: netting, night searches using high powered torches, and egg searches of submerged vegetation. In addition, in 2017 around 25 bottle traps were set overnight to improve capture rates.

The majority of refugia inspections were completed by Aljos Farjon. Other people checking them were Kate Sharma, Margaret Price, Sylfest Muldal and Angela Julian. Aquatic surveys were undertaken by Rosemary Hill, Peter Topley, Angela Julian, Sylfest Muldal and Jim Fairclough, with others helping on occasion. Because Great Crested Newts (Figure 16.1) were known to be present, aquatic searches were completed under a Natural England Great Crested Newt Class Licence (2015-19285-CLS-CLS) held by Dr Angela Julian.

Figure 16.1.   Male Great Crested Newt on rotting oak wood. (Photograph by Ray Hamilton)

16.3. Results

The highly protected Great Crested Newt (*Triturus cristatus*) was found in several ponds and under logs of sawn Ash and a heap of cornbrash fragments (Jurassic limestone) at an old quarry. Great Crested Newts were also recorded in a number of the ponds checked, including a large and heavily vegetated pond adjacent to the road between High Lodge and the Combe Gate, the large pond at the top end of the Palace Vista near High Lodge and the pond near the track to Watermeadow Lodge at the southern end of High Park (near the timber yard). Both adult females and males were found, as well as eggs on folded leaves of aquatic plants, suggesting a healthy breeding population. The majority of records resulted from systematic searches in these ponds by Angela Julian and Sylfest Muldal on 2 June 2021; however, earlier records were from searches and trapping by Jim Fairclough and Angela Julian on 30 April 2017, in the pond between High Lodge and Combe Gate (three adults), and by Steve Gregory on 15 June 2019, also in this pond. The male observed under the cornbrash was identified on 12 March 2018 by Rosemary Hill and the female under Ash logs on 19 July 2020 by Darren Mann (while searching for beetles with Benedict Pollard). It appears therefore that this nationally protected species is well established in High Park. The large pond at the top end of the Palace Vista was restored (by the removal of willows growing in the pond and over-shading planted poplars around it) in 2021 after our surveys for amphibians ended. It is anticipated that this deeper pond may be more attractive to Great Crested Newts, and it is hoped that once the submergent and marginal vegetation is re-established this pond may become an important habitat for amphibians.

Smaller newts were also observed during the evening surveys. The majority of these were positively identified as Smooth Newts, since the adults can be readily distinguished from the Palmate Newt. These were mainly observed in the pond near Watermeadow Lodge where there was a good-sized population. Small newt eggs were also found in the marginal vegetation at this pond (additional searches were not possible at the pond adjacent to the road between High Lodge and the Combe Gate because with Great Crested Newt eggs having been recorded further searches under licence were not permissible). Given the widespread nature of both Smooth and Palmate Newts in this area, with records of both species from Bladon, Long Hanborough and the adjoining Ditchley Park (OxARG database), it is possible that more small newts were present.

Of the amphibians, by far the most common species recorded was the Common Toad. At least 50 adults and many juveniles were found under the refugia during the five years of the survey. On 17 June 2020 and again on 3 and 9 July 2021 thousands of very small metamorph toadlets were spotted migrating along grassy rides and among Bracken with grass in the area to the north of High Lodge where springs cause seepage and moist ground. They did not appear to be moving in a particular direction but were probably dispersing from the site(s) where they had originated as spawn. Based on the presence of adult and young toads during torchlight surveys conducted by Angela Julian and Jim Fairclough during April 2017, and Angela Julian and Sylfest Muldal during June 2021, they are presumed to breed in the multiple small ponds near High Lodge, where they are able to tolerate the presence of fish.

It took a long time before the first (dead) specimen of Common Frog was found – not under a refuge. On 25 March 2021 spawn with larvae in the early 'comma stage' was found by Aljos Farjon, comprising >1,000 individuals, in the pond along the road between the Combe Gate and High Lodge. This proves that Common Frog is breeding in High Park. On 21 July 2019, 10 juveniles were seen crossing the road near Combe Gate by Benedict Pollard. This is just inside High Park. Adults were seen only twice, one dead on 25 March 2019 and one alive on 22 July 2020, with two further juveniles on 17 June 2020 and 22 July 2020. None were found under the mats. It is possible, but unproven, that the abundant presence of pheasants in summer has a negative impact on frogs, which are more active by day than are toads.

Common Lizards are uncommon in High Park. Only four individuals were seen over the duration of the study, the first in May 2016 basking on a fallen oak limb by Aljos Farjon, who also recorded the second on 14 June 2017 crossing the road between High Lodge and the Combe Gate (both were female adults). A third was seen on 14 October 2020 by Benedict Pollard, also on a fallen oak branch (its sex was not recorded) and the fourth, a juvenile, was found on 29 August 2020 under a reptile refugium by Aljos Farjon and Sylfest Muldal. These sightings were well spread throughout High Park, indicating that this small lizard occurs wherever there is more or less open terrain among the trees. They are a diurnal species, becoming active after warming up in the sunshine, so they are often to be found basking on top of logs or timber structures during the early part of the day, and may be quick to move away if disturbed, which may account for the comparatively few sightings.

Slow-worms were found regularly under the refugia except when the weather was very warm. In total 59 'individuals' were seen, but since many of these were repeatedly found under the same refugium (e.g. on 11 occasions under No. 18 between 17 April and 22 June 2018) and they have a relatively small range, it seems likely that some of them may have been repeated observations. Slow-worms tend to be most active at dusk, or after heavy rainfall, when their prey is most abundant, and are often cryptic during the daytime, having a fossorial or burrowing habit. However, this also encourages them to shelter under objects – and the corrugated onduline refugia were ideal for this purpose – during the day. Their relatively wide distribution throughout large parts of High Park indicates that they are a fairly common species, and frequently more than one individual was found under a refuge at the same instant. On one occasion (12 June 2018) refugium No. 18 yielded an adult female, a sub-adult and a juvenile (Figure 16.2). Most records of Slow-worms were by Aljos Farjon, with additional finds by Kate Sharma and Margaret Price. Only one animal was not under a refugium but under a piece of fallen oak bark, spotted by Benedict Pollard when looking for beetles.

Figure 16.2.   Female (top), sub-adult (bottom) and juvenile Slow-worms found sheltering under a reptile refuge in High Park. (Photograph by Aljos Farjon)

Grass Snakes were also most often found under refugia, though some, in particular juveniles, were seen moving through grass. There were eight sightings in total, five under the refugia. One of these, on 9 July 2019 under No. 14, was a large (probably female) snake about 120cm long. It cohabited under this corrugated refugium with two adult male Slow-worms. All Grass Snakes were found in an area to the south of High Lodge, which has glades of open grassland among the oaks, as well as piles of fallen, rotting oak wood mixed with dead Bracken fronds. The low number of specimens seen makes it difficult to draw conclusions about the spread of Grass Snakes in High Park, however, they are a highly mobile species with a large home range, and it is possible that the same animal was not observed more than once.

Despite the presence of refugia, no Grass Snakes were recorded in the vicinity of the lakeshore even though the adjacent dry open grassland and scattered oak trees appeared to provide ideal habitat. Save for one, found by Anthony Cheke, all snakes were recorded by Aljos Farjon.

16.4. Habitat for amphibians and reptiles in High Park

Blenheim High Park provides suitable terrestrial habitat for the majority of our widespread native amphibian and reptile species. The smaller ponds present in High Park also appear to be suitable for amphibians, and two native species of newt as well as Common Frog and Common Toad were observed. It is possible that Palmate Newts are present since they have been recorded in the neighbouring Ditchley Estate (Angela Julian, pers. comm.), but required additional survey effort. Common Toads are tolerant to the presence of fish since adults, metamorphs and tadpoles contain bufotoxin, which makes them unpalatable to fish, and therefore are likely to be able to exploit a greater number of ponds, as well as potentially the lakeside.

On the face of it, the lack of Adders is puzzling since they should be well suited to High Park with its diversity of habitats, including grassland with scrub and open woodland on both sand and limestone. Indeed, there are anecdotal records from the nearby village of Stonesfield from the early 2000s (Stuart Hamilton, pers comm.), but since then they have not been found and are in fact almost certainly absent from the whole of Oxfordshire, and much of the Midlands, where they are in greater decline than elsewhere in the UK (Baker et al. 2004). It seems likely that historic factors, including persecution and perhaps unsympathetic habitat management, such as destruction of hibernation sites, are mainly responsible, since there are no historical records of Adders in the greater Blenheim Park.

It is clear that the waterbodies present in Blenheim High Park play an essential role in supporting the amphibians and reptiles we find there. High Park also borders the long-established Great Lake formed by damming the River Glyme during Capability Brown's remodelling of the Park in the 1760s (Chapter 1). Although the vegetation zone along its shore was included in some of the surveys, no amphibian or reptile species were observed there. It is probable that the large numbers of waterfowl and fish, the latter introduced for licensed angling, have made the habitat less desirable for these species. Inside High Park there are several semi-permanent ponds. While they tend to be mostly overshadowed by surrounding trees and willows growing in the ponds, these still support a number of amphibian species, which in turn provide prey for the Grass Snakes. A pond restoration programme has removed the trees from one large pond, but the impact of this could not be determined within the timescale of the project. The other ponds in which the newts were found all had a thick layer of leaf detritus at the bottom and tend to more or less dry out at the end of summer, filling up again with autumn and winter rains. This may in fact favour the newts by reducing the persistence of fish and large aquatic invertebrate larvae, which may be why the Great Crested Newts are breeding successfully in High Park.

This is why care needs to be taken before additional pond restoration is conducted. Lastly, there are two artificial deep ponds at High Lodge, one inside the fenced private compound and one outside. We had access to the latter, but an investigation by Jim Fairclough and Angela Julian in 2017 established the presence of large numbers of the American Signal Crayfish (*Pacifastacus leniusculus*) as well as a few Common Toads. Because these highly invasive crustaceans not only compete for the same invertebrate food, but also predate directly on the eggs, larvae and adults of amphibians, we excluded this pond from future surveys.

Common Toads appear to be thriving and are breeding successfully (Figure 16.3). They have a preference for relatively deep ponds in which to breed, which they migrate to annually, showing a high level of fidelity. The small toadlets disperse after metamorphosis, when weather conditions are suitable, and if there has been a prolonged dry spell this may result in mass waves of migration soon after a heavy shower, as observed in two successive waves of young toads in 2020 and 2021. Once terrestrially based, toads can migrate considerable distances from water, being more tolerant of dry conditions than frogs, only returning to ponds during the spring breeding season once adult. Toads are nocturnal and hunt by night for small invertebrates, and they appear to be finding abundant prey in the open woodland and glades of High Park. The apparent scarcity of the Common Frog may be due to the generally dry nature of High Park, particularly where ponds are drying out later in the summer, which may confine them to a few damp flushes. Habitat unsuitability rather than general scarcity – this species is of Least Concern in the IUCN Red List for England (Foster et al. 2021) – seems to be the most likely explanation.

While the habitat for the Common Lizard is present in High Park, it is perhaps not optimal. Whereas in the past, the pasture woodland with veteran oaks was more open, as

Figure 16.3.   Common Toad on leaf litter found under a reptile refuge in High Park. (Photograph by Aljos Farjon)

Figure 16.4.   Grass Snake found under a reptile refugium. (Photograph by Nicola Devine)

aerial photographs from as late as the Second World War have shown (see Figure 1.19 on page 34), infill by younger trees, in part planted oaks but mostly naturally generated, has closed the canopy in many places. Thinning, especially of Ash, in recent years has reduced this over-shading in many areas, but Common Lizards are not particularly mobile and once a population has been lost it will be very slow to recolonize a new area. Another major problem for the Common Lizard is likely to be the large numbers of gamebirds, with mass releases of pheasants each summer from pens inside High Park. These are likely to prey on the basking lizards, and particularly on the neonates born during the late summer breeding season, thereby reducing the ability of the small Common Lizard population present to regenerate. Unless pheasants can be specifically excluded from key Common Lizard sites, it will be difficult for this population to survive.

By contrast, their cousins, the Slow-worms, are thriving and were found to be widespread across the Park occurring in dense vegetation, piles of wood or other vegetative debris, but also finding open sunny but secluded spots to bask. The success of this legless lizard is possibly due to its fossorial day-time habit, it lives partly under and partly above ground, where it can hide from avian predators, appearing only during the evenings to hunt.

Only a single snake species – the Grass Snake – is found in High Park (Figure 16.4). This highly mobile snake is almost always associated with water, in the form of ponds, lakes or streams, as well as grassland and open woodland. Here, it hunts for its favoured amphibian prey, mainly Common Frogs, but also small fish and occasionally other live prey as it is an excellent swimmer. Although the bulk of Grass Snake sightings were several hundred meters from waterbodies, the number of recordings was too low to determine a definitive range within High Park and it is likely that the Grass Snake could be found elsewhere, for instance near the lakeshore.

16.5. Amphibians and reptiles in Oxfordshire – the bigger picture

Finally, we review the bigger picture for amphibians and reptiles in Oxfordshire to see how representative our records from Blenheim High Park are. Associated with the county's extensive water-catchments are all of our widespread amphibians, with an abundance of Common Frog, Common Toad, Great Crested Newt, Smooth Newt and even records of the Palmate Newt across the county. Common Toads are particularly successful in Oxfordshire, with some of the Froglife toad patrols reporting many thousands of animals during their spring migrations. Oxfordshire also represents a stronghold for the Great Crested Newt, which while enjoying the highest level of protection under the Wildlife and Countryside Act (1981) and being protected under Habitat and Species regulations for its rarity nationally, is widespread across the lowland parts of the county. With a preference for deep, well-vegetated and fish-free ponds, the ponds at Blenheim provide a perfect habitat for our biggest native newt, and may support an important metapopulation. Finally, we turn to our least abundant native newt – the Palmate Newt. Although not recorded during the current study, they have been recorded close by in the village of Long Hanborough and on the neighbouring Ditchley Estate, where a total of nine Palmate Newts were observed exhibiting courtship behaviour in a deep wheel rut (Angela Julian, pers. comm.). This suggests that additional careful searching in the smaller waterbodies may reveal the presence of Palmate Newts, which are slighter in build and perhaps less competitive than their cousin the Smooth Newt.

Similarly for reptiles with the exception of the Adder, all of our native widespread species are found across the county. Most common are Slow-worms, which are regularly reported from gardens and allotments, even in major urban settlements. Similarly, our largest native snake, the Grass Snake, is frequently encountered in the vicinity of many of the water courses that criss-cross the landscape, including in the Thames itself, and the Oxfordshire Amphibian and Reptile Group has hundreds of records from both the wider countryside and urban settings. It is believed that the main obstacle to Grass Snake expansion is lack of egg-laying sites, such as manure piles, and it may be that the relative scarcity in Blenheim High Park reflects a lack of suitable egg-laying spots.

Sadly, the Common Lizards bely their name as they are no longer as common as they were. While some small colonies still persist even to the edges of the City of Oxford, larger populations are observed only in relatively large or undisturbed habitats, for example in the nearby Wychwood Forest, in the western part of the county on the Ridgeway and on the Chilterns to the east. Common Lizards are among the least mobile of our native herpetofauna and most susceptible to predation, and once gone it is hard for a colony to re-establish itself.

Perhaps our biggest mystery is the Adder. While small populations can still be observed in the neighbouring counties of Berkshire, Buckinghamshire and Gloucestershire, following the extirpation of the last known colony at the BBOWT Warburg reserve on the Chilterns around 10 years ago it appears that there are no indigenous populations left in the county. Why the Adder disappeared from Oxfordshire is less well established. While the clay soils and river valleys that dominate the central zone are considered unsuitable for them, the calcareous soils of the Cotswolds to the west and the Chilterns to the east, as well as sandy outcrops such as Shotover SSSI, ought to suit them well. Indeed, there are historical records for Adder in all of these locations, as well as place names such as Adderbury, that suggest they were commonplace. We can only presume that persecution, habitat fragmentation and the widespread releases of predatory game birds reduced perhaps already small populations to the point of unviability.

Oxfordshire is not within the natural range of any of our native rare amphibian and reptile species. The Northern Pool Frog is confined to Norfolk, though undoubtedly there are non-native water frogs in the county, and individuals have been recorded from nearby

Enstone. Sand Lizards and Natterjack Toads tend to be restricted to a few coastal dune sites, with a few colonies persisting on lowland heathland in the southern counties of Dorset, Hampshire, Surrey and West Sussex. While there is a small Natterjack Toad reintroduction site in Oxfordshire, this is on private land, and little is known about the size of the population. Our rarest snake, the Smooth Snake, similarly persists only in very specific habitats on heathland in Dorset, Hampshire, Surrey and West Sussex, with a small reintroduction in Devon. These small and extremely secretive snakes prefer to live in mature stands of heather. There are none in Oxfordshire.

Acknowledgements

The authors thank Linda Losito and Dr John Baker for helpful comments on earlier drafts of this chapter.

CHAPTER 17

Birds

Anthony S. Cheke

> [A] little bird, sometimes seen, but oftener heard in the *Park* at *Woodstock*, from
> the noise it makes commonly called the *Wood-cracker*, described to me (for I had
> not the Happiness to see it) to be about the Bigness of a *Sparrow*, with a *blue* back,
> and a reddish Breast, a *wide* Mouth and a *long* Bill., which it puts into a Crack
> or Splinter of a rotten Bough of a Tree, and makes a noise as if it were rending
> asunder, with that Violence that the noise may be heard at least 240 yards, some
> have ventured to say a *Mile* from that place. (Plot 1677)

17.1. Introduction

It is fitting that the oldest record of a bird from Blenheim, and indeed the only one from
before Vanbrugh's grand Palace was built, is of the Nuthatch (*Sitta europaea*; Figure 17.1).[1] The
nuthatch is perhaps the bird most associated with oaks in Britain, and remains to this day
very common in High Park. It has been suggested (Mavor 1820) that Geoffrey Chaucer, living
at Woodstock, drew inspiration for his poem 'The Cuckoo and the Nightingale' from birds
encountered as he wandered through the Glyme valley below the town, then just a small river.

Apart from the 'wood-cracker', Robert Plot's *Natural history of Oxfordshire* (1677) is largely
silent on birds, outsourcing the class to John Ray, whose pioneering *Ornithology* (1678) hardly
mentions Oxfordshire, and Blenheim not at all. Apart from the odd mention of a heronry,
waterfowl and gamebirds, it is nearly 200 years before Blenheim reappears in the ornithological literature, and even when knowledge of the birds of Oxfordshire was collated by
Oliver Aplin in 1889, entries for Blenheim are few, though waterbirds on the Lake were given
a chapter by Charles Cornish (1895), and in later updates on Oxfordshire birds (below) Aplin
did include Blenheim. Waterbirds aside, Blenheim was overlooked in Jourdain's Oxford area
list in 1926, and barely features in Alexander's discussion in the *Victoria County History* (1939;
VCH) and his updated county list (1947). However, the same Wilfred Alexander was co-author
of the first attempt to cover the estate more broadly (Elliott and Alexander 1937), followed by
annotated lists in booklets by Margaret Pickles (1960) and John Brucker and John Campbell

---

1   Aplin (1889: 121), while accepting that the description was of a Nuthatch, believed the loud noise
     referred to the drumming of the Lesser Spotted or Barred Woodpecker (*Dendrocopos minor*). While
     the distance the sound carries was clearly exaggerated, it could refer to excavating/foraging by
     woodpeckers, or indeed the cracking of nuts by a Nuthatch; it does not seem to be a description
     of drumming.

Anthony S. Cheke, 'Birds' in: *The Natural History of Blenheim's High Park*. Pelagic Publishing (2024). © Anthony
S. Cheke. DOI: 10.53061/CIGW8855

Figure 17.1.   Nuthatch collecting nest material. (Photograph by Anthony Cheke)

(1975, 1987), the 1987 edition being the most detailed. In all cases the main focus was on the Lake's waterbirds, no doubt because they are easy to observe from public paths, and because much of the Park and woodland has been off-limits. In addition, before the creation of numerous gravel pits, still a 'new feature' in the 1960s (Radford 1966), the Farmoor reservoirs (from 1967), and more recently the RSPB reserve at Otmoor, Capability Brown's Great Lake at Blenheim, at 48.4ha (Alexander 1939) was 'the only permanent sheet of water of any size within our limits' (Jourdain 1926), and thus a focus for birdwatchers. Dredging the Lake in the 1950s made conditions unsuitable for some waterbirds but better for others (Brucker and Campbell 1975),[2] and the exceptionally long, cold and snowbound winter of late December 1962–March 1963 had lasting effects on vulnerable species. In 1975 Blenheim was nevertheless 'the best general bird-watching area in the county' (Bruce Campbell, introduction to Brucker and Campbell 1975), exposed mud in the Lake-bed during the long hot summer of 1976 attracting waders and temporarily adding to its appeal (Campbell 1987). There is a general discussion of the estate's wildlife, including birds, in Campbell (1987) and, very briefly, in Banbury et al. (2010).

Over the years 1985–8 in Oxfordshire,[3] birds were surveyed at the 'tetrad' level (2 × 2 km squares of the National Grid), the square SP4214 encompassing all of High Park (approx. 121ha) apart from a sliver of about 8ha in the north, but of course also covering agricultural and village land beyond the Park. The results were published in an annotated atlas (Brucker et al. 1992), which has occasional text notes on Blenheim in addition to the mapped presence in the atlas. Since 1996 bird records submitted to the Oxford Ornithological Society (OOS) have been held digitally in a database, which County Recorder for birds Ian Lewington has kindly provided me for this chapter; most of the records are again for the Lake and its margins to the north, but a good number of records are labelled High Lodge or Combe Gate and these relate specifically to High Park. There was a further, less thorough, tetrad-based atlas survey

---

2   Wintering shallow water feeders (Teal *Anas crecca*, Wigeon *A.penelope*, Dabchick *Tachybaptus ruficollis*) declined, deep water feeders (Great Crested Grebe *Podiceps cristatus*, Tufted Duck *Aythya fuligula*, Pochard *A.ferina*) increased.

3   The county was expanded in 1974 to include the northern part of old Berkshire.

in 2007–12, which doesn't entirely match database information, and indeed some High Park records I submitted in 2011 to the British Trust for Ornithology (as sponsor and database holder for the atlas) were not included in the published data, which is only available online,[4] together with the mapping from the previous atlas.

Although the overall habitat in and around High Park has been generally stable at least since the disturbances of the Second World War, occasional major interventions can have short or longer-term effects. The 1950s dredging and the 1963 winter have been mentioned; more recently the lowering of water level in the Lake in 2018 and 2020 led to unexpected numbers of egrets, Great and Little,[5] taking advantage of the exposed mud (Lewington et al. 2020) but losses among breeders in lakeside vegetation (short term). The removal of mature conifers in the long strip plantation in the centre of High Park in 2019 has impacted numbers of Coal Tits (*Parus ater*) and Goldcrests (*Regulus regulus*), which favour the presence of at least some conifers (see below). The introduction of free-range British White Cattle in 2020 as a means of restoring the area towards the pasture woodland it once was (Chapter 19) has yet to show substantial effects, but if as intended they remain at current densities long term the woodland will be dramatically altered (see Section 17.9). Changes in agricultural practices inside and outside the estate over time will have affected some species, as will the collapse of the formerly important rabbit populations caused by myxomatosis in the mid-1950s.

The OOS has published annual reports since its inception in 1921 (the first back-dated to 1915), but as they are not geographically indexed, searching for references to Blenheim has not been feasible, but I have scoured the books by Aplin (1889), Radford (1966) and Brucker et al. (1992) for records, the latter two having themselves worked through the reports. There are some pertinent comparisons to be had from comprehensive data from Wytham (Gosler 1990, Perrins and Gosler 2010) and Shotover (Whitehead et al. 2003, Whitehead 2018). Overall, it is not until the last few decades that one can focus on birds in High Park itself rather than the estate as a whole.

There has been unprecedented flux in the scientific nomenclature of birds in recent years since the advent of molecular/DNA phylogeny, a failure to define generic limits in phylotrees leading to rampant Victorian-style splitting of long-established genera and species, together with creation of new families and re-attribution of many species. In the interests of stability and continuity I have retained the Linnean names used in the last county avifauna, Brucker et al.'s *Birds of Oxfordshire* (1992), and (largely), as a similar matter of policy, in the annual county bird reports (e.g. Lewington et al. 2022); new names are given in brackets in tables.

17.2. Game birds and shooting

The avian (and mammalian) history and ecology of a shooting estate is in some ways two parallel stories – that of the target animals ('game'), preserved, bred and shot, together with their predators, presumed or real ('vermin', trapped and shot) and all the other animals that go about their normal business more or less unmolested. The two intersect, of course, as the way the wood is managed will influence the wild species inhabiting it (Sage et al. 2020), the removal of predators by gamekeeping may boost numbers of some species and the high density of reared birds released, in Blenheim's case mostly Common Pheasants (*Phasianus colchicus*),

---

4   http://thamesandchilternbirdatlas.org.uk/. The 1985–8 atlas had 70 summering (potentially breeding) species for tetrad SP4214, but the 2007–12 version only 30; in a single visit in May 2011 I saw (and submitted) an additional 12 species that should have been included. It is not just Blenheim: I analysed the tetrads within Oxford city for a talk (unpublished) in 2016, and found a similar decline in records for the great majority of tetrads. The national atlas, for 2007–11 only, was published as a book (Balmer et al. 2013), but the mapping is at the much lower 10 × 10 km square resolution.
5   Respectively *Ardea alba* and *Egretta garzetta*.

impacts terrestrial insects and small vertebrates (e.g. Hall et al. 2021), and thus the food of ground-feeding wild species.

The practice of emparking large areas, enclosing them within a pale (wall or fence), to create deer parks, was originally limited to royalty, but in later medieval times was permitted also to the nobility. What was then called Woodstock Royal Park was enclosed in a stone wall by Henry I in about 1110 (Baggs et al. 1990; Chapter 1) to retain introduced Fallow Deer (*Dama dama*), and probably also the king's collection of exotic animals (Plot 1677), though the latter, which included predators, would not have been given free range (Banbury et al. 2010). There was a mews for the king's hawks and falcons attached to the manor from at least 1250 (Baggs et al. 1990), and gamebirds were generally hunted with falcons (or poached using traps). Queen Mary gave Lord Bedingfield, in charge of Princess Elizabeth under house arrest in Woodstock Manor, a licence in 1554 to 'hawk for your pastime at the partridge,[6] or hunt the hare, within that our manor of Woodstock, or any of our grounds adjoining the same, from time to time' (Mumby 1909). It was not until the advent of convenient, portable and relatively accurate flintlock firearms later in the sixteenth century that shooting birds became sufficiently productive; even so the vagaries of muzzle-loading guns meant that most birds were shot on the ground or at roost in trees (Tapper 1992). The introduction of breech-loading guns in the mid-nineteenth century, and the concomitant development of driven rather than stalked shooting, led to the modern style of shoot by the late nineteenth century (Yardley 2015). Finally mass rearing with incubators instead of using broody hens, also in the late nineteenth century (Yardley 2015), allowed the production of huge numbers to boost game bags, and also to intense competition between shooting estates, including Blenheim, for bag size and 'bird quality' (Ruffer 1977).

Woodstock Park became somewhat neglected in the seventeenth century (Bond 1987) and, apart from Plot's Nuthatch, wildlife reports are lacking until after the estate was granted to John Churchill, recently elevated to Duke of Marlborough, in 1705 (Green 1989). Even then there are no mentions of game birds and shooting until the 1770s, although Lancelot 'Capability' Brown's landscaping over the decade 1763–73 (Bapasola 2009) included pheasant-friendly planting. As Sarah Rutherford (2016) put it, 'game birds favour woodland edges against open land, so narrow belts [of trees] gave plenty of these edges to encourage them for sporting owners; Blenheim has a belt alongside its park wall'. In an unfortunately unreferenced and somewhat over-novelized history, Fowler (1989) described what appear to have been regular shooting parties in the 1770s: 'on 1 September the shooting began. Shots rang out in the crisp, still air of the oak woods and copses, and Pheasants and Partridges thudded to the damp earth.'[7] Shooting must have been much practised, as pheasants have been reared and released at Blenheim since at least the 1780s – John Byng (later Lord Torrington) reporting in August 1787 that 'in various parts of the park are clusters of faggots around a coop, where are hatch'd and rear'd such quantities of Pheasants that I almost trod upon them in the grass' (Andrews 1934). In July 1792 Byng was back: 'Our ride … was to the new Pheasant grounds, defended, and gated, for the Duke's private sport. These grounds … are of a wild, foresty, nature, with wide drives' (Andrews 1934). Mavor (1797) was more specific:

> the Pheasantry, situated on a rising ground near the lower Cascade, is one of the most superb establishments of the kind in this country. The variety and beauty of these birds present an object that cannot fail to please. But though some of the more curious kinds are kept here, they stock the park and the adjoining woods in prodigious numbers.

---

6   Only the Grey Partridge (*Perdix perdix*) was present at the time.
7   The book has chapter bibliographies, but there's no link to episodes in the text, so original sources cannot be checked for further details.

Figure 17.2.   A rare black and purple plumage morph of the Common Pheasant in High Park. (Photograph by Anthony Cheke)

These facilities are probably what Andrews (1985) meant when commenting that 'in the last decade of the eighteenth century, the Duke of Blenheim's [sic] keepers made extensive use of cooped broody hens to incubate collected Pheasants eggs'. Different varieties of Common Pheasant were being introduced (Figure 17.2), and 'at Blenheim, the Duke of Marlborough has bred, in great quantities, both the Chinese or Pencilled, and also the Gold Pheasant; it is supposed that three or four hundred brace of each of these species [sic] are to be found at large within the Park Wall (Daniel 1807)'.[8] James Hamilton (1860) recalls seeing 'golden and silver pheasants' at Blenheim in 1798, but none of these in 1834.[9] On the same 'pheasantry' visit in 1792 Byng also commented that 'the squirrells are destroy'd, as *suckers of pheasants eggs* !!' – the first Blenheim reference to control of (alleged) game predators; the exclamation marks show that Byng was clearly a better naturalist than the Blenheim gamekeepers.[10] By the early nineteenth century waterfowl were also bred, at the north end of the Lake, by the duke's fisherman (Baggs et al. 1990); at least some of these must have been ornamental, as Mavor (1820) noted that 'the Fisherman's House … and the adjoining duckery, usually containing some very curious species, deserve notice'; the fisherman may also have bred ducks for wildfowling.

---

8   The 'Chinese or pencilled variety' is the oriental race *P. c. torquatus* with a white neck-ring, a feature now cross-bred into most UK pheasants (Kirkman and Hutchinson 1924, Snow and Perrins 1998); the context suggests that the 'gold pheasant' was another variety, not the Golden Pheasant *Chrysolophus pictus*, a quite different species, introduced to the UK much later.

9   Hamilton suggested Daniel (1807) had also seen these two varieties, but Daniel's text makes no mention of 'silver' birds. Mavor (1820), however, mentions 'gold and silver pheasant' kept at the aviary.

10   Red Squirrels (*Sciurus vulgaris*), the only species then present, occasionally take arboreal songbird eggs (Krauze-Gryz and Gryz 2015), but pheasant eggs are much too big and Red Squirrels do not forage much on the ground.

After this brief flurry of mentions (including a few for non-game species, next section), all goes quiet in print for birds until the middle of the nineteenth century, and then material is scant. The 7'h Duke's younger son Lord Randolph Churchill's letters from home when at school (Eton) in 1863 included 'accounts of sport, of partridges and pheasants' (Churchill 1906). A decade later there is little change, as Churchill wrote in 1873 from Blenheim to his fiancée in Paris of 'the round of shooting parties, the varying totals of slaughtered hares and pheasants'. As driven game shooting developed, Blenheim's activity shifted from a family affair to being part of an aristocratic circus where estate owners, landed gentry and their friends toured the country for what became competitive events for the size of bag and the grandness of the setting and repasts. 'Thousands of pheasants' eggs were hatched each spring under hundreds of hens, and the chicks were reared in captivity until they were large enough to be turned loose in the coverts' (Fowler 1989). In 1907, under 'Sport' the *Victoria County History* (Page 1907) noted:

> Blenheim is one of the principal shooting estates in the county. Unfortunately, the old game-books were destroyed in a fire which occurred at the palace some years ago, and the present records do not go further back than the season 1870–1. The bag of those days did not differ in any great degree from that of the past season (1905–6), some 9,000 head of game being killed in the first-named season as compared with about 11,000 head in the past winter. The best season here was that of 1896–7, when H.R.H. the Prince of Wales, our present king, paid the Duke of Marlborough a visit, and on 27 November was one of a party who shot 1,334 Pheasants; the total bag for the day being 2,210 head. The guns on this occasion were H.R.H. the Prince of Wales, Lord Chesterfield, Lord Gosford, Viscount Curzon, Sir S. Scott, Major-General Ellis, the Rt. Hon. Henry Chaplin, and Mr. W. H. Grenfell. The total amount of game shot that season was 23,196 head. The partridge shooting on the Blenheim estate has been greatly improved of late years by the turning down of Hungarian birds, and by driving. In the season 1905–6, 1,571 partridges were killed, 193 brace being obtained by seven guns in a single day's driving, and in 1904–5, when the stock had suffered severely from the disastrous season which preceded it, 1,157 birds were killed. Going back to 1877, the game-book shows that 527 partridges were shot in the season—a fair average bag for many Oxfordshire estates at the present time. Undeterred by the example of Eynsham, the duke is now trying the French method of partridge-rearing, and the pens for this purpose are in course of construction.

The staggering totals noted here were by no means a one-off. Over three days in different park habitats on 7–9 October 1898 five 'guns' shot 6,943 rabbits (day 1), 1,700 pheasants (day 2) and 614 partridges (day 3) – with additional smaller numbers, including many hares, when they weren't the main target (Martin 1987). The 9th Duke's wife Consuelo Vanderbilt (Balsan 1953) recalled the same occasion:

> Marlborough and the other guns were always shooting, for there were pheasants, rabbits, duck, woodcock and snipe to be killed. I remember a record shoot one autumn when seven thousand rabbits were bagged by five of the best English shots in one day. They had two loaders and three guns each, and every one of them had a violent headache on reaching home. At least for a time the High Park was free of rabbits.[11]

'Sunny' Spencer-Churchill, the 9th Duke, was keener on Grey Partridges (*Perdix perdix*) than pheasants, and in 1895 and 1906 planted hedges and shelter belts specifically to create

---

11   Later in life, after divorce from the 9th Duke, Consuelo married Jacques Balsan.

A Game-card from the Blenheim archives

Figure 17.3. 'A game card from the Blenheim archives', from Green (1950, and subsequent editions) – showing two Snipe, a hare and three pheasants.

barriers to drive partridges over (Waterhouse and Wiseman 2019). Snipe (*Gallinago gallinago*), marsh birds much more abundant in the nineteenth century than today, were also targeted; 106 killed in one day in 1883 by three 'guns',[12] 97 by two 'guns' on another occasion (Bryden 1904, Martin 1987; see Figure 17.3). Woodcock (*Scolopax rusticola*) were mentioned by Balsan (1953), but there is no other detail about them for Blenheim at this period. Page (1907) noted that for 'sport' (shooting) 'Oxfordshire is not a very good county for woodcock, though the birds are met with in small numbers in nearly every part of it; half a dozen, or perhaps ten, are accounted a good bag'.

For the 10th Duke (John, known as 'Bert'), who succeeded in 1934, 'shooting always took precedence over agriculture', and he was 'as enthusiastic over duck and snipe shooting as over the more orthodox sport offered by pheasant and partridge' (Martin 1987); this is reflected in an increase in the duck bag from 70 to 80 in the early 1930s to 700 to 800 from 1935 to 1940 (Figure 17.4; Blenheim archives). Although in the 1890s duck were 'seldom shot' (Cornish 1895) and were consequently quite relaxed about humans, 'in Bert's day there were several lake shoots each season, duck being put off the water by keepers in boats. Quite a few snipe were then to be found at marshy spots near the Lake, and good mixed bags could be had' (Martin 1987; Figure 17.3).

Unlike many estates where accounts of shoots are numerous and vivid, there are only snippets published for Blenheim, so the illustrated account of one in the mid-1950s (exact

---

12 The Lake keeper told Charles Cornish (1895) that the total was 112 Snipe, plus 40–50 Mallard (*Anas platyrhynchos*) and Teal.

date unspecified; Frissell 1957) is a useful marker in the history of game management and harvesting:

> the 2,700 acres around the Great House still abound with a variety of game, including snipe, woodcock and pheasant. Shooting in the Blenheim manner is an elaborate affair. On a typical morning a small procession of Land Rovers heads out from the palace loaded with titled sportsmen. Once in the field, a platoon of loaders and gamekeepers readies the guests for the shoot. When the guests have been positioned at the various shooting stands, the drive begins. Then across fields and through thickets, an army of beaters moves toward the party, putting up literally hundreds of birds within range of the waiting guns. If the drive is made in the park of the palace, the guns return to the palace for lunch. But if they shoot on outlying farms they stay out all afternoon and a hot lunch is taken out in heated containers to be served by the butler and footman. By teatime, a successful shoot may have produced as many as 700 palace-reared birds, which will be sorted afterwards from Blenheim's cavernous freezer to be given away as holiday gifts or sold to the local markets.

Blenheim's archives currently hold game bag records from 1929 to 1976 for several bird species, also hares and rabbits, but not deer; the post-1870 game books mentioned by Page (1907) were not found by the archivist in 2022, and post-1976 records appear not to have been handed in by the gamekeepers. Figure 17.4 shows, on a log scale so that widely differing numbers can be accommodated on the same graph, the numbers of birds shot from the 1929–30 season through to 1975–6; note that partridges, ducks and pigeons are not identified to species. In the 1930s most partridges will have been Greys, but by the 1970s largely Red-legged (*Alectoris rufa*). Pigeons will have been largely Woodpigeons (*Columba palumbus*), but could include some Stock Doves (*C. oenas*), while 'ducks', mostly Mallard (*Anas platyrhynchos*), may include several other species. In addition, not graphed, there were occasional culls of up to 165 Coot (*Fulica atra*).

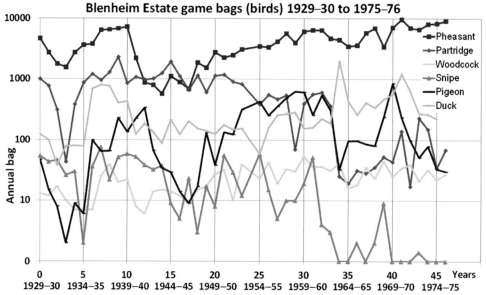

Figure 17.4.   Blenheim Estate game bags 1929–30 to 1975–6, log scale. NB: as zero cannot be registered on a log scale, the zero scores for Snipe in the 1960s–70s are scored as '1'.

When the 11th Duke, also John, took over responsibility for the estate in the 1960s, he 'assisted in modernising the shooting "from the old system of broody hens"' (Martin 1987), in an unspecified way. Martin went on to say that

> the rearing programme is traditional, with all the birds derived from caught-up stock, and the average bag in recent decades has been 500 pheasants per day and 10,000 for the season. Some 25 days are shot throughout the full season (some eleven or twelve let [to outsiders]) during which time the staff are often fully stretched with house parties for up to sixteen guests for a long weekend.

The same year, from a different perspective, Cobham and Hutton (1987) wrote under 'Game':

> The sporting interests have been especially important in recent decades, and this is to a large extent the reason for the retention of woodlands which should ideally have been felled and replanted in rotation years ago, for the delay of thinning, and for some new plantings that are inappropriate from the standpoint of both forestry and amenity. Ironically the existing woodlands have become increasingly unsatisfactory for sporting purposes, because they now lack the understory or shrub layer to provide a suitable habitat for the pheasants. Thinning can only be carried out in a period of six weeks between the end of the shooting season and the beginning of next year's rearing programme. While many of the existing woodlands do need replacement, they have had to be retained as game cover until alternatives can be provided elsewhere. The game department, asked to produce its own ideal management requirements, looked for new woodland planting areas into which the forestry operation could be steered, and drew up its own recommendations for selective felling in the Lince area and phased replanting over a twenty-year period along the western shelter belt. Some of the areas recommended for planting by the game department again proved unacceptable because of their encroachment over the SSSI.

By the early twenty-first century, a management report for High Park commissioned by the estate (Mottram and Kerans 2014) commented under 'Pheasants':

> The Estate runs an active, keepered pheasant shoot that includes High Park. There are pheasant release pens in Compartments 26H and 26N. Pheasants are brought in as poults around June and placed in the release pens until they are able to fly. There is one supplementary feeding area in the south of the site in Compartment 26F but most feeding is done outside High Park. At the feeding station in Compartment 26F there is evidence of ground compaction. This is confined to a relatively small area. During the winter shooting season birds are driven from copses and game strips outside High Park, over the valleys where the guns are stationed and back towards the release pens. These drives, being within the historic parkland, play an important role in the overall sporting strategy for the estate. During several days of field survey for this study [in winter 2012–2013], mostly outside the shooting season a large, though unquantified, number of pheasants were seen throughout High Park. The impact of these birds on the overall ecology of High Park is not known but it seems likely that there will be an impact as they forage on the woodland floor.

Two management reports by Historic Landscape Management Ltd (HLM) for the whole park (2014, 2017) refer to continued shooting of other species too:

> The Palace runs an active shoot with commercial and non-commercial interests. Shooting takes place within the park on 30–34 days per annum, although on most of these days only a few hours will be spent in the park as the guns normally start on the wider estate and move into the park during the day. The shoot allows for 8 guns per day and is based on pheasant and partridge. (HLM 2014)

> Blenheim Palace run an in-house game shoot of pheasant and partridge. A small amount of duck flighting over Mapleton Pond and Bladon Lake takes place, together with a low level of deer stalking, mainly for control purposes. Fishing is also permitted on the Great Lake. Management of the game requires the planting of game cover crops in parts of the park together with the use of some of the woodlands for game rearing and cover. (HLM 2017)

A short shooting holiday on the estate, for early 2020, staying three nights 'all inclusive' at the Palace (and overnight at Claridges) was advertised at $11,000 per couple to wealthy Americans as 'the ultimate sporting experience':

> The estate offers a mixture of Partridge and Pheasant shooting spread out over the Estate and offers both woodland and open ground shooting enabling teams of guns to enjoy a day's challenging and varied sport. A day normally comprises of five of the Duke's favourite drives in the parkland. (Wing and Barrel 2019) [they add '400 birds per day'].

The 'partridge' are nowadays Red-legged Partridges, Greys having more or less vanished from this part of Oxfordshire; they are open country birds, usually seen during the survey in the fields by Combe Gate, but also twice within High Park.

During the High Park Biodiversity Survey (HPBS) 2017–21 very large numbers of pheasants were regularly seen throughout the woodland during the spring and summer, that is, birds from previous releases. However, during 2020 numbers fell dramatically. No releases took place in High Park as shooting was cancelled because of the COVID-19 pandemic, and pheasant numbers were only roughly 10% of 'normal' levels in the winter months 2020–1, though they seemed to have recovered somewhat by the spring. Counting pheasants reliably within the woodland is not feasible, but the High Park birds also feed out on the fields by Combe Gate, especially in the early morning – I have counted up to 52 there with no sign of any diminution within the woodland.

17.3. Gamekeeping and predator control

What is conspicuously absent from these accounts of pheasant management and shoots is any information, historical or recent, about what may have been done to control predators. We know from numerous other sources (e.g. Lovegrove 2007) that the growth of organized sport shooting in the mid- to late eighteenth century led to increasingly zealous gamekeeping, with almost any animal thought to impact on the chosen quarry ('vermin') heavily persecuted – even to killing Kingfishers (*Alcedo atthis*) for taking fish fry. Cornish (1902) remarked on the increase in Kingfishers on the Thames in the 1890s following a ban on shooting on the river by the Thames Conservancy and local county councils. The subsequent disappearance of the heronry in High Park seen in the 1780s and 1790s by Byng (Andrews 1934) may have been a measure to protect fish in the Lake. The last breeding Raven (*Corvus corax*) in Oxfordshire was shot on the estate in 1847 (anon. 1854, Pickles 1960), the last Red Kite (*Milvus milvus*) and Buzzard (*Buteo buteo*) in the county (before the late-twentieth-century reintroduction and a recent return, respectively) were both trapped (alive) at Blenheim in 1851, when also a Peregrine (*Falco peregrinus*) was shot (Powys 1852); Carrion Crows (*Corvus corone*), Rooks and Jays (*Garrulus glandarius*) were reported as being shot in the 1930s and 1950s (Elliott and Alexander 1938, Pickles 1960). There is no mention of systematic predator control for game preservation in the published literature, though it must surely have occurred, and indeed a repurposed 1950s 'vermin' register sheet in the estate archives has the following headings (Figure 17.5): 'Stoats, Weasels, Rats, Cats, Rooks, Jackdaws, Carrion Crows, Jays, Magpies, Hawks, Hedgehogs.'

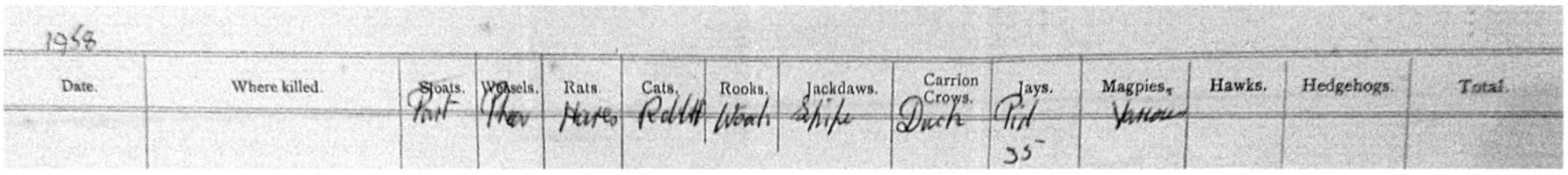

Date.	Where killed.	Stoats.	Weasels.	Rats.	Cats.	Rooks.	Jackdaws.	Carrion Crows.	Jays.	Magpies.	Hawks.	Hedgehogs.	Total.

Figure 17.5. Headings of Blenheim Estate 'vermin register' from the 1950s.

Table 17.1. Numbers of birds regarded as pests recorded as killed on the Blenheim Estate during the 1980s.

Date	Little Owl	Carrion Crow	Jay	Magpie	'hawk'	'various'	TOTAL
1980–81	*56	158	71	76	*10	nl	371
1981–82	*24	112	60	56	*9	nl	261
1983–84	*15	92	111	111	*7	nl	336
1984–85	*15	129	46	113	*9	nl	312
1987–88	nl[6#]	122	70	120	nl[8#]	[$'33']	[326]

Asterisks (*) indicate species later subsumed as 'various', hash signs (#) indicate estimates based on declines in owl and Hedgehog bags over the time series. Dollar sign ($) indicates total of 'various' (including Hedgehogs) as included on original documents. 'nl' = not listed. Totals exclude 'various'.

These records must have been kept by gamekeepers, but most do not seem to have been passed to or retained in the archives. A few records of 'vermin' bags from the 1980s have, however, survived (Table 17.1), including, surprisingly, Little Owls (*Athene noctua*) and with the addition of newly invasive (American) Mink (*Mustela* (=*Neovison*) *vison*). The figures show Little Owl kills dropping dramatically while the other species numbers are maintained. The original typed sheets with added pencil notes that Little Owls, 'hawks' (presumably Sparrowhawks *Accipiter nisus*) and Hedgehogs (*Erinaceus europaeus*) are the species lumped under 'various' in the 1987–8 record. This change probably reflects a very belated realization that killing protected birds deliberately, without specific evidence that they were harming livestock (i.e. pheasants for shooting), was best not explicitly recorded! Little Owls came within the orbit of the Wildlife Protection Act 1954,[13] and Sparrowhawks were protected from 1961 after massive pesticide-related decline (Lovegrove 2007). What the figures do show is the evident abundance of Little Owls at Blenheim in the early 1980s (not reflected in the contemporary bird lists: Brucker and Campbell 1975, 1987) and the relative scarcity of Sparrowhawks. It appears that the owls may have built up prior to 1980,[14] and were subsequently targeted, as Brucker et al. (1992) noted that 'over the years the species has suffered persecution from gamekeepers, and even in the 1980s some estates were known to shoot all Little Owls. This is both illegal and unnecessary since the diet consist primarily of invertebrates and small mammals, and game birds, even young ones, are rarely if ever taken.' Sparrowhawks continue to frequent High Park (details in Section 17.3, below), but there has been no recent sign of Little Owls, which have suffered a countrywide decline in recent decades (Balmer et al. 2013). There were still 'a few pairs' in 1987 (Brucker and Campbell 1987); in the wider estate the last record of a Little Owl was in 2009 (OOS database).

Trapping of Crows and Magpies (*Pica pica*) has certainly taken place in recent years (Aljos Farjon, pers. comm.). As discussed in Section 17.5 below, Jackdaws (*Corvus monedula*) are now abundant in High Park, with smaller numbers of Jays (Figure 6), Carrion Crows, Kites,

13 https://www.legislation.gov.uk/ukpga/1954/30/contents/enacted.

14 They were listed as 'Resident but very seldom recorded' by Pickles (1960) and similarly by Brucker and Campbell (1975).

Figure 17.6.   Typical view of a Jay half obscured by branches. (Photograph by Anthony Cheke)

Buzzards, Sparrowhawks and Ravens, the last four now protected species. Rooks (*Corvus frugilegus*) and Magpies (*Pica pica*) were only recorded peripherally in the survey years.

17.4. The general birdlife of High Park (non-game)

Apart from Plot's (1677) Nuthatch, we know nothing of birds in the old Woodstock Royal Park, and precious little of Blenheim Park in the eighteenth and nineteenth centuries. In the late eighteenth we find John Byng noting wildfowl and seeking out the heronry on two occasions (Andrews 1934):

> [24.7.1784] On the lake are quantities of swans; and flocks of wild geese, who swim in the same particular lines, in which they fly. We next rode to the upper lodge [i.e. High Lodge] (newly built on the same spot, where the old one stood, in which the witty Ld Rochester resided;) then to the new bridge, plantations, &c. &c., and at my desire, to the hernery, where great numbers of herns are yearly bred.[15]

> [Friday July 6, 1792] I took a long evening ride, upon my mare; towards the Hernery; a noble assemblage, and only to be had in old parks, upon very old trees.

His last remark suggests the heronry, apparently large, was in High Park. It was probably a leftover from the time when Grey Herons (*Ardea cinerea*), and thus heronries, were protected as game, targeted by falconers with Peregrines, and the young harvested for food, especially on aristocratic estates (Shrubb 2013). Byng's mention of 'great numbers … yearly bred'

---

15   Hern or hernshaw was a common alternative name for heron well into the nineteenth century (Lockwood 1984). Lockwood, incidentally, did not include Plot's 'wood-cracker' in his extensive listings of British bird names.

suggests harvesting may have still been practised. As we have seen, the old royal manor was used for falconry at least until the mid-sixteenth century, but there is no evidence that the Marlboroughs used falcons; hawking for Herons was in any case dying out by the later eighteenth century. Herons had general protection under a game law of 1603; it was even forbidden (apart from the landowner) to shoot anything else within 600 paces of a heronry (Ray 1678, Marchant and Watkins 1897).[16] This (nominal) protection was lost when Herons went unmentioned in the definitive Game Act of 1831, and persecution to protect fisheries, away from the nesting sites (e.g. Aplin 1889), as well as directly in some cases, appears to have eliminated many heronries as the nineteenth century progressed (Shrubb 2013), continuing into the twentieth (Nicholson 1929).[17] As the Reverend Francis Morris wrote in the 1850s, 'Royal game in the times of falconry, and prized for the table, now-a-days he [the Heron] is the object of all but universal hostility' (Morris 1850–7, Soper 1981), so it is not surprising that the great Blenheim heronry disappeared.

Judging by the rather unclear knowledge of geese in the county in the nineteenth century (Aplin 1889), Byng's January birds would have been Whitefronts (*Anser albifrons*) or Pinkfeet (*A. brachyrhynchus*). Mavor (1797) also noted [Mute] Swans (*Cygnus olor*) 'and other aquatic fowl indigenous as well as exotic'. Byng was distressed by over-enthusiastic pest control in the estate's orchards and fields:

[Thursday July 5, 1792] walk'd to Blenheim kitchen garden … What shocked me much was to hear the firing of guns, and to see a set of Jacobins arm'd against the national guards – the birds – Oh fye! – What, for a few cherries, destroy all the songsters? And here will they come to perish. 'Stretch forth, Marlborough, thy hand of mercy and of pity; and let not infamous slaughter prevail.'

[Friday July 6, 1792] But all such pleasures [observing nature] are only allow'd to an absolute master; for I find that these, here, are nearly destroy'd, as *potent of mischief*! The rooks are destroy'd as *potent of mischief*! The squirrells are destroy'd, as *suckers of pheasants eggs*!! and the singing birds are destroy'd as *destroyers of fruit*! So man, instead of encouraging delights, and the companions of his walk, becomes from ignorance and idleness, the ruin of his own pleasures.

Rooks, some of which had evidently survived, also feature in an extreme weather event reported by William Mavor (1820, anon 1852):

On the evening of the 29th of November, 1797, a freezing rain began to fall, and in the course of the night, incrusted every tree, shrub, and blade of grass, to a thickness almost incredible. In consequence of this, many trees and an immense number of branches were brought to the ground. In Blenheim Park to which, and small surrounding space, this phenomenon was confined, nearly one thousand loads of wood were destroyed. The very rooks had their wings frozen, and fell from their perches.

Mavor revised his guide many times. The 1836 edition includes a note on hirundines; discussing the Grand Bridge the author commented that:

in some of these dark and unexplored **recesses** it is not improbable that one or more species of the swallow tribe find a winter retreat, and lie in a torpid state till

16 Despite quoting the law protecting Herons (2 Jac 1, cap.27), Ray (1678) also included details of a way to trap and kill Herons ('a great devourer of fish') using baited hooks.

17 Although Oxfordshire was not listed among counties known for shooting Herons (some still offering bounties!), neither was it among those protecting them (Nicholson 1929); Nicholson only listed one active heronry in the county. Morris (1850–7) listed all British heronries he knew of, but none in Oxfordshire.

the return of spring. This is certain, that they have been noticed skimming the lake as early as any have been discovered on the sea coasts. One season a white swallow was seen for a considerable time.

Aplin (1914b) commented that the early arrivals were probably Sand Martins (*Riparia riparia*), which to this day nest in holes in the bridge's stonework (pers. obs.). Supposed hibernation of birds that disappeared for the winter was a long-standing myth.

After the glimpses cited, nothing is heard about Blenheim birds until the minutes of the Ashmolean Society (anon. 1854, Pickles 1960) reported,[18] on 6 December 1847, that the 6th Duke had presented '[a] remarkable raven, shot on its return to its eyrie, which it had long occupied on the top of John, Duke of Marlborough's pillar, in Blenheim Park, Dec. 3, 1847,—from the Duke of Marlborough. The expanse of its wings is four feet.'

This was the last recorded breeding of Ravens in Oxfordshire before their return in the twenty-first century. Soon afterwards ornithologist Thomas Powys (later Lord Lilford) mentioned a Kite and a Buzzard trapped, and a peregrine shot, in 1851 (Powys 1852).[19] When he wrote the first county avifauna, Oliver Aplin (1889) had clearly never visited Blenheim, and apart from repeating Powys's remarks, mentions only Coot and Mallard as breeding on the Lake. Cornish (1895), in a previously overlooked book chapter, added Dabchicks (*Tachybaptus ruficollis*), [Mute] Swans and Moorhens (*Gallinula chloropus*) and, in winter, Teal (*Anas crecca*), Wigeon (*A. penelope*), a few Snipe and the odd Heron. By the turn of the century Aplin started visiting Blenheim, reporting Coots again (in 1903; Aplin 1904), adding breeding Great Crested Grebes (*Podiceps cristatus*) on the Lake margin and House Martins (*Delichon urbicum*) under the Grand Bridge, plus Nuthatches, 'a few Redstarts [*Phoenicurus phoenicurus*] about the ancient oaks' in June 1913 (Aplin 1914a, b), and a Moorhen nesting on a boat in 1914 (Aplin 1915). In 1904, alarmingly to twenty-first century sensibilities, he reported (Aplin 1906) that: 'Mr. Darbey, of Oxford, told me he had received many Hawfinches [*Coccothraustes coccothraustes*] to stuff this winter. Also that in the summer, four years ago [i.e. 1902], he had a "basketful" from Blenheim, where they were breeding in the gardens, but "did too much damage to the peas".'

Atlee (1915) noted 'a couple of dozen' Tufted Duck (*Aythya fuligula*) on 11 March 1914, a Greater Black-backed Gull (*Larus marinus*) on 13 April, with a Green Sandpiper (*Tringa ochropus*) and six Great Crested Grebes on 7 November.

I have given the early bird records in full both to indicate the sparseness of published observations prior to the founding of the Oxford Ornithological Society [OOS] in 1921 (Radford 1966), and to pull them together for the first time.

It is only after the founding of the OOS that more systematic observations started to be made at Blenheim, but still concentrated on the Lake and waterbirds. Tucker and Ottley (1924) recorded the first breeding of Pochard (*Aythya ferina*) and Tufted Duck in Oxfordshire on the Great Lake in 1923. It was not until 1975 that another county first, breeding Gadwall (*Anas strepera*), was recorded on the Lake (Brucker and Campbell 1975), followed by a more unexpected invader owing to climate change, Cattle Egret (*Bubulcus ibis*) in 2021 (OOS Bulletin 563, August 2021; pers. obs.). Pochard no longer breed, but the other two ducks nest along the High Park lake edge as well as elsewhere.

Elliott and Alexander's very briefly annotated list (1938) was the first to consider the birds of the estate as a whole, listing all recorded species, gleaned during waterfowl census work

---

18 This story was clearly unknown to Aplin (1889) and Alexander (1939, 1947), and even to Holloway (1996), who all listed the last breeding as in 1834 'near Oxford' (locality unspecified). As she gave no reference, it is unclear whether Pickles (1960) rediscovered it or someone else found it and told her; my thanks to Tosh Moller for tracing the original report.
19 There is, however, no mention of Blenheim in Powys and Lilford's major work *Coloured Figures of the Birds of the British Islands* (London: R.H. Porter, 1885–7; 7 vols), nor, apart from the same brief paper listed among his publications, in the compilation by Trevor-Battye (1903).

by the OOS (dates not specified), and clearly including at least some observations in High Park. Pickles (1960) and Brucker and Campbell (1975, 1987) updated the list, with slightly more annotated detail. The status over time of birds that use High Park are summarized in Appendix 14; more specific information on species of ecological importance to the ancient woodland is discussed in Section 17.5.

17.5. Birds in High Park – 1. Overview of breeding species

While there is no detail on numbers and distribution of birds within High Park prior to the recent biodiversity surveys, the accounts from the 1930s onwards give an overall picture of the avifauna. What is very clear is the number of breeding species that have disappeared since the 1970s or a bit later. Until then, as might be expected in healthy oak woodland, there were nesting Redstarts, Spotted Flycatchers (*Muscicapa striata*), probably Wood Warblers (*Phylloscopus sibilatrix*) (all migrants), plus Lesser Spotted Woodpeckers (*Dendrocopos minor*) and, probably in the areas with ground seepage, Willow Tits (*Parus montanus*), both resident. Hawfinches, a 'scarce resident in the oakwoods' (i.e. High Park) in the 1950s, disappeared earlier, remaining only as occasional winter visitors. The near-absence of Dunnocks (*Prunella modularis*) as egg-hosts may be the reason, in addition to national decline, that Cuckoos (*Cuculus canorus*) do not now frequent High Park, although formerly considered to breed, and still 'frequently heard calling in the wooded areas' in 1987. Also lost as a breeder, although never common, is Woodcock; it is still seen as a scarce winter visitor – one ringed in Perth (Scotland) was shot in November 1931 (Elliott and Alexander 1938). Apart from sporadic breeders such as Crossbills (*Loxia curvirostra*), and the return of long-lost raptors and Raven (discussed later in this section), the only gain has been the Mandarin Duck (*Aix galericulata*; Figure 17.7), an introduced tree-hole-nesting species that appreciates the many ancient hollow trees, first confirmed as breeding in 2020. All the lost species also declined and disappeared in Wytham

Figure 17.7.   Male Mandarin Duck on the pond at the head of the Palace Vista in High Park. (Photograph by Anthony Cheke)

Woods (Overall 1989, Gosler 1990, Perrins and Gosler 2010) and Shotover (Whitehead et al. 2003, Whitehead 2018) during the 1970s and 1980s, and have been declining in England generally for some time (Fuller et al. 2005, Balmer et al. 2013, Harris et al. 2020), particularly in the south, though for some (Willow Tit, Lesser Spotted Woodpecker) the population collapse has been more widespread. In Oxfordshire the Lesser Spotted Woodpecker numbers peaked in the late 1970s taking advantage of dead elms, but had mostly gone by 1983 except in Bagley and Wytham woods (Brucker et al. 1992), surviving, however, until 2009 in Shotover (Whitehead 2018); the last records in High Park were single birds in April 2004 and September 2006 (OOS). As in Wytham (Gosler 1990, pers. obs.) Tree Sparrows (*Passer montanus*) put in an ephemeral appearance, arriving to breed commonly 1961–6, then declining to more or less disappearing by 1975, and have not been reported since.

The reasons for the declines are not fully understood (Fuller et al. 2005, Thaxter et al. 2010), but include local causes as both migrants and residents are affected. Climate change and agricultural-induced insect declines may be a factor for some (Redstart, Wood Warbler, Spotted Flycatcher) whose breeding distribution has retreated north and west, but loss of understorey owing to expanding deer density is also implicated for woodland species (Perrins and Overall 2001, Fuller et al. 2005, Gill and Fuller 2007, Perrins and Gosler 2010, Cooke 2019). Increased deer browsing since the arrival in the county of Muntjac (*Muntiacus reevesi*) and the return of Roe (*Capreolus capreolus*) since the 1970s (Ward 2005) may have had an impact; both, as well as Fallow Deer, are common in High Park (Chapter 18; pers. obs.) Thinning of undergrowth has dramatically increased in High Park with the arrival of British White Cattle in 2020. Several species that persist in High Park, such as Blackbird, Blackcap, Chiffchaff, Song Thrush, and especially Willow and Garden Warblers, are known to have been impacted by high deer density in nearby Wytham Woods (Perrins and Overall 2001, Perrins and Gosler 2010). Though numbers of the first three remain healthy in High Park, where owing to a more open canopy there is perhaps more undergrowth than at the deer density peak in Wytham, it is possible Song Thrush numbers are depressed, and both Willow and Garden warblers are now scarce, but both have suffered national declines (Gill and Fuller 2007), whereas Song Thrush has made some recovery nationally after declining during 1970–95 (Balmer et al. 2013). Turtle Doves (*Streptopelia turtur*), a woodland edge/hedgerow species, another species in catastrophic decline in Britain (Balmer et al. 2013), were recorded breeding 'in the High Lodge region' in 1975 (Brucker and Campbell 1975). By 1987 continued breeding was uncertain (Brucker and Campbell 1987), and there is only a single Blenheim record in the OOS database, from 1999. In this case the loss can be largely related to intensified agriculture and concomitant loss of weed seeds (e.g. Dunn et al. 2017), helped by unsustainable illegal persecution on migration in some Mediterranean countries (Lormée et al. 2020), though disease (Stockdale et al. 2015) and conditions in Africa are also implicated (Ockendon et al. 2012). Woodcock, still visiting in winter, used to breed 'in denser woodland, particularly in the north of the park' (Brucker and Campbell 1975), but have not been recorded 'roding' (their display flight and call) in High Park since 1997; this is another species in long-term decline as a breeder in Britain (Fuller et al. 2005, Heward et al. 2015). Starlings (*Sturnus vulgaris*), a hole-nesting species, were noted as nesting in old trees (Brucker and Campbell 1975, 1987), though where was not mentioned; however, there was no sign of them in High Park during the HPBS period, though some were clearly nesting near the Grand Bridge in 2020. This is another species in long-term decline, so the remaining birds may prefer to nest closer to the best feeding grounds (open close-grazed grassland).

To judge by the older accounts some breeding species have increased: Great Spotted Woodpeckers (*Dendrocopus major*; Figure 17.9) were described as 'a few pairs' in the 1930s and 1960, and 'several pairs' in 1975, Stock Doves (*Columba oenas*; Figure 17.8) were described as 'resident?' in the 1930s, 'breed in old trees' in 1960, but had become 'very common' by 1975, declining in the 1980s (Brucker and Campbell 1987), but unremarkable enough to be

Figure 17.8.   Typical birds of High Park 1: Stock Dove (left); Blackcap (right). (Photographs by Anthony Cheke)

almost unrecorded in the OOS database from 1996. Jackdaws (*Corvus monedula*) went from just 'resident: breeds on palace' (1930s) to 'very common' by 1975 (Elliott and Alexander 1938, Pickles 1960, Brucker and Campbell 1975, 1987). All three are now very common in High Park; indeed Stock Doves so rarely feature in woodland censuses that High Park may be the only woodland in Britain where they outnumber breeding Woodpigeons. The same increase applies to Blackcaps (*Sylvia atricapilla*; Figure 17.8), a summer visitor: 'a few breed' in the 1930s, 'several pairs' in 1975, but relatively 'numerous and widespread' in 1987 and now one of the commonest breeders in High Park. All four, together with Chiffchaffs (*Phylloscopus collybita*),[20] have seen large national increases since 1970 (Burns et al. 2020), though the woodpecker has declined again somewhat in the last decade (Woodward et al. 2020), as it also did in High Park over the years 2019–21. Although Nuthatches have shown a consistent overall increase in England since the mid-1970s (Woodward et al. 2020), and always common at Blenheim, they appear to have suffered a temporary decline here during the 1970s and 1980s (Brucker and Campbell 1975, 1987); they were affected in Oxfordshire by four cold winters between 1971 and 1982 (Overall 1988, Brucker et al. 1992), and no doubt also in 1963.

The other typically woodland birds still present, some 20 species, seem to have been roughly stable in numbers (Appendix 14), though the earlier status of several was rather vague. An example of unclear long-term status is the Willow Warbler (*Phylloscopus trochilus*), a summer migrant species in decline in (especially southern) England (Harris et al. 2020); they were sparse breeders in the 1930s–50s, but considered 'common' in 1975 and 1987, but are now scarce in High Park, with only five or six breeding pairs in 2019–20, declining to just one in 2021. After an increase from the 1930s to the 1970s, Goldcrests have decreased owing to the removal in 2019 of conifers (mostly larch *Larix decidua*) from the former mixed long plantation; Coal Tits, common through to the 1980s, and indeed until 2018, have declined dramatically since for the same reason.[21] Several species of woodland birds are severely knocked back by severe winters – Goldcrest, Green Woodpecker (*Picus viridis*), Long-tailed Tit (*Aegithalos caudatus*), Nuthatch, Treecreeper (*Certhia familiaris*; Figure 17.9), and Wren (*Troglodytes troglodytes*) being the most affected (see Brucker et al. 1992 for Oxfordshire data), but winters have been generally mild in the survey years. Both Green Woodpecker and Mistle Thrush (*Turdus viscivorus*; Figure 17.14) feed largely in grassy areas, the woodpecker largely on ants, so are not very numerous within High Park.

---

20 Past indications for Chiffchaff at Blenheim are too vague to tell whether numbers have changed over time.
21 Both Coal Tits and Goldcrests prefer mixed woodland to pure conifers, and pure conifers to pure broadleaf. The four remaining small conifer plots (in 2021), 3 about 1ha, the other about 2ha, are perhaps too close-planted and dense to suit either species very well.

Figure 17.9. Typical birds of High Park 2: Greater Spotted Woodpecker and its characteristic round excavated nest holes (left); Treecreeper (right). (Photographs by Anthony Cheke)

Figure 17.10. Red Kites: large chick on nest in High Park: note vertebral skeleton of a large fish by the chick (left); adult in flight (right). (Photographs by Anthony Cheke)

Figure 17.11. Buzzard being mobbed by Kestrel over High Park; Raven (right). (Photographs by Anthony Cheke)

Apart from the colourful Mandarin Ducks, the most spectacular changes are seen in the return of three large generalist predator/scavengers – Raven, Red Kite and Buzzard (Figures 10, 11). Wiped out over the whole of south central and eastern England in the nineteenth century by zealous gamekeeping but now very conspicuously back at Blenheim, as also in the rest of Oxfordshire, they nest in High Park but range over the whole estate. All three are absent from Elliott and Alexander's list (1938); the buzzard was a rare winter visitor in Oxfordshire from the 1940s on (Alexander 1947) and appears as an occasional but increasing visitor in the 1960, 1975 and 1987 lists. It became increasingly regular from 1996, and was recorded breeding on the Blenheim Estate in 1999. After a single record in

1996, kites appear regularly from 1999, with nesting behaviour first noted, in High Park, in 2007; these originate from birds reintroduced by the RSPB to the Chilterns initially in 1989 (Wotton et al. 2000). Ravens, like Buzzards spreading naturally from strongholds in the west, first reappeared in 2000, though breeding behaviour was first noticed only in 2011. The other species in this category, Carrion Crow, was heavily persecuted in the past ('much shot': Elliott and Alexander 1938); a few pairs live in High Park.

Tawny Owls (*Strix aluco*) nest in High Park, but the status of two other birds of prey, Barn Owl (*Tyto alba*) and Sparrowhawk (*Accipiter nisus*), is unclear. A Barn Owl roosted in the southern part in early 2019 and another was flushed from a tree not far off in June 2021, but there was no evidence of a nest. No owl activity has been seen around a Barn Owl box on a tree in a field visible from Combe Gate, though a pair of Kestrels (*Falco tinnunculus*) occupied it in 2021; another two owl boxes were put up near High Lodge in early 2021. Remains, mostly of Woodpigeons, apparently Sparrowhawk kills, are seen fairly regularly, but a bird at a nest in May 2019 and one overhead in April 2021 are the only sightings of an actual Sparrowhawk during the survey period.

Finally, there are the waterbirds that use the Lake edge along High Park to nest: one to a few pairs of Great Crested Grebe (Figure 17.13), Mallard (Figure 17.12), Gadwall, Tufted Duck, Coot and Moorhen; the dabbling ducks (Mallard, Gadwall) and Moorhens also

Figure 17.12.   Waterbirds making use of the duck house on the High Lodge ponds: Mallards (left); Grey Heron (right). (Photographs by Anthony Cheke)

Figure 17.13.   Great Crested Grebe nesting among the branches of a tree that had fallen into the Lake on the High Park shoreline. (Photograph by Anthony Cheke)

visit, and may occasionally breed at, the woodland ponds. The appearance in May 2022 of a pair of Egyptian Geese (*Alopochen aegyptiacus*) in an oak by the pond at the head of the Palace Vista may presage breeding in future by this expanding tree-nesting exotic goose. In some years one or two singing Reed Warblers (*Acrocephalus scirpaceus*) occupy lakeside vegetation in the Combe Creek area when it grows up in midsummer, but are apparently non-breeders.

17.6. Birds in High Park – 2. Non-breeding visitors and wood-edge occasional breeders

Some species that live in the surrounding area visit High Park sporadically, either simply flying over, making occasional foraging visits or simply using the peripheral trees for rest or shelter.

The most frequent users are waterbirds: Grey Herons (Figure 17.12), Little Egrets (*Egretta garzetta*), and Greylag and Canada Geese (*Anser anser* and *Branta canadensis*), which breed in other parts of the Lake and River Glyme, and feed along the shore or (geese) graze on the open grassy area off the lower Combe Creek, especially in late summer with their young. Great White Egrets (*Ardea alba*), Black-headed Gulls (*Larus ridibundus*) and Kingfishers also sometimes feed along the shoreline. Herons also visit the woodland ponds. Cormorants (*Phalacrocorax carbo*), which nest with herons on the island in Queen Pool, rest and roost in trees along the High Park shore especially in winter, but necessarily feed only in open water. A few Wigeon sometimes mix with Mallards and Gadwall along the shore vegetation in lower Combe Creek in winter. It is likely that various wader species and other water- or water-edge birds that occur around Queen Pool also occasionally use the shores within High Park, but rarely enough not to have been picked up during the survey period; by chance in 2021 two migrant Yellow Wagtails (*Motacilla flava*) were seen in mid-April and a Grey Wagtail (*M.cinerea*) in September.

The woodland is irregularly host to various winter visitors. The paucity of berries persisting beyond early autumn limits the use by winter thrushes Redwing (*Turdus iliacus*) and Fieldfare (*T. pilaris*) – indeed the latter, although common in hedgerows, appears not to have been seen in High Park. The small finches, Siskin (*Carduelis spinus*) and Redpoll (*C. flammea*), which feed mainly on birch (*Betula* spp.), Alder (*Alnus glutinosa*) and larch seed, are seen fairly regularly in winter, as, more sporadically, are Crossbills, which as conifer cone specialists are likely to be seen less in High Park in future. Hawfinches, tree seed specialists regularly seen in the Palace gardens where they formerly bred, are exceptional in High Park, and Bramblings (*Fringilla montifringilla*), nomadic winter visitors associated with Beech mast, have only been seen once during the survey years.

Hedgerow and woodland edge species – Dunnock (*Prunella modularis*), Goldfinch (*Carduelis carduelis*), Greenfinch (*C. chloris*), Bullfinch (*Pyrrhula pyrrhula*) and Yellowhammer (*Emberiza citrinella*) probably all wander in occasionally. The first three are fairly often seen along the lane just outside Combe Gate, and (especially Goldfinches and Greenfinches) sometimes nest within the High Park boundary in that area, with Yellowhammers a bit further afield nearby. The scarcity of Bullfinches is a bit surprising, as in 1987 they were said to be a 'resident that breeds, more commonly found in the High Park'. When formerly much more common generally, they were notorious for damaging fruit buds in spring (e.g. Perrins and Gosler 2010), and there is plenty of suitable Hawthorn, Blackthorn and Wild Cherry in High Park; however, they are known to be adversely affected by loss of Bramble owing to deer, as are Dunnocks (Gill and Fuller 2007), and have suffered a similar substantial decline in Wytham (Perrins and Overall 2001, Perrins and Gosler 2010). Several other hedgerow and grassland species recorded as breeding in Blenheim Park in the past do not now appear in High Park, and possibly never did; I have excluded these.

Several falcons, Kestrel, Hobby (*Falco subbuteo*) and Peregrine, are occasionally seen around or over High Park, but despite apparently suitable habitat for the first two, and plenty of food (pigeons) for the last, do not nest, though kestrels use a nestbox erected for Barn Owls some 200m from the boundary near Combe Gate. How a kestrel ended up as a skeleton in a cattle trough near High Lodge, found in July 2021, is a mystery; kestrels are open country hunters, but the trough is deep under woodland canopy. A Goshawk (*Accipiter gentilis*), a sub-canopy woodland raptor, was seen in 2019, and the most unlikely British bird saga of 2020 began with my sighting overhead of the subsequently famous Lammergeier (*Gypaetus barbatus*), a huge vulture, over High Park on 23 June,[22] wandering from the Alps to spend the summer in the Pennines.

A few species are simply seen overflying the wood, but not otherwise using it: various gulls (*Larus* spp.), Rooks and in summer, Swifts (*Apus apus*), Swallows (*Hirundo rustica*) and occasional Common Terns (*Sterna hirundo*), are the most frequent. I have once seen a Collared Dove (*Streptopelia decaocto*; an urban species), and Starlings must also do so.

Non-breeding and casual visitors that use the woodland are listed separately in Appendix 14, together with overflying species and water- and water-edge birds.

17.7. Analysis of the breeding birds of High Park

Following 'species present' surveys in 2017, and rather sparsely in 2018, by other observers, over the three years 2019–21 I made an attempt to census the breeding birds of High Park. In each breeding season (March–end-June) I made five to eight visits to the woodland, sometimes accompanied by another birdwatcher, marking bird detections on a map, basically using the 'territory mapping method' (Bibby et al. 1992). As High Park is a rather large area to use this method efficiently, I walked one particular section each visit to get maximum data for that section, and used this as a basis for calculating probable numbers in other areas surveyed less often (methodology to be written up elsewhere). In 2021 I also matched the census made from the public roads by Ben Carpenter in 2017 (Carpenter 2020 and pers. comm.) as a cross check and possible indicator of changes over the interval; additional visits by Ian Lewington in 2017–19 and Russ Hedley in 2017 have added to the overall picture.

For birds with discrete territories that sing a lot the coverage will be good; fortunately, this is most species. Some species, however, are less vocal and also relatively cryptic in behaviour (Marsh Tit *Parus palustris*, Garden Warbler *Sylvia borin*), and some additionally have high-pitched songs and calls outside my hearing range, even with hearing aids (Goldcrest, Treecreeper), though my associates, when present, can hear them – these will all have been under-recorded. Also possibly under-recorded are the most abundant species, particularly Robins (*Erithacus rubecula*; Figure 17.14) and Wrens: they have such small territories that in conservatively assigning territories to clusters in the mapping, some territories may have been combined. The song or advertising (woodpecker drumming) period is not the same for all species – Great Spotted Woodpeckers and Nuthatches are most vocal/noisy in March-April, while for migrants (and Wrens) late April to early June is best. Many woodland birds have only a single brood that ties in with food (mostly caterpillar) abundance, and once breeding is over singing stops, birds moult and become silent and hard to locate, so census work becomes pointless after the end of June. I did not attempt to census Jackdaws as they

---

22 This record was not accepted by the British Birds rarities committee (BBRC; British Birds 114: 626, 2021) as it didn't reach their strict criteria – at the time of the sighting I only knew I had seen, very briefly, a huge but unidentified raptor, only realizing what it was when later seeing photos of the bird – missing its central tail feathers, exactly like the bird I had seen over High Park. It was only the second time this species has been seen in the UK.

Figure 17.14.   Typical High Park birds 3: Mistle Thrush on the Palace Vista grassland (left). (Photograph by Anthony Cheke); Robin, the most abundant woodland species in High Park, with the possible exception of Jackdaws (right). (Photograph by Jackie Ingram)

are both abundant and not territorial; the population is probably of the order of 100+ pairs, concentrated in the south-west part of High Park. Large non-territorial species (raptors, corvids) were estimated based on sightings clusters, but in some cases birds that forage in High Park were nesting outside it – that is, their home ranges extended beyond the High Park boundaries; one Raven pair often seen in 2019 nested in the conifers just north of High Park. Breeding waterbird nests were counted directly, apart from the tree-nesting Mandarin Ducks whose actual breeding numbers are somewhat uncertain.

I had hoped to compare breeding bird densities in High Park with those in other woodlands, but it turns out that this is a lot harder than expected. Widely differing methodologies make comparisons difficult, and the fairly rapid nationwide changes in abundance over time for many woodland species invalidate studies as comparators even when only a couple of decades old.

17.8. Breeding trends during 2019–21

Population estimates for the 15 commonest species are shown in Figure 17.15.

While it is hard to generalize from three years' data, some patterns are clear. Chaffinch, Willow Warbler and Great Spotted Woodpecker all declined as per national trends (Woodward et al. 2020), though the drop in woodpeckers was sharper than expected. Tits notoriously fluctuate from year to year, and the coincident drop in both Blue and Great Tits in 2020, reflected in national data (*Bird Table* 104: 6-7, 2020), suggests a poor breeding season in 2019, a year when there was such severe defoliation of High Park and other local oaks by caterpillars that post-fledging food may have been hard to find; however, the bounce-back seen in High Park in 2021 is not reflected nationally (*Bird Table* 107: 9, 2021). The same pattern is more weakly echoed in Chiffchaffs and Nuthatches, which also mostly feed in the canopy, but also in Song Thrush, which doesn't. The two pigeons (Woodpigeon and Stock Dove) show a completely opposite pattern, peaking in 2020, echoed also by Carrion Crow. Owing to the COVID-19 pandemic, 2020 was a year without pheasant releases or shooting, which may have led to reduced control of crows, but should not have benefited pigeons. The other species graphed (Blackbird, Blackcap, Wren) have been fairly constant over the three years, apart from Robin, which has consistently increased.

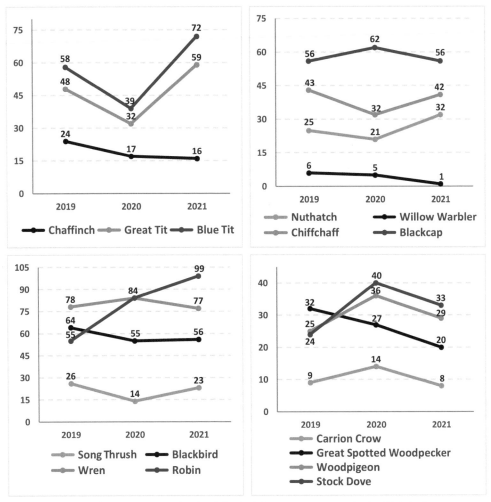

Figure 17.15.   Population estimates (territorial pairs) for 15 common species in High Park, 2019–21.

17.9. Future prospects

While it is likely that Willow Warblers will be lost, Chaffinches and Great Spotted Woodpeckers are resilient species that will probably stabilize at a lower density before recovery. Chaffinches and also Greenfinches have been afflicted by a serious infection, *Trichomonas gallinae*, which appears to be the main cause of their national declines (Lawson et al. 2011). In the other direction it is likely that the oakwoods may be colonized by introduced Ring-necked Parakeets (*Psittacula krameri*); this attractive but aggressive hole-nesting parrot spreading out of London reached Oxford in 2018 and first bred there in 2019 (Evans and Jennings 2022) after a period of sporadic records – so may well soon find High Park to its liking, although it has so far tended to prefer areas with human habitation. Competition for nest sites with Nuthatches has been reported in Belgium (Strubbe et al. 2010), and tree-hole usurpation and killing of Greater Noctule Bats (*Nyctalus lasiopterus*) in Spain (Hernandez-Brito et al. 2018), but Stock Doves seem more likely to be potentially affected in High Park.

Although breeding in low vegetation or on the ground, Blackcaps and Chiffchaffs respectively have increased or maintained numbers in High Park (as at Shotover, Whitehead 2018), but numbers collapsed at peak deer density in Wytham (Perrins and Gosler 2010), which was around 2900 kg/km² MWD,[23] around the year 2000 (calculated from deer numbers in Buesching et al. 2010); this crash could also happen in High Park given the likely longer-term effects of English White Cattle as part of the intended recreation of pasture woodland (Chapter 19). At about 20 animals in 2021 (reduced from 45 in 2020), cattle MWD was about 2000kg/km², which with deer added is likely to result in similar understorey changes as seen in Wytham. Blackbirds and Song Thrushes also do better where there is dense low vegetation as their nests can be better hidden from predators, declining in Wytham at peak deer numbers (Perrins and Gosler 2010). Other birds vulnerable to loss of understorey vegetation, Nightingale, Redstart and possibly Willow Tit, have already long gone from High Park. On the other hand, opening up more grassy area in pasture woodland is likely to favour Mistle Thrush and Green Woodpecker, and might attract Starlings, Rooks and Kestrels back.

It is difficult to predict how far climate change will affect woodland birds, though some trends are clear. Greater storm frequency may lead to more ancient oaks being overthrown, and could thus lead to pressure on nest-sites. Spring rainfall seems to be less predictable (which affects leaf-burst and insect abundance), and a number of studies have noted an increasing mismatch in bird breeding seasons (relatively inflexible) and the earlier cater-pillar peak owing to warmer springs, leading to lower breeding success (e.g. Visser et al. 2006).

17.10. Acknowledgements

Data from 2017–18 come largely from fieldwork by Ian Lewington and Russ Hedley, and an independent survey by Ben Carpenter (summarized in Carpenter 2020), with some input 2019–20 from Ian Lewington. Gerry Tissier and Ruth Ashcroft's help with fieldwork on occasion was much appreciated; extra records were supplied by several people, notably Ivan Wright, Graham Collins and Aljos Farjon. Alexa Frost kindly gave me access to game books and vermin records in the estate archives. Steve Holliday sourced some important documents inaccessible from libraries during the COVID-19 pandemic restrictions. Ben Carpenter and Aljos Farjon made helpful comments and suggestions that have improved the text.

---

23 MWD (metabolic weight density) is defined by Kleiber's Law as mass$^{0.75}$ (Ramirez et al. 2018), which means the larger the animal, the less relative impact it has; the MWD of one cow 600$^{0.75}$ = 121 = roughly 5 Fallow Deer (65kg), 11 Roe (25kg) and 17 Muntjac (14kg).

CHAPTER 18

Mammals

Ray Heaton

18.1. Introduction

The mammal surveys at High Park identified most species that we expected to find in Oxfordshire within this site of mixed woodland that contains a high density of ancient oaks. In the UK there are no mammal species that actually depend on ancient oaks as a habitat requirement, but three of the species of bat that were found are woodland specialists, and one of these has a high affinity and some reliance for ancient oaks. This species, the Barbastelle Bat (*Barbastella barbastellus*), uses cavities and spaces behind areas of shrinking dead bark on large old trees, often oaks, as roosting and breeding sites. Of the other species, the Noctule Bat (*Nyctalus noctula*), known as the woodland bat, roosts and breeds in tree cavities and often forages at or close to the canopy top. The Brown Long-eared Bat (*Plecotus auritus*) forages often below the canopy at the 'browse line' created by deer reaching up to eat young leaves, and often actually removes insect prey from the leaves, a process known as hawking. The nearby Great Lake and other smaller waterbodies are probably the reason for the occurrence of some other bat species at High Park, such as the Daubenton's Bat (*Myotis daubentonii*) and Nathusius' Pipistrelle (*Pipistrellus nathusii*).

Among the other mammal species found some are woodland specialists and some are certainly often found in woodlands. The clue is in the name for the Wood Mouse (*Apodemus sylvaticus*), while the Grey Squirrel (*Sciurus carolinensis*) is mostly dependent on mixed deciduous woodland with mature 'fruiting' trees. Most of the other mammal species identified can be found in a variety of habitats as well as in ancient woodland.

18.2. Survey methods

The presence, in High Park, of twenty-six species of mammal was demonstrated by a variety of techniques including direct observation, sometimes by looking under refugia mats, tracks and signs, live trapping and by sounds and calls made by the mammals. Traces of the mammals were found such as snagged hair and shed antlers from deer species and also the remains of dead animals, including bones in bird pellets. Observations were also made of species in lethal traps from the gamekeeping activity associated with the pheasant shoot.

To discover and identify small mammal species such as mice, voles and shrews, live trapping was undertaken. Aluminium Longworth and similar style traps were used across High Park, in order to sample species from a variety of habitats. Traps were left out overnight for 24-hour periods and longer and were checked regularly at between 8 and 12 hours. When the inspection

Ray Heaton, 'Mammals' in: *The Natural History of Blenheim's High Park*. Pelagic Publishing (2024). © Ray Heaton. DOI: 10.53061/KVXM4486

Figure 18.1.   Fallow Deer (*Dama dama*) grazing under veteran oaks. (Photograph by Aljos Farjon)

found the trap door down the whole trap was placed in a large unsealed clear plastic bag and opened, allowing the animal to emerge and be observed, identified and then released quickly in the same spot. Each trap was provided with dry bedding and vegetable food, for example, fresh carrot, celery and seeds. Insects were also provided as live food for shrews in the form of fly larvae and pupae, in sufficient quantity for an animal to survive between inspections. Shrew trapping licences were held by the operators. Refugia mats originally set out to attract reptiles were found to be a good way to observe and identify some small mammals. These mats are half metre square corrugated sheets of composite material that are laid on the ground on top of short vegetation where the sun can warm them. Voles and mice particularly often shelter under the mats (as reptile species and invertebrates also do) and even build nests. When one carefully lifts the mat, in a door-like fashion, the animals present often freeze for a moment allowing species identification.

Larger species such as deer were observed in the wood and even nocturnal species such as Badger (*Meles meles*) were seen, sometimes by surveyors working on other taxa such as moths and beetles. Tracks and signs such as footprints, snagged hair and feeding evidence were important indicators of the mammal species along with droppings and sometimes the remains of the animals themselves were found. Droppings are referred to as scats while Otter droppings are known as spraint. Identification can be found in guides such as Bang and Dahlstrom (1974) but was often provided by the specialist Bob Cowley.

Bat species were discovered by several methods, including the use of ultrasonic detectors that allow the bat calls to be heard and the species to be identified. Each bat species has a different call that can be distinguished by the trained human ear listening to samples of the call on a heterodyne detector; calls are also recorded and presented digitally on a sonogram. The detectors were used on transects when routes were walked with a hand-held ultrasonic detector that picked up and recorded the calls, and sometimes simultaneous observations of the bats could be made. Bat calls are made faster and at a higher frequency than the human ear can normally listen to, so the full calls and behaviours cannot be matched unless samples of the calls in real time with the behaviours seen are listened to on a heterodyne detector, as was done on the transects. Bat behaviour can also be inferred from sonogram traces. Other types of static fixed detectors such as Anabats were used to record calls at different sites within the wood, including near small waterbodies and by the Lake. Bats, including species not necessarily associated with waterbodies, will often fly to ponds, pools and lakes when they are emerging from day roosts in order to drink before they start foraging; therefore, most bat species can be encountered near water at such times. Direct observations of bats in their roosts, for example, tree holes, were made by optical aids such as an endoscope and torch light. Live trapping and release of flying bats was also carried out using mist nets and a harp trap. All such invasive sampling techniques were carried out under licence.

18.3. Species accounts

The species found arranged by groups following the Mammal Society list are listed below.

18.3.1. Insectivores

Hedgehog (*Erinaceus europaeus*)

Once a common and widespread species on the UK mainland, the Hedgehog is now in decline and has achieved Vulnerable status under IUCN (International Union for Conservation of Nature) Red Listing criteria. There is good habitat at High Park and plenty of resting, breeding and hibernation places. They eat mostly invertebrate prey but will take larger items such as eggs and chicks. Hedgehogs were recorded as dead animals, and live ones were also observed in the Park.

European Mole (*Talpa europaea*)

This is a widespread common subterranean UK species that is rarely seen above ground. An earthworm-eating specialist, moles leave distinctive mounds of loose soil known as molehills that can indicate foraging and breeding behaviour. A number of mole predating species are present in High Park such as Tawny Owl, Buzzard and Stoat. Molehills were observed, including fresh ones, within the Park.

Common Shrew (*Sorex araneus*)

A small mouse-like animal with a long snout-like nose. This species has a pelage of tricoloured dense velvety fur, darker brown on the back, lighter brown on the flanks with greyish white underparts. It lives a fast life of rarely more than twelve months and eats large quantities of mostly arthropods, worms and molluscs. Shrews use burrows in the undergrowth, often those vacated by other small animals. On-site predators include owls, Stoat, Red Fox and Kestrel, though the shrew carcase is often left abandoned apparently owing to the foul-tasting skin. Common Shrews were recorded in live mammal traps, under refugia mats and by their remains in bird pellets at High Park. A licence is necessary for live-trapping shrews and live food and bedding must be provided in the traps and regular inspections of them made.

Pygmy Shrew (*Sorex minutus*)

A smaller species than the Common Shrew, it has a thicker tail (Figure 18.2). The fur is light brown on the back and flanks with greyish white underparts. It has the distinctive long snout-like nose. It is widespread in the UK. Pygmy Shrews also live a fast life and need to consume up to 125% of their body weight daily; this is why great care needs to be taken when live trapping by providing live food to avoid deaths from starvation. This species was found as skeletal remains in bird pellets at High Park.

Water Shrew (*Neomys fodiens*)

The largest British shrew, which is widespread on the mainland, mainly in aquatic habitats but also recorded from mixed woodland. It has the distinctive long snout-like nose particularly prehensile in this species. The fur is a dense velvety jet black with white underparts. It forages under water for aquatic invertebrates and small fish and also terrestrially feeds on

Figure 18.2.   Pygmy Shrew (*Sorex minutus*) with a 'blue bottle' fly as prey. (Photograph by Pat Morris)

earthworms and amphibians. The animal seen in High Park by Anthony Cheke was found under a refugium mat. Its presence may be due in part to the proximity of the large lake bordering much of the north side of the Park and other nearby aquatic areas.

18.3.2. Lagomorphs

European Rabbit (*Oryctolagus cuniculus*)
This species was probably introduced to Britain (Chapter 1) and is now widespread and introduced to many offshore islands. Its use as a human food and fur source resulted in breeding colonies kept protected from predators and poachers in warrens. Introductions to the wild occurred as a result of warreners' activities or most likely a lack of them. Rabbits form their own colonies in the wild by digging a complex of tunnels in which they live, with numerous entrances onto grazing areas. They are smaller than most cats and have a bobbing hopping gait, brown fur and distinctive long ears. Rabbits like to feed on short grass, which their grazing can create, and in doing so they maintain a lawn-like habitat that can be important for many other species of fauna and flora. In High Park this species was recorded even though there is an absence of rabbit-grazed areas. Live animals were observed and scats and latrines found. In the surrounding park land and countryside there is much suitable habitat. Rabbits are an important prey item for Red Foxes and Stoats, in spite of the Stoat being much smaller in size. Young rabbits are predated by Buzzards, Badgers and Weasels.

Brown Hare (*Lepus europaeus*)
This species was introduced during the Iron Age from northern Europe. Larger than the rabbit and with longer black-tipped ears, the Brown Hare is an agile fast running species often seen running and leaping. Hares live above ground in the open, sheltering in vegetation and in depressions they make called forms. They feed mainly on grass and are found on open grasslands and agricultural areas throughout Britain. In north-west and highland areas of Scotland it is replaced by the smaller Mountain Hare (*Lepus timidus*). Hares are particularly predated by the larger birds of prey and Red Foxes. Young hares known as Leverets are taken

by smaller predators. Adult Brown Hares were seen within the wood and in open areas in High Park but are much more likely to reside in the surrounding park land and agricultural areas.

18.3.3. Rodents

Grey Squirrel (*Sciurus carolinensis*)

The Grey Squirrel is an introduced species from North America, released from 1876 into the UK in Cheshire. Since then, Grey Squirrels have become a highly visible and widespread member of the British fauna. Evidence from squirrel damage on planted oaks indicates its arrival after 1910 in High Park (Chapter 3). They have a habitat preference for deciduous woodlands and feed particularly on large seeds such as oak, walnut and chestnut. Often spending time on the ground to cache large nuts in the soil they may inadvertently 'plant' seeds that germinate into seedling trees. The large stick and leaf nests, known as dreys, which are usually situated high up in the crooks of tree branches, make the presence of this species particularly evident. The live animals were observed in a variety of tree species in High Park and also dead ones in the lethal traps around the woodland. Goshawks and Pine Martens, also Stoats and Red Foxes are known to prey on Grey Squirrels. In the Park there is much evidence of tree damage particularly in the growth form of emergent oaks, caused by the feeding behaviour of Grey Squirrels on the bark and buds.

Bank Vole (*Myodes glareolus*)

The Bank Vole is a mouse-like rodent with a rounded snout and a tail that is about half the length of the body (Figure 18.3). The fur is brown and the small dark eyes are obvious. They are found throughout the UK. Bank Voles are vegetarian and feed on green leaves, seeds, fruits and flowers and to a lesser degree on fungi, roots and mosses. They occupy a wide variety of habitat types including gardens, hedgerows, deciduous and mixed woodland, heathland and arable areas. In areas of outgrown grassland vole runs, open tunnels within the vegetation, may be evident upon parting the vegetation; they also make nests and tunnels in the base of

Figure 18.3.   Bank Vole (*Myodes glareolus*). (Photograph by Pat Morris)

tussocks and under sheet building materials. There is vole habitat in the Park in the grassy glades centrally, along the road and in an area of tussock in the southern part. Voles are important prey for many avian and mammalian predators. Bank Voles were recorded in live traps, under refugia mats, from feeding signs on cherry stones and from their remains in bird pellets at High Park.

Field Vole (*Microtus agrestis*)
A common and widespread species in the UK, Field Voles have grey-brown fur and a short tail. They specialize in feeding on grass species but also eat other vegetation and have a wide habitat preference including gardens, grassland, mixed and deciduous woodland, heathland and arable areas. Nests of leaves and grass stems are built in the base of grass tussocks, in small underground tunnels and even under corrugated sheet building material. Field Voles are important quarry for many bird of prey species and mammal predators such as Red Foxes, Stoats and Weasels. Field Voles were recorded in live traps at High Park, under refugia mats and from skeletal remains in bird pellets.

Wood Mouse (*Apodemus sylvaticus*)
Britain's most common and widespread rodent, this mouse occurs throughout the mainland and on many of the islands (Figure 18.4). It has large ears, a long tail and sandy brown fur. It eats seeds, fruit and invertebrates and makes food caches in tree stumps and even bird nests. It lives in underground burrows and is largely nocturnal. Wood Mice are important prey for many predator species. At High Park it is recorded from live traps and was originally noted from holes gnawed into Hazel nuts found by Ray Heaton beneath the only known patch of Hazel coppice in the wood. Bob Cowley also found cherry stones nibbled in the characteristic manner and skeletal remains were found in bird pellets. This animal is very similar to the Yellow-necked Mouse (*Apodemus flavicallis*) and the nibbled nuts could have been from either species. High Park is on the boundary of Yellow-necked Mouse distribution but they were not recorded. The Yellow-necked Mouse spends much time in tree canopy, has a large home range

Figure 18.4. Wood Mouse (*Apodemus sylvaticus*) sitting on a bracket fungus. (Photograph by Pat Morris)

and is notoriously difficult to live-trap. Yellow-necked mice also live at very low densities, and in the summer a regular seasonal die-off means there are even fewer individuals per hectare, which might explain its absence from this survey.

High Park is within the UK range of the Hazel Dormouse (*Muscardinus avellanarius*) but no evidence was found from nut collection, a standard survey technique. There is a paucity of Hazel trees in the woods of High Park, with just one 20 x 30m patch of self-coppiced Hazels and only a scattering of solitary trees elsewhere (Chapter 3), which suggest a population is unlikely.

Also not recorded but expected was the Brown Rat (*Rattus norvegicus*). It is inconceivable that this species does not occur in High Park as it is thoroughly widespread across all habitats in the UK. None were recorded or seen, or their tracks and signs noted.

18.3.4. Carnivora
Red Fox (*Vulpes vulpes*)
The Red Fox is a medium dog-sized carnivore, a famous icon in the UK and very widespread. They live in burrows which they can excavate, but they often use abandoned Badger holes. Red Foxes are scavengers in urban areas and top predators in the countryside and agricultural areas. Adult Red Foxes were seen in the Park and their scats found, including some containing feathers.

Badger (*Meles meles*)
Badgers are low-slung medium dog-sized animals. They spend a lot of time feeding on earthworms in grassland but feed on other animal prey opportunistically. They have a striped black and white head and a grey brindled body. They live in family groups in an excavated tunnel network known as a sett. Badger evidence was found in High Park in the form of feeding signs, and droppings including fresh latrines and adult animals were observed in spite of their nocturnal nature. An adult was even seen emerging from a hollow oak tree. No Badger setts were found in the Park, but setts are known from elsewhere in Blenheim Park and the nearby countryside.

Otter (*Lutra lutra*)
This species was recorded from convincing footprints in a mud patch well away from the Lake, photographed by Aljos Farjon, and spraint with the sweet, fishy, but not unpleasant, smell was located by Bob Cowley near the Lake. The species was unexpected within the Park as it is a highly aquatic specialized predator. However, Otters are spreading and increasing their populations across Britain. It also forages for and eats terrestrial prey, but good aquatic habitat in the Lake is close by.

Stoat (*Mustela erminea*)
One of our smaller carnivores, Stoats are wonderful long and sinuous animals (Figure 18.5). They have short legs and reddish-brown fur with white underparts and a white lower neck, throat and chin. They have a distinctive black tip to the tail, which along with the larger size helps distinguish them from the Weasel. They occur throughout the UK in a wide variety of habitats including woodland. They mainly predate small mammals but will eat worms and fruit and birds. Stoats are noted for catching prey larger than themselves, particularly rabbits, which they fascinate and 'hypnotize' before biting on the neck. A Stoat was found in a lethal trap set for Grey Squirrels in High Park and scats were also found.

It is likely that other smaller carnivore species are present in High Park, though these animals are wide-ranging and can occur at low density. Gamekeeping activities associated with the pheasant shoot may have suppressed numbers. Weasel (*Mustela nivalis*) is the most widespread and numerous British carnivore common in woodlands, but was not recorded. Polecat

(*Mustela putorius*) was also not recorded, even though this species is currently spreading in central England and has wide habitat and food requirements. American Mink (*Neovison vison*), an introduced and very common and widespread carnivore, also with wide food and habitat needs, was not recorded. Weasel, Polecat and American Mink have been recorded in the local area. No Ferrets (*Mustela putorius furo*) or their feral hybrids, which can survive well in the wild, were picked up, and neither were Pine Marten (*Martes martes*). Not finding this latter species could be excused as the nearest populations, though spreading within the UK, are quite distant. Both Pine Marten and Polecat are afforded legal protection in the UK.

18.3.5. Ungulates

Fallow Deer (*Dama dama*)

Fallow Deer is a most common deer species in England; they were a Norman introduction to English deer parks from southern Europe in the twelfth century. Originally an Asian species, it was introduced to

Figure 18.5.   Stoat (*Mustela erminea*). (Public Domain – Steve Hillebrand, USFWS)

Mediterranean Europe much earlier. A popular species in English deer parks owing to its reasonable size, it is smaller than the Red Deer (*Cervus elaphus*) and somewhat easier to manage in park herds. This attractive animal spends part of the year in single-sex herds. The bucks (males) have palmate (partly flattened) antlers and both sexes sport a tan coat with white spots; they constantly flick their tails, which seems to enhance their pelage. Fallow Deer occur in black and white colour forms and some herds have been selectively bred to have these colours as a decorative aspect, at the whim of the owners. Animals seen at High Park were of the darker form (Figure 18.1), which is more likely to be the natural wild colour in UK habitats. They feed mainly on grass with a proportion of browse in the form of leaves, tree seeds and berries. Although once contained within Woodstock (Blenheim) Park, the current herd roams freely across the countryside and is part of the species population that occupies much of southern England and lowland parts of Scotland. This deer was often sighted in High Park in small groups and in herds of up to 70 individuals, but on many days none were seen. Within the central clearings of the Park there are high hides, with salt lick blocks on the ground close by; these attract individuals that can then be harvested.

Roe Deer (*Capreolus capreolus*)

This deer is now widespread across the UK, being Britain's most common deer species. Roe Deer became extinct in the eighteenth century, apart from in the north of Scotland. The species has now spread south and from introduced populations in southern England to occupy most of the mainland. They live as individuals or pairs or small groups. The males (bucks) have short bi-pointed antlers that can be lethal to other males during courtship behaviour. Their diet includes much woodland vegetation, including deciduous tree buds and leaves, shrubs and creepers, herbs and grasses. A single buck, occasional adults and a female with a fawn were observed in High Park, and Roe Deer presence was recorded in other areas of the Park by scats and footprints in soft earth.

Reeves's Muntjac (*Muntiacus reevesi*)

This small fox-sized deer (Figure 18.6) is a native of south-east China and was deliberately introduced into the UK in 1894; it has now spread through much of England. Occurring singly or in family groups, the males mark their territories by rubbing their forehead scent glands onto trees and shrubs. The males have small single-spiked backward-pointing antlers and well-developed canines. This species, although susceptible to cold winters, can achieve a high population density, particularly in south-east England. Mainly feeding on woodland herbs and shrubby vegetation, Muntjac can cause damage by grazing important native wild flower populations including Bluebells and orchids, and at high density can remove the woodland understorey. Reeves's Muntjac were recorded at High Park from numerous field observations, from scats, from footprints and from antlers and bones discovered by Bob Cowley. This animal is often elusive, but the distinctive barks and calls it makes were often heard on our visits to High Park. Anthony Cheke was able to identify Muntjac 'territories' from this barking behaviour. A male barks, and there can be reciprocal barking from another male or a female. Females bark and make other calls immediately after giving birth in order to attract a male to mate straight away. In this way females can be continually pregnant and rearing young; this explains how high-density populations have occurred. Muntjac, except for their calls, are very cryptic and secretive and therefore difficult to observe; they are often silent and then invariably quietly move away through the undergrowth.

There is much evidence of deer browsing and grazing within High Park, especially on shrubs such as bramble and Blackthorn and on trees. It is possible that deer numbers within the Park from across the three species are influencing the vegetation growth and the regeneration of tree species and other plants.

There are three other UK deer species. Two species, Red Deer (*Cervus elaphus*), a native, and Sika Deer (*Cervus nippon*), an introduced species, were not recorded although both have

Figure 18.6.   Reeves's Muntjac (*Muntiacus reevesi*) in High Park. (Photograph by Anthony Cheke)

populations in central England. The third, Chinese Water Deer (*Hydropotes inermis*), another introduced species, was observed once as a single individual near the Lake in 2021. This small deer species prefers wetland habitats and has previously been recorded from individual sightings on Blenheim Estate before the current biodiversity survey began.

Another species in the ungulate group, a pig, the Wild Boar (*Sus scrofa*), has spreading populations to the west and south-east, and may eventually colonize woodlands in Oxfordshire.

18.3.6. Bats

Daubenton's Bat (*Myotis daubentonii*)
Known as the water bat, this medium-sized UK species is most often seen flying purposefully just over the surface of ponds, lakes and rivers. They can trawl over the water surface using their big feet to catch and pick up prey above, below and on the surface. They take mostly insect prey but are believed to also take small fish; so far this has only been shown in captivity. Their brown furred bodies show up as they cross a torch beam shone just above and across the water. A large colony occupies the now partially flooded and abandoned rooms in the bridge parapets over the Lake in front of Blenheim Palace. Daubenton's Bat is in the *Myotis* genus of species, which contains a number of similar species that are similar in their ultrasonic calls, appearance and behaviour, and hence are often difficult to distinguish. In High Park they were recorded on a stationary Anabat Express detector over a small pond; two were seen hunting over the middle of the pond at the top of the Palace Vista. Other individuals were observed, using an illuminated endoscope, inside a tree roost caused by a wound in a small Sycamore. They were also detected on transects, standardized linear locations within the Park, which were walked with a bat detector.

Natterer's Bat (*Myotis nattereri*)
This medium-sized bat is also in the *Myotis* genus of species. This species is associated with broadleaved woodland, trees in parkland and tree-lined river corridors. Natterer's Bat was recorded in High Park on an Anabat Express detector and seen in a tree roost using an endoscope. They were also encountered on transect detector walks.

Noctule Bat (*Nyctalus noctula*)
The Noctule is one of Britain's larger bat species, also known as the woodland bat (Figure 18.7). Noctules roost in holes in trees that often result from dead or damaged large branches. They are often seen flying straight along or just below canopy level in mature woodlands. This is

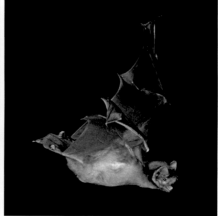

Figure 18.7.   Noctule Bat (*Nyctalus noctula*). (Photographs by Bat Conservation Trust/Hugh Clark)

one of only two mammal species found in High Park that can be said to be reliant on mature woodland; the other is also a bat species, the Barbastelle bat (*Barbastella barbastellus*), which is even more dependent on ageing large trees in mature woodland; a third woodland species, the Brown Long-eared Bat, is less dependent on mature woodland. In High Park Noctule Bats were recorded over a small pond and over the edge of the Lake, using fixed Anabat detectors. Noctule bats were also watched and heard on detectors over a valley in the Park and encountered elsewhere in the Park on bat transects.

Common Pipistrelle Bat (*Pipistrellus pipistrellus*)

The Common Pipistrelle is Britain's most frequently recorded, most numerous and one of our smallest bat species. Until recently considered as one species, pipistrelles have been split into two with the discovery of differences in their echolocation call frequency when heard on bat detectors and inferred from molecular analysis. This second cryptic species, the Soprano Pipistrelle (*Pipistrellus pygmaeus*), is an even smaller species and is found alongside the Common Pipistrelle populations. Common Pipistrelle calls were recorded on Anabat detectors over a small pond and on the edge of the Lake and when walking bat transects. Their social calls were also recorded. Encounters with this species being numerous, they were also seen to make repeated hunting flights around the same circuit through the trees, a well-known behaviour of pipistrelles. Using live traps, under licence, one individual was caught in a mist net and two others in a harp trap; all were released.

Soprano Pipistrelle Bat (*Pipistrellus pygmaeus*)

This bat species was discovered as a cryptic species alongside the Common Pipistrelle population. Fittingly, the Soprano has a higher frequency echolocation call and is a smaller animal. They were frequently encountered on bat transects and their social calls noted. Stationary Anabat detectors picked up this species over small ponds in the Park and over the edge of the Lake. They were also caught in nets, identified and released.

Nathusius' Pipistrelle Bat (*Pipistrellus nathusii*)

Although a threatened species under IUCN Red Listing criteria (Near Threatened in Europe) this bat is widespread in parts of the UK but frequently encountered as single individuals. They are long-distance migrants, having colonized here from mainland Europe, and climate change may well be assisting their spread. It is a medium-sized pipistrelle with dark brown fur and a habitat preference for woodlands and large waterbodies, so well suited to High Park. Nathusius' Pipistrelle was recorded on an Anabat detector over the edge of the Lake.

Barbastelle Bat (*Barbastella barbastellus*)

This medium-sized dark-coloured bat has large broad ears and is sometimes described by bat workers as having wellies (wellington boots) on its head (Figure 18.8). It is IUCN Red Listed as Vulnerable in the UK. It is found throughout the southern half of England and has a habitat preference for broadleaved woodlands, meadows, floodplains and even improved grasslands. The maternity roosts are almost always in trees and more particularly in oaks, sometimes in ancient woodland or parkland, making this the mammal species that has a highest affinity for ancient oaks. Barbastelle bats are known to live at low population density and were only recorded three times during the High Park survey. A flying bat was picked up on a bat detector, an Anabat recording was made over a small pond and this species was encountered once during a bat transect.

Brown Long-eared Bat (*Plecotus auratus*)

Brown Long-eared Bats are a woodland species that prefer broad-leaved, deciduous and mixed woodlands (Figure 18.9). As the name suggests, they have very long ears that are often folded

Figure 18.8.   Barbastelle Bat (*Barbastella barbastellus*). (Photograph by Pat Morris)

Figure 18.9.   Brown Long-eared Bat (*Plecotus auratus*). (Photographs by Pat Morris (left) and Bat Conservation Trust/Hugh Clark (right))

down the back when at rest, leaving the inner part of the ear (tragus) sticking up – and when seen this still looks long. It can hover and feed on larger insects, often taking them from within the canopy or gleaning them from the ground. This bat makes extensive use of old trees and tree cover. Brown Long-eared Bats ultrasound rather quietly and at a low frequency (the call can even be heard by younger people), and this makes them difficult to record on a bat detector. The records from High Park are from two individuals in a tree cavity (both of which were then tagged and returned to the cavity) and a third from trapping; all were caught under licence and released.

18.4. Concluding remarks

A good range of UK mammal species was found within High Park during this five-year long biodiversity survey. There were restrictions on working within the Park, owing to the activities of the pheasant shoot, over several months annually. This problematic access at seasonally key times for some animal populations, along with COVID-19 pandemic restrictions, attenuated and affected the field work. There were some notable species exceptions such as Rats and Weasels, their presence not proven. Mammal species, even larger species, can be notoriously cryptic in their appearance and behaviour, including some species adopting highly secretive nocturnal lifestyles. After all, the ancient ancestors of modern mammals (ours too) spent many more millions of years successfully hiding from the dinosaur fauna than modern species have needed to spend hiding from humans. However, a greater survey effort could have proven the presence of more mammal species.

Acknowledgements

Anthony Cheke is thanked for making considerable mammal observations and discoveries while also working on the bird survey of High Park. Bob Cowley of the Oxfordshire Mammal Group is thanked for stepping in when it looked as if other methods of recording were yielding few results, and for discovering auspicious numbers of tracks and signs and mammal remains, including their bones in bird pellets.

CHAPTER 19

Review of Management Practices

Torsten Moller, Aljos Farjon and Anthony S. Cheke

As described in Chapter 1, the woodland of High Park has been managed for more than a millennium, initially as a royal hunting forest and deer park enjoyed by visiting monarchs and latterly as part of a ducal estate with an increasingly diverse set of priorities. The purpose of this review is to focus on the most recent decades during which the day-to-day stewardship of the Blenheim Estate, including High Park, has been passed to a multi-disciplinary management team challenged with balancing the, often conflicting, demands of different interest groups.

Previous chapters amply demonstrate the complex pasture woodland character of High Park. The interaction of plants, fungi and animals creating a mature quasi-natural environment has evolved during centuries of management action. Arguably the most important imperative underpinning continued interventions is to maintain a historical balance that is undeniably artificial, but without which a unique environment would inevitably revert to a different ecological state.

19.1. Framework for park management interventions

The conservation values of High Park are enshrined in its notification in 1956 as a Site of Special Scientific Interest (SSSI) and its renotification in 1986, followed by the designation in 1987 of Blenheim Palace with its parkland and designed landscapes as a UNESCO World Heritage Site. In the mid-1990s the estate management team started to realize the national and international significance of High Park, which led to the first significant survey of the veteran trees, marking their position along with an abundance of associated information. This baseline survey provided the backdrop for a rolling programme of management plans for High Park, helping to define what long-term actions can be taken to implement corrective measures within a desired conservation framework. Most recently this programme was endorsed through a Higher Level Stewardship agreement between Natural England and Blenheim Estate for the period 2012–22, resulting in a comprehensive analysis of woodland management tools made by the Wychwood Project,[1] entitled *A Management Plan for High Park* ('The Plan') (Mottram and Kerans 2014). This is integrated within the current Management Plan Review for the Blenheim Palace World Heritage Site (MPR 2017).

The Plan provides a benchmark allowing comparisons to be made with conditions prevailing in High Park before and after management procedures have been implemented. Earlier surveys from the 1980s onwards have clearly demonstrated the ecological complexity of the site and

---

1   The Wychwood Project is a local conservation initiative that seeks to inspire and support people to preserve the landscapes and habitats in the former royal hunting forest of Wychwood: www.wychwoodproject.org.

Torsten Moller, Aljos Farjon and Anthony S. Cheke, 'Review of Management Practices' in: *The Natural History of Blenheim's High Park*. Pelagic Publishing (2024). © Torsten Moller, Aljos Farjon and Anthony S. Cheke. DOI: 10.53061/ILNU8469

the need for careful monitoring of the impact and effectiveness of control measures. Good use has been made of photographs to illustrate particular issues and record the condition of specific sites. The Plan includes a set of detailed recommendations to be considered year by year in designated areas defined by their individual woodland/grassland character.

An integral part of this initiative has been the central storage of acquired ecological data by Thames Valley Environmental Records Centre (TVERC) for future reference purposes. Specific surveys targeted the size distribution of ancient and veteran oaks, while other short-term investigations of particular groups of species as well as observations made during the implementation of the Plan have all generated data feeding into the TVERC database (Alexander 2002; Wychwood Project 2008).

19.2. Baseline for management actions

The benefit of early surveys and interventions in High Park was to highlight the sensitivity of different components to ecological change, whether natural or induced by management actions. Examples included the inexorable growth of vegetation encroaching on veteran trees, vistas, grassland and even particular locations of decomposing wood that are strongly influenced by shading and changes in their microclimate. Such changes called for 'haloing' around veteran trees, thinning, pruning and reducing ground vegetation such as Bracken, as well as considering measures to control the depredations of squirrels, deer and rabbits.

A conspicuous outcome of the long-term regime in place in High Park has been the flourishing biota associated with decomposing wood – a renewable resource directly related to the preponderance of veteran trees. It was recognized that dead and fallen trees should be left *in situ* and allowed to decay naturally, so as to provide an undisturbed habitat for a diverse community of saproxylic species (Figure 19.1). The ecological need to retain dead and dying

Figure 19.1.   A fallen veteran oak in High Park. (Photograph by T.H. Moller, 7 April 2019)

trees *in situ* was often in conflict with a desire for tidiness, with the assumption that this meant good management. Although compromises inevitably had to be made, the amount of deadwood habitat in High Park has increased significantly over the last two decades as greater understanding has led to updated management decisions.

A light touch was favoured with respect to regeneration of key tree species, while striving to preserve the mosaic of pasture woodland and grassland and the perimeter belts of trees with a more ornamental character serving *inter alia* as a backdrop to the designed eighteenth-century landscape around Blenheim Palace. However, the coniferous plantations introduced during the tenure of the 10th Duke (Spencer-Churchill 1931–1972) were singled out for removal and replacement with a mixture of broadleaved trees more in keeping with the dominant pasture woodland theme of High Park.

A programme of continuous monitoring was envisaged to identify the impact of interventions and to fill gaps in the knowledge of grassland ecology, pest control and diseases affecting oak, Ash, Beech, Horse Chestnut and other trees. A lack of information on the flora and fauna of the several ponds around High Lodge was noted. Attention was drawn to the sporting use of the Park involving the rearing of pheasants, some of which survive the hunting season and represent a significant component of the resident bird population. It was recognized that sporting interests were more amenable to short-term adjustments than ecological and landscape objectives.

19.3. Implementation of the Management Plan for High Park

Given that the primary objective of the Plan was to maintain the ecological value of the site, it is natural that the aim of safeguarding the veteran oaks has taken centre stage. Halo thinning around specimen ancient oaks to prevent them being shaded out has been a priority over the last two decades but in particular throughout the duration of the Plan. The importance of conserving the surrounding environment by minimizing soil compaction and protecting nearby insect habitats in thorns and various flowering plants has been emphasized. Consideration has been given to the deliberate girdling (ring-barking) of surplus medium-sized oaks as a means of artificially increasing the amount of standing deadwood available for saproxylic species to colonize. Candidate oaks suitable for such treatment were identified in the large northern plantation created by the 9th Duke (Spencer-Churchill 1892–1931), where the crowns of the regularly spaced trees are closing up.

A better understanding of the genetic role in arboreal disease resistance and fitness to the local environment has emerged in recent years. With this knowledge has come the realization that the promotion of natural oak regeneration should be tempered with the opportunity for propagating seeds from veteran trees of known provenance. Recently, about 3,000 acorns have been gathered from the ancient oaks in High Park and are being nursed into saplings for future planting (Rachel Furness-Smith, pers. comm.). Trees from the Middle Ages will have grown naturally, whereas younger trees are more likely planted and potentially less fit to survive. Grey squirrel damage is commonplace in High Park and will remain a problem for the foreseeable future when the most that can be achieved is keeping numbers down. Natural regeneration of oaks has been sporadic after the 1980s, probably because of an increasing population of feral Fallow Deer.

The recommendations for removing the coniferous plantations have indeed been actioned (Figure 19.2). Selective felling has proceeded over the course of several years with a long-term view to replacement with mixed broadleaf planting varying from closed-canopy to more open pasture woodland. Some historic vistas and open ground have been reinstated, including the area adjoining High Lodge to the south. An overgrown glade adjoining the top of the vista towards Blenheim Palace is destined to be opened up, with the ultimate intention of revealing distant hills (Figure 19.3).

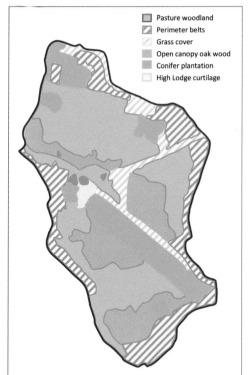

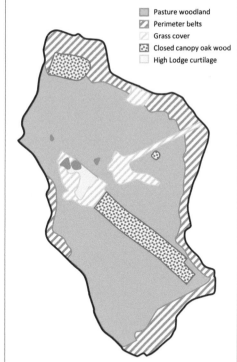

Figure 19.2.   Habitat distribution in 2013.

Figure 19.3.   Aspirational habitat distribution. (Redrawn from Mottram and Kerans, 2014)

19.4. Evaluation of management strategies

Figure 19.2 is redrawn from the Plan and summarizes the patchwork of different woodland and grassland habitats in High Park, plotted by TVERC as of 2013, at the beginning of the recently expired ten-year management plan. Since then, the conifer plantations have been partly removed, thereby reversing the policy introduced by the 10th Duke for exploiting timber. By comparison with Figure 19.3, the overall management strategy can be recognized as one of rationalization and simplification. The central pasture woodland areas are to be amalgamated and extended following the removal of conifer plantations. The perimeter belts are mostly composed of closed-canopy broadleaf woodland. The tussock grassland areas harbouring abundant ant colonies and freed of earlier encroachment are to be retained. The area surrounding High Lodge to the south formerly comprised more open grassland, and the ambition expressed in the Plan is for such open ground to be reinstated.

The Estate's long-standing resistance to the introduction of cattle to High Park was reversed in 2019. A small herd of British White Cattle – a heritage breed – has been hosted on a trial basis to see if the cattle's trampling and grazing pressure is beneficial in discouraging Bracken and maintaining or recreating the pasture woodland character of past centuries (Figure 19.4). The introduction of cattle, albeit on a trial basis, represents an acknowledgement of the challenge of achieving and then maintaining the historically relevant state of pasture woodland. The desired outcome is a mosaic of open woodland with free-growing trees in association with a diverse understorey of shrubs and grassland. This is a difficult ecological balance to achieve using heavy mechanical equipment, which is cumbersome, prone to cause undesirable soil compaction and incompatible with the policy for fallen deadwood to be left undisturbed to decompose *in situ*.

Figure 19.4.   British White Cattle in High Park. (Photograph by T.H. Moller, 7 February 2020)

The Plan recognizes that climate change is likely to entail a greater risk of summer drought and storm damage in autumn/winter. Large trees growing on the thin soils that characterize significant portions of the High Park plateau are prone to increased wind-throw. Consideration is given to planting a wider range of species in the protective boundary belts as a response to predicted climate change.

A brief comment in the Plan refers to the benefits of using a Geographical Information System as a powerful tool for a more visual display of collected data and the facility for easier identification of significant correlations between parameters with a common geographical link.

Maximizing the survival and regeneration potential of the ancient oaks in High Park is rightly prioritized. Measures to check the impact of vegetative growth, disease and vermin all form part of an overarching strategy that seeks to maintain an ecological status quo that is faithful to the historical legacy of the managed environment. These objectives are not compatible with unrestricted public access to the Park, which in turn means that the substantial ongoing management costs for High Park must be met through external financial support such as income streams from other parts of Blenheim Estate.

Blenheim Palace and Blenheim Estate strive to operate together as a largely self-sufficient financial enterprise, while enjoying capital tax exemptions through meeting government conditions regarding public access and maintenance of national heritage assets. The Management Plan Review (MPR 2017) sets out the interests of stakeholders and provides guidance for managers within an agreed framework attempting to reconcile the many and often conflicting objectives of the business. The stakeholders include UK statutory bodies as well as the members of a Steering Group comprising the owner and representatives of Blenheim Palace, International Council on Monuments and Sites, Historic England, Natural England, Oxfordshire County Council and West Oxfordshire District Council. The managers responsible for implementing the Management Plan are mainly senior staff members of Blenheim Palace and Blenheim Estate working in conjunction with the ducal family trustees and a range of other partners to maintain good stewardship.

19.5. Grazing with cattle in High Park

19.5.1. General aims of conservation grazing
The primary aim of introducing grazing in pasture woodland reserves is to restore biodiversity in the landscape and 'maintain the woodpasture' (National Trust: 'Conservation Grazing'; Woodland Trust: 'Conservation grazing in woodland management';[2] Corporation of London: 'Cattle Grazing').[3] Grazed woodland using cattle and sheep is thought to create a close approximation to a natural ecosystem that was widespread in prehistoric times (National Trust). The Woodland Trust also mentions the effects of grazing on woodland in pre-Neolithic times and refers more specifically to animals then present in Britain: Red Deer, Aurochs, Wild Boar and Beaver. Both organizations acknowledge that realizing this aim through grazing requires careful management informed by monitoring developments in the vegetation.

Management involves animal densities, timing and choice of grazers. Lack of grazers has led to 'overgrown and shady' woods, which has 'led to a serious decline in many species that require disturbance and more open conditions, including rare lichens and ground flora' (Woodland Trust). 'Overstocking' of woodland grazers can lead to 'overgrazing and soil compaction preventing natural regeneration' of trees. The Woodland Trust compares various animals, with cattle being more suitable in rough vegetation, breaking open dense stands of Bracken (but not eating it) and creating an uneven sward by trampling the ground. Unlike other grazers, they can consume dead plant material not palatable to other animals. English Longhorns are considered most suitable in this environment by several grazing managers, but other breeds are also used (see Table 19.1). The woodland structure that grazing by domestic ungulates, mainly cattle, is meant to create and maintain is a more or less park-like landscape with (thorny) shrubs, trees and patches of grassland: pasture woodland. It is not closed-canopy wood or forest, to which many pasture woodland sites have reverted in the absence of such grazing.

19.5.2. The situation in High Park
In High Park this transition to closed forest was under way, with Ash as the dominant infill of trees among the ancient oaks, despite a high deer density more recently. Prior to the introduction of cattle in 2019, thinning of the Ash trees had opened the canopy between many of the oaks. The grassy glades were gradually disappearing under cover of spreading Bracken fronds where soil conditions were acidic, while a dense ground cover of False Brome – not grazed by deer – prevailed where the limestone is closer to the surface. Owing to the presence of much fallen dead wood – a deliberate policy – mowing and rolling of Bracken was limited to wood-free areas. The introduction of cattle seeks to overcome these limitations in a more 'natural' way. Cattle grazing was not envisaged in the High Park Management Plan in 2014 (Mottram and Kerans 2014) but was introduced in 2019. Apart from the more general aims of grazing in pasture woodland mentioned above, it is at present not clear what past situation the Blenheim Estate is hoping to bring back. It may be the landscape as perceived to have existed in the medieval hunting park or a landscape envisaged to have existed in prehistoric time under the 'Vera Model' (Vera, 2000). Alternatively, the grazing may merely be the pragmatic choice for maintenance of open grassland glades, given the difficulties encountered with mechanical methods.

---

2   https://www.woodlandtrust.org.uk/media/1824/wood-wise-woodland-conservation-grazing.pdf.
3   https://democracy.cityoflondon.gov.uk/documents/s127406/Appendix%201.%20Final%20 Draft%20Plan.pdf.

Table 19.1. Sites in England with ancient oaks where grazing with cattle ('conservation grazing') has been introduced. The average density of grazing animals is expressed in Livestock Units per hectare, where one cow or heifer or pony is equivalent to 0.8 LSU and a one-year-old calf = 0.7 LSU.[4]

Site name	Cattle breed	Year introduced	Area grazed	Density average
Epping Forest	Red Poll / Longhorn	2016	120 ha	0.23 LSU/ha
Burnham Beeches	British White/ponies	1993/2012	123 ha	0.14 LSU/ha
Ashtead Common	Red Sussex / Belted Galloways	2010	c. 45 ha	0.22 LSU/ha
Cranbourne Park (WGP)	English Longhorn	2003	34.5 ha	0.55 LSU/ha
Sherwood Forest	English Longhorn	2005	84/91.5 ha	0.13 LSU/ha
Grimsthorpe Park	English Longhorn	1998	c. 50 ha	0.44 LSU/ha
High Park (Blenheim)	British White	2019	100 ha	0.20 LSU/ha
Savernake Forest	Hereford	2018	338 ha	0.11 LSU/ha

19.5.3. Pasture woodland sites with conservation grazing

Sites in England with ancient and veteran oaks on which grazing with cattle has been introduced are listed in Table 19.1. The list is not complete, but all sites were visited by AF in the course of his studies of ancient oaks (Farjon 2022).

In Epping Forest, Red Poll cattle were recently replaced by Longhorns; the former breed tending to prefer grazing the grasslands of the plains and avoiding the wooded areas in a 120ha grazing area in the southern part of Epping Forest known as Fairmead (Dennis 2016). Long-term grazing by one type (here cattle) alone is said to be insufficient to create and maintain plant biodiversity. Owing to a high footfall of people (with dogs) from surrounding urban areas, the deer population in Epping Forest is limited, with Fallow Deer and Reeve's Muntjac present, but an 'increase [of] the populations of different sized wild deer' is nevertheless recommended (Dennis 2016). The grazing programme at Epping Forest operates under a Higher Level Stewardship scheme and involves periodic assessment of the management aims and outcomes by an Independent Grazing Assessor.

In Burnham Beeches grazing started on a very small scale in 1992/3 with two ponies and two British White cows. Over time the grazing area was expanded, as were the number of cattle. In 2012 a fenceless system was introduced, with grazing in 'loops' throughout the year and with varying numbers of animals in. The time that the cows spend in each loop depends on the growing season. As an example, in 2022, 43ha of variable terrain were grazed with four cows and two ponies. 'This area includes heathland and a valley mire system, not just the wood pasture with the old trees' (Helen Read by email 11 August 2022). The total area grazed at various times is 123ha.

On Ashtead Common grazing with cattle started in 2010 with Red Sussex cows; in 2017 these were replaced with Belted Galloways. The total area grazed increased gradually to 45–50ha and eventually the management aims to achieve 'whole site grazing' on the entire common of 200ha (Ashtead Common Management Plan 2021–2031).[5] With the use of temporary cattle fencing a specific area is grazed for a relatively short time, while previously grazed areas are given a respite. The numbers of animals in a grazing area varied in the years 2010–16 from 3–10 when Red Sussex were used; from 2017 to 2022 there were 8–12 head present in a grazing area ranging in size from 2.5–9.3ha at any one grazing period (~34 days). This gives an average

4   https://ec.europa.eu/eurostat/statistics-explained/index.php?title=Glossary:Livestock_unit_(LSU).
5   https://democracy.cityoflondon.gov.uk/documents/s148811/Appendix%201%20Final%20Draft%20Ashtead%20plan.pdf.

density of 0.22 LSU/ha for the total grazed area, but with much variation both in space and time. Much of the grazing was initially limited to areas where large fires had opened up the oak canopy, followed by the invasion of birch.

In Cranbourne Park (now part of Windsor Great Park) grazing with English Longhorn cattle began in 2003; there was just one year when these were replaced by Red Sussex Cattle. The area grazed is 34.5ha in two equal compartments that are grazed in rotation or as one unit with the internal gate left open. In 2022 there were 15 cattle and 9 calves, giving a relatively high density of 0.55 LSU/ha. Initially cattle remained on site year-round, but during wet winters the animals caused 'excessive poaching' and so grazing was later limited to spring/summer only. The habitat has had significant growth of scrub, saplings and Bracken, and some mechanical cutting and Bracken spraying is undertaken to try to maintain the habitat structure and prevent grassland being lost to thorn, bramble and Bracken.

In Sherwood Forest there are two linked enclosures in which cattle are not present at the same time. After an initial trial phase with Hebridean sheep and Dexter Cattle, the grazing has been with English Longhorns since 2005, using the two enclosures alternatively. The total grazing area increased from 84ha to 91.5ha in 2012 and the average number of cattle decreased from 12.6 to 10.5, giving densities of 0.15 and 0.11 LSU/ha respectively (0.13 LSU/ha average).

In Grimsthorpe Park grazing by cattle started in 1998. There are two main areas grazed quite differently: Bracken Beds and High Wood (Gymkhana). In Bracken Beds, about 50ha of pasture woodland with ancient oaks, English Longhorn graze; in High Wood, an old oak plantation, there are Aberdeen Angus cows. In Bracken Beds there are on average 15 adults and 10–15 calves, giving an average density of 0.44 LSU/ha. In High Wood there are on average 40 cattle in a slightly larger area, and although it is part of the SSSI of 110ha it compares less well with the other localities so it is not included in the density calculation. Bracken Beds is grazed on rotation using three compartments separated by movable electric fences. Occasionally the compartments can be grazed simultaneously in different scenarios, depending on season and situation.

In High Park grazing by cattle is guided by moveable electric wire fences so that particular areas can be included or excluded. The grazed areas are divided in two by the road through High Park in a southern and a northern half, and only one half is being grazed at one time (year). Otherwise, the animals have access to most of High Park but are excluded from parts of the grassland in the Palace Vista and from a woodland section between High Lodge and Combe Gate, as well as from two pheasant pens. Introduction was in 2019 initially with 45 British White Cattle (including calves) in the southern half, but the animals were then removed from this part owing to heavy poaching of the soil and the number was reduced to about 20 head grazed in the northern half. This gives a density of 0.20 LSU/ha on average.

In Savernake Forest, the area grazed with Hereford Cattle is known as Red Vein Bottom, between the A4 and the Grand Avenue, 338ha in total. Here remain half open areas with grassland and scrub as well as wooded areas with mainly oak and Beech; there are no conifer plantations. In 2021 there were 28 cows/heifers and 20 calves grazing in this area, which gives a density of 0.11 LSU/ha, similar to Sherwood Forest but with a different breed of cattle. Between 2002 and 2008 there was a five-year trial period; data for 2009–2018 seem to be missing (Hannah Bloxham by email 27 July 2022), but since 2018 cattle have been grazing every year. The aim is to convert to organic status in the next few years.

From the review of these eight sites (for Staverton Park in Suffolk, AF was unable to obtain data) with ancient oaks where grazing with cattle has been introduced, it appears that there is much variation among them. In Table 19.1 only average figures on density are given, calculated from the areas grazed and the number of animals as reported by the managers. In actual fact, both parameters tend to fluctuate from year to year and from one season to another depending on changing circumstances, so that the figures are merely indicative. Moreover, on most sites, grazing was introduced in a small area with few animals and only later expanded. With this expansion in area a rotational system was usually adopted, whereby parts of the total area grazed were given a rest when the cows were moved into another part. In High Park, 45 head of cattle introduced into the southern half and kept there year-round caused so much poaching of the ground in the wet winter that this area was left ungrazed in the following few years, while the number of animals was reduced to 20, including some calves.

On most sites now, year-round grazing has been replaced, if it occurred, with seasonal grazing in the vegetation period, sometimes for shorter than two months, as on Ashtead Common. While some effects on the vegetation are soon apparent, much of the impact will only become evident over a number of years, and lasting changes on a landscape scale may take even longer. On only two sites in this survey, Burnham Beeches and Grimsthorpe Park, has cattle grazing lasted longer than 20 years at the time of writing. In High Park and Savernake Forest grazing began only a few years ago. Monitoring the effects may be done on other sites, but only for Epping Forest was published evidence found of this.

19.5.4. Long-term effects of grazing with cattle

In order to obtain some understanding of the effects of cattle grazing in the medium to longer term in former pasture woodland with ancient oaks, a visit on 18 September 2022 was made to Cranbourne Park, where grazing began in 2003. AF was accompanied by Anthony Cheke and his wife Ruth Ashcroft, both ecologists. Cranbourne Park, though smaller, is in many ways similar to High Park, with many great ancient oaks, sections with closed-canopy woodland and grassy glades invaded by Bracken and brambles. The effects of grazing were somewhat obscured in several of the glades owing to mowing or flaying of Bracken and Hawthorn shrubs. In a large glade close to the A332 (a site popular with the cattle) young Hawthorn bushes had become established in the grassland (Figure 19.5a, b), some browsed and others untouched. The sward here is kept short by the grazers and there is some light poaching of the ground. This may enable Hawthorn seeds to reach the soil and germinate. Seedlings of oak are soon more or less protected from grazing under and among these small hawthorns.

In other glades, Hornbeam and to a lesser extent birch and Ash had also established within brambles and thorny bushes. When brambles grow and spread here, saplings become more protected and can grow to trees. Elsewhere tree seedlings in grass were soon eaten out by either deer or cows (Reeve's Muntjac are abundant, Roe Deer common, Fallow Deer absent). In several sections with secondary infill of trees and a near-closed canopy the shrub layer appeared to have been reduced, allowing light to reach the forest floor. Here cattle may be responsible for a thin grass cover and a more open aspect at sub-canopy level. Bracken and/or nettles were nearly absent in these areas where formerly they were more abundant (AF, pers. obs.). Where the

Figure 19.5a.   Grazing with Longhorn Cattle in Cranbourne Park, October 2015. (Photograph by Aljos Farjon)

Figure 19.5b.   The same glade in October 2022. (Photograph by Aljos Farjon)

canopy is too dense there is little or no ground layer and the cattle only pass through; most of the small Hawthorns and Holly in these locations were largely untouched. In the glades, the cattle do not appear to push back Bracken, but depending on the frequency of their presence seem to consolidate the boundary with open grassland. A few big oaks formerly engulfed in brambles and Bracken had now become accessible for measurement of their girths, owing to trampling (AF, pers. obs.). Overall, floristic diversity was diminished, exemplified by the scarcity of Bluebells inside the grazing fences compared with outside. Some tall herbs avoided by cattle, such as Creeping Thistle (*Cirsium arvense*), could be seen to be spreading.

The effects of cattle grazing with a relatively high density (Table 19.1) after nearly 20 years are still limited in extent (consolidation of grassy glades rather than expansion) and vegetation structure (some opening of the sub-canopy layer in secondary woodland areas). Interference in the form of mowing and flaying, as well as removal of trees near ancient oaks (here especially birch), is still required and undertaken.

19.5.5. Can conservation grazing restore pasture woodland?

In the grazing programmes described above, the principal aim, with the possible exception of High Park, has been to preserve open glades and their biodiversity within woodlands with otherwise closed or semi-closed canopy – that is, preserving a particular status quo. This is not rewilding, nor is it the same as reverting to a historical condition in an area that has developed a closed canopy or is well advanced in that direction. Furthermore, in most cases the previous conditions over historical time are poorly known, even where the general history of an estate or park is well documented, as at Blenheim. There are in essence two major management questions to be answered: to which period is it the intention to restore, and how best to do this. Given the current primacy of Capability Brown's landscaping, the logical move at Blenheim would probably be to restore High Park to how it was in the late eighteenth/early nineteenth centuries – if we knew what that was. Although adequate written descriptions are lacking, it is clear from contemporary maps that much of High Park was much more open than it is now, and it seems to have changed little between 1771 (Thomas Richardson's map, Figure 19.6) and the six-inch OS map of 1884,[6] with some thinning out of woodland over the century in the bulk of the area both north and south of the road along High Lodge, presumably as ancient oaks died and were not replaced. Even as late as 1961 (see Figure 1.19, page 34) this open aspect still persisted, perhaps maintained through high pre-myxomatosis rabbit numbers,[7] and/or sheep grazing.

This clearly rather open woodland, with some large areas completely treeless, maintained by high deer density, would require many decades, probably a couple of centuries, to allow for gradual tree fall, to restore using conservation grazing alone; as Tree and Burrell (2023) put it, 'it is much harder to rewild mature or even early succession closed-canopy woodland – to create heterogeneity within it – than it is to generate complexity of habitat on over-grazed land'. Indeed, it is unclear if this has ever been seriously attempted, and the age gap between the ancient oaks and the 9th Duke's oak plantings of 1898–c. 1910, some 1980s planting (Chapters 1 and 3) and also the scant natural oak regeneration in the deer-free period (1914–1970s, Cheke 2023), all make it harder to recreate the eighteenth-century woodscape. Conservation grazing, at various different herbivore densities, can manage maintenance of pasture, as seen in Epping and Cranbourne (above), or conversion of arable to various levels of rewilding, but whether grazing with cattle and/or sheep can 'lead to a restoration of the medieval situation is very uncertain' (Tree and Burrell 2023). It certainly cannot in a reasonable timescale reverse pasture

---

6   See also the 1806 map in Chapter 1, Figure1.17 and Thomas Pride's map of 1772 (Bapasola 2009: 76).
7   Owing to the myxomatosis epidemic, game bags of rabbits on the estate plummeted from 5,301 in 1952–3 to 14 in 1955–6, recovered somewhat to 600–700 then crashed again in 1960 (ASC, pers. obs. in Blenheim archives). In October 1898, 6,943 Rabbits were shot on the estate in one day alone! (Cheke 2023).

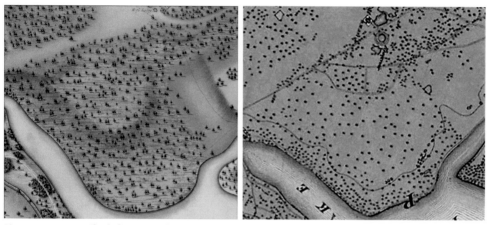

Figure 19.6.   Detail of Thomas Richardson's map of 1771 (left) and the Ordnance Survey 6-inch to 1 mile (1:10,560) of 1884 (right), showing the northern section of High Park (partly inverted) to illustrate apparent tree density decline over the intervening century, and open areas. The Richardson map is in the British Library (Add MS 71602).

woodland that has progressed to or towards closed canopy, so a more drastic approach is needed.

The first stage would require selective felling and thinning of the woodland to the desired tree density (already practised minimally in High Park, Chapter 3), followed by grazing by a mixed guild of herbivores. High Park has numerous small springs and mostly heavy soil, thus easily poached by heavy ungulates, so the ideal post-thinning guild would tend to mimic the eighteenth- and nineteenth century mix of deer and sheep, plus a few cattle and/or ponies to tackle the False Brome and thistles. The best density, probably in the region of 0.5–1.5 LU/ha, might have to be determined by trial and error (see discussion in Tree and Burrell 2023), and would also need to be periodically reduced sharply to allow pulses of natural regeneration or planting. This would mimic natural 'boom and bust' cycles caused by drought or severe winters, though it might be possible to maintain a stocking figure that exactly balances the need for some, but limited, tree replacements – oaks being very long-lived don't need many recruits through time to maintain continuity (Farjon 2022). Clearly, however, this degree of precision would not have been practised in past centuries, hence the apparent slow but steady thinning of High Park's woodland seen in the map series.

Appendix 1. Checklist of vascular plants

Surveyors: David M. Morris, Aljos Farjon,
TVERC Grassland Survey 2015, Martin Corley, Ivan Wright, Russ Hedley

Scientific Name	English Name	DAFOR/Abundance	Ancient woodland indicator
Acer campestre subsp. campestre	Field Maple	locally frequent	
Acer platanoides	Norway Maple	rare	
Acer pseudoplatanus	Sycamore	locally abundant	
Aegopodium podagraria	Cock's-foot	occasional	
Aesculus carnea	Red Horse-chestnut	rare	
Aesculus hippocastanum	Horse-chestnut	locally frequent	
Agrimonia eupatoria	Agrimony	rare	
Agrimonia procera	Fragrant Agrimony	rare	
Agrostis capillaris	Common Bent	locally abundant	
Agrostis stolonifera	Creeping Bent	locally frequent	
Ajuga reptans	Bugle	abundant	
Alnus cordata	Italian Alder	rare	
Alopecurus pratensis	Meadow Foxtail	locally frequent	
Anacamptis pyramidalis	Pyramidal Orchid	rare	
Anemone nemorosa	Wood Anemone	rare	
Angelica sylvestris	Wild Angelica	occasional	
Anisantha sterilis	Barren Brome	locally frequent	
Anthoxanthum odoratum	Sweet Vernal-grass	abundant	
Anthriscus sylvestris	Cow Parsley	locally abundant	
Arctium lappa	Greater Burdock	locally frequent	
Arctium minus	Lesser Burdock	rare	
Arrhenatherum elatius var. elatius	False Oat-grass	locally frequent	
Arum maculatum	Lords-and-Ladies	frequent	
Athyrium filix-femina	Lady-fern	locally frequent	
Atropa belladonna	Deadly Nightshade	rare	
Barbarea vulgaris	Winter-cress	rare	
Bellis perennis	Daisy	rare	
Betula pendula	Silver Birch	frequent	
Betula x aurata	Hybrid Birch	occasional	
Brachypodium rupestre	Tor-grass	locally frequent	
Brachypodium sylvaticum	False Brome	dominant	
Bromopsis ramosa	Hairy-brome	rare	weak
Bromus hordeaceus subsp. hordeaceus	Soft-brome	rare	
Bryonia dioica	White Bryony	rare	
Buxus sempervirens	Box	locally abundant	
Calamagrostis epigejos	Wood Small-reed	locally frequent	weak

Scientific Name	English Name	DAFOR/Abundance	Ancient woodland indicator
Callitriche stagnalis	Common Water-starwort	locally frequent	
Calystegia sepium	Hedge Bindweed	occasional	
Campanula rotundifolia	Harebell	rare	
Cardamine flexuosa	Wavy Bitter-cress	locally frequent	weak
Cardamine hirsuta	Hairy Bitter-cress	frequent	
Cardamine pratensis	Cuckooflower	frequent	
Carex caryophyllea	Spring Sedge	rare	
Carex divulsa subsp. divulsa	Grey Sedge	frequent	
Carex flacca	Glaucous Sedge	abundant	
Carex hirta	Hairy Sedge	frequent	
Carex leporina	Oval Sedge	rare	
Carex panicea	Carnation Sedge	occasional	
Carex pendula	Pendulous Sedge	frequent	strong
Carex remota	Remote Sedge	locally frequent	moderate
Carex riparia	Greater Pond-sedge	locally frequent	
Carex spicata	Spiked Sedge	occasional	
Carex strigosa	Thin-spiked Wood-sedge	rare	strong
Carex sylvatica	Wood-sedge	abundant	strong
Castanea sativa	Sweet Chestnut	rare	
Centaurium erythraea	Common Centaury	rare	
Cerastium fontanum subsp. vulgare	Common Mouse-ear	frequent	
Cerastium glomeratum	Sticky Mouse-ear	locally frequent	
Chamaenerion angustifolium	Rosebay Willowherb	locally frequent	
Circaea lutetiana	Enchanter's-nightshade	locally frequent	
Cirsium arvense	Creeping Thistle	frequent	
Cirsium palustre	Marsh Thistle	frequent	
Cirsium vulgare	Spear Thistle	frequent	
Clematis vitalba	Traveller's-joy	occasional	
Colchicum autumnale	Meadow Saffron	rare	weak
Conium maculatum	Hemlock	rare	
Conopodium majus	Pignut	frequent	moderate
Cornus sanguinea subsp. sanguinea	Dogwood	rare	
Corylus avellana	Hazel	occasional	
Crataegus monogyna	Hawthorn	abundant	
Crataegus x media	Hybrid Hawthorn	frequent	
Crepis vesicaria	Beaked Hawk's-beard	rare	
Cruciata laevipes	Crosswort	abundant	
Cynoglossum officinale	Hound's-tongue	rare	
Cynosurus cristatus	Crested Dog's-tail	locally frequent	
Dactylis glomerata	Cock's-foot	frequent	
Dactylorhiza fuchsii	Common Spotted-orchid	locally frequent	
Dactylorhiza praetermissa	Southern Marsh-orchid	rare	
Deschampsia cespitosa	Tufted Hair-grass	locally frequent	
Digitalis purpurea	Foxglove	occasional	
Dryopteris dilatata	Broad Buckler-fern	locally frequent	
Dryopteris filix-mas	Male-fern	locally abundant	
Elymus caninus	Bearded Couch	rare	moderate
Epilobium hirsutum	Great Willowherb	frequent	
Epilobium parviflorum	Hoary Willowherb	locally frequent	
Epipactis helleborine	Broad-leaved Helleborine	rare	moderate
Ervilia hirsuta	Hairy Tare	frequent	
Ervum tetraspermum	Smooth Tare	rare	

Scientific Name	English Name	DAFOR/Abundance	Ancient woodland indicator
Euonymus europaeus	Spindle	rare	weak
Euphrasia confusa	Eyebright	rare	
Fagus sylvatica	Beech	locally frequent	
Festuca rubra subsp. rubra	Red Fescue	frequent	
Ficaria verna	Lesser Celandine	occasional	
Filipendula ulmaria	Meadowsweet	locally frequent	
Fragaria vesca	Wild Strawberry	occasional	
Fraxinus excelsior	Ash	locally dominant	
Galanthus nivalis	Snowdrop	rare	
Galium album	Hedge Bedstraw	occasional	
Galium aparine	Cleavers	occasional	
Galium palustre subsp. palustre	Common Marsh-bedstraw	occasional	
Galium saxatile	Heath Bedstraw	locally abundant	
Galium uliginosum	Fen Bedstraw	occasional	
Galium verum	Lady's Bedstraw	locally abundant	
Geranium dissectum	Cut-leaved Crane's-bill	occasional	
Geranium robertianum	Herb-Robert	frequent	
Geum urbanum	Wood Avens	abundant	
Glechoma hederacea	Ground-ivy	abundant	
Glyceria fluitans	Floating Sweet-grass	rare	
Glyceria maxima	Reed Sweet-grass	locally abundant	
Hedera helix	Common Ivy	rare	
Helianthemum nummularium	Common Rock-rose	occasional	
Heracleum sphondylium subsp. sphondylium var. angustifolium	Hogweed	rare	
Heracleum sphondylium subsp. sphondylium var. sphondylium	Hogweed	occasional	
Hesperis matronalis	Dame's-violet	rare	
Holcus lanatus	Yorkshire-fog	abundant	
Holcus mollis	Creeping Soft-grass	locally abundant	
Hyacinthoides non-scripta	Bluebell	abundant	moderate
Hypericum androsaemum	Tutsan	rare	weak
Hypericum hirsutum	Hairy St John's-wort	occasional	moderate
Hypericum maculatum	Imperforate St John's-wort	rare	
Hypericum perforatum	Perforate St John's-wort	occasional	
Hypericum tetrapterum	Square-stalked St John's-wort	locally frequent	
Hypochaeris radicata	Cat's-ear	rare	
Ilex aquifolium	Holly	rare	
Iris pseudacorus	Yellow Iris	locally frequent	
Jacobaea erucifolia	Hoary Ragwort	occasional	
Jacobaea vulgaris	Common Ragwort	occasional	
Juncus acutiflorus	Sharp-flowered Rush	locally frequent	
Juncus articulatus	Jointed Rush	rare	
Juncus conglomeratus	Compact Rush	locally frequent	
Juncus effusus	Soft-rush	locally frequent	
Juncus inflexus	Hard Rush	locally frequent	
Lamiastrum galeobdolon subsp. montanum	Yellow Archangel	rare	strong
Lamium album	White Dead-nettle	occasional	
Lapsana communis	Nipplewort	occasional	
Larix decidua	European Larch	locally frequent	
Lathyrus pratensis	Meadow Vetchling	locally frequent	

Scientific Name	English Name	DAFOR/Abundance	Ancient woodland indicator
Lemna minor	Common Duckweed	locally abundant	
Linaria vulgaris	Common Toadflax	rare	
Linum catharticum	Fairy Flax	rare	
Lolium perenne	Perennial Rye-grass	occasional	
Lonicera periclymenum	Honeysuckle	rare	
Lotus corniculatus var. corniculatus	Common Bird's-foot-trefoil	frequent	
Lotus pedunculatus	Greater Bird's-foot-trefoil	locally frequent	
Luzula campestris	Field Wood-rush	abundant	
Lycopus europaeus	Gypsywort	locally frequent	
Lysimachia nummularia	Creeping-Jenny	locally frequent	
Medicago lupulina	Black Medick	locally frequent	
Mentha aquatica	Water Mint	locally abundant	
Mentha arvensis	Corn Mint	occasional	
Mercurialis perennis	Dog's Mercury	dominant	weak
Moehringia trinervia	Three-nerved Sandwort	frequent	weak
Myosotis arvensis	Field Forget-me-not	frequent	
Myosotis discolor	Changing Forget-me-not	occasional	
Myosotis laxa	Tufted Forget-me-not	rare	
Myosotis scorpioides	Water Forget-me-not	rare	
Narcissus sp.	Daffodil cultivars	rare	
Narcissus 'Tete-a-Tete'	Daffodil cultivar	rare	
Neottia ovata	Common Twayblade	locally abundant	
Odontites vernus	Red Bartsia	rare	
Ophioglossum vulgatum	Adder's-tongue	rare	moderate
Origanum vulgare	Wild Marjoram	rare	
Oxalis acetosella	Wood-sorrel	rare	strong
Papaver rhoeas	Common poppy	rare	
Pastinaca sativa subsp. sylvestris	Wild Parsnip	rare	
Petasites pyrenaicus	Winter Heliotrope	rare	
Phleum bertolonii	Smaller Cat's-tail	rare	
Picea abies	Norway Spruce	rare	
Pilosella officinarum	Mouse-ear-hawkweed	occasional	
Pinus nigra	Black Pine	locally frequent	
Pinus sylvestris	Scots Pine	locally frequent	
Plantago lanceolata	Ribwort Plantain	locally frequent	
Plantago major	Greater Plantain	locally frequent	
Platanus x hispanica	London Plane	rare	
Poa angustifolia	Narrow-leaved Meadow-grass	rare	
Poa annua	Annual Meadow-grass	locally frequent	
Poa humilis	Spreading Meadow-grass	rare	
Poa pratensis	Smooth Meadow-grass	locally frequent	
Poa trivialis	Rough Meadow-grass	frequent	
Polypodium interjectum	Intermediate Polypody	occasional	
Populus nigra 'Italica'	Lombardy Poplar	rare	
Populus tremula	Aspen	one clump	
Populus x canadensis	Hybrid Black-poplar	locally frequent	
Potentilla anserina	Silverweed	locally frequent	
Potentilla erecta	Tormentil	frequent	
Potentilla reptans	Creeping Cinquefoil	frequent	
Potentilla sterilis	Barren Strawberry	locally abundant	weak
Primula veris	Cowslip	locally frequent	

Scientific Name	English Name	DAFOR/Abundance	Ancient woodland indicator
Primula vulgaris	Primrose	occasional	weak
Prunella vulgaris	Selfheal	frequent	
Prunus avium	Wild Cherry	locally frequent	
Prunus cerasifera var. pissardii	Cherry cultivar	rare	
Prunus laurocerasus	Cherry Laurel	rare	
Prunus spinosa	Blackthorn	locally frequent	
Prunus padus	Bird Cherry	rare	
Pseudotsuga menziesii	Douglas Fir	rare	
Pteridium aquilinum subsp. aquilinum	Bracken	dominant	
Quercus cerris	Turkey Oak	rare	
Quercus robur	Pedunculate Oak	dominant	
Ranunculus acris subsp. acris	Meadow Buttercup	rare	
Ranunculus aquatilis	Common Water-crowfoot	rare	
Ranunculus bulbosus subsp. bulbosus	Bulbous Buttercup	rare	
Ranunculus repens	Creeping Buttercup	abundant	
Reseda luteola	Weld	rare	
Rhamnus cathartica	Buckthorn	rare	
Rhinanthus minor	Yellow-rattle	occasional	
Rhododendron ponticum	Rhododendron	rare	
Ribes nigrum	Black Currant	occasional	
Ribes rubrum	Red Currant	rare	
Ribes uva-crispa	Gooseberry	rare	
Robinia pseudoacacia	False Acacia	occasional	
Rosa canina agg.	Dog-rose	occasional	
Rosa x scabriuscula	Roughish-leaved Dog-rose	rare	
Rubus echinatus	Echinate Bramble	rare	
Rubus fruticosus agg.	Bramble	abundant	
Rubus idaeus	Raspberry	occasional	
Rubus ulmifolius	Elm-leaved Bramble	frequent	
Rubus vestitus	European Blackberry	occasional	
Rumex acetosa subsp. acetosa	Common Sorrel	abundant	
Rumex acetosella	Sheep's Sorrel	rare	
Rumex conglomeratus	Clustered Dock	rare	
Rumex crispus	Curled Dock	occasional	
Rumex obtusifolius	Broad-leaved Dock	occasional	
Rumex sanguineus var. viridis	Wood Dock	frequent	
Salix caprea	Goat Willow	abundant	weak
Salix cinerea subsp. oleifolia	Grey Willow	occasional	
Salix triandra	Almond Willow	rare	
Sambucus nigra	Elder	abundant	
Saxifraga tridactylites	Rue-leaved Saxifrage	occasional	
Schedonorus arundinaceus	Tall Fescue	occasional	
Schedonorus giganteus	Giant Fescue	rare	
Schedonorus pratensis	Meadow Fescue	locally frequent	
Schoenoplectus lacustris	Common Club-rush	locally frequent	
Scrophularia auriculata	Water Figwort	rare	
Scrophularia nodosa	Common Figwort	frequent	weak
Scutellaria galericulata	Skullcap	occasional	weak
Senecio vulgaris	Groundsel	locally frequent	
Silene dioica	Red Campion	rare	
Silene flos-cuculi	Ragged-Robin	rare	
Solanum dulcamara	Bittersweet	rare	

Scientific Name	English Name	DAFOR/Abundance	Ancient woodland indicator
Solanum nigrum	Black Nightshade	rare	
Sonchus asper	Prickly Sow-thistle	rare	
Sorbus aria	Common Whitebeam	occasional	
Sorbus aucuparia	Rowan	rare	
Stachys sylvatica	Hedge Woundwort	frequent	
Stellaria graminea	Lesser Stitchwort	frequent	
Stellaria holostea	Greater Stitchwort	occasional	
Stellaria media	Common Chickweed	locally frequent	
Symphytum officinale	Common Comfrey	rare	
Symphytum x uplandicum	Russian Comfrey	occasional	
Tamus communis	Black Bryony	occasional	
Taraxacum officinale agg.	Dandelion	frequent	
Tilia cordata	Small-leaved Lime	rare	strong
Tilia x europaea	Lime	occasional	
Tragopogon pratensis	Goat's-beard	rare	
Trifolium dubium	Lesser Trefoil	occasional	
Trifolium pratense var. pratense	Red Clover	occasional	
Trifolium repens	White Clover	frequent	
Trisetum flavescens	Yellow Oat-grass	locally frequent	
Tsuga canadensis	Eastern Hemlock-spruce	rare	
Tsuga heterophylla	Western Hemlock-spruce	rare	
Tussilago farfara	Colt's-foot	occasional	
Typha latifolia	Bulrush	locally frequent	
Ulmus glabra	Wych Elm	locally frequent	
Ulmus procera	English Elm	rare	
Urtica dioica subsp. dioica	Common Nettle	abundant	
Verbascum nigrum	Dark Mullein	rare	
Veronica arvensis	Wall Speedwell	frequent	
Veronica chamaedrys	Germander Speedwell	frequent	
Veronica filiformis	Slender speedwell	occasional	
Veronica montana	Wood Speedwell	occasional	strong
Veronica officinalis	Heath Speedwell	rare	
Veronica persica	Common field-speedwell	occasional	
Veronica serpyllifolia subsp. serpyllifolia	Thyme-leaved Speedwell	frequent	
Vicia cracca	Tufted Vetch	rare	
Vicia sativa	Common Vetch	occasional	
Viola hirta subsp. hirta	Hairy Violet	occasional	
Viola odorata	Sweet Violet	rare	
Viola reichenbachiana	Early Dog-violet	rare	weak
Viola riviniana	Common Dog-violet	frequent	
Vulpia bromoides	Squirreltail Fescue	rare	

Appendix 2. Checklist of bryophytes (mosses and liverworts)

Surveyors: Jacqueline Wright, David M. Morris, Aljos Farjon
Taxa in **bold** were recorded in the HPBS 2017–2021;
other entries are from TVERC records prior to 2017.

Mosses

Scientific name	Abundance	Sex/Stage
Amblystegium serpens var. serpens		
Anomodon veticulosus	abundant	
Atrichum tenellum		
Atrichum undulatum		fruiting
Aulacomnium androgynum		
Barbula unguiculata		
Brachytheciastrum velutinum		
Brachythecium rivulare		
Brachythecium rutabulum		
Brachythecium salebrosum		
Bryoerythrophyllum recurvirostrum		
Bryum capillare var. capillare		fruiting
Bryum klinggraeffii		
Bryum radiculosum		
Bryum rubens		
Bryum ruderale		
Bryum subelegans		
Calliergonella cuspidata		
Campylopus flexuosus		fruiting
Campylopus introflexus	locally abundant	
Ceratodon purpureus	locally abundant	fruiting
Cirriphyllum crassinervium		
Cirriphyllum piliferum		
Cratoneuron filicinum		
Cryphaea heteromalla		fruiting
Ctenidium molluscum		
Dicranella heteromalla		
Dicranella varia/howei agg.		
Dicranoweissia cirrata		fruiting
Dicranum scoparium		
Didymodon fallax		
Didymodon luridus		
Didymodon sinuosus		
Didymodon tophaceus		

Scientific name	Abundance	Sex/Stage
Encalypta streptocarpa		
Fissidens bryoides		
Fissidens exilis		
Fissidens incurvus		
Fissidens taxifolius var. **taxifolius**		
Fontinalis antipyretica		
Funaria hygrometica		
Grimmia pulvinata		
Gyroweisia tenuis		
Homalia trichomanoides		
Homalothecium lutescens		
Homalothecium sericeum		fruiting
Hypnum andoi		
Hypnum cupressiforme var. cupressiforme		fruiting
Hypnum cupressiforme var. resupinatum		fruiting
Isothecium alopecuroides		
Isothecium myosuroides		
Kindbergia praelonga		
Leptodictyum riparium		
Leskea polycarpa		
Leucodon sciuroides		
Lewinskya affinis		fruiting
Metzgeria consanguinea		
Microeurhynchium pumilum		
Mnium hornum		
Neckera complanata	abundant	fruiting
Orthodontium lineare	abundant	fruiting
Orthotrichum anomalum		
Orthotrichum diaphanum		fruiting
Orthotrichum pulchellum		fruiting
Orthotrichum stramineum		fruiting
Orthotrichum tenellum		fruiting
Oxyrrhynchium hians		
Plagiomnium affine		
Plagiomnium rostratum		
Plagiomnium undulatum		
Platygyrium repens		
Plenogemma phyllantha		
Pohlia melanodon		
Pseudocrossidium hornschuchianum		
Pseudoscleropodium purum	abundant	
Pulvigera lyellii		
Rhizomnium punctatum		
Rhychostegium riparioides		
Rhynchostegiella tenella		
Rhynchostegium confertum		fruiting
Rhynchostegium murale		
Rhytidiadelphus squarrosus		
Rhytidiadelphus triquetrus		
Schistidium apocarpum		
Pseudoscleropodium purum		
Streblotrichum convolutum var. *convolutum*		
Syntrichia laevipila		

Scientific name	Abundance	Sex/Stage
Syntrichia montana		
Syntrichia ruralis		
Syntrichia virescens		
Thamnobryum alopecurum		
Thuidium tamariscinum		
Tortula acaulon		
Tortula muralis		fruiting
Trichostomum crispulum		
Ulota bruchii		fruiting
Ulota crispa agg.		
Zygodon conoideus		fruiting
Zygodon rupestris		
Zygodon viridissimus		

Liverworts

Scientific name	Abundance	Sex/Stage
Chiloscyphus pallescens		
Frullania dilatata		
Lophocolea bidentata		
Lophocolea heterophylla		fruiting
Metzgeria consanguinea		
Metzgeria furcata		
Metzgeria violacea		
Myriocoleopsis minutissima		fruiting
Plagiochila asplenioides		
Plagiochila porelloides		
Porella platyphylla	abundant	
Radula complanata		

Appendix 3. Checklist of fungi (excluding lichens)

Recorders: Martyn Ainsworth, Richard Fortey, Molly Dewey, Penny Cullington, Wendy MacEachrane, Caroline Jackson-Houlston, Aljos Farjon, Brian Spooner, Alick Henrici, Alona Yu. Biketova (fruit body DNA), Laura M. Suz (ectomycorrhizal root DNA)

'Oak Saprotrophs' are species living on dead wood of ancient oaks listed in Ainsworth (2022) Table 10-1; of a total of 16 species so recognized, 10 have been recorded in High Park. A minimum of eight species is currently required for a site to be considered for SSSI protection based on this assemblage.

Codes prefixed 'SH' are UNITE species hypothesis codes (see https://unite.ut.ee/) and are provided for those species whose DNA was detected in the sampled root tips.

Source	Scientific Name	English Name	Oak Saprotroph
HPBS	Agaricus campestris	Field Mushroom	
HPBS	Agaricus moelleri	Inky Mushroom	
HPBS	Agaricus xanthodermus	Yellow Stainer	
HPBS	Aleuria aurantia	Orange Peel Fungus	
HPBS	Amanita alseides		
HPBS	Amanita ceciliae	Snakeskin Grisette	
HPBS	Amanita fulva	Tawny Grisette	
HPBS	Amanita lividopallescens		
HPBS	Amanita phalloides	Deathcap	
HPBS	Amanita rubescens	Blusher	
HPBS	Amaropostia stiptica	Bitter Bracket	
HPBS	Aphanobasidium pseudotsugae		
HPBS	Apioperdon pyriforme	Stump Puffball	
HPBS	Arachnocrea stipata		
HPBS	Armillaria gallica	Bulbous Honey Fungus	
HPBS	Armillaria mellea	Honey Fungus	
HPBS	Ascocoryne sarcoides	Purple Jellydisc	
HPBS	Ascodichaena rugosa		
HPBS	Ascotremella faginea		
HPBS	Ascozonus woolhopensis		
HPBS	Athelia arachnoidea	Cobweb Duster	
HPBS	Athelia bombacina		
HPBS	Athelia decipiens		
HPBS	Athelopsis glaucina		
HPBS	Auricularia auricula-judae	Jelly Ear	
HPBS	Auricularia mesenterica	Tripe Fungus	
HPBS	Basidiodendron spinosum		
HPBS	Biscogniauxia anceps		

Source	Scientific Name	English Name	Oak Saprotroph
HPBS	Biscogniauxia nummularia	Beech Tarcrust	
HPBS	Bjerkandera adusta	Smoky Bracket	
HPBS	Bolbitius titubans	Yellow Fieldcap	
HPBS	Botryobasidium aureum		
HPBS	Botryobasidium laeve		
HPBS	Botryobasidium pruinatum		
HPBS	Botryobasidium subcoronatum		
HPBS	Brevicellicium olivascens		
HPBS	Britzelmayria multipedata	Clustered Brittlestem	
HPBS	Buglossoporus quercinus	Oak Polypore	√
HPBS	Bulbillomyces farinosus	Couscous Crust	
TVERC	Bulgaria inquinans	Batchelor's Buttons	
HPBS	Calvatia gigantea	Giant Puffball	
HPBS	Calycina citrina	Lemon Disco	
HPBS	Calycina claroflava	Sulphur Disco	
HPBS	Caudospora taleola		
HPBS	Cenococcum geophilum SH1639590.08FU		
HPBS	Cerioporus mollis	Common Mazegill	
HPBS	Cerioporus squamosus	Dryad's Saddle	
HPBS	Cerioporus varius	Blackfoot Polypore	
HPBS	Ceriporia reticulata		
HPBS	Ceriporia viridans		
HPBS	Cerrena unicolor		
HPBS	Cheilymenia granulata	Cowpat Gem	
HPBS	Chlorociboria aeruginascens	Green Elfcup	
HPBS	Ciboria batschiana		
HPBS	Clavaria acuta	Pointed Club	
HPBS	Clavulina cinerea	Grey Coral	
HPBS	Clavulina coralloides SH1546094.08FU	Crested Coral	
HPBS	Clavulina rugosa	Wrinkled Club	
HPBS	Clavulinopsis helvola	Yellow Club	
HPBS	Clitocybe fragrans	Fragrant Funnel	
HPBS	Clitocybe nebularis	Clouded Funnel	
HPBS	Clitopilus hobsonii	Miller's Oysterling	
HPBS	Clitopilus prunulus	The Miller	
HPBS	Coleosporium tussilaginis		
HPBS	Collybia cookei	Splitpea Shanklet	
HPBS	Collybiopsis confluens	Clustered Toughshank	
HPBS	Collybiopsis peronata	Wood Woollyfoot	
HPBS	Colpoma quercinum		
HPBS	Coniophora arida	Dry Duster	
HPBS	Coniophora olivacea	Olive Duster	
HPBS	Coniophora puteana	Wet Rot	
HPBS	Conocybe juniana		
HPBS	Conocybe microspora var. brunneola		
HPBS	Conocybe siennophylla		
HPBS	Conocybe subovalis		
HPBS	Coprinellus disseminatus	Fairy Inkcap	
HPBS	Coprinellus domesticus	Firerug Inkcap	
HPBS	Coprinellus impatiens		
HPBS	Coprinellus micaceus	Glistening Inkcap	
HPBS	Coprinopsis lagopus	Hare's Foot Inkcap	

Source	Scientific Name	English Name	Oak Saprotroph
HPBS	Coprinus comatus	Shaggy Inkcap	
HPBS	Cortinarius alboadustus		
HPBS	Cortinarius azureovelatus	Yellow Capped Webcap	
HPBS	Cortinarius collocandoides		
HPBS	Cortinarius flexipes	Pine Pelargonium Webcap	
HPBS	Cortinarius cf. glandicolor		
HPBS	Cortinarius incisior		
HPBS	Cortinarius luridus		
HPBS	Cortinarius megacystidiosus SH1283009.09FU		
HPBS	Cortinarius pilatii SH1282901.09FU		
HPBS	Cortinarius semiodoratus		
HPBS	Cortinarius triumphans	Birch Webcap	
HPBS	Craterellus undulatus	Sinuous Chanterelle	
HPBS	Crepidotus applanatus	Flat Oysterling	
HPBS	Crepidotus autochthonus	Grounded Oysterling	
HPBS	Crepidotus cesatii	Roundspored Oysterling	
HPBS	Crepidotus luteolus	Yellowing Oysterling	
HPBS	Crepidotus mollis	Peeling Oysterling	
HPBS	Cudoniella acicularis	Oak Pin	
HPBS	Cuphophyllus pratensis	Meadow Waxcap	
HPBS	Cuphophyllus virgineus	Snowy Waxcap	
HPBS	Cyanosporus subcaesius	Blueing Bracket	
HPBS	Cyathicula coronata		
HPBS	Cyclocybe cylindracea	Poplar Fieldcap	
HPBS	Cylindrobasidium laeve	Tear Dropper	
HPBS	Cystoderma amianthinum	Earthy Powdercap	
HPBS	Cytospora nivea		
HPBS	Dacrymyces stillatus	Common Jelly Spot	
HPBS	Daedaleopsis confragosa	Blushing Bracket	
HPBS	Daldinia concentrica	King Alfred's Cakes	
HPBS	Dasyscyphella nivea		
HPBS	Dendrothele acerina		
HPBS	Dendrothele citrisporella		
HPBS	Diatrype bullata	Willow Barkspot	
HPBS	Diatrype stigma	Common Tarcrust	
HPBS	Diatrypella quercina	Oak Blackhead	
HPBS	Entoloma hirtipes		
HPBS	Entoloma lividoalbum		
HPBS	Entoloma sericellum	Cream Pinkgill	
HPBS	Eonema pyriforme		
HPBS	Epichloe typhina	Choke	
HPBS	Eriopezia caesia		
HPBS	Erysiphe alphitoides	Oak Mildew	
HPBS	Euepixylon udum		
HPBS	Eutypa maura		
HPBS	Eutypa spinosa	Spiral Tarcrust	
HPBS	Exidia glandulosa	Witches' Butter	
HPBS	Fistulina hepatica	Beefsteak Fungus	√
HPBS	Flagelloscypha minutissima		
HPBS	Flammula alnicola	Alder Scalycap	
HPBS	Flammulaster muricatus		
HPBS	Fomitopsis betulina	Birch Polypore	

Source	Scientific Name	English Name	Oak Saprotroph
HPBS	Fuscoporia ferrea	Cinnamon Porecrust	
HPBS	Galerina marginata s. lat.	Funeral Bell	
HPBS	Ganoderma adspersum	Southern Bracket	
HPBS	Geastrum fimbriatum	Sessile Earthstar	
HPBS	Genea hispidula SH1539219.08FU		
HPBS	Gliophorus psittacinus	Parrot Waxcap	
HPBS	Gloeocystidiellum porosum		
HPBS	Grifola frondosa	Hen Of The Woods	√
HPBS	Gymnopilus junonius	Spectacular Rustgill	
HPBS	Gymnopus dryophilus	Russet Toughshank	
HPBS	Gymnopus erythropus	Redleg Toughshank	
HPBS	Gymnopus fusipes	Spindle Toughshank	√
HPBS	Hapalopilus rutilans	Cinnamon Bracket	
HPBS	Hebeloma birrus		
HPBS	Hebeloma theobrominum		
HPBS	Helvella crispa	White Saddle	
HPBS	Hemimycena cephalotricha		
HPBS	Hemimycena cucullata		
HPBS	Hemimycena tortuosa	Dewdrop Bonnet	
HPBS	Henningsomyces candidus	White Tubelet	
HPBS	Heteroradulum deglubens	Blushing Crust	
HPBS	Hodophilus atropunctus	Dotted Fanvault	
HPBS	Hortiboletus engelii		
HPBS	Hortiboletus cf. engelii SH1610853.08FU		
HPBS	Humaria sp. SH1608584.08FU		
HPBS	Hyalopeziza millepunctata		
HPBS	Hyalorbilia inflatula		
HPBS	Hyaloscypha daedaleae		
HPBS	Hyaloscyphaceae sp. SH1522937.08FU		
HPBS	Hydropus floccipes		
HPBS	Hygrocybe ceracea	Butter Waxcap	
HPBS	Hygrocybe chlorophana	Golden Waxcap	
HPBS	Hygrocybe glutinipes	Glutinous Waxcap	
HPBS	Hygrocybe miniata	Vermilion Waxcap	
HPBS	Hygrocybe quieta	Oily Waxcap	
HPBS	Hygrocybe reidii	Honey Waxcap	
HPBS	Hygrophoropsis aurantiaca	False Chanterelle	
HPBS	Hygrophoropsis rufa		
HPBS	Hygrophorus discoxanthus	Yellowing Woodwax	
HPBS	Hymenochaete rubiginosa	Oak Curtain Crust	√
HPBS	Hymenopellis radicata	Rooting Shank	
HPBS	Hymenoscyphus imberbis		
HPBS	Hyphoderma cremeoalbum		
HPBS	Hyphoderma roseocremeum		
HPBS	Hyphodontia alutaria		
HPBS	Hyphodontia arguta		
HPBS	Hypholoma fasciculare	Sulphur Tuft	
HPBS	Hypochnicium erikssonii		
HPBS	Hypocrea sp.		
HPBS	Hypoxylon fragiforme	Beech Woodwart	

Source	Scientific Name	English Name	Oak Saprotroph
HPBS	Hypoxylon fuscum	Hazel Woodwart	
HPBS	Hypoxylon petriniae		
HPBS	Hysterium acuminatum		
HPBS	Infundibulicybe geotropa	Trooping Funnel	
HPBS	Infundibulicybe gibba	Common Funnel	
HPBS	Inocybe asterospora	Star Fibrecap	
HPBS	Inocybe corydalina	Greenflush Fibrecap	
HPBS	Inocybe erinaceomorpha		
HPBS	Inocybe flocculosa	Fleecy Fibrecap	
HPBS	Inocybe fuscidula	Darkish Fibrecap	
HPBS	Inocybe geophylla	White Fibrecap	
HPBS	Inocybe grammata		
HPBS	Inocybe jucunda SH1251230.09FU		
HPBS	Inocybe lilacina s. lat.	Lilac Fibrecap	
HPBS	Inocybe pseudodestricta		
HPBS	Inocybe sp. SH125776.09FU		
HPBS	Inocybe syringae (SH currently unavailable)		
TVERC	Inonotus hispidus	Shaggy Bracket	
HPBS	Inosperma adaequatum		
HPBS	Inosperma bongardii	Fruity Fibrecap	
HPBS	Inosperma cookei	Straw Fibrecap	
HPBS	Inosperma maculatum	Frosty Fibrecap	
HPBS	Jaapia ochroleuca		
HPBS	Jackrogersella cohaerens		
HPBS	Jackrogersella multiformis	Birch Woodwart	
HPBS	Kretzschmaria deusta	Brittle Cinder	
HPBS	Kuehneromyces mutabilis	Sheathed Woodtuft	
HPBS	Kurtia argillacea		
HPBS	Laccaria amethystina	Amethyst Deceiver	
HPBS	Laccaria laccata	Deceiver	
HPBS	Lachnum bicolor		
HPBS	Lachnum brevipilosum		
HPBS	Lachnum rhytismatis		
HPBS	Lachnum tenuissimum		
HPBS	Lachnum virgineum	Snowy Disco	
HPBS	Lacrymaria lacrymabunda	Weeping Widow	
HPBS	Lactarius azonites		
TVERC	Lactarius blennius	Beech Milkcap	
HPBS	Lactarius deterrimus	False Saffron Milkcap	
HPBS	Lactarius quietus	Oakbug Milkcap	
HPBS	Lactarius cf. subdulcis	Mild Milkcap	
HPBS	Lactarius cf. subumbonatus		
HPBS	Lactifluus vellereus	Fleecy Milkcap	
HPBS	Laetiporus sulphureus	Chicken Of The Woods	√
HPBS	Leccinum scabrum	Brown Birch Bolete	
HPBS	Lepiota cristata	Stinking Dapperling	
HPBS	Lepiota felina	Cat Dapperling	
HPBS	Lepista nuda	Wood Blewit	
HPBS	Lepista sordida	Sordid Blewit	
HPBS	Leptosphaeria acuta	Nettle Rash	
HPBS	Leucoagaricus badhamii	Blushing Dapperling	
HPBS	Lindtneria leucobryophila		

Source	Scientific Name	English Name	Oak Saprotroph
HPBS	Lycoperdon nigrescens	Dusky Puffball	
HPBS	Lycoperdon perlatum	Common Puffball	
HPBS	Macrolepiota konradii		
HPBS	Macrolepiota procera	Parasol	
HPBS	Marasmius chordalis		
HPBS	Marasmius epiphyllus	Leaf Parachute	
HPBS	Marasmius rotula	Collared Parachute	
HPBS	Marssonina betulae		
HPBS	Melampsora caprearum		
HPBS	Melampsora populnea	Dog's Mercury Rust	
HPBS	Melanoleuca cognata	Spring Cavalier	
HPBS	Melanoleuca polioleuca	Common Cavalier	
TVERC	Meripilus giganteus	Giant Polypore	
HPBS	Meruliporia pulverulenta		
HPBS	Microscypha grisella		
HPBS	Mollisia cinerea	Common Grey Disco	
HPBS	Mollisia ligni		
HPBS	Monilinia johnsonii	Haw Goblet	
HPBS	Mucidula mucida	Porcelain Fungus	
HPBS	Mucronella calva	Swarming Spine	
HPBS	Mutinus caninus	Dog Stinkhorn	
HPBS	Mycena aetites	Drab Bonnet	
HPBS	Mycena arcangeliana	Angel's Bonnet	
HPBS	Mycena filopes	Iodine Bonnet	
HPBS	Mycena galericulata	Common Bonnet	
HPBS	Mycena galopus	Milking Bonnet	
HPBS	Mycena haematopus	Burgundydrop Bonnet	
HPBS	Mycena inclinata	Clustered Bonnet	√
HPBS	Mycena leptocephala	Nitrous Bonnet	
HPBS	Mycena olivaceomarginata	Brownedge Bonnet	
HPBS	Mycena pelianthina	Blackedge Bonnet	
HPBS	Mycena polygramma	Grooved Bonnet	
HPBS	Mycena pura	Lilac Bonnet	
HPBS	Mycena rosea	Rosy Bonnet	
HPBS	Mycena smithiana		
HPBS	Mycena stylobates	Bulbous Bonnet	
HPBS	Mycena tenerrima	Frosty Bonnet	
HPBS	Mycena vitilis	Snapping Bonnet	
HPBS	Mycetinis alliaceus	Garlic Parachute	
HPBS	Mycoacia aurea		
HPBS	Mycosphaerella aspidii		
HPBS	Myxarium nucleatum	Crystal Brain	
HPBS	Nectria cinnabarina	Coral Spot	
HPBS	Neoboletus luridiformis	Scarletina Bolete	
HPBS	Oidiodendrum maius SH1564437.08FU		
HPBS	Orbilia xanthostigma	Common Glasscup	
HPBS	Oxyporus populinus	Poplar Bracket	
HPBS	Panaeolus cinctulus	Banded Mottlegill	
HPBS	Panaeolus papilionaceus	Petticoat Mottlegill	
HPBS	Panellus stipticus	Bitter Oysterling	
HPBS	Panus conchatus	Lilac Oysterling	
HPBS	Parasola leiocephala	Bald Inkcap	
HPBS	Parasola plicatilis	Pleated Inkcap	
HPBS	Paxillus involutus	Brown Rollrim	

Source	Scientific Name	English Name	Oak Saprotroph
HPBS	Peniophora boidinii		
HPBS	Peniophora incarnata	Rosy Crust	
HPBS	Peniophora lycii		
HPBS	Peniophora quercina		
HPBS	Peniophora reidii		
HPBS	Peniophorella guttulifera		
HPBS	Peniophorella praetermissa		
HPBS	Peniophorella pubera		
HPBS	Pezizaceae sp. SH2729448.08FU		
HPBS	Phaeoclavulina minutispora		
HPBS	Phaeohelotium flexuosum		
HPBS	Phaeohelotium nobile		
HPBS	Phaeophlebiopsis ravenelii		
HPBS	Phaeotremella frondosa		
HPBS	Phanerochaete sordida		
HPBS	Phanerochaete velutina		
HPBS	Phellinus pomaceus	Cushion Bracket	
HPBS	Phlebia lilascens		
HPBS	Phlebia radiata	Wrinkled Crust	
HPBS	Phlebia rufa		
HPBS	Phlebia tremellosa	Jelly Rot	
HPBS	Phloeomana hiemalis		
HPBS	Phloeomana speirea	Bark Bonnet	
HPBS	Pholiota gummosa	Sticky Scalycap	
HPBS	Pholiota tuberculosa		
HPBS	Pholiotina arrhenii	Ringed Conecap	
HPBS	Pholiotina rugosa		
HPBS	Pholiotina velata	Veiled Conecap	
HPBS	Phragmidium violaceum	Violet Bramble Rust	
HPBS	Physisporinus sanguinolentus	Bleeding Porecrust	
HPBS	Physisporinus vitreus		
HPBS	Picipes badius	Bay Polypore	
TVERC	Pleurotus ostreatus	Oyster Mushroom	
HPBS	Pluteus cervinus	Deer Shield	
HPBS	Pluteus salicinus	Willow Shield	
HPBS	Pluteus thomsonii	Veined Shield	
HPBS	Podoscypha multizonata	Zoned Rosette	√
HPBS	Podospora longicollis		
HPBS	Polydesmia pruinosa		
HPBS	Polyporus tuberaster	Tuberous Polypore	
HPBS	Postia ceriflua		
HPBS	Propolis farinosa		
HPBS	Protomerulius dubius		
HPBS	Prunulus diosmus		
HPBS	Psathyrella candolleana	Pale Brittlestem	
HPBS	Psathyrella cf. corrugis	Red Edge Brittlestem	
HPBS	Psathyrella obtusata		
HPBS	Psathyrella piluliformis	Common Stump Brittlestem	
HPBS	Psathyrella pseudogracilis		
HPBS	Psathyrella spadiceogrisea	Spring Brittlestem	
HPBS	Pseudoclitocybe cyathiformis	Goblet	
HPBS	Pseudoinonotus dryadeus	Oak Bracket	√
HPBS	Pseudosperma rimosum	Split Fibrecap	
HPBS	Pseudosperma aff. rimosum		

Source	Scientific Name	English Name	Oak Saprotroph
HPBS	Puccinia glechomatis	Ground Ivy Rust	
HPBS	Puccinia pygmaea var. pygmaea		
HPBS	Radulomyces confluens		
HPBS	Radulomyces molaris	Oak Tooth Crust	
HPBS	Ramularia rubella		
HPBS	Resupinatus applicatus	Smoked Oysterling	
HPBS	Rhizomarasmius setosus	Hairy Stem Parachute	
HPBS	Rhizomarasmius undatus		
HPBS	Rhodocollybia butyracea	Butter Cap	
HPBS	Rhopographus filicinus	Bracken Map	
HPBS	Rhytisma acerinum	Sycamore Tarspot	
HPBS	Rickenella fibula	Orange Mosscap	
HPBS	Rickenella swartzii	Collared Mosscap	
HPBS	Riopa metamorphosa		√
HPBS	Roridomyces roridus	Dripping Bonnet	
HPBS	Russula atropurpurea	Purple Brittlegill	
HPBS	Russula cyanoxantha	Charcoal Burner	
HPBS	Russula cf. graveolens		
HPBS	Russula ochroleuca SH1509013.08FU	Ochre Brittlegill	
HPBS	Russula odorata SH1509034.08FU		
HPBS	Russula pseudointegra	Scarlet Brittlegill	
HPBS	Russula vesca	The Flirt	
HPBS	Russula virescens	Greencracked Brittlegill	
HPBS	Russula sororia SH1569748.08FU	Sepia Brittlegill	
HPBS	Russula sp. SH1528314.08FU		
HPBS	Ruzenia spermoides		
HPBS	Sarcoscypha austriaca	Scarlet Elfcup	
HPBS	Sawadaea bicornis	Maple Mildew	
HPBS	Schizophyllum commune	Splitgill	
HPBS	Scleroderma cepa		
HPBS	Scleroderma verrucosum SH1526179.08FU	Scaly Earthball	
HPBS	Scopuloides rimosa		
HPBS	Scutellinia scutellata	Common Eyelash	
HPBS	Scytinostroma portentosum	Mothball Crust	
HPBS	Sebacina incrustans	Enveloping Crust	
HPBS	Sebacina sp. 01 SH1561852.08FU		
HPBS	Sebacina sp. 02 SH1561859.08FU		
HPBS	Serpula himantioides		
HPBS	Sidera vulgaris		
HPBS	Simocybe centunculus	Dingy Twiglet	
HPBS	Simocybe sumptuosa	Velvet Twiglet	
HPBS	Singerocybe phaeophthalma	Chicken Run Funnel	
HPBS	Sistotrema brinkmannii		
HPBS	Sistotrema brinkmannii agg.		
HPBS	Skeletocutis nivea agg.	Hazel Bracket	
HPBS	Spinellus fusiger	Bonnet Mould	
HPBS	Spongiporus floriformis		
HPBS	Steccherinum ochraceum		
HPBS	Stereum gausapatum	Bleeding Oak Crust	

Source	Scientific Name	English Name	Oak Saprotroph
HPBS	Stereum hirsutum	Hairy Curtain Crust	
HPBS	Stereum ochraceoflavum		
HPBS	Stereum rugosum	Bleeding Broadleaf Crust	
HPBS	Stereum subtomentosum	Yellowing Curtain Crust	
HPBS	Stilbella fimetaria		
HPBS	Stropharia caerulea	Blue Roundhead	
HPBS	Stropharia pseudocyanea	Peppery Roundhead	
HPBS	Stypella grilletii		
HPBS	Subulicystidium longisporum		
HPBS	Suillellus luridus	Lurid Bolete	
HPBS	Szczepkamyces campestris	Hazel Porecrust	
HPBS	Thelephora wakefieldiae		
HPBS	Thelephoraceae sp. 01 SH1528519.08FU		
HPBS	Thelephoraceae sp. 02 SH1528482.08FU		
HPBS	Thelephoraceae sp. 03 SH1502188.08FU		
HPBS	Tomentella crinalis		
HPBS	Tomentella italica		
HPBS	Tomentella lateritia		
HPBS	Tomentella punicea		
HPBS	Tomentella sp. 01 SH2738806.08FU		
HPBS	Tomentella sp. 02 SH1528605.08FU		
HPBS	Tomentella sp. 03 SH1502265.08FU		
HPBS	Tomentella sp. 04 SH1502560.08FU		
HPBS	Tomentella sublilacina SH1502543.08FU		
HPBS	Tomentellopsis sp.		
HPBS	Tomentellopsis aff. zygodesmoides		
TVERC	Trametes gibbosa	Lumpy Bracket	
HPBS	Trametes hirsuta	Hairy Bracket	
HPBS	Trametes versicolor	Turkeytail	
HPBS	Trechispora cohaerens		
HPBS	Trechispora dimitica		
HPBS	Trechispora farinacea		
HPBS	Trechispora nivea		
HPBS	Trechispora stellulata		
HPBS	Trechispora stevensonii		
HPBS	Trichoderma pulvinatum	Ochre Cushion	
HPBS	Tricholoma album SH1647759.08FU	White Knight	
HPBS	Tricholoma cingulatum	Girdled Knight	
HPBS	Tricholoma fulvum	Birch Knight	
HPBS	Tricholoma lascivum	Aromatic Knight	
HPBS	Tricholoma saponaceum	Soapy Knight	
HPBS	Tricholoma sejunctum	Deceiving Knight	
HPBS	Tricholoma stiparophyllum	Chemical Knight	
HPBS	Tricholoma sulphureum	Sulphur Knight	
HPBS	Trimmatostroma betulinum		
HPBS	Tubaria dispersa	Hawthorn Twiglet	
HPBS	Tubaria furfuracea	Scurfy Twiglet	
HPBS	Typhula juncea	Slender Club	
HPBS	Typhula quisquiliaris	Bracken Club	
HPBS	Ustilago avenae		

Source	Scientific Name	English Name	Oak Saprotroph
HPBS	Vanderbylia fraxinea		
HPBS	Vitreoporus dichrous	Bicoloured Bracket	
HPBS	Volvopluteus gloiocephalus	Stubble Rosegill	
HPBS	Vuilleminia comedens	Waxy Crust	
HPBS	Vuilleminia cystidiata		
HPBS	Xenasmatella vaga	Yellow Cobweb	
HPBS	Xerocomellus cisalpinus	Bluefoot Bolete	
HPBS	Xerocomellus porosporus SH1508917.08FU	Sepia Bolete	
HPBS	Xerocomellus pruinatus	Matt Bolete	
HPBS	Xerocomus subtomentosus	Suede Bolete	
HPBS	Xerula pudens		
HPBS	Xylaria carpophila	Beechmast Candlesnuff	
HPBS	Xylaria hypoxylon	Candlesnuff Fungus	
HPBS	Xylaria longipes	Dead Moll's Fingers	
HPBS	Xylodon asper		
HPBS	Xylodon paradoxus	Split Porecrust	
HPBS	Xylodon raduloides		
HPBS	Xylodon sambuci	Elder Whitewash	

Appendix 4. Checklist of lichens

Recorders: Brian Coppins, Mark Powell, Paula Shipway,
Neil Sanderson, Pat Wolseley, Sandy Coppins, Andy Cross

The codes for small-scale habitats are used in standard Excel recording spreadsheets issued by the British Lichen Society (BLS) where CQ stands for Cortex (bark) of Quercus (oak) and LQ for Lignum (wood) of Quercus (oak). The codes are explained in the BLS Lichen Recording Guidelines on the BLS website.

Taxon name	Small-scale habitats
Agonimia allobata	CQ
Agonimia flabelliformis	CQ
Agonimia tristicula	CFx, CQ
Alyxoria culmigena	CFx
Alyxoria ochrocheila	CFx, SSm, LFg, LTr
Alyxoria varia	CFx, CFg
Amandinea punctata	LQ, LDs, CQ
Andreiomyces obtusaticus	CQ
Anisomeridium biforme	CFx
Anisomeridium polypori	CSm, CQ, CU
Anisomeridium sp.	CCt
Aquacida viridifarinosa	CQ
Arthonia atra	CQ,CTw, CFx, CFg
Arthonia didyma	CTb, CQ, CCo
Arthonia punctiformis	CFg, CTb, CCt
Arthonia radiata	CQ,CTb, Co, CPt
Arthopyrenia analepta	CCt
Arthopyrenia punctiformis	CCt
Bacidia delicata	CAc
Bacidia friesiana	CSm
Bacidia phacodes	CFx
Bacidia rubella	CAc, CQ, CFx
Bactrospora corticola	CQ, LQ, LDf
Biatora veteranorum	LQ, LDf
Buellia griseovirens	LQ, LTb, LQ
Buellia hyperbolica	LQ, LDf, LTr, LDs
Buellia schaereri	LQ, LDs, LDf
Calicium abietinum	LQ
Calicium glaucellum	LQ, LDs, LDf, CQ
Calicium viride	CQ
Caloplaca cerinella	CPt,CTb
Caloplaca chlorina	CQ
Caloplaca chrysodeta	CFx

Taxon name	Small-scale habitats
Caloplaca flavocitrina	CQ
Caloplaca lucifuga	CQ
Caloplaca obscurella	CQ, CSm, CFg
Caloplaca ulcerosa	CAc, CFx, CSm
Candelaria concolor	CQ,CTb
Candelariella vitellina f. *vitellina*	PFr
Candelariella xanthostigmoides	CQ, CSx, DFx, CFg, CTb
Catillaria fungoides	CAp, CTw
Catillaria nigroclavata	CFx, CTw
Chaenotheca brachypoda	CFx, CSm, CQ
Chaenotheca brunneola	LQ, LDs, CQ
Chaenotheca chrysocephala	LQ, LDs, LDf, CQ
Chaenotheca ferruginea	LQ, LDf, LDs, CQ
Chaenotheca furfuracea	LQ, LDf
Chaenotheca hispidula	CQ
Chaenotheca stemonea	CQ
Chaenotheca trichialis	CQ, LQ, LDs, LTr
Chaenothecopsis nigra	LQ, LDs, LDf, CQ, LFx, LTr
Chaenothecopsis pusilla	LQ, LDf, LTr
Chaenothecopsis savonica	CQ, LQ
Chrysothrix candelaris	CQ, LQ, LDf, CTr
Chrysothrix flavovirens	LQ, LTr, LDs, LDf, CQ
Cladonia chlorophaea s. lat.	LQ, LDf, CQ
Cladonia coniocraea	QL, LTr, LDs, LDf, LTb
Cladonia cryptochlorophaea	LQ
Cladonia cyathomorpha	CQ
Cladonia digitata	LQ, LDf, LDs, CQ, LTr
Cladonia fimbriata	CQ, LQ, LDf, LDs
Cladonia glauca	CQ
Cladonia humilis	LQ, LDf
Cladonia macilenta	LQ, LDf, CQ
Cladonia ochrochlora	LQ
Cladonia parasitica	LQ, LDf, LDs, LTb, LTr
Cladonia polydactyla var. *polydactyla*	LQ, LDs, CQ, LDf, LTr
Cladonia pyxidata	CQ, LQ, LDf
Cladonia ramulosa	CQ, LQ, LDf
Cliostomum griffithii	CQ, CFx
Clypeococcum hypocenomycis	Z0578, LQ, LDs, LDf, LTr
Coenogonium pineti	CQ
Coenogonium tavaresianum	CQ
Cresponea premnea	CQ, CAp, LQ
Cyphelium sessile	CQ, Z1064
Cyrtidula quercus	CQ, CTw
Dactylospora parasitica	CQ,CFx, Z1076
Diarthonis spadicea	CQ
Diploicia canescens	CAc, CQ, CFx, LQ
Enterographa crassa	CQ, CAp, LQ
Evernia prunastri	CTw, CQ, CTb, CSx, LQ, CPt
Fellhanera bouteillei	CTw
Flavoparmelia caperata	CQ, LQ, CFx, CSx
Fuscidea lightfootii	LQ, LDf, CFg, CTb
Graphis scripta	CCo
Gyalecta flotowii	CQ
Gyalecta truncigena	CFx

Taxon name	Small-scale habitats
Halecania viridescens	Cort
Heterocephalacria physciacearum	Z1120
Hyperphyscia adglutinata	CSm
Hypocenomyce scalaris	LQ, LDs, LTr, LDf
Hypogymnia physodes	CQ, CTb, LQ
Hypogymnia tubulosa	
Hypotrachyna afrorevoluta	CAp, CTb, CSx
Hypotrachyna revoluta s. str.	CQ,CTb
Illosporiopsis christiansenii	
Imshaugia aleurites	LQ, LDf
Jamesiella anastomosans	LQ
Lecanactis abietina	CQ
Lecania cyrtella	CFg, CTr, CQ
Lecania cyrtellina	CFg, CTr, CAc
Lecania naegelii	CAp, CSm, CFx
Lecanidion atratum	LQ
Lecanographa lyncea	CQ
Lecanora albellula var. albellula	LQ, LDf
Lecanora argentata	CFx, CQ
Lecanora carpinea	CPt, CTb
Lecanora chlarotera	CQ, Lx, CPt
Lecanora conizaeoides f. conizaeoides	LQ, LDf
Lecanora dispersa	LWT
Lecanora expallens	LQ, LDs, CQ, LDf
Lecanora hagenii	CPt
Lecanora hybocarpa	CFg, CTw
Lecanora hypoptoides	LQ, LDf
Lecanora muralis	LWT
Lecanora phaeostigma	LQ, LDs, LDf
Lecanora pulicaris	LQ
Lecanora quercicola	CQ
Lecanora saligna	LQ, LDs
Lecanora sublivescens	CQ
Lecanora symmicta	LQ, LDf, LTr
Lecidea nylanderi	LQ, LDf, LTb, LDs
Lecidella elaeochroma f. elaeochroma	CQ, CFx
Lepraria finkii	CQ, LQ, LDf
Lepraria incana s. lat.	LQ, LDs, CQ
Lepraria incana s. str.	LQ, LDs, CQ, LDf, LTr
Lepraria sylvicola	CQ
Lepraria vouauxii	CQ
Leptorhaphis atomaria	CPt, CTb
Leptorhaphis maggiana	CCo
Lichenochora obscuroides	CTb, CSm, Z1107
Lichenoconium erodens	CTw, Z0511, CQ, Z1087
Lichenodiplis pertusariicola	CFx, Z1079
Lichenodiplis sp.	Z0685
Lichenotubeufia heterodermiae	CTw, Z1120
Melanelixia glabratula	CQ, LQ, Cfx
Melanelixia subaurifera	CLx, LTb, LQ
Melanohalea elegantula	CQ, CTw, CAp, CTb, CFx
Melaspilea sp.	CQ
Micarea denigrata	LQ, LDs, LDf
Micarea globulosella	LQ, LDf

Taxon name	Small-scale habitats
Micarea melaena	LQ, LDf
Micarea misella	LQ, LDf, LDS
Micarea nitschkeana	LQ
Micarea peliocarpa	LQ, LDf
Micarea prasina s. lat.	LQ, LDs, LDf, LTr, CQ
Microcalicium ahlneri	LQ, LDs, LTr
Microcalicium disseminatum	LQ, LDs, CQ
Milospium graphideorum	CQ, Zo600
Mycocalicium subtile	LQ
Navicella pileata	CFx
Normandina pulchella	CQ
Ochrolechia androgyna	LQ, LDf
Ochrolechia arborea	CLx, LQ
Ochrolechia microstictoides	LQ
Ochrolechia subviridis	CFx, CQ
Ochrolechia turneri s. str.	CQ, CFx
Opegrapha niveoatra	CQ
Opegrapha vermicellifera	CQ, CAc, CFx
Opegrapha vulgata	CAp, CQ, CFg
Pachnolepia pruinata	CQ, LQ
Pachyphiale carneola	CQ, CCo
Parmelia saxatilis	LQ, CFx, CQ
Parmelia sulcata	CQ, LTb, LQ
Parmeliopsis ambigua	LQ
Parmotrema perlatum	CFx, CSx, CQ
Pertusaria albescens var. *albescens*	CFx
Pertusaria albescens var. *corallina*	CQ
Pertusaria amara f. *amara*	CQ, CFx
Pertusaria coccodes	CQ, LQ, CFx, LDf
Pertusaria flavida	CQ, CFx
Pertusaria hymenea	CQ, CFx
Pertusaria leioplaca	CTb, CCt, CQ, CFx, CCo
Pertusaria pertusa	CQ, CFx
Phaeographis smithii	CQ, CTw
Phaeophyscia orbicularis	CAp, CSm, CTb, CQ
Phlyctis argena	CQ, CSx, CCo, CFx, CU
Phoma sp.	LQ, LDs, Zo578
Phylloblastia fortuita	Flx
Phylloblastia inexpectata	Flx
Physcia aipolia	LWT
Physcia tenella	
Physcia tribacia	CQ
Physconia grisea	CQ, CAp, CFx
Physconia perisidiosa	CFx
Placynthiella icmalea	LQ, LDs, LDf
Platismatia glauca	LQ
Porina borreri	CQ
Porina byssophila	CAp, CU, CQ
Protoparmelia oleagina	LQ, LDf
Pseudoschismatomma rufescens	CFx
Psoroglaena stigonemoides	CAc, CQ, CSm
Punctelia jeckeri	CQ, CTb, CLx, CAp, CTw, CFx
Punctelia subrudecta s. str.	CQ, CTb, CLx, CAp, LQ
Pycnora sorophora	LQ

Taxon name	Small-scale habitats
Pyrenula chlorospila	CFx
Pyrrhospora quernea	CQ, CFx, LQ
Ramalina farinacea	CQ, CTb, LQ
Ramalina fastigiata	CQ
Rinodina exigua	CQ
Rinodina furfuracea	CFx, CTr
Rinodina roboris var. roboris	CQ
Schismatomma decolorans	CQ, CAc, CSm, CFx, LQ
Scoliciosporum chlorococcum	Cort
Scoliciosporum pruinosum	CQ
Sphinctrina anglica	LQ, Z1793
Sphinctrina turbinata	CQ, Lic
Sporodophoron cretaceum	CQ
Stigmidium eucline	Z1075, CQ
Strigula jamesii	CU, CQ, CFg
Strigula phaea	CQ
Strigula taylorii	CFx, CSm, CAc, CAp, CQ, CFg
Syncesia myrticola (sorediate morph)	CQ
Taeniolella phaeophysciae	Z1107, CAp, CTb
Thelopsis corticola	CAc, CQ
Thelopsis rubella	CQ
Trapeliopsis flexuosa	LQ, LDs, LDf, CQ
Trapeliopsis granulosa	LQ, LDf, LDs
Tremella sp.	CQ, CTw, Z1989
Unguiculariopsis lesdainii	CQ, Z0675
Unguiculariopsis thallophila	Lic
Usnea subfloridana	CTb
Varicellaria hemisphaerica	CQ, LQ, LDf, CFx, CSx
Verrucaria elaeina	Sax
Violella fucata	LQ, LDf, LTb
Xanthoria parietina	CQ, CTb, CFx, CPt
Xanthoria polycarpa	CQ, CTw
Xanthoriicola physciae	Z1530
Xylopsora friesii	LQ, LDf

Appendix 5. Checklist of snails, slugs and bivalves (Mollusca)

Surveyors: Rosemary Hill, Rosemary Winnall, Peter Topley, Tom Walker

Taxon name
Acanthinula aculeata
Aegopinella nitidula
Aegopinella pura
Anisus leucostoma
Arianta arbustorum
Arion ater
Arion circumscriptus
Arion distinctus
Arion fasciatus
Arion flagellus
Arion hortensis
Arion intermedius
Arion owenii
Arion rufus
Arion subfuscus
Ashfordia granulata
Balea perversa
Bithynia tentaculata
Boettgerilla pallens
Carychium minimum
Carychium tridentatum
Cepaea hortensis
Cepaea nemoralis
Clausilia bidentata
Cochlicopa lubrica
Cochlicopa lubricella
Cochlodina laminata
Columella edentula
Cornu aspersum
Deroceras invadens
Deroceras laeve
Deroceras reticulatum
Discus rotundatus
Euconulus alderi
Euconulus fulvus
Galba truncatula
Gyraulus crista
Helicigona lapicida

Taxon name
Helix pomatia
Hygromia cinctella
Lauria cylindracea
Lehmannia marginata
Limacus maculatus
Limax maximus
Lymnaea stagnalis
Merdigera obscura
Monacha cantiana
Nesovitrea hammonis
Oxychilus alliarius
Oxychilus cellarius
Oxychilus draparnaudi
Oxychilus navarricus helveticus
Oxychius cellarius
Pisidium obtusale
Pisidium personatum
Pomatias elegans
Potamopyrgus antipodarum
Punctum pygmaeum
Pyramidula pusilla
Radix balthica
Succinea putris
Tandonia budapestensis
Tandonia sowerbyi
Testacella haliotidea
Trochulus hispidus
Trochulus striolatus
Vallonia costata
Vallonia excentrica
Vallonia pulchella
Vertigo pygmaea
Vertigo substriata
Vitrea contracta
Vitrea crystallina
Vitrina pellucida
Zenobiella subrufescens

Appendix 6. Checklist of Arachnids

Recorders: Bill Parker, OBRC (= TVERC), Steve Gregory,
Keith Alexander, Lawrence Bee, Brian Spooner, Ivan Wright

Status (UK): NR = Nationally Rare, NT = Near Threatened (IUCN), 'declining' is here recorded if >10% and 'amber' >40% according to Harvey et al. (2017); Source: HPBS = High Park Biodiversity Survey, TVERC = records prior to 2017 held by Thames Valley Environmental Records Centre; Moccas: X = also recorded at Moccas Park, Herefordshire. *Pardosa lugubris* is marked as a doubtful identification with ?

Group	Family	Scientific name	Status	Source	Moccas
harvestmen	Nemastomatidae	*Nemastoma bimaculatum*	common	TVERC	
	Phalangiidae	*Dicranopalpus ramosus*	spreading	HPBS	
		Lacinius ephippiatus	common	TVERC	
		Mitopus morio	common	TVERC	
mites	Eriophyidae	*Eriophyes rubicolens*	rare	HPBS	
pseudoscorpions	Chernetidae	*Dendrochernes cyrneus*	rare	HPBS, TVERC	
spiders	Agelenidae	*Agelena labrynthica*	common	HPBS	
		Coelotes atropos	declining	TVERC	
		Eratigena gigantea	common	HPBS	
		Tegenaria silvestris	common	HPBS	
		Tegenaria sp.	?	HPBS	
	Amaurobiidae	*Amaurobius fenestralis*	common	HPBS, TVERC	X
		Amaurobius similis	common	TVERC	X
		Amaurobius sp.	?	HPBS	
	Anyphaenidae	*Anyphaena accentuata*	common	HPBS, TVERC	X
	Araneidae	*Araneus diadematus*	common	HPBS, TVERC	X
		Araneus marmoreus	common	TVERC	
		Araneus triguttatus	rare	TVERC	X
		Araniella cucurbitina s. str.	common	HPBS	X
		Gibbaranea gibbosa	common	TVERC	X
		Larinioides cornutus	common	HPBS	X
		Leviellus (Stroemiellus) stroemi	NR, NT	TVERC	X
		Mangora acalypha	common	HPBS	
		Nuctenea umbratica	common	HPBS	X
		Zygiella atrica	common	TVERC	X
		Zygiella x-notata	common	TVERC	X
	Clubionidae	*Clubiona brevipes*	common	TVERC	X
		Clubiona comta	common	HPBS	X

Group	Family	Scientific name	Status	Source	Moccas
		Clubiona corticalis	common	HPBS, TVERC	
		Clubiona lutescens	common	TVERC	
		Clubiona reclusa	common	HPBS, TVERC	
	Dictynidae	*Lathys humilis*	common	TVERC	X
	Dysderidae	*Harpactea hombergi*	common	HPBS, TVERC	X
	Gnaphosidae	*Zelotes latreillei*	local	TVERC	
	Linyphiidae	*Agyneta (Meioneta) innotabilis*	declining	TVERC	
		Agyneta (Meloneta) innotabilis	declining	TVERC	
		Agyneta (Meloneta) rurestris	declining	TVERC	
		Bathyphantes gracilis	common	TVERC	X
		Bathyphantes nigrinus	declining	TVERC	X
		Bathyphantes parvulus	declining	TVERC	
		Centromerus sylvaticus	declining	TVERC	
		Collinsia inerrans	uncommon	TVERC	X
		Dicymbium nigrum	declining	TVERC	X
		Diplocephalus picinus	common	TVERC	X
		Diplostyla concolor	declining	HPBS, TVERC	X
		Dismodicus bifrons	declining	TVERC	X
		Erigone atra	common	TVERC	X
		Erigone dentipalpis	common	TVERC	X
		Gonatium rubens	declining	TVERC	X
		Hylyphantes graminicola	common	HPBS, TVERC	
		Hypomma cornutum	common	TVERC	X
		Kaestneria dorsalis	declining	TVERC	
		Linyphia triangularis	common	HPBS, TVERC	X
		Micrargus herbigradus	declining	TVERC	
		Neriene clathrata	common	HPBS	X
		Neriene peltata	common	TVERC	X
		Palliduphantes ericaeus	declining	TVERC	
		Palliduphantes pallidus	declining	TVERC	
		Pocadicnemis juncea	common	TVERC	
		Porrhomma pygmaeum	common	TVERC	X
		Savignia frontata	declining	TVERC	X
		Sintula corniger	NS, amber	TVERC	
		Tenuiphantes cristatus	declining	TVERC	X
		Tenuiphantes mengei	declining	TVERC	X
		Tenuiphantes tenuis	common	HPBS, TVERC	X
		Tenuiphantes zimmermanni	common	TVERC	X
		Walckenaeria acuminata	declining	TVERC	X
		Walckenaeria atrotibialis	declining	TVERC	
	Lycosidae	*Alopecosa pulverulenta*	common	TVERC	
		Pardosa amentata	common	HPBS, TVERC	X
	?	*Pardosa lugubris*	rare	TVERC	X
		Pardosa prativaga	common	HPBS, TVERC	

Group	Family	Scientific name	Status	Source	Moccas
		Pardosa pullata	common	TVERC	X
		Pardosa saltans	common	HPBS	
		Piratula hygrophila	common	HPBS, TVERC	X
		Trochosa terricola	common	TVERC	
	Philodromidae	Philodromus albidus	common	HPBS, TVERC	
		Philodromus aureolus	common	TVERC	X
		Tibellus oblongus	common	TVERC	
	Pisauridae	Pisaura mirabilis	common	HPBS, TVERC	X
	Salticidae	Evarcha falcata	common	HPBS	
		Heliophanus flavipes	common	TVERC	
		Neon reticulatus	declining	HPBS	
		Pseudeuophrys erratica	declining	HPBS	
	Segestriidae	Segestria senoculata	common	TVERC	X
	Tetragnathidae	Metellina mengei	common	HPBS, TVERC	X
		Metellina segmentata	common	HPBS, TVERC	X
		Metellina sp.	?	HPBS	
		Pachygnatha clercki	common	TVERC	
		Pachygnatha degeeri	common	HPBS, TVERC	X
		Tetragnatha extensa	common	TVERC	X
		Tetragnatha montana	common	HPBS, TVERC	X
		Tetragnatha sp.	?	HPBS	
	Theridiidae	Anelosimus vittatus	common	HPBS, TVERC	X
		Enoplognatha ovata	common	HPBS, TVERC	X
		Neottiura bimaculata	common	HPBS, TVERC	
		Paidiscura pallens	common	TVERC	
		Parasteatoda lunata	uncommon	HPBS	
		Parasteatoda simulans	uncommon	TVERC	
		Phylloneta sisyphia	common	HPBS, TVERC	
		Platnickina tincta	common	HPBS	X
		Steatoda bipunctata	common	HPBS	X
	Thomisidae	Diaea dorsata	local	HPBS, TVERC	X
		Xysticus cristatus	common	TVERC	X
		Xysticus lanio	declining	HPBS	
		Xysticus sp.	?	HPBS	
	Zoridae	Zora spinimana	common	TVERC	X

Appendix 7. Checklist of Diptera

Recorders: Peter Chandler, Jon Cooter, Ivan Wright,
Benedict Pollard, Graham Collins, Peter Hall, Jovita Kaunang

Conservation status: NS = Nationally Scarce, pNS = provisionally NS, NT = Near Threatened (IUCN), VU = Vulnerable (IUCN), pVU = provisionally VU, RDB2 = Red Data Book status 2 (JNCC), RDB3 = RDB status 3, Notable = as in Falk (1991), DD = Data Deficient (IUCN)

Family	Species name	Saproxylic	Conservation status
Agromyzidae	Chromatomyia milii		
	Napomyza scrophulariae		
	Phytomyza glechomae		
Anisopodidae	Sylvicola cinctus	X	
	Sylvicola punctatus		
Anthomyiidae	Adia cinerella		
	Alliopsis billbergi		
	Anthomyia confusanea		
	Anthomyia liturata		
	Anthomyia mimetica		
	Anthomyia pluvialis		
	Anthomyia procellaris	X	
	Botanophila dissecta		
	Botanophila fugax		
	Botanophila laterella		pNS
	Botanophila phrenione		
	Botanophila striolata		
	Chirosia albitarsis		
	Chirosia crassiseta		
	Chirosia grossicauda		
	Chirosia histricina		
	Chirosia nigripes		
	Delia florilega		
	Delia platura		
	Delia radicum		
	Egle inermis		pNS
	Egle parvaeformis		pNS
	Egle rhinotmeta		
	Eustalomyia festiva	X	
	Eustalomyia hilaris	X	pNS
	Eustalomyia vittipes	X	pNS
	Hydrophoria ruralis		
	Hylemya nigrimana	X	
	Hylemya urbica		

Family	Species name	Saproxylic	Conservation status
	Hylemya vagans		
	Hylemya variata		
	Hylemyza partita		
	Lasiomma picipes		
	Lasiomma seminitidum		
	Lasiomma strigilatum		
	Leucophora obtusa		
	Mycophaga testacea		
	Paradelia intersecta		
	Paregle audacula		
	Pegomya bicolor		
	Pegomya geniculata	X	
	Pegomya solennis		
	Pegoplata aestiva		
	Pegoplata annulata		
	Pegoplata infirma		
	Phorbia bartaki		
	Phorbia fumigata		
	Phorbia sepia		
Asilidae	Choerades marginatus	X	
	Dioctria baumhaueri		
	Dioctria linearis		
	Dioctria rufipes		
	Leptogaster cylindrica		
	Machimus atricapillus		
Asteiidae	Asteia amoena	X	
Atelestidae	Atelestus pulicarius		
Bibionidae	Bibio lanigerus		
	Bibio leucopterus		
	Bibio marci		
	Bibio nigriventris		
	Bibio reticulatus		
	Bibio varipes		
	Dilophus bispinosus		
	Dilophus febrilis		
	Dilophus femoratus		
Bolitophilidae	Bolitophila cinerea	X	
	Bolitophila occlusa	X	
	Bolitophila pseudohybrida	X	
	Bolitophila saundersii	X	
Bombylidae	Bombylius major		
Brachystomatidae	Trichopeza longicornis		
Calliphoridae	Bellardia viarum		
	Bellardia vulgaris		
	Calliphora vicina		
	Calliphora vomitoria		
	Lucilia ampullacea		
	Lucilia caesar		
	Lucilia illustris		
	Melanomya nana		
	Melanophora roralis	X	
	Melinda gentilis		
	Melinda viridicyanea		
	Paykullia maculata	X	

Family	Species name	Saproxylic	Conservation status
	Protocalliphora azurea		
	Rhinophora lepida	X	
	Tricogena rubricosa	X	
Campichoetidae	Campichoeta punctum		
Cecidomyiidae	Dasineura crataegi		
	Dasineura urticae		
	Hartigiola annulipes		
	Iteomyia major		
Ceratopogonidae	Sphaeromias fasciatus		
Chironomidae	Brillia bifida		
	Chironomus commutatus		
	Cricotopus annulator		
	Cricotopus bicinctus		
	Kiefferulus tendipediformis		
	Microchironomus tener		
	Microtendipes pedellus		
	Paraphaenocladius impensus		
	Paratendipes albimanus		
	Polypedilum nubeculosum		
Chloropidae	Chlorops hypostigma		
	Dicraeus vagans		
	Thaumatomyia notata		
	Tricimba cincta		
Clusiidae	Clusia flava	X	
	Clusiodes gentilis	X	
	Clusiodes verticalis	X	
Conopidae	Myopa pellucida		RDB3
	Myopa testacea		
	Sicus ferrugineus		
	Thecophora atra		
Culicidae	Dahliana geniculata	X	
Diadocidiidae	Diadocidia ferruginosa	X	
Diastatidae	Diastata fuscula		
Ditomyiidae	Symmerus annulatus	X	
Dolichopodidae	Argyra leucocephala		
	Campsicnemus curvipes		
	Campsicnemus loripes		
	Campsicnemus scambus		
	Chrysotus blepharosceles		
	Chrysotus gramineus		
	Chrysotus laesus		
	Chrysotus neglectus		
	Dolichopus festivus		
	Dolichopus griseipennis		
	Dolichopus plumipes		
	Dolichopus popularis		
	Dolichopus ungulatus		
	Dolichopus virgultorum		NS
	Dolichopus wahlbergi		
	Gymnopternus aerosus		
	Gymnopternus cupreus		
	Gymnopternus metallicus		
	Hercostomus nigrilamellatus	X	NS
	Hercostomus parvilamellatus		

Family	Species name	Saproxylic	Conservation status
	Medetera saxatilis	X	
	Medetera truncorum	X	
	Neurigona pallida	X	
	Neurigona quadrifasciata	X	
	Poecilobothrus nobilitatus		
	Rhaphium appendiculatum		
	Scellus notatus		
	Sciapus platypterus	X	
	Sybistroma obscurellum		
	Sympycnus pulicarius		
	Syntormon bicolorellum		
	Syntormon pallipes		
	Thrypticus bellus		
	Xanthochlorus galbanus		
	Xanthochlorus ornatus		
Drosophilidae	*Chymomyza fuscimana*	X	
	Drosophila phalerata	X	
	Drosophila subobscura	X	
	Drosophila suzukii		
	Hirtodrosophila cameraria	X	
	Hirtodrosophila confusa	X	
	Leucophenga maculata	X	
	Phortica variegata	X	pVU
	Scaptomyza flava		
	Scaptomyza graminum		
	Scaptomyza pallida		
Dryomyzidae	*Dryomyza anilis*		
	Dryope flaveola		
Empididae	*Chelifera precatoria*		
	Clinocera nigra		
	Dolichocephala oblongoguttata		
	Empis aestiva		
	Empis caudatula		
	Empis femorata		
	Empis grisea		
	Empis livida		
	Empis longipes		
	Empis lutea		
	Empis nigripes		
	Empis nigritarsis		
	Empis nuntia		
	Empis pennipes		
	Empis planetica		
	Empis scutellata		
	Empis stercorea		
	Empis tessellata		
	Empis trigramma		
	Hilara anglodanica		
	Hilara cornicula		
	Hilara fuscipes		
	Hilara lurida	X	
	Hilara ternovensis		
	Hilara thoracica		
	Phyllodromia melanocephala		

Family	Species name	Saproxylic	Conservation status
	Rhamphomyia caliginosa		NS
	Rhamphomyia crassirostris		
	Rhamphomyia hybotina		
	Rhamphomyia lamellata		NS
	Rhamphomyia sulcata		
	Rhamphomyia tarsata		
Ephydridae	*Discomyza incurva*		
	Hydrellia albiceps		
	Philygria interstincta		
Fanniidae	*Fannia aequilineata*	X	pNS
	Fannia armata		
	Fannia canicularis	X	
	Fannia fuscula		
	Fannia mollissima		
	Fannia monilis	X	
	Fannia pallitibia		
	Fannia rondanii		
	Fannia serena		
	Fannia sociella		
Heleomyzidae	*Eccoptomera obscura*		
	Heteromyza rotundicornis		
	Suillia affinis		
	Suillia atricornis	X	
	Suillia bicolor	X	
	Suillia pallida		
	Suillia variegata	X	
	Tephrochlamys flavipes	X	
Hippoboscidae	*Lipoptena cervi*		
Hybotidae	*Bicellaria nigra*		
	Bicellaria vana		
	Elaphropeza ephippiata		
	Hybos culiciformis		
	Hybos femoratus		
	Leptopeza flavipes	X	
	Ocydromia glabricula		
	Oedalea holmgreni	X	
	Oedalea stigmatella	X	
	Platypalpus agilis		
	Platypalpus annulipes		
	Platypalpus candicans		
	Platypalpus ciliaris		
	Platypalpus exilis		
	Platypalpus longicornis		
	Platypalpus mikii	X	NS
	Platypalpus minutus		
	Platypalpus optivus		
	Platypalpus pallidiventris		
	Platypalpus parvicauda		
	Platypalpus pectoralis		
	Tachypeza fuscipennis	X	NS
	Tachypeza nubila	X	
	Trichina elongata		
	Trichinomyia flavipes		
Keroplatidae	*Cerotelion striatum*	X	

Family	Species name	Saproxylic	Conservation status
	Keroplatus testaceus	X	NS
	Macrocera angulata	X	
	Macrocera centralis	X	
	Macrocera phalerata		
	Macrocera fasciata	X	
	Macrocera lutea		
	Macrocera nigricoxa		
	Macrocera phalerata		
	Macrocera stigmoides	X	
	Macrocera vittata	X	
	Macrorrhyncha flava	X	
	Monocentrota lundstromi		
	Neoplatyura modesta		
	Orfelia fasciata	X	
	Orfelia lugubris		
	Orfelia nemoralis	X	
	Orfelia ochracea	X	
	Platyura marginata		
	Pyratula zonata		
Lauxaniidae	*Calliopum aeneum*		
	Calliopum simillimum		
	Calliopum tuberculosum		
	Homoneura interstincta		
	Meiosimyza decipiens		
	Meiosimyza platycephala		
	Meiosimyza rorida		
	Minettia fasciata		
	Minettia longipennis		
	Poecilolycia vittata		
	Pseudolyciella pallidiventris	X	DD
	Pseudolyciella stylata	X	
	Sapromyza sexpunctata		
	Tricholauxania praeusta		
Limoniidae	*Austrolimnophila ochracea*	X	
	Dicranomyia autumnalis		
	Dicranomyia chorea		
	Dicranomyia mitis		
	Dicranomyia modesta		
	Epiphragma ocellare	X	
	Erioptera lutea		
	Ilisia occoecata		
	Limonia nubeculosa		
	Limonia phragmitidis	X	
	Molophilus cinereifrons		
	Molophilus griseus		
	Molophilus ochraceus		
	Neolimonia dumetorum	X	
	Ormosia nodulosa		
	Phylidorea ferruginea		
	Pilaria discicollis		
	Rhipidia maculata		
Lonchaeidae	*Lonchaea patens*	X	
	Lonchaea tarsata		
Lonchopteridae	*Lonchoptera lutea*		

Family	Species name	Saproxylic	Conservation status
Muscidae	*Azelia cilipes*		
	Coenosia agromyzina		
	Coenosia albicornis		
	Coenosia infantula		
	Coenosia mollicula		
	Coenosia testacea		
	Coenosia tigrina		
	Eudasyphora cyanella		
	Eudasyphora cyanicolor		
	Graphomya maculata		
	Hebecnema vespertina		
	Helina abdominalis	X	pNS
	Helina depuncta		
	Helina evecta		
	Helina impuncta		
	Helina lasiophthalma		
	Hydrotaea cyrtoneurina		
	Hydrotaea dentipes		
	Hydrotaea irritans		
	Limnophora triangula		
	Lispocephala falculata		pNS
	Mesembrina meridiana		
	Musca autumnalis		
	Muscina levida	X	
	Muscina prolapsa		
	Mydaea ancilla		
	Mydaea electa		
	Myospila meditabunda		
	Neomyia cornicina		
	Neomyia viridescens		
	Phaonia errans		
	Phaonia pallida	X	
	Phaonia palpata	X	
	Phaonia rufiventris	X	
	Phaonia subventa	X	
	Phaonia tuguriorum		
	Phaonia valida		
Mycetophilidae	*Acnemia amoena*	X	NT
	Acnemia nitidicollis	X	
	Allodia ornaticollis	X	
	Apolephthisa subincana	X	
	Boletina gripha	X	
	Boletina sciarina		
	Brachycampta neglecta	X	NS
	Brachypeza radiata	X	
	Brevicornu fissicauda		
	Brevicornu griseicolle		
	Brevicornu proximum		
	Brevicornu sericoma	X	
	Coelophthinia thoracica		
	Coelosia flava		
	Cordyla brevicornis	X	
	Cordyla crassicornis		
	Cordyla fissa		

Family	Species name	Saproxylic	Conservation status
	Cordyla semiflava		
	Cordyla sp near *murina*		
	Docosia fumosa		
	Docosia gilvipes	X	
	Docosia moravica		
	Docosia sciarina		
	Dynatosoma fuscicorne	X	
	Ectrepesthoneura hirta	X	
	Epicypta aterrima	X	
	Epicypta fumigata	X	
	Exechia bicincta		
	Exechia contaminata		
	Exechia dorsalis	X	
	Exechia festiva		
	Exechia fusca	X	
	Exechia neorepanda		
	Exechia parva	X	
	Exechiopsis intersecta		
	Exechiopsis leptura	X	
	Grzegorzekia collaris	X	NS
	Leia bimaculata	X	
	Leia crucigera	X	
	Leia fascipennis		
	Leptomorphus walkeri	X	
	Megalopelma nigroclavatum		
	Megophthalmidia crassicornis		
	Monoclona rufilatera	X	
	Mycetophila adumbrata	X	
	Mycetophila alea		
	Mycetophila britannica	X	
	Mycetophila curviseta		
	Mycetophila deflexa		DD
	Mycetophila dentata	X	
	Mycetophila edwardsi		
	Mycetophila formosa	X	
	Mycetophila fraterna	X	
	Mycetophila fungorum	X	
	Mycetophila ichneumonea	X	
	Mycetophila luctuosa	X	
	Mycetophila lunata	X	
	Mycetophila marginata	X	
	Mycetophila occultans		
	Mycetophila ocellus	X	
	Mycetophila ornata	X	
	Mycetophila perpallida		
	Mycetophila pictula	X	
	Mycetophila pumila	X	
	Mycetophila ruficollis	X	
	Mycetophila sepulta		
	Mycetophila signatoides		
	Mycetophila sordida		
	Mycetophila stolida		
	Mycetophila stylatiformis		
	Mycetophila tridentata	X	

Family	Species name	Saproxylic	Conservation status
	Mycetophila vittipes	X	
	Mycomya cinerascens	X	
	Mycomya circumdata	X	
	Mycomya flavicollis		
	Mycomya insignis	X	NS
	Mycomya marginata	X	
	Mycomya occultans	X	NS
	Mycomya tenuis		
	Mycomya trilineata	X	
	Mycomya tumida	X	
	Mycomya wankowiczii	X	
	Mycomya winnertzi	X	
	Neoempheria striata	X	VU
	Palaeodocosia vittata		
	Phronia biarcuata	X	
	Phronia braueri	X	
	Phronia cinerascens	X	
	Phronia conformis	X	
	Phronia coritanica	X	
	Phronia forcipata		
	Phronia nigricornis	X	
	Phronia notata		
	Phronia signata		
	Phronia strenua	X	
	Phronia tenuis	X	
	Phthinia humilis	X	
	Phthinia mira	X	
	Platurocypta punctum	X	
	Platurocypta testata	X	
	Polylepta guttiventris		
	Pseudobrachypeza helvetica	X	
	Rondaniella dimidiata	X	
	Rymosia fasciata		
	Rymosia spinipes		NS
	Sceptonia cryptocauda		
	Sceptonia fumipes		
	Sceptonia nigra		
	Sciophila baltica		
	Sciophila fenestella		
	Sciophila hirta	X	
	Stigmatomeria crassicornis	X	
	Synapha vitripennis		
	Synplasta gracilis	X	
	Tetragoneura sylvatica	X	
	Trichonta atricauda		
	Trichonta foeda	X	
	Trichonta fragilis		NS
	Trichonta melanura	X	
	Trichonta vitta	X	
	Zygomyia humeralis	X	
	Zygomyia valida		
	Zygomyia vara	X	
Opetiidae	*Opetia nigra*	X	
Opomyzidae	*Opomyza florum*		

Family	Species name	Saproxylic	Conservation status
	Opomyza germinationis		
Pallopteridae	*Palloptera quinquemaculata*		
	Palloptera scutellata		
	Palloptera ustulata	X	
Pediciidae	*Dicranota bimaculata*		
	Ula mollissima	X	
	Ula sylvatica	X	
Phoridae	*Chaetopleurophora erythronota*		
	Megaselia picta		
	Spiniphora maculata		
Piophilidae	*Prochyliza nigrimana*		
Pipunculidae	*Chalarus pughi*		
	Chalarus spurius		
	Eudorylas zonellus		
	Pipunculus campestris		
	Tomosvaryella geniculata		
Platypezidae	*Agathomyia unicolor*	X	
	Bolopus furcatus	X	
	Callomyia amoena	X	
	Paraplatypeza bicincta	X	NS
	Polyporivora picta	X	
	Protoclythia modesta	X	
	Protoclythia rufa	X	
Platystomatidae	*Platystoma seminationis*		
Polleniidae	*Pollenia amentaria*		
	Pollenia angustigena		
	Pollenia pediculata		
	Pollenia rudis		
Psychodidae	*Pericoma pilularia*		
	Pericoma trivialis		
	Philosepedon humeralis		
	Trichomyia urbica	X	
Ptychopteridae	*Ptychoptera albimana*		
Rhagionidae	*Chrysopilus asiliformis*		
	Chrysopilus cristatus		
	Ptiolina obscura		
	Rhagio scolopaceus		
Sarcophagidae	*Macronychia dolini*		
	Macronychia polyodon		pNS
	Nyctia halterata		
	Ravinia pernix		
	Sarcophaga aratrix		
	Sarcophaga carnaria		
	Sarcophaga coerulescens		
	Sarcophaga crassimargo		
	Sarcophaga dissimilis		
	Sarcophaga haemorrhoa		
	Sarcophaga incisilobata		
	Sarcophaga melanura		
	Sarcophaga nigriventris		
	Sarcophaga teretirostris		
	Sarcophaga vagans		
	Sarcophaga variegata		
Scathophagidae	*Cleigastra tibiella*		

Family	Species name	Saproxylic	Conservation status
	Cordilura albipes		
	Megaphthalma pallida		
	Norellisoma spinimanum		
	Scathophaga furcata		
	Scathophaga inquinata		
	Scathophaga lutaria		
	Scathophaga stercoraria		
Scatopsidae	*Apiloscatopse flavicollis*		
	Coboldia fuscipes		
Sciaridae	*Austrosciara hyalipennis*		
	Bradysia fungicola		
	Bradysia nitidicollis	X	
	Bradysia placida		
	Bradysia trivittata		
	Corynoptera flavicauda		
	Corynoptera forcipata	X	
	Leptosciarella fuscipalpa		
	Leptosciarella rejecta		
	Leptosciarella subpilosa		
	Leptosciarella trochanterata		
	Phytosciara flavipes		
	Schwenckfeldina carbonaria	X	
	Scythropochroa quercicola	X	
	Trichosia caudata	X	
	Trichosia confusa		
	Trichosia splendens	X	
Sciomyzidae	*Euthycera fumigata*		
	Hydromya dorsalis		
	Pherbellia dubia		
	Pherbellia ventralis		
	Sepedon sphegea		
	Tetanocera ferruginea		
Sepsidae	*Nemopoda nitidula*		
	Sepsis fulgens		
	Themira annulipes		
Sphaeroceridae	*Copromyza equina*		
	Copromyza stercoraria		
	Crumomyia fimetaria		
	Crumomyia nitida		
	Gigalimosina flaviceps		
	Limosina silvatica		
	Lotophila atra		
Stratiomyidae	*Beris chalybata*		
	Beris fuscipes		
	Beris vallata		
	Chloromyia formosa		
	Chorisops nagatomii	X	
	Chorisops tibialis	X	
	Eupachygaster tarsalis	X	NS
	Pachygaster atra	X	
	Pachygaster leachii	X	
	Sargus bipunctatus		
	Sargus iridatus		
Syrphidae	*Anasimyia lineata*		

Family	Species name	Saproxylic	Conservation status
	Baccha elongata		
	Brachyopa bicolor	X	NS
	Brachyopa insensilis	X	
	Brachypalpoides lentus	X	
	Brachypalpus laphriformis	X	
	Chalcosyrphus nemorum	X	
	Cheilosia albitarsis		
	Cheilosia bergenstammi		
	Cheilosia fraterna		
	Cheilosia impressa		
	Cheilosia pagana		
	Chrysotoxum bicinctum		
	Chrysotoxum festivum		
	Criorhina berberina	X	
	Dasysyrphus albostriatus		
	Epistrophe eligans		
	Episyrphus balteatus		
	Eristalinus sepulchralis		
	Eristalis intricaria		
	Eristalis nemorum		
	Eristalis pertinax		
	Eristalis tenax		
	Eumerus ornatus		
	Eupeodes luniger		
	Helophilus pendulus		
	Leucozona glaucia		
	Leucozona lucorum		
	Melangyna cincta		
	Melanogaster hirtella		
	Melanostoma mellinum		
	Melanostoma scalare		
	Meliscaeva auricollis		
	Meliscaeva cinctella		
	Merodon equestris		
	Myathropa florea	X	
	Neoascia podagrica		
	Parasyrphus annulatus		
	Parhelophilus frutetorum		
	Pipiza noctiluca		
	Platycheirus albimanus		
	Platycheirus clypeatus		
	Platycheirus manicatus		
	Platycheirus peltatus		
	Platycheirus sticticus		NS
	Pocota personata	X	NS
	Rhingia campestris		
	Riponnensia splendens		
	Scaeva pyrastri		
	Sphaerophoria batava		
	Sphaerophoria scripta		
	Syrphus nitidifrons		
	Syrphus ribesii		
	Syrphus vitripennis		
	Volucella bombylans		

Family	Species name	Saproxylic	Conservation status
	Volucella inanis		
	Volucella inflata	X	
	Volucella pellucens		
	Xanthogramma pedissequum		
	Xylota segnis	X	
	Xylota sylvarum	X	
Tabanidae	Chrysops caecutiens		
	Chrysops relictus		
	Haematopota pluvialis		
	Hybomitra bimaculata		
	Hybomitra distinguenda		
	Tabanus bromius		
Tachinidae	Carcelia rasa		
	Eriothrix rufomaculata		
	Gonia picea		
	Gymnocheta viridis		
	Ocytata pallipes		
	Pales pavida		
	Phasia obesa		
	Ramonda spathulata		
	Redtenbacheria insignis		RDB2
	Tachina fera		
	Thelaira nigrina		
Tephritidae	Chaetostomella cylindrica		
	Goniglossum wiedemanni		Notable
	Tephritis formosa		
	Tephritis hyoscyami		
	Terellia longicauda		
	Terellia tussilaginis		
	Urophiora cardui		
Therevidae	Thereva nobilitata		
Tipulidae	Ctenophora pectinicornis	X	Notable
	Dictenidia bimaculata	X	
	Nephrotoma appendiculata		
	Nephrotoma flavescens		
	Nephrotoma quadrifaria		
	Tipula lunata		
	Tipula oleracea		
	Tipula pagana		
	Tipula paludosa		
	Tipula scripta	X	
	Tipula signata		
	Tipula varipennis		
Trichoceridae	Cladoneura hirtipennis	X	NS
	Trichocera annulata	X	
	Trichocera major		
	Trichocera parva		
	Trichocera regelationis		
Ulidiidae	Otites guttatus		
Xylomyidae	Solva marginata	X	
	Xylophagus ater	X	

Appendix 8. Checklist of Hymenoptera

Recorders: Ivan Wright, Jon Cooter, Martyn Ainsworth, Anthony Cheke, Graham Collins, Martin Corley, Aljos Farjon, Steve Gregory, Peter Hall, Russell Hedley, Jovita F. Kaunang, Benedict John Pollard, Martin Townsend, Rosemary Winnall

AAgroup	Family	Species Name
Aculeate (ant)	Formicidae	*Lasius brunneus*
Aculeate (ant)	Formicidae	*Lasius flavus*
Aculeate (ant)	Formicidae	*Lasius mixtus*
Aculeate (ant)	Formicidae	*Lasius niger*
Aculeate (ant)	Formicidae	*Lasius niger* s. lat.
Aculeate (ant)	Formicidae	*Myrmica ruginodis*
Aculeate (ant)	Formicidae	*Stenamma debile*
Aculeate (ant)	Formicidae	*Temnothorax nylanderi*
Aculeate (bee)	Andrenidae	*Andrena bicolor*
Aculeate (bee)	Andrenidae	*Andrena bucephala*
Aculeate (bee)	Andrenidae	*Andrena chrysosceles*
Aculeate (bee)	Andrenidae	*Andrena cineraria*
Aculeate (bee)	Andrenidae	*Andrena dorsata*
Aculeate (bee)	Andrenidae	*Andrena flavipes*
Aculeate (bee)	Andrenidae	*Andrena fucata*
Aculeate (bee)	Andrenidae	*Andrena fulva*
Aculeate (bee)	Andrenidae	*Andrena fulvago*
Aculeate (bee)	Andrenidae	*Andrena haemorrhoa*
Aculeate (bee)	Andrenidae	*Andrena helvola*
Aculeate (bee)	Andrenidae	*Andrena labialis*
Aculeate (bee)	Andrenidae	*Andrena minutula*
Aculeate (bee)	Andrenidae	*Andrena nigroaenea*
Aculeate (bee)	Andrenidae	*Andrena nitida*
Aculeate (bee)	Andrenidae	*Andrena praecox*
Aculeate (bee)	Andrenidae	*Andrena scotica*
Aculeate (bee)	Andrenidae	*Andrena semilaevis*
Aculeate (bee)	Andrenidae	*Andrena subopaca*
Aculeate (bee)	Apidae	*Anthophora plumipes*
Aculeate (bee)	Apidae	*Apis mellifera*
Aculeate (bee)	Apidae	*Bombus barbutellus*
Aculeate (bee)	Apidae	*Bombus campestris*
Aculeate (bee)	Apidae	*Bombus hortorum*
Aculeate (bee)	Apidae	*Bombus hypnorum*
Aculeate (bee)	Apidae	*Bombus lapidarius*
Aculeate (bee)	Apidae	*Bombus lucorum* s. lat.
Aculeate (bee)	Apidae	*Bombus pascuorum*
Aculeate (bee)	Apidae	*Bombus pratorum*

AAgroup	Family	Species Name
Aculeate (bee)	Apidae	*Bombus ruderarius*
Aculeate (bee)	Apidae	*Bombus rupestris*
Aculeate (bee)	Apidae	*Bombus sylvestris*
Aculeate (bee)	Apidae	*Bombus vestalis*
Aculeate (bee)	Apidae	*Melecta albifrons*
Aculeate (bee)	Apidae	*Nomada fabriciana*
Aculeate (bee)	Apidae	*Nomada facilis*
Aculeate (bee)	Apidae	*Nomada flava*
Aculeate (bee)	Apidae	*Nomada flavoguttata*
Aculeate (bee)	Apidae	*Nomada goodeniana*
Aculeate (bee)	Apidae	*Nomada hirtipes*
Aculeate (bee)	Apidae	*Nomada lathburiana*
Aculeate (bee)	Apidae	*Nomada leucophthalma*
Aculeate (bee)	Apidae	*Nomada panzeri*
Aculeate (bee)	Apidae	*Nomada ruficornis*
Aculeate (bee)	Colletidae	*Hylaeus brevicornis*
Aculeate (bee)	Colletidae	*Hylaeus communis*
Aculeate (bee)	Colletidae	*Hylaeus confusus*
Aculeate (bee)	Colletidae	*Hylaeus dilatatus*
Aculeate (bee)	Colletidae	*Hylaeus hyalinatus*
Aculeate (bee)	Halictidae	*Halictus rubicundus*
Aculeate (bee)	Halictidae	*Halictus tumulorum*
Aculeate (bee)	Halictidae	*Lasioglossum albipes*
Aculeate (bee)	Halictidae	*Lasioglossum calceatum*
Aculeate (bee)	Halictidae	*Lasioglossum fulvicorne*
Aculeate (bee)	Halictidae	*Lasioglossum lativentre*
Aculeate (bee)	Halictidae	*Lasioglossum leucopus*
Aculeate (bee)	Halictidae	*Lasioglossum leucozonium*
Aculeate (bee)	Halictidae	*Lasioglossum malachurum*
Aculeate (bee)	Halictidae	*Lasioglossum minutissimum*
Aculeate (bee)	Halictidae	*Lasioglossum morio*
Aculeate (bee)	Halictidae	*Lasioglossum parvulum*
Aculeate (bee)	Halictidae	*Lasioglossum pauxillum*
Aculeate (bee)	Halictidae	*Lasioglossum puncticolle*
Aculeate (bee)	Halictidae	*Lasioglossum smeathmanellum*
Aculeate (bee)	Halictidae	*Lasioglossum villosulum*
Aculeate (bee)	Halictidae	*Sphecodes ephippius*
Aculeate (bee)	Halictidae	*Sphecodes geoffrellus*
Aculeate (bee)	Halictidae	*Sphecodes hyalinatus*
Aculeate (bee)	Halictidae	*Sphecodes monilicornis*
Aculeate (bee)	Megachilidae	*Chelostoma campanularum*
Aculeate (bee)	Megachilidae	*Chelostoma florisomne*
Aculeate (bee)	Megachilidae	*Coelioxys inermis*
Aculeate (bee)	Megachilidae	*Heriades truncorum*
Aculeate (bee)	Megachilidae	*Megachile centuncularis*
Aculeate (bee)	Megachilidae	*Megachile ligniseca*
Aculeate (bee)	Megachilidae	*Megachile versicolor*
Aculeate (bee)	Megachilidae	*Osmia bicolor*
Aculeate (bee)	Megachilidae	*Osmia bicornis*
Aculeate (bee)	Megachilidae	*Osmia caerulescens*
Aculeate (bee)	Megachilidae	*Osmia leaiana*
Aculeate (bee)	Megachilidae	*Osmia spinulosa*
Aculeate (wasp)	Chrysididae	*Chrysis ignita* s. lat.

AAgroup	Family	Species Name
Aculeate (wasp)	Chrysididae	Chrysis illigeri
Aculeate (wasp)	Chrysididae	Chrysis terminata
Aculeate (wasp)	Chrysididae	Omalus aeneus
Aculeate (wasp)	Crabronidae	Arachnospila spissa
Aculeate (wasp)	Crabronidae	Argogorytes mystaceus
Aculeate (wasp)	Crabronidae	Crossocerus annulipes
Aculeate (wasp)	Crabronidae	Crossocerus cetratus
Aculeate (wasp)	Crabronidae	Crossocerus elongatulus
Aculeate (wasp)	Crabronidae	Crossocerus megacephalus
Aculeate (wasp)	Crabronidae	Crossocerus walkeri
Aculeate (wasp)	Crabronidae	Didineis lunicornis
Aculeate (wasp)	Crabronidae	Dipagon subintermedius
Aculeate (wasp)	Crabronidae	Ectemnius cephalotes
Aculeate (wasp)	Crabronidae	Ectemnius continuus
Aculeate (wasp)	Crabronidae	Ectemnius lituratus
Aculeate (wasp)	Crabronidae	Evagetes crassicornis
Aculeate (wasp)	Crabronidae	Mellinus arvensis
Aculeate (wasp)	Crabronidae	Mimumesa dahlbomi
Aculeate (wasp)	Crabronidae	Mimumesa unicolor
Aculeate (wasp)	Crabronidae	Nysson spinosus
Aculeate (wasp)	Crabronidae	Nysson trimaculatus
Aculeate (wasp)	Crabronidae	Passaloecus corniger
Aculeate (wasp)	Crabronidae	Passaloecus insignis
Aculeate (wasp)	Crabronidae	Passaloecus singularis
Aculeate (wasp)	Crabronidae	Pemphredon inornata
Aculeate (wasp)	Crabronidae	Pemphredon lugubris
Aculeate (wasp)	Crabronidae	Pemphredon morio
Aculeate (wasp)	Crabronidae	Priocnemis fennica
Aculeate (wasp)	Crabronidae	Priocnemis hyalinata
Aculeate (wasp)	Crabronidae	Rhopalum clavipes
Aculeate (wasp)	Crabronidae	Rhopalum coarctatum
Aculeate (wasp)	Crabronidae	Spilomena curruca
Aculeate (wasp)	Crabronidae	Stigmus solskyi
Aculeate (wasp)	Crabronidae	Trypoxylon attenuatum
Aculeate (wasp)	Dryinidae	Lonchodryinus ruficornis
Aculeate (wasp)	Mutillidae	Myrmosa atra
Aculeate (wasp)	Pompilidae	Anoplius caviventris
Aculeate (wasp)	Pompilidae	Anoplius nigerrimus
Aculeate (wasp)	Pompilidae	Arachnospila anceps
Aculeate (wasp)	Pompilidae	Arachnospila spissa
Aculeate (wasp)	Pompilidae	Auplopus carbonarius
Aculeate (wasp)	Pompilidae	Caliadurgus fasciatellus
Aculeate (wasp)	Pompilidae	Priocnemis confusor
Aculeate (wasp)	Pompilidae	Priocnemis exaltata
Aculeate (wasp)	Pompilidae	Priocnemis fennica
Aculeate (wasp)	Pompilidae	Priocnemis hyalinata
Aculeate (wasp)	Pompilidae	Priocnemis parvula
Aculeate (wasp)	Pompilidae	Priocnemis perturbator
Aculeate (wasp)	Tiphiidae	Tiphia minuta
Aculeate (wasp)	Torymidae	Torymidae sp.
Aculeate (wasp)	Vespidae	Ancistrocerus nigricornis
Aculeate (wasp)	Vespidae	Ancistrocerus trifasciatus
Aculeate (wasp)	Vespidae	Dolichovespula media

AAgroup	Family	Species Name
Aculeate (wasp)	Vespidae	*Dolichovespula saxonica*
Aculeate (wasp)	Vespidae	*Gymnomerus laevipes*
Aculeate (wasp)	Vespidae	*Symmorphus bifasciatus*
Aculeate (wasp)	Vespidae	*Vespa crabro*
Aculeate (wasp)	Vespidae	*Vespula germanica*
Aculeate (wasp)	Vespidae	*Vespula vulgaris*
Parasitica	Agaomidae	*Otitesellinae* sp.
Parasitica	Cynipidae	*Andricus fecundator*
Parasitica	Cynipidae	*Andricus kollari*
Parasitica	Cynipidae	*Andricus lignicola*
Parasitica	Cynipidae	*Andricus quercuscalicis*
Parasitica	Cynipidae	*Andricus quercuscalicis* f. *agamic*
Parasitica	Cynipidae	*Biorhiza pallida*
Parasitica	Cynipidae	*Diplolepis rosae*
Parasitica	Cynipidae	*Neuroterus albipes*
Parasitica	Cynipidae	*Neuroterus anthracinus*
Parasitica	Cynipidae	*Neuroterus numismalis*
Parasitica	Cynipidae	*Neuroterus quercusbaccarum*
Parasitica	Cynipidae	*Neuroterus quercusbaccarum* f. *agamic*
Parasitica	Diapriidae	*Psilus frontalis*
Parasitica	Diapriidae	*Spilomicrus stigmaticalis*
Parasitica	Gasteruptiidae	*Gasteruption jaculator*
Parasitica	Ichneumonidae	*Baranisobas ridibundus*
Parasitica	Ichneumonidae	*Barichneumon ridibundus*
Parasitica	Ichneumonidae	*Chasmia palidulus*
Parasitica	Ichneumonidae	*Cratichneumon flavifrons*
Parasitica	Ichneumonidae	*Ichneumon gracilicornis*
Parasitica	Ichneumonidae	*Ichneumon oblongus*
Parasitica	Ichneumonidae	*Melanichneumon leucochielus*
Parasitica	Ichneumonidae	*Ophion minutus*
Parasitica	Ichneumonidae	*Ophion obscuratus*
Parasitica	Ichneumonidae	*Ophion parvulus*
Parasitica	Ichneumonidae	*Pimpla contemplator*
Parasitica	Ichneumonidae	*Pimpla flavicoxis*
Parasitica	Ichneumonidae	*Pimpla turionellae*
Parasitica	Ichneumonidae	*Pseudorhyssa alpestris*
Parasitica	Ichneumonidae	*Rhembobius*
Parasitica	Ichneumonidae	*Rhyssa persuasoria*
Parasitica	Ichneumonidae	*Syspasis lineator*
Parasitica	Ichneumonidae	*Trychosis legator*
Parasitica	Ichneumonidae	*Ophion ventricosus*
Parasitica	Ormyridae	*Ormyrus nitidulus*
Symphyta	Argidae	*Arge berberidis*
Symphyta	Argidae	*Arge gracilicornis*
Symphyta	Cephidae	*Cephus pygmaeus*
Symphyta	Cephidae	*Cephus spinipes*
Symphyta	Tenthredinidae	*Aglaostigma aucupariae*
Symphyta	Tenthredinidae	*Ametastegia albipes*
Symphyta	Tenthredinidae	*Ametastegia equiseti*
Symphyta	Tenthredinidae	*Ametastegia glabrata*
Symphyta	Tenthredinidae	*Aneugmenus furstenbergensis*
Symphyta	Tenthredinidae	*Aneugmenus padi*
Symphyta	Tenthredinidae	*Apethymus filiformis*

AAgroup	Family	Species Name
Symphyta	Tenthredinidae	Apethymus serotinus
Symphyta	Tenthredinidae	Athalia bicolor
Symphyta	Tenthredinidae	Athalia circularis
Symphyta	Tenthredinidae	Athalia cordata
Symphyta	Tenthredinidae	Athalia liberta
Symphyta	Tenthredinidae	Athalia rosae
Symphyta	Tenthredinidae	Athalia scutellariae
Symphyta	Tenthredinidae	Dolerus fumosus
Symphyta	Tenthredinidae	Dolerus nigratus
Symphyta	Tenthredinidae	Eutomostethus aphippium
Symphyta	Tenthredinidae	Eutomostethus nigrans
Symphyta	Tenthredinidae	Euura obducta
Symphyta	Tenthredinidae	Hoplocampa ariae
Symphyta	Tenthredinidae	Hoplocampa pectoralis
Symphyta	Tenthredinidae	Macrophya annulata
Symphyta	Tenthredinidae	Nematus crassus
Symphyta	Tenthredinidae	Pachygaster leachii
Symphyta	Tenthredinidae	Periclista albida
Symphyta	Tenthredinidae	Pristiphora
Symphyta	Tenthredinidae	Rhogogaster chambersi
Symphyta	Tenthredinidae	Rhogogaster viridis
Symphyta	Tenthredinidae	Selandria serva
Symphyta	Tenthredinidae	Stromboceros delicatulus
Symphyta	Tenthredinidae	Strongylogaster lineata
Symphyta	Tenthredinidae	Strongylogaster multifasciata
Symphyta	Tenthredinidae	Strongylogaster xanthocera
Symphyta	Tenthredinidae	Tenthredo arcuata
Symphyta	Tenthredinidae	Tenthredo atra
Symphyta	Tenthredinidae	Tenthredo celtica
Symphyta	Tenthredinidae	Tenthredo mesomela
Symphyta	Tenthredinidae	Tenthredo zona
Symphyta	Tenthredinidae	Tomostethus nigritus

Appendix 9. Checklist of butterflies (Lepidoptera)

Recorders: Caroline Steel, Margaret Price, Phil Cribb, Anthony Cheke, Graham Collins, Martin Corley, Bob Cowley, Aljos Farjon, Ray Heaton, Russell Hedley, Jovita Kaunang, Linda Losito, Sylfest Muldal, Bill Parker, Chris Perrins, Benedict Pollard, Paulo Salbany

Scientific Name	English Name	Source
Aglais urticae	Small Tortoiseshell	TVERC
Anthocharis cardamines	Orange Tip	HPBS
Apatura iris	Purple Emperor	HPBS
Aphantopus hyperantus	Ringlet	HPBS
Argynnis paphia	Silver-washed Fritillary	HPBS
Aricia agestis	Brown Argus	HPBS
Celastrina argiolus britanna	Holly Blue	TVERC
Erynnis tages	Dingy Skipper	TVERC
Favonius quercus	Purple Hairstreak	HPBS
Gonepteryx rhamni	Brimstone	HPBS
Inachis io	Peacock	HPBS
Lasiommata megera	Wall	TVERC
Lycaena phlaeas	Small Copper	HPBS
Maniola jurtina	Meadow Brown	HPBS
Melanargia galathea	Marbled White	HPBS
Ochlodes sylvanus	Large Skipper	HPBS
Pararge aegeria	Speckled Wood	HPBS
Pieris brassicae	Large White	HPBS
Pieris napi	Green-veined White	HPBS
Pieris rapae	Small White	HPBS
Polygonia c-album	Comma	HPBS
Polyommatus icarus	Common Blue	HPBS
Pyrgus malvae	Grizzled Skipper	TVERC
Pyronia tithonus	Gatekeeper	HPBS
Thymelicus lineola	Essex Skipper	HPBS
Thymelicus sylvestris	Small Skipper	HPBS
Vanessa atalanta	Red Admiral	HPBS
Vanessa cardui	Painted Lady	TVERC

Appendix 10. Checklist of Moths (Lepidoptera) 2017–2021

Recorders: Martin Corley, Martin Townsend, Marc Botham, BNHS,
Graham Collins, Jon Cooter, Mary Elford, Peter Hall, Julian Howe,
Steve Nash, Pedro Pires, Benedict Pollard, Ivan Wright

Status: RDB1,2,3 = British Red Data Book status categories, pRDB1,2,3 = potentially Red
Databook status categories (JNCC)

Family	Scientific name	English name	Records	Inds	Status
Adelidae	Adela croesella		1	1	
	Adela reaumurella		2	2	
	Cauchas fibulella		2	8	
	Cauchas rufimitrella		1	1	
	Nematopogon metaxella		5	12	
	Nematopogon swammerdamella		7	7	
	Nemophora degeerella		4	8	
Argyresthiidae	Argyresthia albistria		2	3	
	Argyresthia bonnetella		2	2	
	Argyresthia brockeella		9	21	
	Argyresthia glaucinella		1	1	
	Argyresthia goedartella		13	81	
	Argyresthia pruniella	Cherry Fruit Moth	1	1	
	Argyresthia pygmaeella		2	2	
	Argyresthia retinella		5	8	
	Argyresthia semifusca		1	1	
	Argyresthia semitestacella		1	1	
	Argyresthia spinosella		1	1	
Batrachedridae	Batrachedra pinicolella agg.		1	1	
	Batrachedra praeangusta		1	1	
Blastobasidae	Blastobasis adustella		23	311	
	Blastobasis lacticolella		13	23	
Bucculatricidae	Bucculatrix frangutella		1	1	
	Bucculatrix nigricomella		1	1	
	Bucculatrix ulmella		2	6	
Chimabachidae	Diurnea fagella		2	6	
Choreutidae	Anthophila fabriciana		5	6	
	Choreutis pariana	Apple-leaf Skeletoniser	1	1	
	Prochoreutis myllerana		1	1	
Coleophoridae	Coleophora albidella		2	2	
	Coleophora albitarsella		1	1	

Family	Scientific name	English name	Records	Inds	Status
	Coleophora alcyonipennella		1	1	
	Coleophora alticolella		8	22	
	Coleophora anatipennella		1	1	
	Coleophora argentula		1	1	
	Coleophora betulella		3	7	
	Coleophora caespititiella		5	21	
	Coleophora conyzae		1	1	
	Coleophora deauratella		1	1	
	Coleophora flavipennella		7	16	
	Coleophora glaucicolella		2	2	
	Coleophora ibipennella		1	2	
	Coleophora kuehnella		1	2	
	Coleophora lusciniaepennella		4	12	
	Coleophora lutipennella		5	16	
	Coleophora mayrella		2	7	
	Coleophora otidipennella		1	2	
	Coleophora serratella		3	3	
	Coleophora striatipennella		1	1	
Cosmopterigidae	*Limnaecia phragmitella*		1	2	
Cossidae	*Zeuzera pyrina*	Leopard Moth	3	4	Common
Crambidae	*Acentria ephemerella*	Water Veneer	20	201	
	Agriphila geniculea		4	18	
	Agriphila selasella		4	8	
	Agriphila straminella		16	251	
	Agriphila tristella		13	61	
	Anania fuscalis		5	6	
	Anania hortulata	Small Magpie	9	23	
	Anania perlucidalis		3	3	
	Calamotropha paludella		2	6	
	Cataclysta lemnata	Small China-mark	2	2	
	Catoptria falsella		5	8	
	Catoptria pinella		8	13	
	Chrysoteuchia culmella	Garden Grass-veneer	25	303	
	Crambus lathoniellus		12	72	
	Donacaula forficella		3	4	
	Elophila nymphaeata	Brown China-mark	4	6	
	Eudonia lacustrata		17	146	
	Eudonia mercurella		22	102	
	Eudonia pallida		4	4	
	Eudonia truncicolella		4	5	
	Nymphula nitidulata	Beautiful China-mark	6	7	
	Ostrinia nubilalis	European Corn Borer	1	1	
	Parapoynx stratiotata	Ringed China-mark	16	34	
	Patania ruralis	Mother of Pearl	20	119	
	Pediasia contaminella		1	2	
	Pyrausta aurata		2	3	
	Pyrausta purpuralis		6	9	
	Scoparia ambigualis		19	63	
	Scoparia basistrigalis		17	100	
	Scoparia pyralella		10	39	
	Scoparia subfusca		1	1	
	Udea ferrugalis	Rusty-dot Pearl	1	2	Migrant
	Udea olivalis		13	50	
	Udea prunalis		4	4	

Family	Scientific name	English name	Records	Inds	Status
Depressariidae	Agonopterix arenella		3	3	
	Agonopterix heracliana		6	8	
	Agonopterix liturosa		5	5	
	Agonopterix ocellana		3	4	
	Depressaria radiella	Parsnip Moth	1	5	
Drepanidae	Achlya flavicornis	Yellow Horned	1	5	Common
	Cilix glaucata	Chinese Character	3	4	Common
	Cymatophorina diluta	Oak Lutestring	1	1	Local
	Drepana falcataria	Pebble Hook-tip	7	11	Common
	Habrosyne pyritoides	Buff Arches	7	20	Common
	Ochropacha duplaris	Common Lutestring	2	4	Common
	Polyploca ridens	Frosted Green	14	133	Local
	Tethea ocularis	Figure of Eighty	2	11	Common
	Thyatira batis	Peach Blossom	4	4	Common
	Watsonalla binaria	Oak Hook-tip	10	18	Common
	Watsonalla cultraria	Barred Hook-tip	6	10	Local
Elachistidae	Elachista albifrontella		2	2	
	Elachista apicipunctella		1	1	
	Elachista argentella		2	2	
	Elachista atricomella		6	11	
	Elachista canapennella		4	4	
	Elachista freyerella		1	1	
	Elachista gangabella		6	15	
	Elachista humilis		6	15	
	Elachista obliquella		1	1	
	Elachista subnigrella		1	1	
	Elachista subocellea		1	1	
Epermeniidae	Epermenia chaerophyllella		1	1	
Erebidae	Atolmis rubricollis	Red-necked Footman	2	3	Local
	Callimorpha dominula	Scarlet Tiger	6	7	Local
	Calliteara pudibunda	Pale Tussock	8	54	Common
	Catocala fraxini	Clifden Nonpareil	6	7	Migrant
	Catocala nupta	Red Underwing	8	11	Common
	Catocala sponsa	Dark Crimson Underwing	1	1	RDB2
	Eilema complana	Scarce Footman	14	57	Local
	Eilema depressa	Buff Footman	14	71	Local
	Eilema griseola	Dingy Footman	19	52	Common
	Eilema lurideola	Common Footman	12	47	Common
	Eilema sororcula	Orange Footman	15	48	Local
	Euclidia glyphica	Burnet Companion	2	2	Common
	Euproctis similis	Yellow-tail	15	95	Common
	Herminia grisealis	Small Fan-foot	3	5	Common
	Herminia tarsipennalis	Fan-foot	2	2	Common
	Hypena proboscidalis	Snout	32	104	Common
	Hypenodes humidalis	Marsh Oblique-barred	1	1	
	Laspeyria flexula	Beautiful Hook-tip	9	32	Local
	Leucoma salicis	White Satin	1	1	Local
	Lithosia quadra	Four-spotted Footman	4	6	Ireland only
	Lygephila pastinum	Blackneck	1	1	Local
	Lymantria monacha	Black Arches	19	234	Local
	Miltochrista miniata	Rosy Footman	4	5	Local
	Nudaria mundana	Muslin Footman	7	29	Local
	Orgyia antiqua	Vapourer	3	3	Common

Family	Scientific name	English name	Records	Inds	Status
	Parascotia fuliginaria	Waved Black	1	1	
	Phragmatobia fuliginosa	Ruby Tiger	10	48	Common
	Rivula sericealis	Straw Dot	42	344	Common
	Scoliopteryx libatrix	Herald	8	17	Common
	Spilosoma lubricipeda	White Ermine	11	45	Common
	Spilosoma lutea	Buff Ermine	13	93	Common
	Thumatha senex	Round-winged Muslin	3	3	Local
	Tyria jacobaeae	Cinnabar	3	4	Common
Eriocraniidae	Dyseriocrania subpurpurella		8	11	
Ethmiidae	Ethmia dodecea		2	2	
Gelechiidae	Acompsia cinerella		5	5	
	Anacampsis blattariella		3	3	
	Anacampsis populella		1	2	
	Aproaerema larseniella		3	3	
	Brachmia blandella		2	2	
	Bryotropha senectella		1	1	
	Bryotropha terrella		4	4	
	Carpatolechia fugitivella		3	7	
	Carpatolechia notatella		1	1	
	Dichomeris alacella		5	7	
	Eulamprotes atrella		2	3	
	Gelechia sororculella		2	2	
	Helcystogramma rufescens		2	3	
	Monochroa cytisella		1	1	
	Parachronistis albiceps		2	2	
	Psoricoptera gibbosella		13	117	
	Scrobipalpa acuminatella		3	4	
	Scrobipalpa costella		4	4	
	Stenolechia gemmella		5	10	
	Teleiodes luculella		11	31	
	Teleiodes vulgella		2	3	
Geometridae	Abraxas sylvata	Clouded Magpie	1	1	Local
	Acasis viretata	Yellow-barred Brindle	3	4	Local
	Agriopis leucophaearia	Spring Usher	3	27	Common
	Agriopis marginaria	Dotted Border	3	9	Common
	Alcis repandata	Mottled Beauty	14	93	Common
	Alsophila aescularia	March Moth	3	5	Common
	Anticlea derivata	Streamer	1	1	Common
	Apocheima hispidaria	Small Brindled Beauty	3	54	Local
	Biston betularia	Peppered Moth	8	33	Common
	Biston strataria	Oak Beauty	4	15	Common
	Cabera exanthemata	Common Wave	13	21	Common
	Cabera pusaria	Common White Wave	13	29	Common
	Campaea margaritata	Light Emerald	14	103	Common
	Camptogramma bilineata	Yellow Shell	9	9	Common
	Chiasmia clathrata	Latticed Heath	1	1	
	Chloroclysta siterata	Red-green Carpet	12	25	Common
	Chloroclystis v-ata	V-Pug	4	11	Common
	Cidaria fulvata	Barred Yellow	4	7	Common
	Colostygia pectinataria	Green Carpet	25	131	Common
	Colotois pennaria	Feathered Thorn	8	24	Common
	Comibaena bajularia	Blotched Emerald	6	30	Local
	Cosmorhoe ocellata	Purple Bar	3	3	Common
	Crocallis elinguaria	Scalloped Oak	3	6	Common

Family	Scientific name	English name	Records	Inds	Status
	Cyclophora linearia	Clay Triple-lines	5	9	Local
	Cyclophora punctaria	Maiden's Blush	24	57	Local
	Deileptenia ribeata	Satin Beauty	2	4	Common
	Dysstroma truncata	Common Marbled Carpet	14	35	Common
	Earophila badiata	Shoulder Stripe	1	1	Common
	Ecliptopera silaceata	Small Phoenix	25	59	Common
	Ectropis crepuscularia	Engrailed	9	29	Common
	Electrophaes corylata	Broken-barred Carpet	4	6	Common
	Ennomos alniaria	Canary-shouldered Thorn	2	2	Common
	Ennomos erosaria	September Thorn	21	184	Common
	Ennomos fuscantaria	Dusky Thorn	9	19	Common
	Ennomos quercinaria	August Thorn	1	1	Local
	Epirrhoe alternata	Common Carpet	20	37	Common
	Epirrita dilutata	November Moth	7	60	Common
	Erannis defoliaria	Mottled Umber	6	7	Common
	Eupithecia abbreviata	Brindled Pug	15	209	Common
	Eupithecia assimilata	Currant Pug	3	3	Common
	Eupithecia centaureata	Lime-speck Pug	7	7	Common
	Eupithecia dodoneata	Oak-tree Pug	11	29	Common
	Eupithecia exiguata	Mottled Pug	4	4	Common
	Eupithecia haworthiata	Haworth's Pug	4	4	Local
	Eupithecia inturbata	Maple Pug	2	4	Local
	Eupithecia subfuscata	Grey Pug	7	13	Common
	Eupithecia tantillaria	Dwarf Pug	2	2	Common
	Eupithecia tenuiata	Slender Pug	3	6	Common
	Eupithecia tripunctaria	White-spotted Pug	4	5	Local
	Eupithecia vulgata	Common Pug	8	19	Common
	Gandaritis pyraliata	Barred Straw	6	10	Common
	Geometra papilionaria	Large Emerald	4	11	Common
	Gymnoscelis rufifasciata	Double-striped Pug	8	11	Common
	Hemistola chrysoprasaria	Small Emerald	3	6	Local
	Hemithea aestivaria	Common Emerald	9	43	Common
	Horisme tersata	Fern	2	2	Common
	Horisme vitalbata	Small Waved Umber	1	1	Common
	Hydrelia flammeolaria	Small Yellow Wave	1	1	Common
	Hydria undulata	Scallop Shell	1	2	Common
	Hydriomena furcata	July Highflyer	19	109	Common
	Hypomecis punctinalis	Pale Oak Beauty	3	9	Common
	Idaea aversata	Riband Wave	25	117	Common
	Idaea biselata	Small Fan-footed Wave	17	108	Common
	Idaea dimidiata	Single-dotted Wave	12	20	Common
	Idaea emarginata	Small Scallop	2	2	Local
	Idaea fuscovenosa	Dwarf Cream Wave	1	2	Local
	Idaea rusticata	Least Carpet	4	5	Local
	Idaea trigeminata	Treble Brown Spot	7	7	Local
	Lampropteryx suffumata	Water Carpet	2	2	Common
	Ligdia adustata	Scorched Carpet	4	4	Local
	Lobophora halterata	Seraphim	5	5	Local
	Lomaspilis marginata	Clouded Border	20	195	Common
	Lomographa bimaculata	White-pinion Spotted	5	13	Common
	Lomographa temerata	Clouded Silver	10	28	Common
	Lycia hirtaria	Brindled Beauty	4	12	Common

Family	Scientific name	English name	Records	Inds	Status
	Macaria alternata	Sharp-angled Peacock	1	1	Local
	Macaria liturata	Tawny-barred Angle	9	15	Common
	Macaria notata	Peacock Moth	3	4	Local
	Melanthia procellata	Pretty Chalk Carpet	1	1	Common
	Menophra abruptaria	Waved Umber	5	6	Common
	Odezia atrata	Chimney Sweeper	2	2	Common
	Odontopera bidentata	Scalloped Hazel	1	1	Common
	Operophtera brumata	Winter Moth	3	3	Common
	Opisthograptis luteolata	Brimstone Moth	33	99	Common
	Ourapteryx sambucaria	Swallow-tailed Moth	7	27	Common
	Pasiphila chloerata	Sloe Pug	1	1	Common
	Pasiphila rectangulata	Green Pug	7	27	Common
	Peribatodes rhomboidaria	Willow Beauty	17	30	Common
	Perizoma albulata	Grass Rivulet	2	8	Local
	Perizoma alchemillata	Small Rivulet	1	1	Common
	Petrophora chlorosata	Brown Silver-line	22	121	Common
	Phigalia pilosaria	Pale Brindled Beauty	4	151	Common
	Philereme transversata	Dark Umber	3	4	Local
	Philereme vetulata	Brown Scallop	8	23	Local
	Plagodis dolabraria	Scorched Wing	9	55	Local
	Plemyria rubiginata	Blue-bordered Carpet	4	6	Common
	Pterapherapteryx sexalata	Small Seraphim	7	13	Local
	Scopula imitaria	Small Blood-vein	2	2	Common
	Scotopteryx chenopodiata	Shaded Broad-bar	3	5	Common
	Selenia dentaria	Early Thorn	5	6	Common
	Selenia tetralunaria	Purple Thorn	7	12	Common
	Thera britannica	Spruce Carpet	9	10	Common
	Thera obeliscata	Grey Pine Carpet	3	3	Common
	Timandra comae	Blood-vein	21	47	Common
	Xanthorhoe designata	Flame Carpet	11	15	Common
	Xanthorhoe fluctuata	Garden Carpet	4	4	Common
	Xanthorhoe montanata	Silver-ground Carpet	14	44	Common
	Xanthorhoe quadrifasiata	Large Twin-spot Carpet	10	17	Local
	Xanthorhoe spadicearia	Red Twin-spot Carpet	22	54	Common
Glyphipterigidae	Acrolepia autumnitella		1	1	
	Glyphipterix fuscoviridella		2	2	
	Glyphipterix simpliciella	Cocksfoot Moth	5	16	
	Orthotelia sparganella		1	1	
Gracillariidae	Caloptilia alchimiella		1	1	
	Caloptilia betulicola		2	2	
	Caloptilia robustella		10	24	
	Caloptilia rufipennella		1	1	
	Caloptilia semifascia		1	3	
	Caloptilia stigmatella		4	4	
	Cameraria ohridella		15	405	
	Euspilapteryx auroguttella		2	5	
	Gracillaria syringella		2	3	
	Parornix anglicella		4	17	
	Parornix betulae		1	1	
	Parornix devoniella		1	1	
	Parornix finitimella		1	1	
	Parornix torquillella		2	2	
	Phyllonorycter acerifoliella		2	7	
	Phyllonorycter coryli	Nut Leaf Blister Moth	3	53	

Family	Scientific name	English name	Records	Inds	Status
	Phyllonorycter corylifoliella		5	7	
	Phyllonorycter dubitella		1	1	
	Phyllonorycter esperella		1	6	
	Phyllonorycter geniculella		1	4	
	Phyllonorycter harrisella		3	3	
	Phyllonorycter maestingella		3	16	
	Phyllonorycter mespilella		1	4	
	Phyllonorycter nicellii		1	2	
	Phyllonorycter oxyacanthae		1	1	
	Phyllonorycter quercifoliella		1	1	
	Phyllonorycter salicicolella		1	2	
	Phyllonorycter schreberella		1	2	
	Phyllonorycter spinicolella		1	3	
	Phyllonorycter tenerella		1	1	
	Phyllonorycter trifasciella		1	4	
	Phyllonorycter ulmifoliella		1	3	
Hepialidae	Hepialus humuli	Ghost Moth	2	2	Common
	Korscheltellus lupulina	Common Swift	14	123	Common
	Phymatopus hecta	Gold Swift	4	11	Local
	Triodia sylvina	Orange Swift	9	43	Common
Incurvariidae	Incurvaria masculella		2	3	
	Incurvaria oehlmanniella		2	2	
Lasiocampidae	Euthrix potatoria	Drinker	1	1	Common
	Poecilocampa populi	December Moth	3	12	Common
Lyonetiidae	Leucoptera malifoliella	Pear Leaf Blister Moth	1	1	
	Lyonetia clerkella	Apple Leaf Miner	2	2	
Lypusidae	Agnoea josephinae		1	1	
Micropterigidae	Micropterix aruncella		2	2	
	Micropterix calthella		3	9	
Momphidae	Mompha langiella		1	1	
Nepticulidae	Ectoedemia albifasciella		4	14	
	Ectoedemia argyropeza		1	5	
	Ectoedemia intimella		1	3	
	Ectoedemia quinquella		1	3	
	Ectoedemia subbimaculella		2	7	
	Etainia decentella		1	1	
	Stigmella aceris		1	1	pRDB2
	Stigmella aurella		1	3	
	Stigmella catharticella		1	2	
	Stigmella continuella		1	1	
	Stigmella floslactella		1	1	
	Stigmella hemargyrella		1	2	
	Stigmella hybnerella		1	1	
	Stigmella lemniscella		1	7	
	Stigmella microtheriella		2	12	
	Stigmella oxyacanthella		1	1	
	Stigmella perpygmaeella		1	5	
	Stigmella plagicolella		1	1	
	Stigmella salicis		2	5	
	Stigmella samiatella		1	1	pRDB3
	Stigmella speciosa		1	1	
	Stigmella tiliae		1	6	
	Stigmella tityrella		1	4	
	Stigmella trimaculella		1	2	

Family	Scientific name	English name	Records	Inds	Status
	Stigmella ulmivora		1	1	
	Zimmermannia atrifrontella		1	1	
Noctuidae	*Abrostola tripartita*	Spectacle	10	18	Common
	Acronicta aceris	Sycamore	3	3	Local
	Acronicta alni	Alder Moth	3	22	Local
	Acronicta leporina	Miller	2	2	Common
	Acronicta psi	Grey Dagger	3	3	Common
	Acronicta rumicis	Knot Grass	5	7	Common
	Acronicta tridens	Dark Dagger	1	1	Common
	Agrotis clavis	Heart and Club	5	16	Common
	Agrotis exclamationis	Heart and Dart	11	48	Common
	Agrotis puta	Shuttle-shaped Dart	3	4	Local
	Agrotis segetum	Turnip Moth	3	6	Common
	Allophyes oxyacanthae	Green-brindled Crescent	5	9	Common
	Amphipyra berbera	Svensson's Copper Underwing	19	128	Common
	Amphipyra pyramidea	Copper Underwing	15	30	Common
	Amphipyra tragopoginis	Mouse Moth	7	7	Common
	Anaplectoides prasina	Green Arches	4	14	Common
	Anchoscelis litura	Brown-spot Pinion	2	2	Common
	Anchoscelis lunosa	Lunar Underwing	6	21	Common
	Anorthoa munda	Twin-spotted Quaker	4	15	Common
	Apamea anceps	Large Nutmeg	3	4	Local
	Apamea crenata	Clouded-bordered Brindle	5	9	Common
	Apamea epomidion	Clouded Brindle	8	17	Common
	Apamea lithoxylaea	Light Arches	4	6	Common
	Apamea monoglypha	Dark Arches	19	38	Common
	Apamea remissa	Dusky Brocade	2	3	Common
	Apamea scolopacina	Slender Brindle	8	25	Local
	Apamea sordens	Rustic Shoulder-knot	1	2	Common
	Apamea sublustris	Reddish Light Arches	2	2	Local
	Apamea unanimis	Small Clouded Brindle	1	1	Common
	Aporophyla lutulenta	Deep-brown Dart	1	1	Common
	Aporophyla nigra	Black Rustic	3	5	Common
	Asteroscopus sphinx	Sprawler	3	5	Common
	Atethmia centrago	Centre-barred Sallow	5	31	Common
	Autographa gamma	Silver Y	10	14	Migrant
	Autographa pulchrina	Beautiful Golden Y	6	7	Common
	Axylia putris	Flame	5	20	Common
	Brachylomia viminalis	Minor Shoulder-knot	11	103	Common
	Bryophila domestica	Marbled Beauty	1	1	Common
	Bryopsis muralis	Marbled Green	2	4	Local
	Caradrina clavipalpis	Pale Mottled Willow	1	1	Common
	Caradrina morpheus	Mottled Rustic	1	2	Common
	Cerapteryx graminis	Antler	1	2	Common
	Cerastis leucographa	White-marked	2	2	Local
	Cerastis rubricosa	Red Chestnut	6	9	Common
	Charanyca trigrammica	Treble Lines	13	48	Common
	Cirrhia icteritia	Sallow	6	12	Common
	Colocasia coryli	Nut-tree Tussock	21	70	Common
	Conistra ligula	Dark Chestnut	10	19	Common
	Conistra vaccinii	Chestnut	17	38	Common

Family	Scientific name	English name	Records	Inds	Status
	Cosmia affinis	Lesser-spotted Pinion	4	4	Local
	Cosmia pyralina	Lunar-spotted Pinion	5	15	Local
	Cosmia trapezina	Dun-bar	27	281	Common
	Craniophora ligustri	Coronet	20	116	Local
	Cryphia algae	Tree-lichen Beauty	1	1	Rare migrant
	Cucullia verbasci	Mullein	1	1	Common
	Denticucullus pygmina	Small Wainscot	1	1	Common
	Diachrysia chrysitis	Burnished Brass	6	10	Common
	Diarsia brunnea	Purple Clay	6	14	Common
	Diarsia mendica	Ingrailed Clay	12	89	Common
	Diarsia rubi	Small Square-spot	12	37	Common
	Dryobotodes eremita	Brindled Green	10	48	Common
	Eugnorisma glareosa	Autumnal Rustic	4	13	Common
	Euplexia lucipara	Small Angle Shades	7	23	Common
	Eupsilia transversa	Satellite	8	39	Common
	Fissipunctia ypsillon	Dingy Shears	2	2	Local
	Griposia aprilina	Merveille du Jour	9	24	Common
	Helicoverpa armigera	Scarce Bordered Straw	1	1	Migrant
	Hoplodrina ambigua	Vine's Rustic	4	5	Local
	Hoplodrina blanda	Rustic	6	10	Common
	Hoplodrina octogenaria	Uncertain	13	77	Common
	Lacanobia oleracea	Bright-line Brown-eye	4	7	Common
	Lacanobia thalassina	Pale-shouldered Brocade	5	7	Common
	Leptologia lota	Red-line Quaker	1	2	Common
	Leptologia macilenta	Yellow-line Quaker	3	8	Common
	Leucania comma	Shoulder-striped Wainscot	1	1	Common
	Lithophane hepatica	Pale Pinion	1	1	Local
	Lithophane ornitopus	Grey Shoulder-knot	3	3	Common
	Lithophane semibrunnea	Tawny Pinion	1	1	Local
	Luperina testacea	Flounced Rustic	7	13	Common
	Melanchra persicariae	Dot Moth	2	7	Common
	Mesapamea didyma	Lesser Common Rustic	2	2	Common
	Mesapamea secalis	Common Rustic	12	22	Common
	Mormo maura	Old Lady	3	3	Local
	Mythimna albipuncta	White-point	4	9	Migrant
	Mythimna conigera	Brown-line Bright Eye	2	3	Common
	Mythimna ferrago	Clay	5	10	Common
	Mythimna impura	Smoky Wainscot	18	116	Common
	Mythimna pallens	Common Wainscot	17	45	Common
	Mythimna straminea	Southern Wainscot	1	1	Local
	Noctua comes	Lesser Yellow Underwing	17	24	Common
	Noctua fimbriata	Broad-bordered Yellow Underwing	13	18	Common
	Noctua janthe	Lesser Broad-bordered Yellow Underwing	13	49	Common
	Noctua pronuba	Large Yellow Underwing	35	267	Common
	Ochropleura plecta	Flame Shoulder	27	63	Common
	Oligia fasciuncula	Middle-barred Minor	4	7	Common
	Oligia latruncula	Tawny Marbled Minor	6	14	Common
	Oligia strigilis	Marbled Minor	5	8	Common

Family	Scientific name	English name	Records	Inds	Status
	Oligia versicolor	Rufous Minor	8	11	Local
	Orthosia cerasi	Common Quaker	17	314	Common
	Orthosia cruda	Small Quaker	11	140	Common
	Orthosia gothica	Hebrew Character	11	98	Common
	Orthosia gracilis	Powdered Quaker	2	2	Common
	Orthosia incerta	Clouded Drab	7	35	Common
	Orthosia miniosa	Blossom Underwing	1	2	Local
	Orthosia populeti	Lead-coloured Drab	2	2	Local
	Panolis flammea	Pine Beauty	1	1	Common
	Parastichtis suspecta	Suspected	1	1	Local
	Phlogophora meticulosa	Angle Shades	9	14	Common
	Photedes fluxa	Mere Wainscot	3	3	
	Photedes minima	Small Dotted Buff	12	34	Common
	Plusia festucae	Gold Spot	4	4	Common
	Protodeltote pygarga	Marbled White Spot	23	305	Common
	Rhizedra lutosa	Large Wainscot	3	3	Common
	Rusina ferruginea	Brown Rustic	14	50	Common
	Subacronicta megacephala	Poplar Grey	6	30	Common
	Sunira circellaris	Brick	3	6	Common
	Thalpophila matura	Straw Underwing	3	7	Common
	Tholera decimalis	Feathered Gothic	4	15	Common
	Tiliacea aurago	Barred Sallow	3	3	Common
	Tiliacea citrago	Orange Sallow	1	1	Common
	Xanthia togata	Pink-barred Sallow	6	13	Common
	Xestia c-nigrum	Setaceous Hebrew Character	19	36	Common
	Xestia sexstrigata	Six-striped Rustic	1	1	Common
	Xestia triangulum	Double Square-spot	14	53	Common
	Xestia xanthographa	Square-spot Rustic	20	289	Common
	Xylocampa areola	Early Grey	2	2	Common
Nolidae	Bena bicolorana	Scarce Silver-lines	1	1	Local
	Meganola albula	Kent Black Arches	1	3	
	Nola confusalis	Least Black Arches	6	14	Local
	Nola cucullatella	Short-cloaked Moth	7	16	Common
	Nycteola revayana	Oak Nycteoline	1	1	Local
	Pheosia gnoma	Lesser Swallow Prominent	6	13	Common
	Pseudoips prasinana	Green Silver-lines	4	4	Common
Notodontidae	Cerura vinula	Puss Moth	1	1	Common
	Clostera curtula	Chocolate-tip	1	1	Local
	Drymonia dodonaea	Marbled Brown	10	22	Common
	Drymonia ruficornis	Lunar Marbled Brown	7	34	Local
	Furcula bifida	Poplar Kitten	2	3	Local
	Furcula furcula	Sallow Kitten	3	3	Common
	Notodonta dromedarius	Iron Prominent	8	11	Common
	Notodonta ziczac	Pebble Prominent	7	11	Common
	Peridea anceps	Great Prominent	1	1	Local
	Phalera bucephala	Buff-tip	13	86	Common
	Pheosia tremula	Swallow Prominent	8	15	Common
	Pterostoma palpina	Pale Prominent	13	17	Common
	Ptilodon capucina	Coxcomb Prominent	9	13	Common
	Stauropus fagi	Lobster Moth	10	21	Common
Oecophoridae	Aplota palpellus		4	5	pRDB1
	Batia lunaris		1	2	

Family	Scientific name	English name	Records	Inds	Status
	Crassa tinctella		3	3	
	Crassa unitella		16	114	
	Esperia sulphurella		2	2	
	Hofmannophila pseudospretella	Brown House Moth	6	7	
	Metalampra italica		9	9	
Parametriotidae	Blastodacna hellerella		1	1	
	Dystebenna stephensi		2	2	pRDB3
	Spuleria flavicaput		2	2	
Peleopodidae	Carcina quercana		21	145	
Plutellidae	Plutella xylostella	Diamond-back Moth	11	21	Migrant
Praydidae	Prays fraxinella		3	3	
	Prays ruficeps		2	2	
Psychidae	Luffia lapidella		2	5	
	Psyche casta		1	1	
Pterophoridae	Emmelina monodactyla		1	1	
	Stenoptilia pterodactyla		1	1	
Pyralidae	Acrobasis advenella		11	26	
	Acrobasis consociella		11	24	
	Acrobasis repandana		13	43	
	Aphomia sociella	Bee Moth	11	19	
	Cryptoblabes bistriga		9	14	
	Delplanqueia inscriptella		1	3	
	Elegia similella		1	1	
	Endotricha flammealis		10	42	
	Ephestia woodiella		1	1	
	Euzophera pinguis		11	13	
	Homoeosoma sinuella		5	7	
	Hypsopygia costalis	Gold Triangle	7	10	
	Hypsopygia glaucinalis		2	2	
	Myelois circumvoluta	Thistle Ermine	1	1	
	Nephopterix angustella		1	1	
	Oncocera semirubella		1	1	
	Phycita roborella		20	336	
	Phycitodes binaevella		2	2	
Roeslerstammiidae	Roeslerstammia erxlebella		2	2	
Sesiidae	Sesia apiformis	Hornet Moth	1	5	
	Synanthedon vespiformis	Yellow-legged Clearwing	1	1	
Sphingidae	Deilephila elpenor	Elephant Hawk-moth	8	18	Common
	Deilephila porcellus	Small Elephant Hawk-moth	1	1	Local
	Laothoe populi	Poplar Hawk-moth	14	30	Common
	Mimas tiliae	Lime Hawk-moth	4	6	Common
	Smerinthus ocellata	Eyed Hawk-moth	1	1	Common
	Sphinx ligustri	Privet Hawk-moth	2	2	Common
	Sphinx pinastri	Pine Hawk-moth	3	5	Local
Tineidae	Monopis crocicapitella		1	1	
	Monopis laevigella	Skin Moth	4	4	
	Monopis obviella		1	1	
	Monopis weaverella		5	6	
	Morophaga choragella		2	2	
	Nemapogon cloacella	Cork Moth	2	3	
	Nemapogon koenigi		3	3	

Family	Scientific name	English name	Records	Inds	Status
	Nemapogon variatella		1	1	
	Niditinea striolella		5	5	
	Tinea semifulvella		2	2	
	Tinea trinotella		5	6	
	Triaxomera parasitella		1	1	
Tischeriidae	Coptotriche marginea		2	2	
	Tischeria ekebladella		8	13	
Tortricidae	Acleris cristana		2	3	
	Acleris emargana		8	22	
	Acleris ferrugana		15	28	
	Acleris forsskaleana		6	8	
	Acleris hastiana		1	1	
	Acleris holmiana		2	2	
	Acleris laterana		9	25	
	Acleris literana		2	2	
	Acleris rhombana	Rhomboid Tortrix	6	11	
	Acleris sparsana		4	4	
	Acleris variegana	Garden Rose Tortrix	1	1	
	Aethes beatricella		1	1	
	Aethes cnicana		1	1	
	Aethes rubigana		1	1	
	Aethes smeathmanniana		1	1	
	Agapeta hamana		14	34	
	Aleimma loeflingiana		12	176	
	Ancylis achatana		10	67	
	Ancylis badiana		1	1	
	Ancylis diminutana		5	6	
	Ancylis mitterbacheriana		4	14	
	Apotomis betuletana		18	84	
	Apotomis capreana		4	14	
	Apotomis turbidana		2	6	
	Archips crataegana		4	14	
	Archips podana	Large Fruit-tree Tortrix	19	94	
	Archips xylosteana	Variegated Golden Tortrix	18	1411	
	Bactra lancealana		4	8	
	Celypha lacunana		19	69	
	Choristoneura hebenstreitella		4	17	
	Clepsis consimilana		1	1	
	Clepsis spectrana	Cyclamen Tortrix	3	3	
	Cnephasia asseclana	Flax Tortrix	11	34	
	Cnephasia genitalana		1	1	pRDB2
	Cnephasia incertana	Light Grey Tortrix	2	2	
	Cnephasia pasiuana		3	4	
	Cnephasia stephensiana	Grey Tortrix	8	44	
	Cochylichroa atricapitana		5	5	
	Cochylidia rupicola		1	1	
	Cochylimorpha straminea		1	2	
	Cydia fagiglandana		6	14	
	Cydia illutana		1	1	
	Cydia splendana		14	61	
	Cydia strobilella	Spruce Seed Moth	1	1	
	Ditula angustiorana	Red-barred Tortrix	17	47	
	Eana incanana		1	1	

Family	Scientific name	English name	Records	Inds	Status
	Endothenia quadrimaculana		1	1	
	Epagoge grotiana		3	5	
	Epinotia abbreviana		4	16	
	Epinotia bilunana		5	12	
	Epinotia brunnichana		11	63	
	Epinotia demarniana		2	2	
	Epinotia immundana		3	7	
	Epinotia nisella		11	57	
	Epinotia ramella		5	11	
	Epinotia signatana		1	1	
	Epinotia subocellana		4	7	
	Epinotia tenerana	Nut Bud Moth	1	1	
	Epinotia tetraquetrana		2	2	
	Epiphyas postvittana	Light Brown Apple Moth	2	2	
	Eucosma campoliliana		3	3	
	Eucosma cana		7	12	
	Eudemis profundana		11	83	
	Eupoecilia angustana		1	1	
	Grapholita janthinana		2	3	
	Grapholita jungiella		1	1	
	Gypsonoma aceriana		1	1	
	Gypsonoma dealbana		13	85	
	Gypsonoma oppressana		1	1	
	Gypsonoma sociana		1	3	
	Hedya nubiferana	Marbled Orchard Tortrix	11	67	
	Hedya ochroleucana		1	1	
	Hedya pruniana	Plum Tortrix	11	25	
	Hedya salicella		8	41	
	Lathronympha strigana		4	8	
	Lobesia reliquana		1	1	
	Neocochylis dubitana		3	6	
	Neocochylis molliculana		2	2	
	Notocelia cynosbatella		2	3	
	Notocelia rosaecolana		2	3	
	Notocelia trimaculana		2	3	
	Notocelia uddmanniana	Bramble Shoot Moth	14	36	
	Pammene albuginana		5	7	
	Pammene aurita		1	1	
	Pammene fasciana		7	17	
	Pammene germmana		1	1	
	Pammene regiana		1	1	
	Pandemis cerasana	Barred Fruit-tree Tortrix	15	189	
	Pandemis corylana	Chequered Fruit-tree Tortrix	13	142	
	Pandemis heparana	Dark Fruit-tree Tortrix	11	16	
	Phalonidia manniana		2	2	
	Phtheochroa inopiana		1	2	
	Pseudargyrotoza conwagana		7	14	
	Ptycholoma lecheana		4	12	
	Ptycholomoides aeriferana		1	1	
	Rhopobota naevana	Holly Tortrix	14	28	
	Spilonota laricana		5	10	

Family	Scientific name	English name	Records	Inds	Status
	Spilonota ocellana	Bud Moth	13	44	
	Strophedra nitidana		2	2	
	Strophedra weirana		1	1	
	Syndemis musculana		6	16	
	Thyraylia nana		1	2	
	Tortricodes alternella		4	140	
	Tortrix viridana	Green Oak Tortrix	14	463	
	Zeiraphera isertana		17	138	
	Zeiraphera ratzeburgiana	Spruce Bud Moth	1	1	
	Zelotherses paleana	Timothy Tortrix	1	1	
Yponomeutidae	Paraswammerdamia albicapitella		3	3	
	Paraswammerdamia nebulella		8	14	
	Pseudoswammerdamia combinella		1	1	
	Swammerdamia caesiella		2	2	
	Yponomeuta cagnagella	Spindle Ermine	3	20	
	Yponomeuta evonymella	Bird-cherry Ermine	16	60	
	Yponomeuta malinellus	Apple Ermine	2	2	
	Yponomeuta padella	Orchard Ermine	1	1	
	Yponomeuta rorrella	Willow Ermine	1	2	
Ypsolophidae	Ypsolopha alpella		6	22	
	Ypsolopha parenthesella		10	25	
	Ypsolopha scabrella		7	9	
	Ypsolopha sequella		2	3	
	Ypsolopha sylvella		5	12	
	Ypsolopha ustella		5	5	
	Ypsolopha vittella		2	2	

Appendix 11a. Checklist of Coleoptera

Recorders: Benedict Pollard, Jon Cooter, Ivan Wright, Keith Alexander, A.J. Allen, R.B. Angus, R.J. Barnett, J. Blincow, Marc Botham, J.M. Campbell, Ryan Clark, Keith Cohen, J. Cole, Martin Collier, Graham Collins, D. Copestake, Martin Corley, Jonty Denton, Aljos Farjon, G.L. Finch, R.N. Fleetwood, Steve Gregory, S. Grove, Clive Hambler, Wil Heeney, Emily Hobson, Jovita Kaunang, Linda Losito, Ceri Mann, Darren Mann, R. Morris, T. Newton, Oxfordshire County Museum Records, Keith Porter, G. Prior, Martin Townsend, Wytham Survey

VC 23 = New to VC 23 or first modern record (since 1939); * = new to VC 23, ** = new to VC 23 and modern Oxfordshire, *** = first modern record (since 1939). SQS = Saproxylic Quality Score; IEC = Index of Ecological Continuity (for both see Chapter 15). Not refound during HPBS = earlier records held by TVERC but not recorded in the period 2017–2021.

Family	Species name	VC 23	SQS	IEC	Not refound during HPBS
Aderidae	Euglenes oculatus		8	1	
Anthicidae	Anthicus antherinus				
	Omonadus floralis				
Anthribidae	Anthribus nebulosus				X
	Platyrhinus resinosus		4	1	
	Platystomus albinus		8	1	
	Betulapion simile				
	Ceratapion gibbirostre				X
	Ischnopterapion loti				
	Oxystoma pomonae				
	Perapion curtirostre				
	Protapion apricans				
	Protapion fulvipes				
	Protapion trifolii				
	Pseudapion rufirostre				
Biphyllidae	Biphyllus lunatus		4	1	
	Diplocoelus fagi		8	1	
Buprestidae	Agrilus angustulus		8		
	Agrilus biguttatus		8		
	Agrilus laticornis		8		
	Agrilus sinuatus		4		
	Agrilus sulcicollis		24		
	Agrilus viridis				
	Trachys scrobiculatus				
Byrrhidae	Byrrhus pilula				
	Byturus ochraceus				
	Byturus tomentosus				
Cantharidae	Cantharis cryptica				X

Family	Species name	VC 23	SQS	IEC	Not refound during HPBS
	Cantharis decipiens				
	Cantharis livida				
	Cantharis nigricans				
	Cantharis pallida				
	Cantharis pellucida				
	Cantharis rustica				
	Malthinus balteatus		8		
	Malthinus flaveolus		1		
	Malthinus foveolus				
	Malthinus frontalis		8		
	Malthinus seriepunctatus		2		
	Malthodes crassicornis		24	3	
	Malthodes marginatus		1		
	Malthodes maurus		16		
	Malthodes minimus		1		
	Malthodes pumilus		2		
	Podabrus alpinus				
	Rhagonycha fulva				
	Rhagonycha lignosa				
	Rhagonycha lutea				
	Rhagonycha nigriventris				
	Rhagonycha translucida				
Carabidae	*Abax parallelepipedus*				
	Acupalpus dubius				
	Acupalpus exiguus				
	Agonum emarginatum				X
	Agonum fuliginosum				
	Agonum viduum				
	Amara aenea				
	Amara consularis				
	Amara montivaga				
	Amara ovata				
	Amara similata				
	Anthracus consputus				
	Asaphidion curtum				
	Badister bullatus				
	Bembidion aeneum				
	Bembidion articulatum				
	Bembidion assimile				
	Bembidion biguttatum				
	Bembidion clarkii				
	Bembidion dentellum				
	Bembidion doris				
	Bembidion gilvipes				
	Bembidion guttula				
	Bembidion lunulatum				
	Bembidion octomaculatum				
	Bembidion properans				
	Bembidion quadrimaculatum				
	Bembidion tetracolum				
	Bradycellus harpalinus				
	Bradycellus verbasci				
	Calathus rotundicollis				
	Calodromius spilotus				
	Carabus nemoralis				

Family	Species name	VC 23	SQS	IEC	Not refound during HPBS
	Carabus problematicus				
	Carabus violaceus				
	Curtonotus aulicus				x
	Cychrus caraboides				
	Demetrias atricapillus				x
	Demetrias imperialis				
	Dromius agilis				
	Dromius meridionalis				x
	Dromius quadrimaculatus				
	Elaphrus cupreus				
	Elaphrus riparius				
	Harpalus affinis				
	Harpalus rubripes				
	Harpalus rufipes				
	Leistus ferrugineus				
	Leistus fulvibarbis				
	Leistus rufomarginatus				
	Leistus spinibarbis				
	Limodromus assimilis				
	Loricera pilicornis				
	Microlestes minutulus				
	Nebria brevicollis				
	Nebria salina				
	Notiophilus biguttatus				
	Notiophilus palustris				x
	Notiophilus substriatus				
	Ocys harpaloides				
	Ophonus ardosiacus				
	Ophonus rufibarbis				
	Oxypselaphus obscurus				
	Paradromius linearis				
	Paranchus albipes				
	Pedius longicollis				
	Perigona nigriceps				
	Philorhizus melanocephalus				x
	Poecilus versicolor				
	Pterostichus madidus				
	Pterostichus melanarius				
	Pterostichus minor				
	Pterostichus niger				
	Pterostichus nigrita				
	Pterostichus oblongopunctatus				
	Pterostichus strenuus				
	Pterostichus vernalis				
	Stenolophus mixtus				
	Stomis pumicatus				
	Syntomus foveatus				x
	Trechus quadristriatus				
Cerambycidae	Alosterna tabacicolor		2		
	Anaglyptus mysticus		4		
	Clytus arietis		1		
	Grammoptera abdominalis		24	1	
	Grammoptera ruficornis		1		
	Leiopus linnei				
	Leiopus nebulosus		2		

Family	Species name	VC 23	SQS	IEC	Not refound during HPBS
	Leptura quadrifasciata		2	1	
	Molorchus minor				x
	Phymatodes testaceus		4	1	
	Phytoecia cylindrica				
	Poecilium alni		16		
	Pseudovadernia livida				
	Pyrrhidium sanguineum	*	24	3	
	Rhagium mordax		1		
	Rutpela maculata		1		
	Stenocorus meridianus		2		
	Stenurella melanura		2		
	Stictoleptura rubra	**			
	Tetrops praeustus		2		x
Cerylonidae	*Cerylon ferrugineum*		2		
	Cerylon histeroides		4		
Chrysomelidae	*Altica helianthemi*				
	Altica lythri				
	Altica palustris				x
	Aphthona euphorbiae				
	Aphthona herbigrada				x
	Aphthona nonstriata				
	Batophila rubi				
	Bruchidius varius				
	Bruchus rufimanus				
	Cassida rubiginosa				
	Cassida vibex				x
	Cassida viridis				
	Chaetocnema arida				
	Chaetocnema concinna				
	Chaetocnema hortensis				
	Chrysolina herbacea				x
	Chrysolina polita				
	Chrysolina staphylaea				
	Crepidodera aurata				
	Crepidodera aurea				
	Crepidodera fulvicornis				
	Cryptocephalus labiatus				
	Cryptocephalus pusillus				
	Donacia semicuprea				
	Epitrix atropae				x
	Epitrix pubescens				
	Galerucella sagittariae				
	Gastrophysa polygoni				
	Gastrophysa viridula				
	Hermaeophaga mercurialis				
	Lema cyanella				
	Lochmaea crataegi				
	Longitarsus anchusae				
	Longitarsus dorsalis				
	Longitarsus exsoletus				
	Longitarsus flavicornis				x
	Longitarsus kutscherae				
	Longitarsus luridus				
	Longitarsus lycopi				
	Longitarsus parvulus				
	Longitarsus pellucidus				x

Family	Species name	VC 23	SQS	IEC	Not refound during HPBS
	Longitarsus pratensis				
	Longitarsus rubiginosus				X
	Longitarsus suturellus				
	Neocrepidodera transversa				
	Orsodacne cerasi				
	Orsodacne humeralis				
	Oulema melanopus s. lat.				
	Oulema obscura				X
	Phaedon tumidulus				X
	Phratora vulgatissima				
	Phyllotreta astrachanica				
	Phyllotreta atra				
	Phyllotreta exclamationis				
	Phyllotreta nemorum				
	Phyllotreta nigripes				
	Phyllotreta undulata				
	Phyllotreta vittula				
	Psylliodes affinis				
	Psylliodes chrysocephala				
	Psylliodes luteola				
	Psylliodes napi				
Ciidae	Cis bilamellatus				
	Cis boleti		1		
	Cis castaneus				
	Cis fagi		2		
	Cis festivus		2		
	Cis micans	***	4		
	Cis pygmaeus	***	2		
	Cis vestitus	*	2		
	Cis villosulus		2		
	Ennearthron cornutum		2		
	Octotemnus glabriculus		1		
	Orthocis alni		2		X
	Strigocis bicornis		8		
	Sulcacis nitidus		2		
Clambidae	Clambus pallidulus				
	Clambus simsoni				
Cleridae	Opilo mollis		8	1	
	Thanasimus formicarius		4	1	
	Tillus elongatus		8	1	
Coccinellidae	Adalia bipunctata				
	Adalia decempunctata				
	Anatis ocellata				
	Anisosticta novemdecimpunctata				
	Aphidecta obliterata				
	Calvia quattuordecimguttata				
	Chilocorus bipustulatus				X
	Chilocorus renipustulatus				
	Coccidula scutellata				
	Coccinella septempunctata				
	Exochomus quadripustulatus				
	Halyzia sedecimguttata				
	Harmonia axyridis				
	Hippodamia variegata				X
	Propylea quattuordecimpunctata				
	Psyllobora vigintiduopunctata				

Family	Species name	VC 23	SQS	IEC	Not refound during HPBS
	Subcoccinella vigintiquattuorpunctata				
	Tytthaspis sedecimpunctata				
Corylophidae	Orthoperus aequalis		16		
	Orthoperus corticalis				
	Orthoperus nigrescens				
	Sericoderus brevicornis				
Cryptophagidae	Antherophagus pallens				
	Antherophagus similis				
	Atomaria apicalis				
	Atomaria atricapilla				
	Atomaria clavigera				
	Atomaria lewisi				
	Atomaria linearis				
	Atomaria mesomela				
	Atomaria nigrirostris				
	Atomaria nigriventris				
	Atomaria testacea				
	Cryptophagus dentatus		1		
	Cryptophagus distinguendus				
	Cryptophagus insulicola	**	8		
	Cryptophagus labilis		8		X
	Cryptophagus micaceus		16	3	
	Cryptophagus punctipennis				
	Cryptophagus scanicus				
	Cryptophagus scutellatus	***			
	Cryptophagus setulosus				
	Ephistemus globulus				
	Ootypus globosus				
	Paramecosoma melanocephalum				
	Telmatophilus caricis				
	Telmatophilus typhae				
Cucujidae	Pediacus dermestoides		4	1	
Curculionidae	Acalles misellus		2		
	Anisandrus dispar		8		
	Anthonomus pedicularis				
	Archarius pyrrhoceras				
	Archarius salicivorus				
	Barynotus moerens				
	Barynotus obscurus				
	Brachypera zoilus				
	Ceutorhynchus alliariae				
	Ceutorhynchus obstrictus				
	Ceutorhynchus pallidactylus				
	Ceutorhynchus picitarsis	***			
	Ceutorhynchus typhae				
	Cionus hortulanus				
	Cionus scrophulariae				
	Cionus tuberculosus				
	Cleopus pulchellus				
	Coeliodes rana				
	Coeliodes transversealbofasciatus				
	Curculio glandium				
	Curculio nucum				
	Curculio venosus				
	Curculio villosus				

Family	Species name	VC 23	SQS	IEC	Not refound during HPBS
	Cyclorhipidion bodoanum	**			
	Dorytomus filirostris	**			
	Dorytomus rufatus				
	Dorytomus taeniatus				
	Dryocoetes autographus		2		
	Dryocoetes villosus		2		
	Ernoporicus caucasicus	**	16	2	
	Ernoporicus fagi		8	1	
	Euophryum confine				
	Exomius araneiformis				
	Exomius pellucidus				
	Hylastes attenuatus				
	Hylesinus crenatus		2		
	Hylesinus toranio		2		
	Hylesinus varius		1		
	Kyklioacalles roboris		8		
	Leiosoma deflexum				
	Magdalis armigera		2		
	Magdalis carbonaria		4		
	Magdalis cerasi		4		
	Magdalis memnonia	**			
	Nedyus quadrimaculatus				
	Phloeophagus lignarius		2		
	Phyllobius oblongus				
	Phyllobius pomaceus				
	Phyllobius pyri				
	Phyllobius roboretanus				
	Phyllobius viridiaeris				
	Pissodes castaneus	**	2		
	Platypus cylindrus		8	1	
	Polydrusus cervinus				
	Polydrusus flavipes				
	Polydrusus formosus				
	Polydrusus pterygomalis				
	Rhamphus oxyacanthae				
	Rhamphus pulicarius				
	Rhinoncus leucostigma				
	Rhinoncus pericarpius				
	Rhinoncus perpendicularis				
	Rhynocyllus conicus				
	Sciaphylus asperatus				
	Scolytus intricatus		2		
	Scolytus multistriatus		1		
	Scolytus scolytus		2		
	Sitona hispidulus				
	Sitona lineatus				
	Strophosoma melanogrammum				
	Taphrorychus bicolor		8		
	Trichosirocalus troglodytes				
	Trypodendron domesticum		2	1	
	Trypodendron signatum		8	1	
	Tychius picirostris				
	Xyleborinus saxesenii		4	1	
	Xyleborus dryographus		8	1	
	Xyleborus monographus	**			

Family	Species name	VC 23	SQS	IEC	Not refound during HPBS
	Xylocleptes bispinus				
Dasytidae	*Dasytes aeratus*		2		
Dermestidae	*Anthrenus fuscus*				
	Ctesias serra		4		
	Dermestes murinus				
	Globicornis nigripes		32	3	
	Megatoma undata		8		
	Trinodes hirtus		24	3	
Dytiscidae	*Agabus bipustulatus*				
	Colymbetes fuscus				
	Hydroporus planus				
	Ilybius fuliginosus				X
Elateridae	*Adrastus pallens*				
	Agriotes acuminatus				
	Agriotes lineatus				
	Agriotes obscurus				
	Agriotes pallidulus				
	Agriotes sputator				
	Ampedus balteatus		2		
	Ampedus cardinalis		32	3	
	Ampedus elongantulus		8	1	
	Aplotarsus incanus				
	Athous bicolor				
	Athous haemorrhoidalis				
	Athous vittatus				
	Calambus bipustulatus		8	1	
	Dalopius marginatus				
	Denticollis linearis		1		
	Elater ferrugineus	**	32	3	
	Hemicrepidius hirtus				
	Ischnodes sanguinicollis		16	2	X
	Limonius poneli				
	Melanotus castanipes				
	Melanotus villosus		1		
	Procraerus tibialis		16	3	
	Stenagostus rhombeus		4	1	
Endomychidae	*Endomychus coccineus*		2		
	Symbiotes latus		8	1	
Erotylidae	*Dacne bipustulata*		2		
	Dacne rufifrons		2		
	Triplax aenea		2		
	Triplax russica		4	1	
Eucnemidae	*Epiphanis cornutus*		8		
	Eucnemis capucinus	**	32	3	
	Hylis cariniceps	**	32		
	Hylis olexai	**	24		
	Microrhagus pygmaeus	**	8	1	
Geotrupidae	*Geotrupes spiniger*				
	Geotrupes stercorarius				
Haliplidae	*Haliplus flavicollis*				X
Helophoridae	*Helophorus aequalis*				
	Helophorus brevipalpis				
	Helophorus dorsalis				
	Helophorus grandis				
	Helophorus minutus				

Family	Species name	VC 23	SQS	IEC	Not refound during HPBS
	Helophorus obscurus				
Heteroceridae	Heterocerus fenestratus				
Histeridae	Abraeus granulum		8	3	
	Abraeus perpusillus		4		
	Aeletes atomarius		16	3	
	Dendrophilus punctatus				
	Gnathoncus buyssoni				
	Gnathoncus rotundatus				
	Margarinotus brunneus				
	Margarinotus merdarius				
	Margarinotus striola				
	Margarinotus ventralis				
	Paromalus flavicornis		2		
	Plegaderus dissectus		8	2	
	Saprinus semistriatus				
Hydraenidae	Limnebius truncatellus				
	Ochthebius minimus				
Hydrophilidae	Anacaena lutescens				x
	Cercyon convexiusculus				
	Cercyon haemorrhoidalis				
	Cercyon impressus				
	Cercyon lateralis				
	Cercyon marinus				
	Cercyon melanocephalus				
	Cercyon pygmaeus				
	Cercyon quisquilius				
	Cercyon sternalis				
	Cryptopleurum minutum				
	Enochrus melanocephalus				x
	Hydrobius fuscipes				
	Hydrobius rottenbergii				
	Laccobius colon				x
	Megasternum concinnum				
	Sphaeridium bipustulatum				
	Sphaeridium scarabaeoides				
Kateretidae	Brachypterus glaber				
	Brachypterus urticae				
Laemophloeidae	Cryptolestes ferrugineus		2		
	Laemophloeus monilis		32		
	Leptophloeus alternans	**			
	Notolaemus unifasciatus	***	16	2	
Lampyridae	Lampyris noctiluca				
Latridiidae	Cartodere bifasciata				
	Cartodere nodifer				
	Corticaria crenulata				
	Corticaria punctulata				
	Corticaria serrata				
	Corticarina minuta				
	Corticarina similata				
	Cortinicara gibbosa				
	Dienerella clathrata	*			
	Dienerella ruficollis				
	Dienerella vincenti	***			
	Enicmus brevicornis		8	1	
	Enicmus histrio				

Family	Species name	VC 23	SQS	IEC	Not refound during HPBS
	Enicmus rugosus		8	2	
	Enicmus testaceus		2		
	Enicmus transversus				
	Latridius assimilis				X
	Latridius porcatus				
Leiodidae	*Anisotoma humeralis*		2		
	Anisotoma orbicularis		2		
	Catops coracinus				X
	Catops fuliginosus				
	Catops kirbii				
	Catops nigricans				
	Catops tristis				
	Choleva angustata				
	Choleva glauca				
	Colenis immunda				
	Hydnobius punctatus				X
	Leiodes calcarata				
	Leiodes macropus	**			
	Leiodes strigipennis				
	Leptinus testaceus				
	Liocyrtusa vittata				
	Nargus velox				
	Nargus wilkini				
	Nemadus colonoides		2		
	Ptomaphagus sericatus				
	Ptomaphagus subvillosus				
	Sciodrepoides fumatus				
	Sciodrepoides watsoni				
Lucanidae	*Dorcus parallelipipedus*		2		
	Sinodendron cylindricum		2		
Lycidae	*Platycis minutus*		8	1	
Lymexylidae	*Lymexylon navale*		32	2	
Melandryidae	*Abdera biflexuosa*		8	1	
	Abdera quadrifasciata		16	3	
	Anisoxya fuscula		16	1	
	Conopalpus testaceus		8	1	
	Melandrya caraboides		4	1	X
	Orchesia micans		4		X
	Orchesia minor		8		
	Orchesia undulata		4	1	
	Orchestes hortorum				
	Orchestes pilosus				
	Orchestes quercus				
	Phloiotrya vaudoueri		8	2	
Melyridae	*Axinotarsus marginalis*				
	Malachius bipustulatus		1		
Monotomidae	*Monotoma picipes*				
	Rhizophagus bipustulatus		1		
	Rhizophagus depressus	*	2		
	Rhizophagus dispar		1		
	Rhizophagus ferrugineus		2		
	Rhizophagus nitidulus		4	1	
	Rhizophagus oblongicollis		24	3	
	Rhizophagus perforatus		2		
Mordellidae	*Mordellistena neuwaldeggiana*		16	1	
	Mordellistena variegata		8		

Family	Species name	VC 23	SQS	IEC	Not refound during HPBS
	Mordellochroa abdominalis		4		
	Tomoxia bucephala		16	1	
	Variimorda villosa				
Mycetophagidae	Eulagius filicornis				
	Litargus connexus		2		
	Mycetophagus atomarius		2	1	
	Mycetophagus multipunctatus		2		
	Mycetophagus piceus		4	2	
	Mycetophagus quadriguttatus	*	16	2	
	Mycetophagus quadripustulatus		2		
	Pseudotriphyllus suturalis		4	1	
	Triphyllus bicolor		4	2	
	Typhaea haagi				
	Typhaea stercorea				
Nitidulidae	Carpophilus marginellus				
	Cryptarcha strigata		8		
	Cryptarcha undata		8		
	Epuraea aestiva				
	Epuraea guttata	*	8		
	Epuraea marseuli		1		
	Epuraea melanocephala				
	Epuraea melina				
	Epuraea pallescens		2		
	Epuraea silacea		1		
	Epuraea unicolor				
	Glischrochilus hortensis				
	Meligethes aeneus				
	Meligethes atratus				
	Meligethes carinulatus				
	Meligethes gagathinus				
	Meligethes haemorrhoidalis	**			
	Meligethes matronalis	**			
	Meligethes nigrescens				
	Meligethes ovatus				
	Meligethes viridescens				
	Omosita colon	*			
	Pocadius adustus				
	Pocadius ferrugineus				
	Soronia grisea		2		
Noteridae	Noterus clavicornis				x
Oedemeridae	Ischnomera caerulea		24	3	x
	Ischnomera cyanea		4	1	
	Ischnomera sanguinicollis		8	3	
	Oedemera lurida				
	Oedemera nobilis				
Phalacridae	Olibrus aeneus				
	Olibrus liquidus				
	Phalacrus corruscus				
	Stilbus testaceus				
Phloiophilidae	Phloiophilus edwardsii		8	1	
Ptiliidae	Acrotrichis fascicularis				
	Acrotrichis insularis				x
	Nephanes titan				
	Ptenidium laevigatum				
	Ptenidium nitidum				
	Ptenidium pusillum				

Family	Species name	VC 23	SQS	IEC	Not refound during HPBS
	Ptiliola kunzei				X
	Ptinella aptera		2		
	Ptinella denticollis		8		X
	Ptinella errabunda				
	Acrotrichis sericans				
	Anitys rubens	**	8	3	
	Anobium inexspectatum		8		
	Anobium punctatum		1		
	Dorcatoma chrysomelina		4	1	
	Dorcatoma dresdensis		16	2	
	Dorcatoma flavicornis		8	1	
	Dorcatoma substriata		16	2	
	Dryophilus pusillus		2		
	Ernobius mollis		2		
	Grynobius planus		2		
	Hadrobregmus denticollis	**	8		
	Hemicoelus fulvicornis		1		
	Ochina ptinoides		2		
	Ptilinus pectinicornis		1		
	Ptinomorphus imperialis		8		
	Ptinus fur				
	Ptinus subpilosus	*	8	2	
	Xestobium rufovillosum		4	1	
Pyrochroidae	Pyrochroa coccinea		4	1	
	Pyrochroa serraticornis		1		
Rhynchitidae	Tatianaerhynchites aequatus				
Salpingidae	Salpingus planirostris		1		
	Salpingus ruficollis		1		
	Vincenzellus ruficollis		2		
Scarabaeidae	Acrossus rufipes				
	Agrilinus ater				
	Amphimallon solstitiale				
	Aphodius fimetarius				
	Aphodius pedellus				
	Bodilopsis rufa				
	Calamosternus granarius				
	Colobopterus erraticus				
	Esymus pusillus				
	Limarus zenkeri				
	Melinopterus consputus	**			
	Melinopterus prodromus				
	Melinopterus sphacelatus				
	Melolontha melolontha				
	Nimbus contaminatus				
	Nimbus obliteratus				
	Onthophagus coenobita				
	Onthophagus joannae				
	Onthophagus similis				
	Otophorus haemorrhoidalis				
	Oxyomus sylvestris				
	Phyllopertha horticola				
	Planolinoides borealis				
	Serica brunnea				
	Teuchestes fossor				
	Volinus sticticus				

Family	Species name	VC 23	SQS	IEC	Not refound during HPBS
Scirtidae	Contacyphon coarctatus				
	Contacyphon ochraceus				
	Contacyphon palustris				
	Microcara testacea				
	Prionocyphon serricornis		8	1	
	Scirtes hemisphaericus				
Scraptiidae	Anaspis costai		2		X
	Anaspis fasciata		2		
	Anaspis frontalis		1		
	Anaspis garneysi				
	Anaspis lurida		2		X
	Anaspis maculata				
	Anaspis pulicaria		1		
	Anaspis regimbarti				
	Anaspis rufilabris		1		
	Anaspis thoracica		8		
	Anaspis thoracica var. septentrionalis				
	Scraptia testacea	**	16	3	
Silphidae	Necrodes littoralis				
	Nicrophorus humator				
	Nicrophorus investigator				
	Nicrophorus vespillo				
	Nicrophorus vespilloides				
	Phosphuga atrata				
	Thanatophilus rugosus				
	Thanatophilus sinuatus				
Silvanidae	Psammoecus bipunctatus				
	Silvanus bidentatus		8	2	
	Silvanus unidentatus		4	1	
	Uleiota planatus		16	2	
Sphindidae	Aspidiphorus orbiculatus		2		
	Sphindus dubius		8		
Staphylinidae	Acrolocha sulcula				
	Acrotona muscorum				
	Acrotona parvula				
	Agaricochara latissima		2		
	Aleochara bipustulata				
	Aleochara curtula				
	Aleochara discipennis				
	Aleochara diversa	*			
	Aleochara funebris				X
	Aleochara intricata				
	Aleochara lanuginosa				
	Aleochara lata				
	Aleochara sparsa				
	Aloconota gregaria				
	Amarochara bonnairei	**	24		
	Amischa analis				
	Amischa decipiens				
	Amischa nigrofusca				
	Anomognathus cuspidatus		2		
	Anotylus clypeonitens				
	Anotylus inustus				

Family	Species name	VC 23	SQS	IEC	Not refound during HPBS
	Anotylus mutator				
	Anotylus nitidulus				
	Anotylus rugosus				
	Anotylus sculpturatus				
	Anotylus tetracarinatus				
	Anthobium atrocephalum				
	Anthobium unicolor				
	Astenus lyonessius				
	Atheta amplicollis				
	Atheta aquatica				
	Atheta britanniae				
	Atheta castanoptera				
	Atheta clientula				
	Atheta crassicornis				
	Atheta dadopora				
	Atheta difficilis				
	Atheta fungi				
	Atheta gagatina				
	Atheta indubia				
	Atheta intermedia				
	Atheta ischnocera				
	Atheta laevana				
	Atheta liliputana	*			
	Atheta liturata		2		
	Atheta longicornis				
	Atheta minuscula	*			
	Atheta palustris	*			
	Atheta pervagata	***			
	Atheta scapularis				
	Atheta sordidula				
	Atheta subtilis				
	Atheta taxiceroides				
	Atheta vaga				
	Atrecus affinis		1		
	Autalia rivularis				
	Batrisodes venustus		8	3	
	Bibloporus bicolor		2		
	Bibloporus minutus	*	8	2	
	Bisnius fimetarius				
	Bolitobius castaneus				
	Bolitochara obliqua				
	Bryaxis curtisii				X
	Callicerus obscurus				
	Callicerus rigidicornis				
	Carpelimus corticinus				
	Carpelimus elongatus				
	Carpelimus fuliginosus				
	Carpelimus impressus	***			
	Carpelimus rivularis				
	Cephennium gallicum				
	Coprothassa melanaria				
	Creophilus maxillosus				
	Cypha discoidea				
	Cypha longicornis				
	Cypha pulicaria	***			
	Cyphea curtula		4		

Family	Species name	VC 23	SQS	IEC	Not refound during HPBS
	Dadobia immersa	***	2		
	Dropephylla devillei		2		
	Dropephylla gracilicornis				
	Dropephylla ioptera		1		
	Dropephylla koltzei	***			
	Drusilla canaliculata				
	Erichsonius cinerascens				
	Euplectus bonvouloiri		8		
	Euplectus infirmus	**	2		
	Euplectus karstenii	***	2		
	Euplectus piceus		2		
	Euplectus tholini		24	3	
	Gabrius splendidulus		1		
	Geostiba circellaris				
	Gyrohypnus fracticornis				
	Gyrophaena affinis				
	Gyrophaena fasciata				
	Gyrophaena joyi		8		
	Gyrophaena manca		8		
	Gyrophaena strictula		8		
	Habrocerus capiilaricornis				
	Hapalaraea pygmaea		2		
	Haploglossa villosula				
	Holobus apicatus				
	Homalota plana		2		
	Hygronoma dimidiata				
	Hypnogyra angularis		16	2	
	Hypopycna rufula	**			
	Lathrobium brunnipes				
	Lathrobium geminum				
	Lathrobium impressum				
	Leptusa fumida		1		
	Leptusa pulchella		2		
	Leptusa ruficollis		1		
	Lesteva heeri				
	Lesteva longoelytrata				
	Liogluta longiuscula				
	Lordithon exoletus				
	Lordithon lunulatus				
	Lordithon trinotatus				X
	Megarthrus bellevoyei				
	Megarthrus depressus				
	Megarthrus prosseni				
	Metopsia clypeata				
	Micropelus staphylinoides				
	Micropeplus fulvus				
	Mycetoporus nigricollis				
	Mycetoporus rufescens				
	Neuraphes praeteritus	***			
	Neuraphes ruthenus				
	Nudobius lentus		2		
	Ocypus aeneocephalus				
	Ocypus olens				
	Oligota apicata	***			
	Oligota punctulata				
	Olophrum piceum				X

Family	Species name	VC 23	SQS	IEC	Not refound during HPBS
	Omalium caesum				
	Omalium rivulare				
	Omalium rugatum	**			
	Othius punctulatus				
	Othius subuliformis				
	Oxypoda acuminata				
	Oxypoda brevicornis				
	Oxypoda elongatula				
	Oxypoda haemorrhoa				X
	Oxypoda opaca				
	Oxytelus laqueatus				
	Pachnida nigella				
	Paederus fuscipes				
	Paederus riparius				
	Philonthus addendus				
	Philonthus carbonarius				
	Philonthus cognatus				
	Philonthus concinnus				
	Philonthus decorus				
	Philonthus laminatus				X
	Philonthus marginatus				
	Philonthus parvicornis				
	Philonthus politus				
	Philonthus splendens				
	Philonthus succicola				
	Philonthus tenuicornis				
	Philonthus varians				
	Phloeonomus punctipennis		2		
	Phloeonomus pusillus		2		
	Phloeopora corticalis		8		
	Phloeopora scribae		2		
	Phloeopora testacea		1		
	Placusa pumilio	***			
	Platystethus arenarius				
	Platystethus nitens				
	Plectophloeus nitidus		32	3	
	Proteinus brachypterus				
	Proteinus ovalis				
	Quedius aetolicus		16	1	
	Quedius cinctus				
	Quedius cruentus				
	Quedius curtipennis				
	Quedius fuliginosus				
	Quedius fumatus				
	Quedius invreae				
	Quedius lateralis				
	Quedius levicollis				
	Quedius lucidulus				
	Quedius maurus		4	1	
	Quedius mesomelinus				X
	Quedius molochinus				X
	Quedius picipes				
	Quedius scitus		8	2	
	Quedius truncicola		8	1	X
	Rugilus rufipes				

Family	Species name	VC 23	SQS	IEC	Not refound during HPBS
	Scaphidium quadrimaculatum		2		
	Scydmaenus rufus		24	1	
	Scydmoraphes helvolus	***			
	Scydmoraphes sparshalli				
	Sepedophilus littoreus		2		
	Sepedophilus marshami				
	Sepedophilus nigripennis				
	Sepedophilus testaceus		8		
	Siagonium quadricorne		2		
	Stenichnus collaris				
	Stenichnus godarti	**	24	2	
	Stenichnus scutellaris				
	Stenus bimaculatus				X
	Stenus boops				
	Stenus brunnipes				
	Stenus butrintensis				
	Stenus clavicornis				
	Stenus flavipes				
	Stenus fulvicornis				
	Stenus impressus				
	Stenus juno				
	Stenus latifrons				
	Stenus nitidiusculus				X
	Stenus niveus				
	Stenus picipes				
	Stenus providus				
	Stenus solutus				
	Stichoglossa semirufa	**	24		
	Sunias propinquus				
	Tachinus humeralis				
	Tachinus laticollis				
	Tachinus marginellus				
	Tachinus rufipes				
	Tachinus subterraneus				
	Tachyporus chrysomelinus				
	Tachyporus dispar				
	Tachyporus hypnorum				
	Tachyporus nitidulus				
	Tachyporus obtusus				X
	Tachyporus pallidus				
	Tachyporus solutus				
	Tachyporus tersus				
	Tasgius melanarius				X
	Tasgius morsitans				
	Tetartopeus terminatus				
	Thamiaraea cinnamomea		2		
	Tinotus morion				
	Tychus niger				
	Xantholinus linearis				
	Xantholinus longiventris				
	Xylodromus concinnus				
	Zyras haworthi				
Tenebrionidae	Alphitophagus bifasciatus				
	Eledona agricola		4	1	
	Mycetochara humeralis	**	16	2	
	Nalassus laevioctostriatus				

Family	Species name	VC 23	SQS	IEC	Not refound during HPBS
	Platydema violaceum		32		
	Prionychus ater		8	1	
	Pseudocistela ceramboides		8	2	
Tetratomidae	Hallomenus binotatus		8	1	
	Tetratoma fungorum		2		
Throscidae	Aulonothroscus brevicollis	**	24	3	
	Trixagus carinifrons				
	Trixagus dermestoides				
	Trixagus leseigneuri	**			
	Trixagus meybohmi	**			
	Trixagus obtusus				
Trogidae	Trox scaber				
Trogossitidae	Nemozoma elongatum		24		
Zopheridae	Bitoma crenata		4	1	
	Colydium elongatum		16		
	Synchita separanda		24	1	
	Synchita undata	*			
	Synchita variegata		8	2	
			1,882	**160**	**57**

Appendix 11b. Saproxylic beetles

Current list of saproxylic beetles recorded in High Park with their scores for Saproxylic Quality Index (SQI) and the Index of Ecological Continuity (IEC – 2004)

Family	Species name	GB status	SQI Score	IEC Score
Histeridae	Abraeus granulum	NS (Lane 2017)	8	3
	Abraeus perpusillus		4	
	Plegaderus dissectus		8	2
	Aeletes atomarius	NS (Lane 2017)	16	3
	Gnathoncus buyssoni	NS (Lane 2017)	0	
	Gnathoncus rotundatus		0	
	Dendrophilus punctatus		0	
	Paromalus flavicornis		2	
	Margarinotus merdarius		0	
Ptiliidae	Ptenidium laevigatum		0	
	Ptinella aptera		2	
	Ptinella denticollis	NS (Hyman 1992)	8	
	Ptinella errabunda		0	
Leiodidae	Anisotoma humeralis		2	
	Anisotoma orbicularis		2	
	Nemadus colonoides		2	
Staphylinidae: Omaliinae	Dropephylla devillei		2	
	Dropephylla gracilicornis	NS (Hyman 1992)	0	
	Dropephylla ioptera		1	
	Dropephylla koltzei		1	
	Hapalaraea pygmaea		2	
	Hypopycna rufula		0	
	Phloeonomus punctipennis		2	
	Phloeonomus pusillus		2	
Staphylinidae: Pselaphinae	Bibloporus bicolor		2	
	Bibloporus minutus	NS (Hyman 1992)	8	2
	Euplectus bonvouloiri	NS (Hyman 1992)	8	
	Euplectus infirmus		2	
	Euplectus karstenii		2	
	Euplectus piceus		2	
	Euplectus tholini	RDB3 (Hyman 1992)	24	3
	Plectophloeus nitidus	RDB2 (Hyman 1992)	32	3
	Batrisodes venustus	NS (Hyman 1992)	8	3
Staphylinidae: Tachyporinae	Sepedophilus littoreus		2	
	Sepedophilus testaceus	NS (Hyman 1992)	8	
Staphylinidae: Aleocharinae	Amarochara bonnairei	RDBi (Hyman 1992)	24	
	Phloeopora corticalis	NS (Hyman 1992)	8	
	Phloeopora scribae		2	
	Phloeopora testacea		1	

Family	Species name	GB status	SQI Score	IEC Score
	Stichoglossa semirufa		24	
	Haploglossa villosula		0	
	Leptusa fumida		1	
	Leptusa pulchella		2	
	Leptusa ruficollis		1	
	Bolitochara obliqua		0	
	Gyrophaena fasciata		0	
	Gyrophaena joyi	NS (Hyman 1992)	8	
	Gyrophaena manca	NS (Hyman 1992)	8	
	Gyrophaena affinis		0	
	Gyrophaena strictula	NS (Hyman 1992)	8	
	Agaricochara latissima		2	
	Homalota plana		2	
	Anomognathus cuspidatus		2	
	Cyphea curtula		4	
	Placusa pumilio		2	
	Dadobia immersa		2	
	Atheta liturata		2	
	Thamiaraea cinnamomea		2	
Staphylinidae: Scaphidiinae	*Scaphidium quadrimaculatum*		2	
Staphylinidae: Piestinae	*Siagonium quadricorne*		2	
Staphylinidae: Scydmaeninae	*Neuraphes praeteritus*	NS (Hyman 1992)	0	
	Stenichnus godarti	RDB3 (Hyman 1992)	24	2
	*Scydmaenus rufus**	RDB2 (Hyman 1992)	24	1
Staphylinidae: Staphylininae	*Atrecus affinis*		1	
	Nudobius lentus		2	
	Hypnogyra angularis	NS (Hyman 1992)	16	2
	Quedius aetolicus	NS (Hyman 1992)	16	1
	Quedius maurus		4	1
	Quedius scitus	NS (Hyman 1992)	8	2
	Quedius truncicola	NS (Hyman 1992)	8	1
	Gabrius splendidulus		1	
Lucanidae	*Sinodendron cylindricum*		2	
	Dorcus parallelepipedus		2	
Scirtidae	*Prionocyphon serricornis*	NS (Hyman 1992)	8	1
Buprestidae	*Agrilus angustulus*	NS (Alexander 2014b)	8	
	Agrilus biguttatus		8	
	Agrilus laticornis		8	
	Agrilus sinuatus		4	
	Agrilus sulcicollis		0	
	Agrilus viridis	NS (Alexander 2014b)	24	
Eucnemidae	*Microrhagus pygmaeus*	RDB3 (Hyman 1992)	8	1
	Epiphanis cornutus		8	
	Hylis cariniceps	RDB1 (Hyman 1992)	32	
	Hylis olexai	RDB3 (Hyman 1992)	24	
	Eucnemis capucinus	RDB1 (Hyman 1992)	32	3
Throscidae	*Aulonothroscus brevicollis*	RDB3 (Hyman 1992)	24	3
Elateridae	*Ampedus balteatus*		2	
	Ampedus cardinalis	RDB2 (Hyman 1992)	32	3
	Ampedus elongatulus	NS (Hyman 1992)	8	1
	Ischnodes sanguinicollis	NS (Hyman 1992)	16	2
	Elater ferrugineus	RDB1 (Hyman 1992)	32	3
	Procraerus tibialis	RDB3 (Hyman 1992)	16	3

Family	Species name	GB status	SQI Score	IEC Score
	Melanotus castanipes		1	
	Melanotus villosus		0	
	Denticollis linearis		1	
	Stenagostus rhombeus		4	1
	Calambus bipustulatus	NS (Hyman 1992)	8	1
Lycidae	Platycis minutus		8	1
Cantharidae	Malthinus balteatus		8	
	Malthinus flaveolus		1	
	Malthinus frontalis	NS (Alexander 2014b)	8	
	Malthinus seriepunctatus		2	
	Malthodes crassicornis	NT (Alexander 2014b)	24	3
	Malthodes marginatus		1	
	Malthodes maurus		16	
	Malthodes minimus		1	
	Malthodes pumilus	NS (Alexander 2014b)	2	
Dermestidae	Trinodes hirtus	NT (Alexander 2017)	24	3
	Globicornis nigripes	VU (Alexander 2017)	32	3
	Megatoma undata	NS (Alexander 2017)	8	
	Ctesias serra		4	
Ptinidae	Ptinomorphus imperialis		8	
	Ptinus subpilosus	NS (Alexander 2017)	8	2
	Grynobius planus		2	
	Dryophilus pusillus		2	
	Ochina ptinoides		2	
	Xestobium rufovillosum		4	1
	Ernobius mollis		2	
	Hemicoelus fulvicornis		1	
	Anobium inexspectatum		8	
	Anobium punctatum		1	
	Hadrobregmus denticollis	NS (Alexander 2017)	8	
	Ptilinus pectinicornis		1	
	Dorcatoma chrysomelina		4	1
	Dorcatoma dresdensis	NS (Alexander 2017)	16	2
	Dorcatoma flavicornis	NS (Alexander 2017)	8	1
	Dorcatoma substriata	NS (Alexander 2017)	16	2
	Anitys rubens	NS (Alexander 2017)	8	3
Lymexylidae	Lymexylon navale	NS (Alexander 2014b)	32	2
Phloiophilidae	Phloiophilus edwardsii	NS (Alexander 2014b)	8	1
Trogossitidae	Nemozoma elongatum	VU (Alexander 2014b)	24	
Cleridae	Tillus elongatus	NS (Alexander 2014b)	8	1
	Opilo mollis	NS (Alexander 2014b)	8	1
	Thanasimus formicarius		4	1
Melyridae	Dasytes aeratus		2	
	Axinotarsus marginalis		0	
	Malachius bipustulatus		1	
Sphindidae	Sphindus dubius	NS (Hyman 1992)	8	
	Aspidiphorus orbiculatus		2	
Biphyllidae	Biphyllus lunatus		4	1
	Diplocoelus fagi	NS (Hyman 1992)	8	1
Erotylidae	Dacne bipustulata		2	
	Dacne rufifrons		2	
	Triplax aenea		2	
	Triplax russica		4	1

Family	Species name	GB status	SQI Score	IEC Score
Monotomidae	*Rhizophagus depressus*		2	
	Rhizophagus bipustulatus		1	
	Rhizophagus dispar		1	
	Rhizophagus ferrugineus		2	
	Rhizophagus nitidulus	NS (Hyman 1992)	4	1
	Rhizophagus oblongicollis	RDB1 (Hyman 1992)	24	3
	Rhizophagus perforatus		2	
Cryptophagidae	*Cryptophagus dentatus*		1	
	Cryptophagus insulicola		8	
	Cryptophagus labilis	NS (Hyman 1992)	8	
	Cryptophagus micaceus	RDBk (Hyman 1992)	16	3
	Cryptophagus scanicus		0	
Silvanidae	*Uleiota planatus**		16	2
	Silvanus bidentatus*	NS (Hyman 1992)	8	
	Silvanus unidentatus		4	1
Cucujidae	*Pediacus dermestoides*		4	1
Laemophloeidae	*Laemophloeus monilis*	RDB1 (Hyman 1992)	32	
	Cryptolestes ferrugineus		2	
	Notolaemus unifasciatus	NS (Hyman 1992)	16	2
Nitidulidae	*Epuraea aestiva*		0	
	Epuraea guttata	NS (Hyman 1992)	8	
	Epuraea marseuli		1	
	Epuraea melina		0	
	Epuraea pallescens		2	
	Epuraea silacea		1	
	Epuraea unicolor		0	
	Soronia grisea		2	
	Cryptarcha strigata	NS (Hyman 1992)	8	
	Cryptarcha undata	NS (Hyman 1992)	8	
	Glischrochilus hortensis		0	
Cerylonidae	*Cerylon ferrugineum*		2	
	Cerylon histeroides		4	
Endomychidae	*Symbiotes latus*	NS (Hyman 1992)	8	1
	Endomychus coccineus		2	
Corylophidae	*Orthoperus aequalis*	RDBk (Hyman 1992)	16	
	Orthoperus corticalis		4	
Latridiidae	*Enicmus brevicornis*	NS (Hyman 1992)	8	1
	Enicmus rugosus	NS (Hyman 1992)	8	2
	Enicmus testaceus		2	
Mycetophagidae	*Pseudotriphyllus suturalis*	NS (Alexander et al. 2014)	4	1
	Triphyllus bicolor	NS (Alexander et al. 2014)	4	2
	Litargus connexus		2	
	Mycetophagus atomarius		2	1
	Mycetophagus multipunctatus		2	
	Mycetophagus piceus		4	2
	Mycetophagus quadriguttatus	NS (Alexander et al. 2014)	16	2
	Mycetophagus quadripustulatus		2	
	Eulagius filicornis		0	
	Typhaea stercorea		0	
Ciidae	*Octotemnus glabriculus*		1	
	Sulcacis nitidus		2	
	Strigocis bicornis	NS (Hyman 1992)	8	
	Orthocis alni		2	

Family	Species name	GB status	SQI Score	IEC Score
	Cis bilamellatus		0	
	Cis boleti		1	
	Cis castaneus		2	
	Cis fagi		2	
	Cis festivus	NS (Hyman 1992)	2	
	Cis micans		4	
	Cis pygmaeus		2	
	Cis vestitus		2	
	Cis villosulus		2	
	Ennearthron cornutum		2	
Tetratomidae	Tetratoma fungorum		2	
	Hallomenus binotatus	NS (Alexander et al. 2014)	8	1
Melandryidae	Abdera biflexuosa	NS (Alexander et al. 2014)	8	1
	Abdera quadrifasciata	NS (Alexander et al. 2014)	16	3
	Anisoxya fuscula	NS (Alexander et al. 2014)	16	1
	Phloiotrya vaudoueri	NS (Alexander et al. 2014)	8	2
	Melandrya caraboides		4	1
	Orchesia micans	NS (Alexander et al. 2014)	4	
	Orchesia minor	NS (Alexander et al. 2014)	8	
	Orchesia undulata		4	1
	Conopalpus testaceus		8	1
Mordellidae	Tomoxia bucephala	NS (Alexander et al. 2014)	16	1
	Mordellistena neuwaldeggiana	NS (Alexander et al. 2014)	16	1
	Mordellistena variegata	NS (Alexander et al. 2014)	8	
	Mordellochroa abdominalis		4	
Zopheridae	Colydium elongatum	NS (Alexander et al. 2014)	16	
	Synchita separanda	NS (Alexander et al. 2014)	24	1
	Synchita undata		0	
	Synchita variegata	NS (Alexander et al. 2014)	8	2
	Bitoma crenata		4	1
Tenebrionidae	Eledona agricola		4	1
	Nalassus laevioctostriatus		0	
	Platydema violaceum	NR (Alexander et al. 2014)	32	
	Prionychus ater		8	1
	Pseudocistela ceramboides	NS (Alexander et al. 2014)	8	2
	Mycetochara humeralis	NS (Alexander et al. 2014)	16	2
Oedemeridae	Ischnomera caerulea	NS (Alexander et al. 2014)	24	3
	Ischnomera cyanea		4	1
	Ischnomera sanguinicollis	NS (Alexander et al. 2014)	8	3
Pyrochroidae	Pyrochroa coccinea		4	1
	Pyrochroa serraticornis		1	
Salpingidae	Vincenzellus ruficollis		2	
	Salpingus planirostris		1	
	Salpingus ruficollis		1	
Aderidae	Euglenes oculatus	NS (Alexander et al. 2014)	8	1
Scraptiidae	Scraptia testacea	NS (Alexander et al. 2014)	16	3
	Anaspis costai	NS (Alexander et al. 2014)	2	
	Anaspis fasciata		2	
	Anaspis frontalis		1	
	Anaspis garneysi		0	
	Anaspis lurida		2	
	Anaspis maculata		0	
	Anaspis pulicaria		1	
	Anaspis regimbarti		0	

Family	Species name	GB status	SQI Score	IEC Score
	Anaspis rufilabris		1	
	Anaspis thoracica	NS (Alexander et al. 2014)	8	
Cerambycidae	*Rhagium mordax*		1	
	Stenocorus meridianus		2	
	Grammoptera abdominalis	NS (Hyman 1992)	24	1
	Grammoptera ruficornis		1	
	Leptura quadrifasciata		2	1
	Alosterna tabacicolor		2	
	Rutpela maculata		1	
	Stenurella melanura		2	
	Pyrrhidium sanguineum*	RDB2 (Hyman 1992)	24	3
	Phymatodes testaceus		4	1
	Poecilium alni	NS (Hyman 1992)	16	
	Clytus arietis		1	
	Anaglyptus mysticus	NS (Hyman 1992)	4	
	Leiopus linnei		0	
	Leiopus nebulosus		2	
	Tetrops praeustus		2	
Anthribidae	*Platyrhinus resinosus*	NS (Hyman 1992)	4	1
	Platystomos albinus	NS (Hyman 1992)	8	1
Curculionidae: Platypodinae	*Platypus cylindrus*	NS (Hyman 1992)	8	1
Curculionidae: Scolytinae	*Hylastes attenuatus*		0	
	Hylesinus crenatus		2	
	Hylesinus toranio		2	
	Hylesinus varius		1	
	Ernoporicus caucasicus	NS (Hyman 1992)	16	2
	Ernoporicus fagi	NS (Hyman 1992)	8	1
	Dryocoetes autographus		2	
	Dryocoetes villosus		2	
	Taphrorychus bicolor	NS (Hyman 1992)	8	
	Scolytus intricatus		2	
	Scolytus multistriatus		1	
	Scolytus scolytus		2	
	Anisandrus dispar	NS (Hyman 1992)	8	1
	Xyleborinus saxesenii		4	1
	Xyleborus dryographus	NS (Hyman 1992)	8	1
	Xyleborus monographus		0	
	Trypodendron domesticum		2	1
	Trypodendron signatum	NS (Hyman 1992)	8	1
Curculionidae: Cossoninae	*Euophryum confine*		0	
	Phloeophagus lignarius		?	
Curculionidae: Cryptorhynchinae	*Acalles misellus*		2	
	Kyklioacalles roboris	NS (Hyman 1992)	8	
Curculionidae: Mesoptiliinae	*Magdalis armigera*		2	
	Magdalis carbonaria	NS (Hyman 1992)	4	
	Magdalis cerasi	NS (Hyman 1992)	4	
	Magdalis memnonia		0	
Curculionidae: Molytinae	*Pissodes castaneus*		2	
	TOTALS:	**SCORING SPECIES**	**268**	
		SQS	**1,891**	
		SQI	**722.7**	
		IEC		**161**

Appendix 12a. Checklist of Hemiptera and other insect orders

Recorders: Graham Collins, Ivan Wright, Jovita Kaunang, Keith Alexander, J. Blincow, Marc Botham, Anthony Cheke, Jon Cooter, Martin Corley, Aljos Farjon, R.N. Fleetwood, Russell Hedley, Steve Gregory, Sylfest Muldal, Brian Spooner, Martin Townsend

Order	Family	Species Name	Source
Hemiptera	Aradidae	*Aneurus avenius*	HPBS
Hemiptera	Aradidae	*Aneurus laevis*	TVERC
Hemiptera	Acanthosomatidae	*Acanthosoma haemorrhoidale*	HPBS
Hemiptera	Acanthosomatidae	*Elasmostethus interstinctus*	HPBS
Hemiptera	Acanthosomatidae	*Elasmucha grisea*	HPBS
Hemiptera	Pentatomidae	*Aelia acuminata*	HPBS
Hemiptera	Pentatomidae	*Dolycoris baccarum*	HPBS
Hemiptera	Pentatomidae	*Palomena prasina*	HPBS
Hemiptera	Pentatomidae	*Pentatoma rufipes*	HPBS
Hemiptera	Lygaeidae	*Cymus melanocephalus*	TVERC
Hemiptera	Lygaeidae	*Drymus sylvaticus*	HPBS
Hemiptera	Lygaeidae	*Gastrodes grossipes*	HPBS
Hemiptera	Lygaeidae	*Heterogaster urticae*	TVERC
Hemiptera	Lygaeidae	*Kleidocerys resedae*	HPBS
Hemiptera	Lygaeidae	*Scolopostethus affinis*	TVERC
Hemiptera	Lygaeidae	*Scolopostethus thomsoni*	HPBS
Hemiptera	Lygaeidae	*Stygnocoris sabulosus*	HPBS
Hemiptera	Coreidae	*Coriomeris denticulatus*	HPBS
Hemiptera	Coreidae	*Gonocerus acuteangulatus*	HPBS
Hemiptera	Rhopalidae	*Myrmus miriformis*	TVERC
Hemiptera	Rhopalidae	*Rhopalus subrufus*	HPBS
Hemiptera	Tingidae	*Physatocheila dumetorum*	HPBS
Hemiptera	Tingidae	*Tingis ampliata*	TVERC
Hemiptera	Tingidae	*Tingis cardui*	HPBS
Hemiptera	Miridae	*Acetropis gimmerthalii*	TVERC
Hemiptera	Miridae	*Amblytylus nasutus*	HPBS
Hemiptera	Miridae	*Apolygus lucorum*	TVERC
Hemiptera	Miridae	*Atractotomus magnicornis*	TVERC
Hemiptera	Miridae	*Atractotomus mali*	HPBS
Hemiptera	Miridae	*Blepharidopterus angulatus*	TVERC
Hemiptera	Miridae	*Campyloneura virgula*	HPBS
Hemiptera	Miridae	*Capsus ater*	HPBS
Hemiptera	Miridae	*Charagochilus gyllenhali*	HPBS
Hemiptera	Miridae	*Closterotomus norwegicus*	TVERC
Hemiptera	Miridae	*Cyrtorhinus caricis*	TVERC

Order	Family	Species Name	Source
Hemiptera	Miridae	Deraeocoris flavilinea	HPBS
Hemiptera	Miridae	Deraeocoris lutescens	HPBS
Hemiptera	Miridae	Deraeocoris ruber	HPBS
Hemiptera	Miridae	Dicyphus epilobii	TVERC
Hemiptera	Miridae	Dicyphus errans	TVERC
Hemiptera	Miridae	Dicyphus stachydis	TVERC
Hemiptera	Miridae	Dryophilocoris flavoquadrimaculatus	HPBS
Hemiptera	Miridae	Grypocoris stysi	HPBS
Hemiptera	Miridae	Halticus luteicollis	HPBS
Hemiptera	Miridae	Harpocera thoracica	HPBS
Hemiptera	Miridae	Heterotoma planicornis	HPBS
Hemiptera	Miridae	Leptopterna dolabrata	HPBS
Hemiptera	Miridae	Liocoris tripustulatus	HPBS
Hemiptera	Miridae	Lygocoris pabulinus	TVERC
Hemiptera	Miridae	Lygocoris rugicollis	TVERC
Hemiptera	Miridae	Mecomma ambulans	TVERC
Hemiptera	Miridae	Megaloceroea recticornis	TVERC
Hemiptera	Miridae	Miris striatus	HPBS
Hemiptera	Miridae	Monalocoris filicis	TVERC
Hemiptera	Miridae	Neolygus contaminatus	TVERC
Hemiptera	Miridae	Notostira elongata	TVERC
Hemiptera	Miridae	Orthonotus rufifrons	TVERC
Hemiptera	Miridae	Orthops campestris	TVERC
Hemiptera	Miridae	Orthotylus nassatus	TVERC
Hemiptera	Miridae	Orthotylus tenellus	TVERC
Hemiptera	Miridae	Orthotylus viridinervis	TVERC
Hemiptera	Miridae	Phoenicocoris obscurellus	TVERC
Hemiptera	Miridae	Phylus coryli	TVERC
Hemiptera	Miridae	Phylus melanocephalus	HPBS
Hemiptera	Miridae	Phytocoris longipennis	TVERC
Hemiptera	Miridae	Phytocoris populi	TVERC
Hemiptera	Miridae	Phytocoris tiliae	HPBS
Hemiptera	Miridae	Pinalitus cervinus	TVERC
Hemiptera	Miridae	Pithanus maerkelii	TVERC
Hemiptera	Miridae	Plagiognathus arbustorum	HPBS
Hemiptera	Miridae	Plagiognathus chrysanthemi	HPBS
Hemiptera	Miridae	Psallus ambiguus	TVERC
Hemiptera	Miridae	Psallus assimilis	TVERC
Hemiptera	Miridae	Psallus betuleti	TVERC
Hemiptera	Miridae	Psallus confusus	TVERC
Hemiptera	Miridae	Psallus perrisi	HPBS
Hemiptera	Miridae	Psallus varians	HPBS
Hemiptera	Miridae	Psallus wagneri	TVERC
Hemiptera	Miridae	Rhabdomiris striatellus	HPBS
Hemiptera	Miridae	Stenodema calcarata	TVERC
Hemiptera	Miridae	Stenodema laevigata	HPBS
Hemiptera	Miridae	Stenotus binotatus	HPBS
Hemiptera	Miridae	Trigonotylus ruficornis	TVERC
Hemiptera	Microphysidae	Loricula elegantula	TVERC
Hemiptera	Nabidae	Himacerus apterus	HPBS
Hemiptera	Nabidae	Nabis ferus	TVERC
Hemiptera	Nabidae	Nabis flavomarginatus	TVERC
Hemiptera	Nabidae	Nabis limbatus	HPBS
Hemiptera	Nabidae	Nabis rugosus	HPBS

Order	Family	Species Name	Source
Hemiptera	Anthocoridae	Anthocoris confusus	TVERC
Hemiptera	Anthocoridae	Anthocoris nemoralis	HPBS
Hemiptera	Anthocoridae	Anthocoris nemorum	HPBS
Hemiptera	Anthocoridae	Orius niger	TVERC
Hemiptera	Anthocoridae	Temnostethus gracilis	TVERC
Hemiptera	Anthocoridae	Temnostethus pusillus	HPBS
Hemiptera	Anthocoridae	Xylocoris cursitans	TVERC
Hemiptera	Saldidae	Saldula saltatoria	TVERC
Hemiptera	Cixiidae	Cixius nervosus	HPBS
Hemiptera	Cixiidae	Tachycixius pilosus	HPBS
Hemiptera	Delphacidae	Conomelus anceps	TVERC
Hemiptera	Delphacidae	Dicranotropis hamata	TVERC
Hemiptera	Delphacidae	Ditropis pteridis	HPBS
Hemiptera	Delphacidae	Hyledelphax elegantulus	TVERC
Hemiptera	Delphacidae	Javesella pellucida	TVERC
Hemiptera	Delphacidae	Kosswigianella exigua	TVERC
Hemiptera	Delphacidae	Stenocranus minutus	HPBS
Hemiptera	Cercopidae	Cercopis vulnerata	HPBS
Hemiptera	Aphrophoridae	Aphrophora alni	HPBS
Hemiptera	Aphrophoridae	Neophilaenus lineatus	TVERC
Hemiptera	Aphrophoridae	Philaenus spumarius	HPBS
Hemiptera	Cicadellidae	Acericerus heydenii	HPBS
Hemiptera	Cicadellidae	Agallia consobrina	TVERC
Hemiptera	Cicadellidae	Alebra albostriella	HPBS
Hemiptera	Cicadellidae	Allygus mixtus	HPBS
Hemiptera	Cicadellidae	Allygus modestus	HPBS
Hemiptera	Cicadellidae	Anoscopus albifrons	TVERC
Hemiptera	Cicadellidae	Aphrodes bicinctus	TVERC
Hemiptera	Cicadellidae	Aphrodes makarovi	TVERC
Hemiptera	Cicadellidae	Balclutha punctata	HPBS
Hemiptera	Cicadellidae	Cicadella viridis	TVERC
Hemiptera	Cicadellidae	Empoasca vitis	HPBS
Hemiptera	Cicadellidae	Eupteryx aurata	TVERC
Hemiptera	Cicadellidae	Eupteryx florida	TVERC
Hemiptera	Cicadellidae	Eupteryx urticae	TVERC
Hemiptera	Cicadellidae	Eupteryx vittata	TVERC
Hemiptera	Cicadellidae	Eurhadina pulchella	HPBS
Hemiptera	Cicadellidae	Euscelis incisus	HPBS
Hemiptera	Cicadellidae	Euscelis obsoletus	HPBS
Hemiptera	Cicadellidae	Evacanthus acuminatus	HPBS
Hemiptera	Cicadellidae	Evacanthus interruptus	TVERC
Hemiptera	Cicadellidae	Forcipata forcipata	TVERC
Hemiptera	Cicadellidae	Iassus lanio	HPBS
Hemiptera	Cicadellidae	Idiocerus lituratus	HPBS
Hemiptera	Cicadellidae	Lamprotettix nitidulus	HPBS
Hemiptera	Cicadellidae	Ledra aurita	HPBS
Hemiptera	Cicadellidae	Linnavuoriana sexmaculata	HPBS
Hemiptera	Cicadellidae	Macropsis prasina	HPBS
Hemiptera	Cicadellidae	Macropsis scotti	HPBS
Hemiptera	Cicadellidae	Macrosteles sexnotatus	TVERC
Hemiptera	Cicadellidae	Metidiocerus rutilans	HPBS
Hemiptera	Cicadellidae	Mocydia crocea	HPBS
Hemiptera	Cicadellidae	Oncopsis flavicollis	HPBS
Hemiptera	Cicadellidae	Oncopsis tristis	TVERC

Order	Family	Species Name	Source
Hemiptera	Cicadellidae	*Populicerus confusus*	HPBS
Hemiptera	Cicadellidae	*Recilia coronifera*	TVERC
Hemiptera	Cicadellidae	*Ribautiana tenerrima*	HPBS
Hemiptera	Cicadellidae	*Ribautiana ulmi*	HPBS
Hemiptera	Cicadellidae	*Speudotettix subfusculus*	HPBS
Hemiptera	Cicadellidae	*Streptanus aemulans*	TVERC
Hemiptera	Cicadellidae	*Thamnotettix confinis*	HPBS
Hemiptera	Cicadellidae	*Thamnotettix dilutior*	HPBS
Hemiptera	Cicadellidae	*Verdanus abdominalis*	HPBS
Hemiptera	Membracidae	*Centrotus cornutus*	TVERC
Hemiptera	Triozidae	*Trichochermes walkeri*	HPBS
Hemiptera	Triozidae	*Trioza galii*	HPBS
Hemiptera	Triozidae	*Trioza remota*	HPBS
Dermaptera	Forficulidae	*Forficula auricularia*	HPBS
Ephemeroptera	Ephemeridae	*Ephemera danica*	HPBS
Mecoptera	Panorpidae	*Panorpa communis*	HPBS
Mecoptera	Panorpidae	*Panorpa germanica*	HPBS
Megaloptera	Sialidae	*Sialis lutaria*	HPBS
Neuroptera	Chrysopidae	*Chrysopa perla*	HPBS
Neuroptera	Chrysopidae	*Chrysoperla carnea* group	TVERC
Neuroptera	Chrysopidae	*Chrysotropia ciliata*	TVERC
Neuroptera	Chrysopidae	*Dichochrysa prasina*	HPBS
Neuroptera	Chrysopidae	*Nineta flava*	HPBS
Neuroptera	Hemerobiidae	*Hemerobius humulinus*	HPBS
Neuroptera	Hemerobiidae	*Hemerobius lutescens*	HPBS
Neuroptera	Hemerobiidae	*Hemerobius micans*	HPBS
Neuroptera	Hemerobiidae	*Micromus variegatus*	HPBS
Odonata	Aeshnidae	*Aeshna cyanea*	HPBS
Odonata	Aeshnidae	*Aeshna grandis*	HPBS
Odonata	Aeshnidae	*Aeshna mixta*	TVERC
Odonata	Aeshnidae	*Anax imperator*	HPBS
Odonata	Calopterygidae	*Calopteryx splendens*	HPBS
Odonata	Calopterygidae	*Calopteryx virgo*	HPBS
Odonata	Coenagrionidae	*Enallagma cyathigerum*	HPBS
Odonata	Coenagrionidae	*Erythromma najas*	HPBS
Odonata	Coenagrionidae	*Ischnura elegans*	HPBS
Odonata	Libellulidae	*Libellula depressa*	TVERC
Odonata	Libellulidae	*Orthetrum cancellatum*	HPBS
Odonata	Libellulidae	*Sympetrum sanguineum*	HPBS
Odonata	Libellulidae	*Sympetrum striolatum*	HPBS
Odonata	Platycnemididae	*Platycnemis pennipes*	TVERC
Orthoptera	Acrididae	*Chorthippus brunneus*	HPBS
Orthoptera	Acrididae	*Chorthippus parullelus*	HPBS
Orthoptera	Acrididae	*Omocestus viridulus*	HPBS
Orthoptera	Tetrigidae	*Tetrix subulata*	HPBS
Orthoptera	Tetrigidae	*Tetrix undulata*	HPBS
Orthoptera	Tettigoniidae	*Conocephalus discolor*	HPBS
Orthoptera	Tettigoniidae	*Leptophyes punctatissima*	HPBS
Orthoptera	Tettigoniidae	*Meconema thalassinum*	HPBS
Orthoptera	Tettigoniidae	*Roeseliana roeselii*	HPBS
Prostigmata	Eriophyidae	*Eriophyes rubicolens*	HPBS
Psocoptera	Caeciliusidae	*Caecilius fuscopterus*	HPBS
Psocoptera	Caeciliusidae	*Epicaecilius pilipennis*	HPBS
Psocoptera	Caeciliusidae	*Valenzuela flavidus*	HPBS

Order	Family	Species Name	Source
Psocoptera	Ectopsocidae	*Ectopsocus briggsi*	HPBS
Psocoptera	Ectopsocidae	*Ectopsocus petersi*	HPBS
Psocoptera	Elipsocidae	*Elipsocus hyalinus*	HPBS
Psocoptera	Mesopsocidae	*Mesopsocus immunis*	HPBS
Psocoptera	Mesopsocidae	*Mesopsocus unipunctatus*	HPBS
Psocoptera	Peripsocidae	*Peripsocus milleri*	HPBS
Psocoptera	Philotarsidae	*Philotarsus parviceps*	HPBS
Psocoptera	Psocidae	*Loensia fasciata*	HPBS
Psocoptera	Psocidae	*Loensia pearmani*	HPBS
Psocoptera	Psocidae	*Loensia variegata*	HPBS
Psocoptera	Stenopsocidae	*Graphopsocus cruciatus*	HPBS
Psocoptera	Stenopsocidae	*Stenopsocus immaculatus*	HPBS
Rhapidioptera	Raphidiidae	*Phaeostigma notata*	HPBS
Siphonaptera	Hystrichopsyllidae	*Hystrichopsylla talpae*	HPBS
Trichoptera	Brachycentridae	*Brachycentrus subnubilus*	TVERC
Trichoptera	Hydropsychidae	*Hydropsyche pellucidula*	TVERC
Trichoptera	Leptoceridae	*Athripsodes aterrimus*	TVERC
Trichoptera	Leptoceridae	*Mystacides longicornis*	HPBS
Trichoptera	Limnephilidae	*Glyphotaelius pellucidus*	HPBS
Trichoptera	Limnephilidae	*Limnephilus auricula*	TVERC
Trichoptera	Limnephilidae	*Limnephilus flavicornis*	TVERC
Trichoptera	Limnephilidae	*Limnephilus lunatus*	HPBS
Trichoptera	Limnephilidae	*Potamophylax latipennis*	TVERC
Trichoptera	Phrygraneidae	*Phryganea bipunctata*	HPBS
Trichoptera	Phrygraneidae	*Phryganea grandis*	HPBS

Appendix 12b. Checklist of other invertebrates (worms)

Recorders: ESB from TVERC data

Group	Family	Species name	Source
Ophistopora	Lumbricidae	*Allolobophora chlorotica*	TVERC
	Lumbricidae	*Aporrectodea calliginosa*	TVERC
	Lumbricidae	*Aporrectodea longa*	TVERC
	Lumbricidae	*Bimastos elseni*	TVERC
	Lumbricidae	*Lumbricus castaneus*	TVERC
	Lumbricidae	*Lumbricus terrestris*	TVERC
	Lumbricidae	*Octolasion cyaneum*	TVERC
	Lumbricidae	*Satchellius mammalis*	TVERC

Appendix 13. Checklist of Amphibians and Reptiles

Recorders: Aljos Farjon, Angie Julian, Kate Sharma, Sylfest Muldal, Margaret Price, Martyn Ainsworth, Anthony Cheke, Bob Cowley, Graham Collins, Penny Cullington, Molly Dewey, Jim Fairclough, Sara Gregory, Steve Gregory, Rosemary Hill, Emily Hobson, Keith Kirby, Linda Losito, Darren Mann, Benedict Pollard, Peter Topley, Tom Walker, Rosemary Winnall

Amphibians

Scientific name	English name	Status
Bufo bufo	Common Toad	frequent
Lissotriton vulgaris	Smooth Newt	common in one pond
Rana temporaria	Common Frog	rare
Triturus cristatus	Great Crested Newt	common in ponds

Reptiles

Anguis fragilis	Slow-worm	common
Natrix helvetica	Grass Snake	occasional
Zootoca vivipara	Common Lizard	rare

Appendix 14. Checklist of birds: summary by survey year 2017–2021

codes: **pr** = pair; **t** = territory; **y** = young/juveniles (out of nest); **#** = present, several, no count; **##** = many, no count; **j/o** = just outside; **ov** = flying over
CG = Combe Gate, CGf = Combe Gate field & hedges (i.e. just outside HP); GBr = Grand bridge; HP = High Park, HPb = High Park bank/shore of lake; QP = Queen Pool, QEI = Queen Elizabeth Island, VP = Vista Pond (near High Lodge), WP = Woodyard pond
NB: Territory figures not generally quoted for 2017 as surveys not comparable to 2019–21, but given for a species that subsequently declined, and cryptic species.

Date	2017–2021	2017	2018	2019	2020	2021
Comment green highlight = species formerly present, not seen during survey	B = breeds	* = recorded + max No. if counted [*] seen from, but not within, HP	* = recorded + max No. if counted [# = several, ## = many]	* = recorded + max No. if counted; x[-y]t = No. territories counted [- estimated total for HP]	* = recorded + max No. if counted; x[-y]t = No. territories counted [- estimated total for HP]	* = recorded + max No. if counted; x[-y]t = No. territories counted [- estimated total for HP]
Recorders / Observers		Ben Carpenter [BC], Russell Hedley	Ian Lewington, Anthony Cheke	Anthony Cheke, Ian Lewington	Anthony Cheke, Gerry Tissier, Ian Lewington	Anthony Cheke, Gerry Tissier
Record Type		limited coverage; BC on roads only	limited coverage	walks not standardized; attempt to cover whole area at least twice	part of route followed constant on each visit; rest varied; winter dates roads only	part of route followed constant on each visit; rest varied; winter dates roads only

Species Name	English Name	2017–2021	Status	2017	2018	2019	2020	2021
Accipiter gentilis	Goshawk	*	? passage only	-	-	*1	-	-
Accipiter nisus	Sparrowhawk	*B	resident, breeds	-	-	*1; lt	*[only kill remains found, bird not seen]	*1 [+kill remains]; ?2t
Aegithalos caudatus	Long-tailed Tit	*B	resident, breeds	*3	*	*6; 4t	*2+y; 2–3t	*5; 6t
Aix galericulata	Mandarin Duck	*B	resident, tree hole nester; often on HP ponds; breeds (irregularly ?)	-	-	*5; 1 pr seen in tree	*11; 4+ pr; ♂ + brood seen on VP	*4pr+l
Alauda arvensis	Skylark		adjacent farm fields only	-	-		[*CGf]	-
Alectoris rufa	Red-legged Partridge	*	occ. grassy areas, winter only	-	-	*2	-	*2
Alopochen aegyptiaca	Egyptian Goose	*	future breeder ?; tree-hole nester	-	-	-	-	*pr VP up tree
Anas platyrhynchos	Mallard	*B	some pairs nest in HP	*4	[*]	*#; s	*10+y; 4–5 pr min	*8pr+ br. ponds/HPb

Date		Notes	2017–2021	2017	2018	2019	2020	2021
Anas strepera	Gadwall	a few nest along HPb	*B	-	-	*2;	*8; 3 1 pr nested HPb	*pr; ?2pr br. ponds/HPb
Buteo buteo	Buzzard	resident, breeds; territories extend beyond HP boundaries	*B	*5; 1+t	*6	*6; ? 3t	*8 [?duplicates]; 2–3 pr	*5; 3–4 pr
Carduelis (Linaria) cannabina	Linnet	passage only	*	-	*ov.6	*6	-	-
Carduelis carduelis	Goldfinch	resident, breeds; edge species	*B	*3	*	-; ot	*pr;?1 t	*pr; 1t
Carduelis (Chloris) chloris	Greenfinch	resident, breeds; edge species	*B	-	*	*1	-	*pr; 1t
Carduelis (Spinus) spinus	Siskin	winter visitor	=	*2	-	*2	-	-
Carduelis (Acanthis) flammea	Redpoll	winter visitor	*	*15	-	*1	*2	-
Certhia familiaris	Treecreeper	resident, breeds; cryptic, under-recorded	*B	*7; 4t	*4	*4;?t	*3; 3+t	*pr; 3+t
Columba oenas	Stock Dove	resident, breeds;	*E	*4	*2	*6; 13[-24]t	*20; 30[-40]t	*16; 23[-33]t
Columba palumbus	Wood Pigeon	resident, breeds; only singing birds counted	*B	*14	*	*8; 17[-25]t;	*17; 26[-36]t	*10; 20[-29]t
Corvus corax	Raven	resident, breeds; territories extend beyond HP boundaries	*B	*1	*2	*3; 3t,	*2pr; 2[-?3]t	*?4; 3t
Corvus corone	Carrion Crow	resident, breeds; edge species - trees along farmland & over only	*B	*2	*	*10; 8[-9]t	*7; 1[-14]t	*6; 6[-8]t
Corvus frugilegus	Rook		*	-	-	-;	*1	[*]

Date		2017–2021		2017	2018	2019	2020	2021
Corvus monedula	Jackdaw	*B	abundant resident, breeds; not possible to census; dozens of pairs probable former breeder	*14	*#	*##	##; est. 50–100 prs	##
Cuculus canorus	Cuckoo							
Dendrocopos major	Gr. Spotted Woodpecker	*B	resident, breeds;	*3	*4	*19; 21[–32]t	*10; 17[–27]t	*6; 14[–20]t
Dendrocopos minor	Lesser Spotted Woodpecker		former breeding resident	–	–	–	–	–
Erithacus rubecula	Robin	*B	resident, breeds;	*14	*##	*25; 43[–55]t	*21;56[–84]t	*32; 67[–99]t
Falco peregrinus	Peregrine	*	? 1 local territory sometimes includes HP	–	–	*2	–	[*]
Falco subbuteo	Hobby		summer visitor, to be expected	–	–	–	–	–
Falco tinnunculus	Kestrel	*	resident; no territory in HP; lt j/o	–	–	*1	*1	[*CGf, nest]
Fringilla coelebs	Chaffinch	*B	resident, breeds;	*5; 5+t	*	*12; 19[–24]t	*8; 13[–17]	*5; 1l[–16]t
Fringilla montifringilla	Brambling	*	winter visitor	*2+	–	–	–	–
Garrulus glandarius	Jay	*B	resident, breeds; wide-ranging, hard to census	*1	*3	*8; ?5t	*4; 5–6t	*6; 7t
Loxia curvirostra	Common Crossbill	*	winter visitor, occ. breeder; breeds early, not proven in HP recently	–	–	*3; may have bred	*2 over	–
Luscinia megarhynchos	Nightingale		summer visitor; former breeder	–	–	–	–	–
Milvus milvus	Red Kite	*B	resident, breeds; territories extend beyond HP boundaries	*5	*3	*7;1[–2]t	*4; 2t+ 1 j/o	*4; 2t

Date			2017–2021		2017	2018	2019	2020	2021
Motacilla cinerea	Grey Wagtail		*	winter visitor	-	-	*1	-	*1
Motacilla flava	Yellow Wagtail		*	passage migrant	-	-	-	-	*2
Muscicapa striata	Spotted Flycatcher		*?B	summer visitor, may breed	*1	-	*2; ?1t	-	*1; ?1t
Parus (Periparus) ater	Coal Tit		*B	resident, breeds; under-recorded; inconspicuous in summer,	*15; 1ot	*#.y	*11; 1ot before long conifer plantation cut	*3y; 1+t,	*1; 1+t
Parus (Cyanistes) caeruleus	Blue Tit		*B	resident, breeds;	*25	*#	*43; 38[−58+t]	*13; 25[−39]t	*20; 47[−72]t
Parus major	Great Tit		*B	resident, breeds;	*11	*#	*18; 31[−48]t	*11; 22[−32]t	*18; 36[−59]t
Parus (Poecile) montanus	Willow Tit			resident; former breeder	-	-	-	-	-
Parus (Poecile) palustris	Marsh Tit		*B	resident, breeds; secretive, under-recorded	*4	*4t.y	*6; 6t	*2; 3+t	*1;2+t
Phasianus colchicus	Pheasant		*B	not censused [## birds released for shooting]	*45	*##	*##	*# [no release]	*##
Phylloscopus collybita	Chiffchaff		*E	summer visitor, breeds	*8	*6	*11; 27[−43]t	*10; 22[−32]t	*15; 28[−41]t
Phylloscopus trochilus	Willow Warbler		*B	declining summer visitor, breeds	*3; 3t	*4	*3; 6t	*3; c.5t	*1; 1t
Pica pica	Magpie		*	resident [? controlled]	-	-	*1; ot	-	-
Picus viridis	Green Woodpecker		*B	resident, breeds;	*2	*2	*2; no summer records 2019	*4; c.4t	*3; 5t
Phoenicurus phoenicurus	Redstart		*	passage only	-	-	*1	-	-
Prunella modularis	Dunnock		*B	resident, breeds; edge species	*3	*	*1; ot	[*; 1t j/o]	*1; ?1t[+1t j/o]
Pyrrhula pyrrhula	Bullfinch		*	resident, scarce	*2	-	-	-	-
Regulus regulus	Goldcrest		*B	resident, breeds + winter visitor; hard to find in summer, under-recorded	*6; 5t	*	*3; ?5t before long conifer plantation cut	*1; 1t	*3; ?2t

Date		2017–2021		2017	2018	2019	2020	2021
Scolopax rusticola	Woodcock	*	scarce winter visitor				*1	*2
Sitta europaea	Nuthatch	*B	resident, breeds; rather silent in summer	*4	*4	*15; 17[-25]t,	*7; 15[-21]t	*13;21[-32]t
Streptopelia turtur	Turtle dove		summer visitor; former breeder	-	-		-	-
Strix aluco	Tawny Owl	*B	resident, breeds; nocturnal so under-recorded	*2	*1	-	*1; 1+t	*1+y; 4–5t
Sturnus vulgaris	Starling	*	resident & winter visitor; nest nr. GBr	-	*	-	-	-
Sylvia atricapilla	Blackcap	*B	summer visitor, breeds	*7	*6	*20; 41[-56]t	*18; 41[-62]t	*17: 39[-56]t
Sylvia borin	Garden Warbler	*B	summer visitor, inconspicuous; probably under-recorded	-	*1	*1; 1t	*1; 1t	- [not confirmed]
Sylvia communis	Whitethroat		edge species	-	-	-	-	-
Troglodytes troglodytes	Wren	*B	resident, breeds;	*23	*##	*27; 54[-78]t	*25; 49[-84]t	*19; 51[-77]t
Turdus iliacus	Redwing	*	winter visitor	-	-	-	-	*40+
Turdus merula	Blackbird	*B	resident, breeds;	*15	*#	*21; 43[-64]t	*17; 39[-55]t	*17: 38[-56]t
Turdus philomelos	Song Thrush	*B	resident, breeds;	*10	*6	*11; 19[-26]t	*4; 1 [-14]t	*9; 16[-23]t
Turdus pilaris	Fieldfare		winter visitor	-	-	-	-	-
Turdus viscivorus	Mistle Thrush	*B	resident, breeds;	*2	*3	*2; 1 [-2?]t	*2; 3–4t	*3; ?4t
Tyto alba	Barn Owl	*?B	resident, may breed; also roosting	-	-	*1;0t	-	*1; ?1t
TOTALS		58	37 breeders + 2 possible	37	34	48	40	44

Waterbirds, water related, aerial feeders and others that use HP or fly over only. [*] = seen on lake from HP (not included in totals)

Date		2017–2021	status	2017	2018	2019	2020	2021
Acrocephalus scirpaceus	Reed Warbler	*	summer visitor, reedy HPb only	-	*1	*2; 1[–2]t ?	*1; 1t [? not breeding]	-
Acrocephalus schoenobaenus	Sedge Warbler		summer visitor; potential breeder HPb reedy edge	-	-	-	-	-
Alcedo atthis	Kingfisher	*	no suitable nesting bank	[*1]	-	[*1]	-	-
Anas penelope	Widgeon	*	winter visitor; occ uses HPb	-	-	-	-	*3
Anser anser	Greylag Goose	*	no nest in HP; feed on bank	-	-	[*]	*12+15y	*6+y
Apus apus	Swift	*	overhead or lake only, rare at HP	-	*	-	*3 over	[*]
Ardea cinerea	Grey Heron	*	no nest in HP (colony QEI); feed on bank, & rest on trees	-	*	*2	*6	*4
Ardea alba	Great White Egret	*	feed on bank, & rest on trees	-	-	-	*1	*1
Aythya fuligula	Tufted Duck	*	occ. nest HPb	-	-	[*]	*pr; ?1 pr nested HBp	[*]
Branta canadensis	Canada Goose	*	no nest in HP; occ. feed on bank	-	[*]	[*]	*1	*4+5y
Cygnus olor	Mute Swan	*	no nest in HP; occ. feed on bank	-	-	[*]	[*]	*1
Egretta garzetta	Little Egret	*	no nest in HP (colony QEI); feed on bank, & rest on trees	-	-	[*1]	*1	*2
Fulica atra	Coot	*	resident; nest along HP lake edge	-	*	*2	5; 2–3 pr HPb	*3; no confirmed nests
Gallinula chloropus	Moorhen	*	resident; nest along HP lake edge; once seen on internal pond (VP)	[*1]	*3	*3	3; 1+ pr HPb	*2: no confirmed nests
Hirundo rustica	Swallow	*	summer visitor; >2019 nested at East End Farm nr.CG	-	-	[*]	*1 over	[*]
Larus fuscus	Lesser Black-backed Gull		no nest in HP; occ. over, or on HPb	-	-	-	-	-
Larus ridibundus	Black-headed Gull	*	no nest in HP (colony QEI); rests on trees HPb	-	-	[*]	*1 + 5 over	*3 over
Phalacrocorax carbo	Cormorant	*	resident; breeds; nest HPb	-	-	*1	*14	*1+2 over
Podiceps cristatus	Great Crested Grebe	*	could nest on HPb	-	[*]	[*]	*3 nests HPb	[*]
Podiceps ruficollis	Dabchick		colony on GBr, but not in HP	-	-	[*]	-	-
Riparia riparia	Sand Martin	*	no nest in HP	*1 over	-	[*]	[*]	[*]
Sterna hirundo	Common Tern	*	resident; nest Woodstock etc.	-	-	[*]	[*]	[*]
Streptopelia decaocto	Collared Dove	*		-	-	-	*2 over	-
		19		1	4	5	15	11

Appendix 15. Checklist of mammals

Recorders: Ben Carpenter, Anthony Cheke, Keith Cohen, Bob Cowley, Chloe Dalglish, Aljos Farjon, Ray Heaton, Russell Hedley, Stuart Jenkins, Ian Lewington, Sylfest Muldal, Benedict Pollard, Helen Simmons

Species Name	English Name
Apodemus sylvaticus	Wood Mouse
Barbastella barbastellus	Barbastelle
Capreolus capreolus	Roe Deer
Dama dama	Fallow Deer
Erinaceus europaeus	Hedgehog
Lepus europaeus	Brown Hare
Lutra lutra	Otter
Meles meles	Badger
Microtus agrestis	Field Vole
Muntiacus reevesi	Muntjac
Mustela erminea	Stoat
Myodes glareolus	Bank Vole
Myotis daubentoni	Daubenton's Bat
Myotis nattereri	Natterer's Bat
Neomys fodiens	Water Shrew
Nyctalus noctula	Noctule Bat
Oryctolagus cuniculus	Rabbit
Pipistrellus nathusii	Nathusius's Pipistrelle
Pipistrellus pipistrellus	Common Pipistrelle
Pipistrellus pygmaeus	Soprano Pipistrelle
Plecotus auritus	Brown Long-eared Bat
Sciurus carolinensis	Grey Squirrel
Sorex araneus	Common Shrew
Sorex minutus	Pygmy Shrew
Talpa europaea	European Mole
Vulpes vulpes	Red Fox

References

Agassiz, D.J.L., S.D. Beavan and R.J. Heckford. 2013. *A Checklist of the Lepidoptera of the British Isles*. 2013. London: Royal Entomological Society. i–iv, 1–206.

Agassiz, D.J.L., S.D. Beavan and R.J. Heckford. 2022. Fifth update to A Checklist of the Lepidoptera of the British Isles, 2013 on account of subsequently published data. The *Entomologist's Record and Journal of Variation* 134: 1–5.

Agassiz, D., 2022. Natural History Museum 2022. Data Portal Query on "Agassiz Lepidoptera 20220628a.xlsx" created at 2022-07-01 07:24:22.187770 PID https://doi.org/10.5519/qd.zg8ojon9 Subset of 'Checklist of the Lepidoptera of the British Isles – Data' (dataset) PID https://doi.org/10.5519/0093915

Ainsworth, A.M. 2005. Identifying important sites for beech deadwood fungi. *Field Mycology* 6: 41–61. https://doi.org/10.1016/S1468-1641(10)60303-9

Ainsworth, A.M. 2010. *Survey for oak polypore at Epping Forest (2009) and updated records for zoned rosette and other fungi of conservation importance.* Unpublished report for the City of London Corporation.

Ainsworth, A.M. 2017. Non-lichenised fungi. In A. Farjon, *Ancient Oaks in the English Landscape*, pp. 274–85. Royal Botanic Gardens, Kew: Kew Publishing.

Ainsworth, A.M. 2022. Non-lichenised fungi. In A. Farjon, *Ancient Oaks in the English Landscape*, second edition, pp. 279–89. Royal Botanic Gardens, Kew: Kew Publishing.

Ainsworth, A.M. and A. Henrici. 2023. Checklist of the British and Irish Basidiomycota, Update 11. https://fungi.myspecies.info/content/checklists, accessed 23 January 2023.

Ainsworth, A.M. and K. Liimatainen. 2022. Meruliporia (Serpula, Leucogyrophana) pulverulenta: an overlooked saprotroph of hollow oak trunks with a stronghold in Windsor Great Park. *Field Mycology* 23 (2): 57–62.

Ainsworth, G.C. 1950. The Oxford foray, 12–19 September 1949. *Transactions of the British Mycological Society* 33 (3–4): 372–7. https://doi.org/10.1016/S0007-1536(50)80095-5

Alexander, K. 2002a. *High Park, Blenheim Estate Veteran Tree Survey – Survey of Saproxylic Invertebrates.* Cirencester: Keith Alexander.

Alexander, K. 2002b. The invertebrates of living & decaying timber in Britain and Ireland – a provisional annotated checklist. *English Nature Research Report* No. 467. Peterborough: English Nature (Natural England).

Alexander, K. 2017. Invertebrates. In A. Farjon, *Ancient Oaks in the English Landscape*. Royal Botanic Gardens, Kew: Kew Publishing.

Alexander, K. 2022. Invertebrates. In A. Farjon, *Ancient Oaks in the English Landscape*, second edition, pp. 296–301. Royal Botanic Gardens, Kew: Kew Publishing.

Alexander, K.N.A., 1988. The development of an index of ecological continuity for deadwood associated beetles. In R.C. Welch, Insect indicators of ancient woodland (East Region Regional News). *Antenna* 12: 69–70.

Alexander, K.N.A., 2002. The invertebrates of living and decaying timber in Britain and Ireland. *English Nature Research Reports* No 47. Peterborough: English Nature.

Alexander, K.N.A. 2003a. *Globicornis rufitarsis* (Panzer) at Blenheim Park, Oxfordshire. The *Coleopterist* 12: 96.

Alexander, K.N.A. 2003b. *Provisional atlas of the Cantharoidea and Buprestoidea (Coleoptera) of Britain and Ireland.* Huntingdon: Biological Records Centre.

Alexander, K.N.A. 2004. Revision of the Index of Ecological Continuity as used for saproxylic beetles. *English Nature Research Reports* No. 574.

Alexander, K.N.A. 2008. Tree biology and saproxylic Coleoptera: issues of definitions and conservation language. In V. Vignon and J.F. Asmodé (eds) *Proceedings of the 4th Symposium and Workshop on the Conservation of Saproxylic Beetles*, pp. 9–13. Held in Vivoin, Sarthe, France, 27–9 June 2006. *Revue d'Écologie (Terre Vie)*, supplement 10. https://doi.org/10.3406/revec.2008.1455

Alexander, K.N.A. 2009. The status of *Uleiota planatus* (Linnaeus) (Silvanidae) in Britain – long-established native or importation? The *Coleopterist* 18: 49–52.

Alexander, K.N.A. 2010. Saproxylic Beetles. In A.C. Newton (ed.) *Biodiversity in the New Forest*, pp. 46–53. Newbury: Pisces Publications.

Alexander, K.N.A. 2011. The European Red List of Saproxylic Beetles – the status of species occurring in Britain and Ireland. The *Coleopterist* 20: 55–61.

Alexander, K.N.A. 2014a. Saproxylic Coleoptera from Calke, Hardwick and Kedleston Parks, Derbyshire: additional records including two species new to the county list and reassessment of site quality and condition. The *Coleopterist* 23: 72–7.

Alexander, K.N.A. 2014b. A review of the scarce and threatened beetles of Great Britain. Buprestidae, Cantharidae, Cleridae, Dasytidae, Drilidae, Lampyridae, Lycidae, Lymexylidae, Malachiidae, Phloiophilidae and Trogossitidae. Species Status No. 16. *Natural England Commissioned Report NECR* 134.

Alexander, K.N.A. 2016. The British status of *Silvanus bidentatus* Fabricius (Silvanidae). The *Coleopterist* 25: 135–8.

Alexander, K.N.A. 2017. A review of the status of the beetles of Great Britain – the wood-boring beetles, spider beetles, woodworm, false powder-post beetles, hide beetles and their allies – Derodontidoidea (Derodontidae) and Bostrichoidea (Dermestidae, Bostrichidae and Ptinidae) *Natural England Commissioned Reports*, No. 236.

Alexander, K.N.A. 2019. A review of the status of the beetles of Great Britain – Longhorn beetles (Cerambycidae). *Natural England Commissioned Reports* No. 272: 1–75.

Alexander, K.N.A. in prep. *Updating the Index of Ecological Continuity as Used for Site Quality Assessment for Saproxylic Beetles.*

Alexander, K.N.A. and P.J. Chandler. 2021. *Dilophus bispinosus* Lundström (Diptera, Bibionidae) taken in numbers at its only known Welsh locality and new sites in Gloucestershire, Oxfordshire and Middlesex. *Dipterists Digest (Second Series)* 28: 97–101.

Alexander, K.N.A., S. Dodd and J.S. Denton. 2014. A review of the scarce and threatened beetles of Great Britain. The darkling beetles and their allies. Aderidae, Anthicidae, Colydiidae, Melandryidae, Meloidae, Mordellidae, Mycetophagidae, Mycteridae, Oedemeridae, Pyrochroidae, Pythidae, Ripiphoridae, Salpingidae, Scraptiidae, Tenebrionidae & Tetratomidae (Tenebrionoidea less Ciidae). *Species Status No.18. Natural England Commissioned Report NECR148.*

Alexander, K.N.A., M. Smith, R. Stiven and N. Sanderson. 2003. Defining 'old growth' in the UK context. *English Nature Research Reports* 49.

Alexander, W.B. 1939. Birds. In L.F. Salzman (ed.) *The Victoria History of the County of Oxford* Volume 1, pp. 199–216. Oxford and London: Oxford University Press/University of London Institute of Historical Research [Reprinted 1970, Folkestone: Dawsons/ULIHR. NB: the faunal chapters (to p. 223) of this volume are the only parts of the Oxfordshire VCH not available online].

Alexander, W.B. 1947. *A Revised List of the Birds of Oxfordshire.* Oxford: Oxford Ornithological Society.

Allen, A.J., M.G. Telfer and W.H. Rücker. 2021. Description of *Corticaria horreum* sp. Nov. (Coleoptera: Latridiidae). *The Coleopterist* 30: 65–71.

Andrews, C.B. (ed.), 1934. *The Torrington Diaries, Containing the Tours through England and Wales of the Hon. John Byng, between the Years 1781 and 1794,* 3 vols. London: Eyre and Spottiswoode.

Andrews, S., 1975. *Eighteenth-Century Europe: the 1600s to 1815.* London: Longmans.

Anon., 1852. *A History, Gazetteer and Directory of the County of Oxford.* Peterborough: Robert Gardner. 863 pp.

Anon., 1854. [meeting of December 6 1847]. *Abstracts of the Proceedings of the Ashmolean Society for 1843–58.* 2: 175.

Anonymous, 2009. Checklist of the British & Irish Basidiomycota Update 4. https://fungi.myspecies.info/content/checklists, accessed 18 October 2022.

Aplin, O.V., 1889. *The Birds of Oxfordshire.* Oxford: Clarendon Press. https://doi.org/10.5962/bhl.title.62783

Aplin, O.V., 1903. Notes on the Ornithology of Oxfordshire (1899–1901), *Zoologist* 4 (7): 5–59.

Aplin, O.V., 1906. Notes on the Ornithology of Oxfordshire (1904), *Zoologist* 4 (10): 410–45.

Aplin, O.V., 1914a. Early breeding of the Crested Grebe, *Zoologist* 4 (18): 235.

Aplin, O.V., 1914b. Notes on the Ornithology of Oxfordshire (1913), *Zoologist* 4 (18): 401–13.

Aplin, O.V., 1915. Notes on the Ornithology of Oxfordshire (1914), *Zoologist* 4 (19): 201–12.

Appleton, D., 2004. Scarcer Coleoptera in Hampshire and Isle of Wight 1964–2001. *The Coleopterist* 13: 41–80.

Arkell, W.J., 1947. A Palaeolith from the Hanborough Terrace. *Oxoniensia* (11–12): 1–4.

Asher, J., N. Bowles, J. Haseler, G. Hawker, P. Ogden, D. Roy, R. Soulsby and M. Wilkins. 2016. *Atlas of Butterflies in Berkshire, Buckinghamshire & Oxfordshire.* Pisces Publications, Butterfly Conservation.

Attlee, H.G., 1915. Bird-notes for Oxford district in 1913 and 1914, *Zoologist* 4 (19): 109–12.

Aubrook, E.W., 1939. Coleoptera. In B.M. Hobby (ed.) *Zoology of Oxfordshire.* In L.F. Salzman (ed.), *The Victoria History of the County of Oxford.* Volume 1. Oxford University Press.

Baggs, A.P., C. Colvin, H.M. Colvin, J. Cooper, C.J. Day, N. Selwyn and A. Tomkinson. 1983. *A History of the County of Oxford: Volume 11, Wootton Hundred (Northern Part),* ed. A. Crossley. New York: Oxford University Press.

Baggs, A.P., W.J. Blair, E. Chance, C. Colvin, J. Cooper, C.J. Day, N. Selwyn and S.C. Townley. 1990. *A History of the County of Oxford: Volume 12, Wootton Hundred (South) including Woodstock,* ed. A. Crossley and C.R. Elrington. New York: Oxford University Press. http://www.british-history.ac.uk/vch/oxon/vol12

Baker, J., J. Suckling and R. Carey. 2004. Status of the adder *Vipera berus* and slow-worm *Anguis fragilis* in England. *English Nature Research Reports* No. 546. Peterborough.

Baldock, D.W. 2010. *Wasps of Surrey.* Pirbright: Surrey Wildlife Trust.

Ball, S.G. and R. Morris. 2014. A Review of the Scarce and Threatened Flies of Great Britain. Part 6 Hoverflies family Syrphidae. *Species Status* No. 9. Peterborough: Joint Nature Conservation Committee.

Ballard, A., 1908. Woodstock Manor in the thirteenth century. *Vierteljahrschrift für Social- und Wirtschaftsgeschichte* 6: 424–59.

Balmer, D.E., S. Gillings, B.J. Caffrey, R.L. Swann, I.S. Downie and R.J. Fuller. 2013. *Bird Atlas 2007–11: The Breeding and Wintering Birds of Britain and Ireland.* Thetford: British Trust for Ornithology.

Balsan, C.V. 1953. *The Glitter and the Gold. The American Duchess – in her own words.* London and New York: Harper & Heinemann. 336 pp. [reissued 2011 by Hodder & Stoughton, London].

Banbury, J., R. Edwards, E. Poskitt and T. Nutt (eds). 2010. *Woodstock and the Royal Park. Nine hundred years of history.* Oxford: Woodstock & the Royal Park 900 Year Association/Chris Andrews Publications.

Bang, P. and P. Dahlstrom. 1974. *Collins Guide to Animal Tracks and Signs: British and European Mammals and Birds.* London: Harper Collins.

Bapasola, J. 2009. *The Finest View in England. The Landscape and Gardens at Blenheim Palace.* Blenheim Palace, Woodstock.

BAS (British Arachnological Society). 2010–23. Spider and Harvestmen Recording Scheme https://srs.britishspiders.org.uk

Bat Conservation Trust online species accounts. www.bats.org.uk

Bee, L., G. Oxford, H. Smith and N. Baker. 2017. *Britain's Spiders: a field guide.* Wild Guides of Britain and Europe, Vol. 21. Princeton, NJ: Princeton University Press.

Beebee, T.J. and R.A. Griffiths. 2000. *Amphibians and Reptiles.* London: Harper Collins.

Bibby, C., N.D. Burgess and D.A. Hill. 1992. *Bird Census Techniques.* London: Academic Press.

Bintley, M.D.J. and M.G. Shapland (eds). 2013. Trees and timber in the Anglo-Saxon world. In *Medieval History and Archaeology,* ed. J. Blair and H. Hamerow.

Oxford: Oxford University Press, 336 pp. https://doi.org/10.1093/acprof:oso/9780199680795.001.0001

Blackwell, E. 2000. Fungi. In P.T. Harding and T. Wall (eds) *Moccas, an English Deer Park*, pp. 137–143. Peterborough: English Nature.

Blair, J. 2010. The Anglo-Saxon minsters. In *An Historical Atlas of Oxfordshire*, ed. K. Tiller and G. Darkes, Oxfordshire Record Society, 67: 26–7.

Blincow, J. 2023. Yardley Chase Training Area, Northamptonshire: a hidden gem. *British Wildlife* 34 (4): 260–9.

Bond, C.J. 1966. *Changes in Land Use and Settlement in Woodstock and Blenheim, Oxfordshire*. BA dissertation, University of Birmingham.

Bond, C.J. 1981. Woodstock Park under the Plantagenet kings – exploitation of wood and timber in medieval deer park. *Arboricultural Journal* 5: 201–213. https://doi.org/10.1080/03071375.1981.9746522

Bond, J. 1987. The park before the palace; the sixteenth and seventeenth centuries. In J. Bond and K. Tiller (eds) *Blenheim – Landscape for a Palace*, Chapter 5, pp. 55–66. Gloucester: Alan Sutton Publishing.

Bond, J. and K. Tiller (eds) 1987. *Blenheim – Landscape for a Palace*. Gloucester: Alan Sutton Publishing.

Booth, P. 1999. Ralegh Radford and the Roman villa at Ditchley: a review. *Oxoniensia* 64: 39–49.

Booth, P. 2011. Romano-British trackways in the Upper Thames Valley. *Oxoniensia* 76: 1–13.

Bosanquet, S.D.S., A.M. Ainsworth, S.P. Cooch, D.R. Genney and T.C. Wilkins. 2018. *Guidelines for the Selection of Biological SSSIs. Part 2: Detailed guidelines for habitats and species groups. Chapter 14 Non-lichenised fungi*. Peterborough: Joint Nature Conservation Committee.

British Mycological Society. 2022. English names for fungi. April 2022. https://www.britmycolsoc.org.uk/resources/english-names, accessed 15 October 2022.

Brock, P.D. 2011. *A Photographic Guide to Insects of the New Forest and Surrounding Area*. Newbury: Pisces Publications.

Brock, P.D. and A.J. Allen. 2022. An overview of Coleoptera of the New Forest. *The Coleopterist* 31: 66–91.

Brown, D., and T. Williamson. 2016. *Lancelot Brown and the Capability Men: Landscape revolution in eighteenth-century England*. London: Reaktion Books.

Brucker, J.W. and J.M. Campbell. 1975. *The Birds of Blenheim Park*. Museum Service Publication No. 7. Oxford: Oxfordshire County Council.

Brucker, J.W. and J.M. Campbell. 1987. *The Birds of Blenheim Palace* [cover], … *Park* [title page]. Museum Service Publication No. 7, Oxford: Oxfordshire County Council [revised from 1975 edition].

Brucker, J.W., A.G. Gosler and A.R. Heryet. 1992. *Birds of Oxfordshire*. Pisces Publications, Newbury.

Bryden, H.A. 1904. *Nature & Sport in Britain*. London: Grant Richards. Xii.

Buesching, C.D., J.R. Clarke, S.A. Ellwood, C. King, C. Newman and D.W. Macdonald. 2010. The mammals of Wytham Woods. In P. A. E. Savill, , C.M. Perrins, K. Kirby and N. Fisher (eds), *Wytham Woods: Oxford's ecological laboratory*, pp. 173–96. Oxford: Oxford University Press. 282 pp. https://doi.org/10.1093/acprof:osobl/9780199605187.003.0010

Burns, F., M.A. Eaton, D.E. Balmer, A. Banks, R. Caldow, J.L. Donelan, A. Douse, C. Duigan, S. Foster, T. Frost, P.V. Grice, C. Hall, H.J. Hanmer, S.J. Harris, I. Johnstone, P. Lindley, N. McCulloch, D.G. Noble, K. Risely, R.A. Robinson and S. Wotton. 2020. The *State of the UK's Birds* 2020. Sandy: RSPB, BTO, WWT, DAERA, JNCC, NatureScot, NE & NRW.

Campbell, J. 1987. Some aspects of the natural history of Blenheim Park. In J. Bond and K. Tiller (eds) *Blenheim – Landscape for a Palace*, Chapter 12, pp. 151–62. Gloucester: Alan Sutton Publishing.

Cannon, P.F., B. Aguirre-Hudson, M.C. Aime, A.M. Ainsworth, M.I. Bidartondo, E. Gaya, D. Hawksworth, P. Kirk, I.J. Leitch and R. Lücking. 2018. Definition and diversity. In K.J. Willis (ed.), *State of the World's Fungi* 2018, pp. 4–11. Royal Botanic Gardens, Kew: Kew Publishing.

Carpenter, B. n.d. [2020]. Oxfordshire woodland bird study 2017. In I. Lewington, G. Coleman and J. Uren (eds) *Birds of Oxfordshire* 2017, pp. 93–113. [Oxford]: Oxford Ornithological Society.

Carreck, N.L. 2008. Are honey bees (*Apis mellifera* L.) native to the British Isles? *Journal of Apicultural Research and Bee World* 47: 318–22. https://doi.org/10.1080/00218839.2008.11101482

Cartwright, K.St.G. 1951. *Polyporus quercinus* on *Fagus sylvatica*. *Transactions of the British Mycological Society* 34 (4): 604–6. https://doi.org/10.1016/S0007-1536(51)80045-7

Chandler, P.J. 2010. Associations with fungi and Mycetozoa. In P.J. Chandler (ed) *A Dipterist's Handbook*, second edition, pp. 417–41. Orpington: The Amateur Entomologists' Society. 525 pp.

Chandler, P.J. 2015. Diptera recording at Bushy Park, Middlesex. *Dipterists Digest (Second Series)* 23: 69–110.

Chandler, P.J. 2021. The two-winged flies (Diptera) of Windsor Forest and Great Park. *Dipterists Digest (Second Series)* 28: Supplement: 1–126; Appendix: 127–232.

Chandler, P.J. 2023. An update of the 1998 Checklist of Diptera of the British Isles [updated 20 March 2023]. *Dipterists Forum*. Available from https://www.dipterists.org.uk

Cheffings, C.M., L. Farell, T.D. Dines, R.A. Jones, S.J. Leach, D.R. McKean, C.D. Pearman, C.D. Preston, F.J. Rumsey and I. Taylor. 2005. *The Vascular Plant Red Data List for Great Britain Species Status* 7: 1–116. Peterborough: Joint Nature Conservation Committee.

Cheke, A.S. 2023. The Changing Functions and Perceptions of a Deer Park – the history of deer in the Woodstock Royal Park, later Blenheim Park, from emparking to the present. *Fritillary* 11: 22–47.

Churchill, W.S. 1906. *Lord Randolph Churchill*. 2 vols. London and New York: Macmillan.

Clark, R. 2017. Saproxylic Coleoptera of Blenheim Palace. *The Coleopterist* 26: 53–64.

Cobham, R. and P. Hutton. 1987. Present management and future restoration. In J. Bond and K. Tiller (eds) *Blenheim – Landscape for a Palace*, Chapter 11, pp. 133–50. Gloucester: Alan Sutton Publishing.

Combe, W. 1794. *An History of the River Thames*. An history of the principal rivers of Great Britain, Vol. 1. John and Josiah Boydell.

Cooke, A. 2019. *Muntjac and Water Deer. Natural history, environmental impact and management*. Exeter: Pelagic Publishing. https://doi.org/10.53061/URGZ9475

Cooke, A.S. 2006. *Monitoring Muntjac deer Muntiacus reevesi at Monks Wood National Nature Reserve*. Peterborough: English Nature.

Cooper, N. 1999. *Houses of the Gentry 1480–1680*. London: Yale University Press.

6

Cooter, J. 2022a. There is no 2022b. A tale of two staphs (Col., Staphylinidae: Aleocharinae); correcting the record. *The Coleopterist* 31: 21–26.

Copeland, T. 1988 The North Oxfordshire Grim's Ditch: a fieldwork survey. *Oxoniensia* 53: 277–92.

Coppins, A.M. and B.J. Coppins. 2002. *Indices of Ecological Continuity for Woodland Epiphytic Lichen Habitats in the British Isles.* London: British Lichen Society.

Cornish, C.J. 1895. *Wild England of Today and the Wild Life in it.* London: Thomas Nelson & Sons.

Cornish, C.J. 1902. *The Naturalist on the Thames.* London: Seeley & Co.

Crawley, M.J. 2005. *The Flora of Berkshire.* Harpenden: Brambleby Books.

Creighton, O.H. 2013. *Designs upon the Land. Elite landscapes of the Middle Ages.* Woodbridge: The Boydell Press.

Crombie, J.M. 2008. On the lichens of Dillenius's 'Historia Muscorum,' as illustrated by his Herbarium. *Botanical Journal of the Linnean Society* 17: 553–81. https://doi.org/10.1111/j.1095-8339.1880.tb01244.x

Dalton, C. 2009. Sir John Vanbrugh and the Vitruvian Landscape. *Garden History Journal* 37(10): 3–38.

Dalton, C. 2012. *Sir John Vanbrugh and the Vitruvian Landscape.* Abingdon: Routledge.

Daniel, W.B. 1807. *Rural Sports.* 3 vols. London: Longman, Hurst, Rees & Orme.

DEFRA, 2018. National Statistics Release: Emissions of air pollutants in the UK, 1970 to 2016.

Dennis, P., 2016. *Grazing Assessor's Report on Status of Cattle Grazing and Associated Habitat Monitoring across Epping Forest.* Bangor: Institute of Biological, Environmental and Rural Sciences (IBERS), Aberystwyth University.

Denton, J. 1999. Is *Scydmaenus rufus* Müller & Kunze (Scydmaenidae) really Vulnerable? *The Coleopterist* 8: 89.

Denton, J. and J.M. Campbell. in prep. *Beetles of Oxfordshire.*

Denwood, P., 2017. The Dark Bee *Apis mellifera mellifera* in the United Kingdom. Bee Improvement and Bee Breeders Association website, accessed January 2023. https://bibba.com

Donisthorpe, H.St.J.K. 1927. *The Guests of British Ants.* London: George Routledge & Sons Ltd.

Duchesne, T., E. Graitson, O. Lourdais, S. Ursenbacher and M. Dufrêne. 2022. Fine-scale vegetation complexity and habitat structure influence predation pressure on a declining snake. *Journal of Zoology.* Corpus ID: 2511754875 https://doi.org/10.1111/jzo.13007

Duff, A.G. 2018. *Checklist of Beetles of the British Isles,* third edition. Iver: Pemberley Books (Publishing).

Dunn, J.C., A.J. Morris and P.V. Grice. 2017. Post-fledging habitat selection in a rapidly declining farmland bird, the European turtle dove Streptopelia turtur. *Bird Conservation International* 27 (1): 45–57. https://doi.org/10.1017/S0959270916000022

Dury, G. 1958. Tests of a general theory of misfit streams. *Transactions and Papers,* Institute of British Geographers, 25: 105–18. https://doi.org/10.2307/621181

Easton, A.M. 1947. The Coleoptera of flood-refuse; a comparison of samples from Surrey, and Oxfordshire. *Entomologist's Monthly Magazine* 83: 113–15.

Elliott, H.F.I. and W.B. Alexander. n.d. [1938]. Systematic list of the birds of Blenheim Park, Woodstock, Oxford. *Report of the Oxford Ornithological Society for 1937:* 45–48.

Ellis, C.J. 2015. Ancient woodland indicators signal the climate change risk for dispersal-limited species. *Ecological Indicators* 53: 106–14. https://doi.org/10.1016/j.ecolind.2015.01.028

Else, G.R. and M. Edwards. 2018. *Handbook of the Bees of the British Isles.* London: The Ray Society.

Emery, F.V. 1974. *The Making of the English Landscape: The Oxfordshire landscape.* London: Hodder and Stoughton.

Erskine, S., H.J. Killick, C.R.L. Lambrick and E.M. Lee. 2018. *Oxfordshire's Threatened Plants.* Newbury: Pisces.

Evans, T. and S. Jennings. 2022. The arrival of Ring-necked Parakeets in Oxford – a request for information. *OOS Bulletin* 569 (March 2022): 4.

Falk, S.J. 1991. *Review of the Scarce and Threatened Flies of Great Britain.* (Research & Survey in Nature Conservation). Joint Nature Conservation Committee (JNCC).

Falk, S.J., and A.C. Pont. 2017. *A Provisional Assessment of the Status of Calypterate flies in the UK.* Natural England. Commissioned Reports, No. 234.

Farjon, A. 2017. *Ancient Oaks in the English Landscape.* Royal Botanical Gardens, Kew: Kew Publishing.

Farjon, A. 2022. *Ancient Oaks in the English Landscape,* second edition. Royal Botanic Gardens, Kew: Kew Publishing.

Fletcher, J. 2011. *Gardens of Earthly Delight. The history of deer parks.* Oxford: Windgather Press. 284 pp. https://doi.org/10.2307/j.ctv13gvfzg

Foster, J., D. Driver, R. Ward and J. Wilkinson. 2021. *IUCN Red List Assessment of Amphibians and Reptiles at Great Britain and Country Scale.* Report to Natural England, July 2021. Bournemouth: Amphibian and Reptile Conservation.

Foundation for Common Land. http://www.foundationforcommonland.org.uk/rights-of-common

Fowler, M. 1989. *Blenheim: Biography of a palace.* London: Viking. 272 pp.

Fowles, A.P. 2023. updated continuously. The Saproxylic Quality Index: evaluated sites ranked by SQI. https://khepri.uk, accessed 8 June 2023.

Fowles, A.P., K.N.A. Alexander and R.S. Key. 1999. The Saproxylic Quality Index: evaluating wooded habitats for the conservation of the dead-wood Coleoptera. *The Coleopterist* 8: 121–141.

Fraiture, A. and P. Otto. 2015. *Distribution, Ecology and Status of 51 Macromycetes in Europe.* Meise: Meise Botanic Garden.

Freschet, G. T., J. T. Weedon, R. Aerts, J. R. van Hal and J.H.C. Cornelissen. 2012. Interspecific differences in wood decay rates: Insights from a new short-term method to study long-term wood decomposition. *Journal of Ecology* 100 (1): 161–70. https://doi.org/10.1111/j.1365-2745.2011.01896.x

Fuller, R.J., D.G. Noble, K.W. Smith and D. Vanhinsberg. 2005. Recent declines in populations of woodland birds in Britain: a review of possible causes. *British Birds* 98: 116–43.

Gardes, M. and T.D. Bruns. 1993. ITS primers with enhanced specificity for basidiomycetes – application to the identification of mycorrhizae and rusts. *Molecular Ecology* 2: 113–18. https://doi.org/10.1111/j.1365-294X.1993.tb00005.x

Gardes, M. and T.D. Bruns. 1997. Community structure of ectomycorrhizal fungi in a Pinus muricata forest: above- and below-ground views. *Canadian Journal of Botany* 74 (10): 1572–83. https://doi.org/10.1139/b96-190

GBIF. updated continuously. https://gbif.org/species/4447680, accessed 22 April 2023.

Gill, R.M.A. and R.J. Fuller. 2007. The effects of deer browsing on woodland structure and songbirds in lowland Britain. *Ibis* 149 (Suppl. 2): 119–27. https://doi.org/10.1111/j.1474-919X.2007.00731.x

Gosler, A.G. 1990. The birds of Wytham – an historical survey. *Fritillary* 1: 29–74.

Green, D. 1950. *Blenheim Palace, Oxfordshire.* Woodstock: Blenheim Estate Office.

Green, D. 1951. *Blenheim Palace.* London: Country Life Limited.

Green, E.E. 2010. The importance of open-grown trees; from acorn to ancient. *British Wildlife* 21(3): 334–8.

Green, H., F. Llewellyn, R. Pumphrey and H. Milner (eds). 2022. *Shrawley Wood History and Natural History, a nationally important Small-leaved Lime wood in Worcestershire.* Malvern: Greyhound Self Publishing.

Green, J. 1987. Blenheim: the palace and gardens under Vanbrugh, Hawkmoor and Wise. In J. Bond and K. Tiller (eds) *Blenheim – Landscape for a Palace,* Chapter 6, pp. 67–79. Gloucester: Alan Sutton Publishing.

Green, T. 2010. The importance of open-grown trees. From acorn to ancient. *British Wildlife* 21: 334–38.

Grigson, C. 2016. *Menagerie: The history of exotic animals.* Oxford: Oxford University Press.

Hall, A., R.A. Sage and J.R. Madden. 2021. The effects of released pheasants on invertebrate populations in and around woodland release sites. *Ecology and Evolution,* 11, 13559–69. https://doi.org/10.1002/ece3.8083

Hallé, F. 2002. *In Praise of Plants.* Portland, OR: Timber Press.

Hamilton, J.P. 1860. *Reminiscences of an Old Sportsman,* 2 vols. London: Longman, Green, Longman & Roberts.

Hanss, J.-M. and P.-A. Moreau. 2020 [2017]. Une révision des amanites 'vaginées' (Amanita sect. Vaginatae) en Europe, 1re partie: quelques amanites argentées. *Bulletin de la Société mycologique de France* 133 (1–2): 67–141.

Hardarker, T. 1992. Neolithic communities of the Evenlode Valley. *Lithics* 13: 45–51.

Hardaker, T. 2004. Present-day Lower Palaeolithic land surfaces in Britain: two examples from the Upper Thames. *Lithics* 25: 22–38.

Harden, D.B. 1937. Excavations on Grim's Dyke, North Oxfordshire. *Oxoniensia* 2: 74–92.

Harden, D.B. 1947. A ring-ditch at Long Hanborough, Oxon. *Oxoniensia* 11–12: 175.

Harding, P.T. 1973–75. A preliminary list of the fauna of Staverton Park, Suffolk. *Suffolk Natural History* 16(4): 232–8; 16(5): 287–304 (with R.C. Welch); 16(6): 399–419.

Harding, P.T. and F. Rose. 1986. *Pasture Woodlands in Lowland Britain: A review of their importance for wildlife conservation.* Huntingdon: Institute of Terrestrial Ecology.

Harding, P.T. and K.N.A. Alexander. 1993. The Saproxylic invertebrates of historic parklands: progress and problems:. In K.J. Kirby and C.M. Drake (eds) *Dead Wood Matters: The ecology and conservation of invertebrates in Britain,* pp. 58–73. Proceedings of a British Ecology Society Meeting held at Dunham Massey Park on 24 April 1992. English Nature Science, No. 7.

Harding, P.T. and K.N.A. Alexander. 1994. The use of Saproxylic invertebrates in the selection and evaluation of areas of relic forest in pasture-woodlands. *British Journal of Entomology and Natural History* 7 (Suppl. 1): 21–26.

Harding, P.T. and T. Wall (eds). 2000. *Moccas: An English deer park.* English Nature (Natural England), Peterborough.

Hardy, T.D. 1790. *The Deeds of the Kings of the English by William of Malmsbury, Monk.* London: Thomas Duftus Hardy, London Sumptibus Societatis.

Hardy, T.D. 1840. *William of Malmesbury.* Gesta Regum Anglorum Vol. 2: 959–1390.

Harris, S.J., D. Massimino, D.E. Balmer, M.A. Eaton, D.G. Noble, J.W. Pearce-Higgins, P. Woodcock, and S. Gillings. 2020. *The Breeding Bird Survey 2019.* BTO Research Report 726. Thetford: British Trust for Ornithology.

Harrison, T. 2012. A fourth British record for *Hylis cariniceps* (Reitter) (Eucnemidae). *The Coleopterist* 21: 49.

Harvey, D.J., H. Harvey, M.C. Larsson, G.P. Svensson, E. Hedenström, P. Finch and A.C. Gange. 2017. Making the invisible visible: determining an accurate National distribution of *Elater ferrugineus* in the United Kingdom using pheromones. *Insect Conservation and Diversity* 10: 283–93. https://doi.org/10.1111/icad.12227

Harvey, D.J., J. Vuts., A. Hooper, P. Finch, C.M. Woodcock, J.C. Caulfield, M. Kadej, A. Smolis, D.M. Withall, S. Henshall, J.A. Pickett, A.C. Gange and M.A. Birkett. 2018. Environmentally vulnerable noble chafers exhibit unusual pheromone-mediated behaviour. *PLoS ONE* 13. https://doi.org/10.1371/journal.pone.0206526

Harvey, J. 1990. *Mediaeval Gardens.* London: B.T. Batsford Ltd. 199 pp.

Harvey, P., M. Davidson, I. Dawson, A. Fowles, G. Hitchcock, P. Lee, P. Merrett, A. Russell-Smith and H. Smith. 2017. *A Review of the Scarce and Threatened Spiders (Araneae) of Great Britain.* Species Status No. 22. British Arachnological Society, NRW Evidence Report No. 11. National Resources Wales, Cardiff.

Hawkins, R.D. 2003. *Shieldbugs of Surrey.* Pirbright: Surrey Wildlife Trust.

Henderson, P. 2005. *The Tudor House and Garden Architecture and Landscape in the Sixteenth and Early Seventeenth Centuries.* New Haven, CT, and London: Paul Mellon Centre, Yale University Press. 288 pp.

Henry, C.S., S.J. Brooks, P. Duelli, and J.B. Johnson. 2002. Discovering the true Chrysoperla carnea (Insecta: Neuroptera: Chrysopidae) using song analysis, morphology, and ecology. *Annals of the Entomological Society of America* 95: 172–91. https://doi.org/10.1603/0013-8746(2002)095[0172:DTTCCI]2.0.CO;2

Henwood, B. and P. Sterling. 2020. *Field Guide to the Caterpillars of Great Britain and Ireland.* London: Bloomsbury.

Hernández-Brito, D., M. Carrete, C. Ibáñez, J. Juste and J.L. Tella. 2018. Nest-site competition and killing by invasive parakeets cause the decline of a threatened bat population. Royal Society Open Science. https://doi.org/10.1098/rsos.172477

Heward, C.J., A.N. Hoodless, G.J. Conway, N.J. Aebischer, S. Gillings and R.J. Fuller 2015. Current status and recent trend of the Eurasian Woodcock Scolopax rusticola as a breeding bird in Britain. *Bird Study* 62 (4): 535–51. https://doi.org/10.1080/00063657.2015.1092497

Hill, M.O., C.D. Preston and D.B. Roy. 2004. *PLAN-TATT: attributes of British and Irish plants.* Cambridge: Biological Records Centre (NERC).

Hill, M.O., T.H. Blackstock, D.G. Long and G.P. Rothero. 2008. *A Checklist and Census Catalogue of British and Irish Bryophytes.* London: British Bryological Society.

Historic England. https://historicengland.org.uk /listing/the-list/map-search?clearresults=True

Historic Landscape Management Ltd. 2014. *Blenheim Palace and Park, Oxfordshire. Parkland management plan.* Lyng: HLM Ltd.

Historic Landscape Management Ltd. 2017. *Blenheim Palace World Heritage Site. World Heritage and National Heritage management plan review.* Lyng: HLM. 82pp + apps.

Holloway, S, 1996. *The Historical Atlas of Breeding Birds in Britain and Ireland 1875–1900.* London: T & AD Poyser.

Hoskins, W.G. 1977. *The Making of the English Landscape.* London: Hodder and Stoughton.

Hyman, P.S. (revised by M.S. Parsons). 1992. *A Review of the Scarce and Threatened Coleoptera of Great Britain. Part 1.* UK Nature Conservation: 3. Peterborough: Joint Nature Conservation Committee.

Hyman, P.S. (revised by M.S. Parsons). 1994. *A Review of the Scarce and Threatened Coleoptera of Great Britain. Part 2.* UK Nature Conservation: 12. Peterborough: Joint Nature Conservation Committee.

Index/Species Fungorum. 2022. http://www.index fungorum.org/names/Names.asp, accessed 27 December 2021.

Inns, H. 2011. *Britain's Reptiles and Amphibians* (Vol. 23). Princeton, NJ: Princeton University Press. https:// doi.org/10.1515/9780691206813

James, P.W, D.L. Hawksworth and F. Rose. 1977. Lichen communities in the British Isles: a preliminary conspectus. In M.R.D. Seaward (ed.) *Lichen Ecology,* pp. 295–413. London: Academic Press.

Jourdain, F.C.R. 1926. The birds of the Oxford district. In J.J. Walker (ed.) *The Natural History of the Oxford District,* pp. 128–60. Oxford: Oxford University Press.

Jørgensen, D., 2010. The roots of the English royal forest. In *Anglo-Norman Studies XXXII – Proceedings of The Battle Conference 2009,* ed. C.P. Lewis, pp. 114–28.

Krahl-Urban, J. 1959. *Die Eichen: forstliche Monographie der Traubeneiche und der Stieleiche.* Hamburg and Berlin: Paul Parey.

Kautmanova, I., A.M. Ainsworth, I. Olariaga, A. Mešić and J. Jordal, 2021. *Buglossoporus quercinus* The IUCN Red List of Threatened Species 2021: e.T70401726A204094456. https://dx.doi.org/10.2305 /IUCN.UK.2021-2.RLTS.T70401726A204094456.en, accessed 12 October 2022.

Keczyński, A., ed., 2017. *The Forests of the Strict Reserve of Białowieża National Park.* Białowieża: Białowieski Park Narodowy.

Kirby, K. and R.C. Thomas. 2000. Changes in the ground flora in Wytham Woods, southern England, from 1974 to 1991 – implications for nature conservation. *Journal of Vegetation Science* 11 (6): 871–80. https://doi .org/10.2307/3236557

Kirby, K. 2001. The impact of deer on the ground flora of British broadleaved woodland. *Forestry (An International Journal of Forest Research)* 74 (3): 219–29. https:// doi.org/10.1093/forestry/74.3.219

Kirby, K. 2020. *Woodland Flowers, Colourful Past, Uncertain Future.* London: Bloomsbury Wildlife.

Kõljalg, U., R.H. Nilsson, K. Abarenkov, L. Tedersoo, A.F.S. Taylor, M. Bahram, S.T. Bates, T.D. Bruns, J. Bengtsson-Palme, T.M. Callaghan, B. Douglas, T. Drenkhan, U. Eberhardt, M. Dueñas, T. Grebenc, G.W. Griffith, M. Hartmann, P.M. Kirk, P. Kohout, E. Larsson, B.D. Lindahl, R. Lücking, M.P. Martín, P.B. Matheny, N.H. Nguyen, T. Niskanen, J. Oja, K.G. Peay, U. Peintner, M. Peterson, K. Põldmaa, L. Saag, I. Saar, A. Schüßler, J.A. Scott, C. Senés, M.E. Smith, A. Suija, D.L. Taylor, M.T. Telleria, M. Weiss and K.-H. Larsson. 2013. Towards a unified paradigm for sequence-based identification of fungi. *Molecular Ecology* 22: 5271–7. https://doi.org/10.1111/mec.12481

Krauze-Gryz, D. and J. Gryz. 2015. a review of the diet of the red squirrel in different habitats. Pp. 51–65 in Shuttleworth C.M., Lurz, P.W.W. and Hayward M.W. (eds) *Red squirrels: Ecology, Conservation & Management in Europe.* European Squirrel Initiative, Stoneleigh Park, Warwickshire.

Læssøe, T. and J.H. Petersen. 2019. *Fungi of temperate Europe.* Princeton University Press, Princeton NJ.

Lambrick, G. 2013. Prehistoric Oxford. *Oxoniensia* 78: 1–48.

Lane, S.A. 2017. A review of the status of the beetles of Great Britain – The clown beetles and false clown beetles – Histeridae and Sphaeritidae. *Natural England Commissioned Reports Number 235. JNCC.*

Lane, S.A., A.L. Drewitt and A.J. Allen. 2019. IUCN threat status and British rarity status for British Coleoptera: Part 1. *The Coleopterist* 28: 71–100.

Lane, S.A., C.B.H. Lucas and A.L. Whiffin. 2020. *The Histeridae, Sphaeritidae and Silphidae of Britain and Ireland.* Shrewsbury: FSC Publications.

Langmaid, J.R., S.M. Palmer and M.R. Young. 2018. *A Field Guide to the Smaller Moths of Great Britain and Ireland.* Reading: The British Entomological and Natural History Society.

Langton, J. 2014. Forest law in the landscape: not the clearing of the woods, but the running of the deer? In *Deer and People,* ed. K. Baker, R. Carden and R. Madgwick, pp. 218–30. Macclesfield: Windgather Press. https://doi.org/10.2307/j.ctv13gvgms.22

Lasdun, S. 1991. *The English Park Royal, Private and Public.* London: Andre Deutsch.

Lawson, B., R.A. Robinson, A. Neimanis, B. Lawson (corresp. author), E.O. Agren, M. Bennett, J. Chantrey, A.A. Cunningham, I. Hammes, K. Handeland, L.A. Hughes, M. Isomursu, S.K. John, J.K. Kirkwood, B. Lawson, A. Neimanis, K.M. Peck, T.W. Pennycott, R.A. Robinson, V.R. Simpson, M.P. Toms and K.M. Tyler. 2011. Evidence of spread of the emerging infectious disease, Finch Trichomonosis, by migrating birds. *EcoHealth* 8: 143–53. https://doi.org/10.1007/ s10393-011-0696-8

Liddiard, R. 2003. The deer parks of Domesday Book. *Landscapes* 4: 4–23. https://doi.org/10.1179/lan.2003.4.1.4

Liddiard, R. (ed.). 2007. *The Medieval Park. New Perspectives – Introduction.* Macclesfield: Windgather Press.

Linde, van der S., L.M. Suz, C.D.L. Orme, F. Cox, H. Andreae, E. Asi, B. Atkinson, S. Benham, C. Carroll, N. Cools, B. De Vos, H.-P. Dietrich, J. Eichorn, J. Gehrmann, T. Grebenc, H.S. Gweon, K. Hansen, F. Jacob, F. Kristöfel, P. Lech, M. Manninger, J. Martin, H. Meesenburg, P. Merilä, M. Nicolas, P. Pavlenda, P. Rautio, M. Schaub, H.-W. Schröck, W. Seidling, V. Šrámek, A. Thimonier, I.M. Thomsen, H. Titeux, E. Vanguelova, A. Verstraeten, L. Vesterdal, P. Waldner,

S. Wijk, Y. Zhang, D. Žlindra and M.I. Bidartondo. 2018. Environment and host as large-scale controls of ectomycorrhizal fungi. *Nature* 558: 243–8. https://doi.org/10.1038/s41586-018-0189-9

Liu, W., D. Schaefer, L. Qiao and X. Liu. 2013. What controls the variability of wood-decay rates? *Forest Ecology and Management* 310: 623–31. https://doi.org/10.1016/j.foreco.2013.09.013

Lockwood, W.B. 1984. *The Oxford Book of British Bird Names*. Oxford: Oxford University Press.

Lormée, H., C. Barbraud, W. Peach, C. Carboneras, J.D. Lebreton, L. Moreno-Zarate, L. Bacon and C. Eraud. 2020. Assessing the sustainability of harvest of the European Turtle-dove along the European western flyway. *Bird Conservation International* 30: 506–21. https://doi.org/10.1017/S0959270919000479

Losito, L. 2018. Reptiles and amphibians. In I. Wright and J. Wright (eds) *Shotover, the Life of an Oxfordshire Hill*. Newbury: Pisces Publications.

Lovegrove, R. 2007. *Silent Fields. The long decline of a nation's wildlife*. Oxford: Oxford University Press.

Luff, M.L. and M. Towns. 2018. Trunk pitfall traps for catching arboreal beetles. *The Coleopterist* 27: 67–9.

Macky, J. 1722. *A Journey through England*, Vol. II. London: J. Pemberton.

Mammal Society. Online species hub and species data sheets. www.mammal.org.uk/species-hub

Marchant, J.R.V. and W. Watkins. 1897. *Wild Birds Protection Acts 1880–1896*. London: R.H. Porter.

Marren, P. 1997. The devil's bolete, *Boletus satanas* Lenz. Unpublished Plantlife 'Back from the Brink' report No. 88. London: National History Museum.

Marren, P. 2000. Surveying the royal and devil's boletes. *Field Mycology* 1(3): 94–8. https://doi.org/10.1016/S1468-1641(10)60049-7

Marren, P. 2012. *Mushrooms*. Milton on Stour: British Wildlife Publishing.

Marshall, E. 1873. *The Early History of Woodstock Manor*. Oxford and London: James Parker & Co.

Marshall, S.A. 2018. *Beetles: The natural history and diversity of Coleoptera*. Buffalo, NY: Firefly Books.

Martin, B.P. 1987. *The Great Shoots. Britain's premier sporting estates.*, Newton Abbot: David & Charles.

Mavor, W.F. 1797. *A New Description of Blenheim, the Seat of the Duke of Marlborough…* fourth edition, enlarged. London: Cadell & Davies.

Mavor, W.F. 1820. *A New Description of Blenheim, the Seat of the Duke of Marlborough …* 11th edition, improved. Oxford: Henry Slatter.

Mavor, W.F. 1836. *A Description of Blenheim, the Seat of his Grace the Duke of Marlborough: Containing a full and accurate account of the painting, tapestry, and furniture: a picturesque tour of the gardens and park: a general delineation of the China gallery, private gardens, &c.: to which are also added: an itinerary: an account of the Roman villa, near Northleigh, &c. &c.: with a preliminary essay on landscape gardening*. Oxford: Henry Slatter.

McLean, T. 1989. *Medieval Gardens*. London: Barrie and Jenkins.

Moller, T.H. 2019. Estimating the age of ancient oaks. *Landscapes* 19 (2): 91–110. https://doi.org/10.1080/14662035.2019.1684737

Morris, F.O. 1850–1857. *A History of British Birds*, 6 vols. London: Groombridge & Sons. https://doi.org/10.5962/bhl.title.29060

Mottram, N and J. Kerans. 2014. *A Management Plan for High Park, the Blenheim Estate*. Eynsham: The Wychwood Project with Thames Valley Environmental Records Centre.

MPR. 2017. *Management Plan Review – Blenheim Palace, Blenheim Palace World Heritage Site*. Norfolk: Historic Landscape Management Ltd.

Mumby, F.A. 1909. *The Girlhood of Queen Elizabeth. A narrative in contemporary letters*. London: Constable & Co.

Muñoz, I., D. .Henriques, L. Jara, J.S. Johnston, J. Chávez-Galarza, P. De la Rúa and M.A. Pinto. 2017. SNPs selected by information content outperform randomly selected microsatellite loci for delineating genetic identification and introgression in the endangered dark European honey bee (*Apis mellifera mellifera*). *Molecular Ecology Resources* 17: 783–95. https://doi.org/10.1111/1755-0998.12637

Murton, R.K., N.J. Westwood and A.J. Isaacson. 1974. A study of Wood-Pigeon shooting: the exploitation of a natural animal population. *Journal of Applied Ecology* 11: 61–81. https://doi.org/10.2307/2402005

National Biodiversity Network Atlas Partnership. 2017. https://nbnatlas.org/

Newton, A.C. (ed.). 2010. *Biodiversity in the New Forest*. Newbury: Pisces Publications.

Nicholson, E.M. 1929. Report on the 'British Birds' census of heronries 1928. *British Birds* 22: 270-323, 334–72.

Nieto, A. and K.N.A. Alexander. 2010. *European Red List of Saproxylic Beetles*. Luxembourg: Publications Office of the European Union.

Notton, D.G. and H. Norman. 2017. Hawk's-beard Nomad Bee, *Nomada facilis*, new to Britain (Hymenoptera: Apidae). *British Journal of Entomology and Natural History* 30: 201–14.

Oberle, B., K. Ogle, A. E. Zanne and C. W. Woodall. 2018. When a tree falls: controls on wood decay predict standing dead tree fall and new risks in changing forests. *PLoS ONE* 13 (5): e0196712. https://doi.org/10.1371/journal.pone.0196712

Ockendon, N., C.M. Hewson, A. Johnston and P.W. Atkinson. 2012. Declines in British-breeding populations of Afro-Palaearctic migrant birds are linked to bioclimatic wintering zone in Africa, possibly via constraints on arrival time advancement, *Bird Study* 59 (2): 111–25. https://doi.org/10.1080/00063657.2011.645798

Overall, R. 1989. Wytham Woods, Oxford, 1971–1987. *OOS Report [for]1988*: 28–30.

Owen, J.A. 1992. Catching beetles with a simple flight-interception trap. *The Coleopterist* 1: 22–6.

Owen, J.A. 1995. A pitfall trap for repetitive sampling of hypogean arthropod faunas. *Entomologist's Record and Journal of Variation* 107: 225–8.

Owen, J.A. 2000. Beetles occurring underground at the roots of old trees. *Entomologist's Gazette* 51: 239–56.

Page, C.N. 1997. *The Ferns of Britain and Ireland*. Cambridge: Cambridge University Press.

Page, W. (ed.). 1907. *A History of the County of Oxford: Volume 2*, London: Victoria County History.

Parker, M. 2011. *Syrphus nitidifrons* Becker (Diptera, Syrphidae) new to Great Britain. *Dipterists Digest (Second Series)* 17(2010): 145–6.

Parminter, J. (2002). *Coarse wood debris decomposition – principles, rates and models*. (https://www.for.gov.bc.ca/hre/pubs/docs/nivma.pdf) Presented to Northern Interior Vegetation Management Association (NIVMA) and Northern Silviculture Committee (NSC) Winter Workshop: Optimizing wildlife trees and coarse woody debris retention at the stand and landscape level. 22–4 January 2002. Prince George, BC.

Parsons, A.J. 2008. *Hylis cariniceps* (Reitter) (Eucnemidae) in Somerset. *The Coleopterist* 17: 82.

Paulson, R. 1929 Lichens of the Oxford foray. *Transactions of the British Mycological Society* 14: 183–5 https://doi.org/10.1016/S0007-1536(29)80002-2

Perrins, C.M., and A. Gosler. 2010. Birds. Savill P. A. E., Perrins, C.M., Kirby, K. and Fisher, N. (eds), *Wytham Woods: Oxford's ecological laboratory*, pp. 145–71. Oxford: Oxford University Press. 282 pp. https://doi.org/10.1093/acprof:osobl/9780199605187.003.0009

Perrins, C.M. and R. Overall. 2001. Effect of increasing numbers of deer on bird populations in Wytham Woods, central England. *Forestry* 74: 299–309. https://doi.org/10.1093/forestry/74.3.299

Perry, I. and P.H. Langton. 2000. Diptera. In L. Friday and B. Harley (eds) *Checklist of the Flora and Fauna of Wicken Fen*, pp 64–80. Harley Books.

Peterken, G.F. 1974. A method for assessing woodland flora for conservation using indicator species. *Biological Conservation* 6 (4): 239–45. https://doi.org/10.1016/0006-3207(74)90001-9

Peterken, G.F. 1993. *Woodland Conservation and Management*, second edition. London: Chapman & Hall.

Peterken, G.F. 2000. Identifying ancient woodland using vascular plant indicators. *British Wildlife* 11 (3): 153–8.

Peterken, G.F. 2017. Recognising wood-meadows in Britain? *British Wildlife* 23 (3): 155–65.

Pickles, M.M. 1960. *The Birds of Blenheim Park*. Oxford: Oxford Ornithological Society.

Piper, R. and A.J. Allen. 2020. Beetles collected in vane traps from King's Beeches, Berkshire. *The Coleopterist* 29: 33–40.

Plot, R. 1677. *The Natural History of Oxford-Shire*. London: Charles Brome & John Nicholson, [also second revised edition 1705; reprinted 1972: Paul Minet, Chicheley].

PMP. 2014. *Blenheim Palace Parkland Management Plan*. Historic Landscape Management Ltd, Norfolk.

Pollard, B.J. 2021. *Elater ferrugineus* Linnaeus (Elateridae) and *Laemophloeus monilis* (Fabricius) (Laemophloeidae) in High Park, Blenheim Palace, Oxfordshire (VC 23). *The Coleopterist* 30: 134–6.

Pollard, B.J., J. Cooter, S.A. Lane, J. Denton, I. Wright, L. Losito and D.J. Mann. 2022. Forty-five species newly recorded for VC 23 (Oxfordshire), from High Park, Blenheim Palace, with notes on other significant finds. *The Coleopterist* 31: 119–38.

Pollard, B.J., H. Mendel and L. Losito. 2023. *Eucnemis capucinus* (Eucnemidae) new to Oxfordshire (VC 23), with further notes on High Park, Blenheim Palace including corrections to Pollard et al. (2022). *The Coleopterist* 32: 111–13.

Powys, T.L, 1852. Note on the kite and buzzard trapped at Blenheim. *Zoologist* 10: 3388.

Preston, C.D., D.A. Pearman and T.D. Dines. 2002. *New Atlas of the British and Irish Flora*. Oxford: Oxford University Press.

Pritchard, D. 2009. Is the dark bee really native to Britain and Ireland? *Bee Improvement and Conservation* 30: 11–17.

Rackham, O. 1997. *The History of the Countryside*. London: Weidenfeld & Nicolson.

Rackham, O. 2003. *Ancient Woodland, its History, Vegetation and Uses in England*, new edition. Colvend: Castlepoint Press.

Rackham, O. 2011. Forest and upland. In *A Social History of England 900–1200*, ed. J. Crick and E. van Houts, pp. 46–55. Cambridge: Cambridge University Press. https://doi.org/10.1017/CBO9780511976056.006

Radford, C.A.R. 1936. The Roman villa at Ditchley Oxon. *Oxoniensia* 1: 24–69.

Radford, M.C. 1966. *The Birds of Berkshire and Oxfordshire*. London: Longmans, Green & Co.

Randle, Z., L.J. Evans-Hill, M.S. Parsons, A. Tyner, N.A.D. Bourn, A.M. Davis, E.B. Dennis, M. O'Donnell, T. Prescott, G.M. Tordoff and R. Fox. 2019. *Atlas of Britain & Ireland's Larger Moths*. Newbury: Pisces Publications.

Ray, J. 1678. *The Ornithology of Francis Willughby*. London: John Martyn [reprinted 1972: Paul Minet, Chicheley].

Read, H. 1999. *Veteran Trees: A guide to good management*. Peterborough: English Nature (Natural England).

Reid, D.A. 1966. Coloured icones of rare and interesting fungi Part 1. *Nova Hedwigia* 11 (Suppl.): 1–32.

Richardson, D.H.S. 1992: *Pollution Monitoring with Lichens*. Naturalists' Handbooks, 19. Slough: Richmond Publishing Co. Ltd. 76 pp.

Rivet, A.L.F. 1964. *Town and Country in Roman Britain*. London: Hutchinson & Co. Ltd. 196 pp.

Roberts, P.J. 2002. *Report on the oak polypore* Piptoporus quercinus *(syn.* Buglossoporus quercinus; B. pulvinus*)*. English Nature Research Reports 458. Peterborough: English Nature.

Rodwell, J.S. (ed.). 1991–2000. *British Plant Communities*. Volumes I–V. Cambridge: Cambridge University Press. https://doi.org/10.1017/9780521235587

Rose, F. 1976. Lichenological indicators of age and environmental continuity in woodlands. In D. H. Brown, D. L. Hawksworth and R. H. Bailey (eds): *Lichenology: Progress and Problems*, pp. 279–307. London: Academic Press.

Rose, F. 1993. Ancient British woodlands and their epiphytes. *British Wildlife* 5 (1): 83–93.

Rose, F. 1999. Indicators of ancient woodland – the use of vascular plants in evaluating ancient woods for nature conservation. *British Wildlife* 10 (4): 241–251.

Rowe, A. 2019. *Medieval Parks of Hertfordshire*. Hatfield: University of Hertford Press.

Ruffer, J.G. 1977. *The Big Shots. Edwardian shooting parties*. London: Debrett's Peerage.

Rutherford, S. 2016. *Capability Brown and his Landscape Gardens*. London: National Trust/Pavilion Books.

Sage, R.B., A.N. Hoodless, M.I.A. Woodburn, R.A.H. Draycott, J.R. Madden and N.W. Sotherton. 2020. Summary review and synthesis: effects on habitats and wildlife of the release and management of Pheasants and Red-legged Partridges on UK lowland shoots. *Wildlife Biology* 2020 (4), 1–12. https://doi.org/10.2981/wlb.00766

Salway, P. 1999. Roman Oxfordshire. *Oxoniensia* 64: 1–22.

Sanderson, N.A. and Wolseley, P. 2001. Management of pasture woodlands for lichens. In *Habitat Management for Lichens*, ed. A. Fletcher, pp. 05-1–05-25. London: British Lichen Society.

Sanderson, N.A. 2018a. *A Review of Woodland Epiphytic Lichen Habitat Quality Indices in the UK*. A report by Botanical Survey and Assessment for Natural England, June 2019. Lichen Survey of Moccas NNR, Herefordshire, British Lichen Society 23.

Sanderson, N.A. 2018b. *The Development of TNTN Lichen Assemblage Scoring*. A report by Botanical Survey and Assessment for Natural England.

Sanderson, N.A., T. Wilkins, S. Bosanquet and D. Genney. 2018. *Guidelines for the Selection of Biological SSSIs. Part 2:*

Detailed Guidelines for Habitats and Species Groups. Chapter 13: Lichens and associated microfungi. Peterborough: Joint Nature Conservation Committee 2018.

Savill, P, C. Perrins, K. Kirby and N. Fisher (eds). 2011. *Wytham Woods, Oxford's ecological laboratory.* Oxford: Oxford University Press. https://doi.org/10.1093/acprof:osobl/9780199605187.001.0001

Schulz, T.E. 1938. The Woodstock glove industry. *Oxoniensia* 3: 139–52.

Schumer, B. 1984. *The Evolution of Wychwood to 1400: Pioneers, frontiers and forests.* 3rd Series No. 6, ed. Alan Everitt. Leicester: Leicester University Press.

Schumer, B. 1999. *Wychwood. The evolution of a wooded landscape.* Charlbury: The Wychwood Press.

Schumer, B. 2004. The 1298/1300 Perambulations of Wychwood Forest – and after. *Oxoniensia,* 69: 1–28.

Shapland, M.G. 2012. Buildings of secular and religious lordship: Anglo-Saxon tower-nave churches. PhD thesis, University College London. 822 pp.

Shelmerdine, J.M. 1951. *Introduction to Woodstock.* The Woodstock Society, Samson Press.

Shigo, A.L. 1985. Compartmentalization of decay in trees. *Scientific American* 252 (4): 96–103. https://doi.org/10.1038/scientificamerican0485-96

Shirt, D.B. 1987. *British Red Data Books: 2 Insects.* Peterborough: Nature Conservancy Council.

Shotton, F.W., A.S. Goudie, D.J. Briggs and H.A. Osmaston, 1980. Cromerian interglacial deposits at Sugworth, near Oxford, England, and their relation to the plateau drift of the Cotswolds and the terrace sequence of the Upper and Middle Thames Source. *Philosophical Transactions of the Royal Society of London. Series B, Biological Sciences,* 289 (1034): 55–86. https://doi.org/10.1098/rstb.1980.0027

Shrub, M. 2013. *Feasting, Fowling and Feathers. A history of the exploitation of wild birds.* London: Bloomsbury (T & AD Poyser).

Soper, T. (ed. and introduction). 1981. *British Birds [by] the Rev. F.O. Morris. A selection from the original work.* London: Webb & Bower.

Spake, R., S. van der Linde, A.C. Newton, L.M. Suz, M.I. Bidartondo and C.P. Doncaster. 2016. Similar biodiversity of ectomycorrhizal fungi in set-aside plantations and ancient old-growth broadleaved forests. *Biological Conservation* 194: 71–79. https://doi.org/10.1016/j.biocon.2015.12.003

Speake, G. 2012. An early Romano-British villa at Combe East End. *Oxoniensia* 77: 1–90.

Spencer-Churchill, C.R.J. 1892–1931. Trees planted at Blenheim by 9th Duke of Marlborough. Blenheim Archives, E/A/193.

Spooner, B. and P. Roberts. 2005. *Fungi.* London: Collins.

Stace, C. 2019. *New Flora of the British Isles,* fourth edition. Middlewood Green: C & M Floristics.

Sterling, P.H. 1987. *Aplota palpella* (Haworth) (Lep.: Oecophoridae) rediscovered in Britain. *The Entomologist's Record and Journal of Variation* 99: 275–6.

Sterling, P.H. 1988. *Aplota palpella* (Haworth) (Lep.: Oecophoridae), further notes on the biology and habitat requirements. *The Entomologist's Record and Journal of Variation* 100: 203–4.

Stockdale, J.E., J.C. Dunn, S.J. Goodman, A.J. Morris, D.K. Sheehan, P.V. Grice and K.C. Hamer. 2015. The protozoan parasite *Trichomonas gallinae* causes adult and nestling mortality in a declining population of European Turtle Doves, *Streptopelia turtur. Parasitology* 142 (3): 490–8. https://doi.org/10.1017/S0031182014001474

Streeter, D.T. 1974. Ecological aspects of oak woodland conservation. In M.G. Morris and F. H. Perring (eds) *The British Oak,* pp. 341–354. Faringdon: E. W. Classey, Ltd.

Stroh, P.A., S.J. Leach, T.A. August, K.J. Walker, D.A. Pearman, F.J. Rumsey, C.A. Harrower, M.F. Fay, J.P. Martin, T. Pankhurst, C.D. Preston and I. Taylor. 2014. *A Vascular Plant Red List for England.* Bristol: Botanical Society of Britain and Ireland.

Strubbe, D., E. Matthysen and C.H. Graham. 2010. Assessing the potential impact of invasive ring-necked parakeets *Psittacula rameria* on native nuthatches *Sitta europeae* in Belgium. *Journal of Applied Ecology* 47: 549–57. https://doi.org/10.1111/j.1365-2664.2010.01808.x

Suz, L.M., N. Barsoum, S. Benham, H.-P. Dietrich, K.D. Fetzer, R. Fischer, P. García, J. Gehrman, F. Kristöfel, M. Manninger, S. Neagu, M. Nicolas, J. Oldenburger, S. Raspe, G. Sánchez, H.W. Schröck, A. Schubert, K. Verheyen, A. Verstraeten and M.I. Bidartondo. 2014. Environmental drivers of ectomycorrhizal communities in Europe's temperate oak forests. *Molecular Ecology* 23: 5628–44. https://doi.org/10.1111/mec.12947

Taigel, A. and T. Williamson. 1993. *Parks and Gardens. Know the landscape.* London: Batsford.

Tapper, S. 1992. *Game Heritage. An ecological review from shooting and gamekeeping records.* Fordingbridge: Game Conservancy.

Taunt, H.W. 1914. *Blenheim & Woodstock. Their story and some of the scenes around them.* Oxford: Henry W. Taunt & Co.

Taunt, H.W. 1925. *Alden's Complete Guide to Blenheim and Woodstock.* Oxford: Woodstock Town Council and Alden & Co.

Telfer, M.G. 2020. Saproxylic invertebrate survey of Petworth Deer Park. Report to Buglife: the Invertebrate Conservation Trust.

Thaxter, C.B., A.C. Joys, R.D. Gregory, S.R. Baillie and D.G. Noble. 2010. Hypotheses to explain patterns of population change among breeding bird species in England. *Biological Conservation* 143: 2006–19. https://doi.org/10.1016/j.biocon.2010.05.004

Thomas, C. 1971. *Britain and Ireland in Early Christian Times, AD 400–800.* London: Thames and Hudson Limited.

Thomas, J. and R. Lewington. 2019. *The Butterflies of Britain and Ireland,* third edition. London: Bloomsbury Publishing.

Tree, I. and C. Burrell, 2023. *The Book of Wilding. A practical guide to rewilding big and small.* London: Bloomsbury Publishing.

Trevor-Battye, A.B.R. (ed.). 1903. *Lord Lilford on Birds: Being a collection of informal and unpublished writings by the late president of the British Ornithologists' Union. With contributed papers upon falconry and otter hunting, his favourite sports.* London: Hutchinson.

Trinder, A.N. 1983. *Woodstock.* Oxford: Oxford Polytechnic.

Tubbs, C.R. 1986. *The New Forest, a Natural History.* The New Naturalist Library, London: William Collins Sons & Co.

Tucker, B.W. and M.P.B. Ottley. 1924. Breeding of pochard and tufted duck in Oxfordshire. *British Birds* 17: 311.

Turner-Bennett, E. 1829. *The Tower Menagerie comprising The Natural History of the Animals Contained in that Establishment with Anecdotes of their Character and*

History. London: Robert Jennings. 264 pp. https://doi.org/10.5962/bhl.title.44488

TVERC. 2015. *Grassland Survey 2015 High Park, The Blenheim Estate*. Oxford: Thames Valley Environmental Record Centre.

Tyler, M. 2008. *British Oaks, a Concise Guide*. Ramsbury: The Crowood Press.

Vera, F.W.M. 2000. *Grazing Ecology and Forest History*. Wallingford: CABI Publishing. https://doi.org/10.1079/9780851994420.0000

Vincent, J. 1897. *Guide to Blenheim*. London: Marshall, Hamilton, Kent & Co.

Visser, M., L. Holleman and P. Gienapp. 2006. Shifts in caterpillar biomass phenology due to climate change and its impact on the breeding biology of an insectivorous bird. *Oecologia* 147: 164–72. https://doi.org/10.1007/s00442-005-0299-6

Walker, J.J. 1907–37. *Preliminary List of Coleoptera Observed in the Neighbourhood of Oxford from 1819 to 1907*. With additional reports and supplements to 1937. Compiled for the Ashmolean Natural History Society of Oxfordshire, from all available records. Reprinted from the Ashmolean Natural History Society Reports.

Ward, A.I. 2005. Expanding ranges of wild and feral deer in Great Britain. *Mammal Review* 35: 165–73. https://doi.org/10.1111/j.1365-2907.2005.00060.x

Waring, P. 2022. Moths. In G. Freeman, Wildlife reports. *British Wildlife* 33 (4): 290–2.

Watanabe M., H. Sato, H. Matsuzaki, T. Kobayashi, N. Sakagami, Y. Maejima, H. Ohta, N. Fujitake and S. Hiradate. 2007. 14C ages and delta δ13C of sclerotium grains found in forest soils. *Soil Science and Plant Nutrition* 53: 125–131. https://doi.org/10.1111/j.1747-0765.2007.00121.x

Waterhouse, M. and K. Wiseman. 2019. *The Churchill who Saved Blenheim. The life of Sunny, 9th Duke of Marlborough*. London: Unicorn Publishing.

Watkins, C. 2014. *Trees, Woods and Forests: A social and cultural history*. London: Reaktion Books Ltd.

Welden, N.A., P.A. Wolseley and M.R. Ashmore. 2018. [2017]: Citizen science identifies the effects of nitrogen deposition, climate and tree species on epiphytic lichens across the UK. – *Environmental Pollution* 232: 80–9. https://doi.org/10.1016/j.envpol.2017.09.020

Westwood, B., P. Shirley, R. Winnall and H. Green (eds). 2015. *The Nature of Wyre: A wildlife-rich forest in the heart of Britain*. Newbury: Pisces Publications.

White, J. 1998. Estimating the age of large and veteran trees in Britain. *Forestry Commission Information Note* 12. Edinburgh: Forestry Commission.

Whitehead, 'T.' [=A.], I. Wright and A. Gosler. 2003. *The Birds of Shotover*. Oxford: Oxford Ornithological Society/Shotover Wildlife.

Whitehead, A. 2018. Birds. In I. and J. Wright (eds) *Shotover. The life of an Oxfordshire hill*, pp. 201–17. Newbury: Pisces Publications.

Whiteman, C.A. and J. Rose. 1997. Early-Middle Pleistocene beheading of the River Thames. *Géographie physique et Quaternaire* 51 (3): 327–36. https://doi.org/10.7202/033131ar

Williamson, T. 2007. *Rabbits, Warrens & Archaeology*. Stroud: Tempus Publishing Ltd.

Williamson, T., 2010. *The Origins of Hertfordshire*. Manchester: Manchester University Press.

Windsor Great Park website. 2023, accessed 29 March 2023. https://www.windsorgreatpark.co.uk/the-estate/environment/conservation-stewardship/wildlife/

Wing and Barrel Ranch. n.d. [2019]. The ultimate sporting experience. January 20–25, 2020. publicity brochure online at https://www.wingandbarrelranch.com/s/WBR_Blenheim_2020_PDFforweb

Wolseley, P. 2022. Lichens. In A. Farjon, *Ancient Oaks in the English Landscape*, second edition, pp. 290–295. Royal Botanic Gardens, Kew: Kew Publishing.

Wolseley, P., N. Sanderson, H. Thüs, D. Carpenter and P. Eggleton. 2017. [2016] Patterns and drivers of lichen species composition in a NW-European lowland deciduous woodland complex. *Biodiversity and Conservation* 26(2): 401–419. https://doi.org/10.1007/s10531-016-1250-3

Wolton, R.J., H. Bentley, P.J. Chandler, C.M. Drake, J. Kramer, A.R. Plant and A.E. Stubbs. 2014. The diversity of Diptera associated with a British hedge. *Dipterists Digest* (Second Series) 21: 1–36.

Wood, T., and D. Baldock. 2016. *Chrysis terminata* Dahlbohm, 1854 from Farnham Heath, Surrey, *BWARS Newsletter* Spring 2016, 26–7. https://doi.org/10.1016/S0262-1762(16)30273-5

Woodward, F. 1982. *Oxfordshire Parks*. Abingdon: Abbey Press.

Woodward, I.D., D. Massimino, M.J. Hammond, L. Barber, C. Barimore, S.J. Harris, D.I. Leech, D.G. Noble, R.H. Walker, S.R. Baillie and R.A. Robinson. 2020. *BirdTrends 2020: Trends in numbers, breeding success and survival for UK breeding birds*. BTO Research Report 732. Thetford: BTO. www.bto.org/birdtrends

Wotton, S.R., I. Carter, A.V. Cross, B. Etheridge, N. Snell, K. Duffy, R. Thorpe and R.D. Gregory. 2002. Breeding status of the Red Kite *Milvus milvus* in Britain in 2000. *Bird Study* 49 (3): 278–86. https://doi.org/10.1080/00063650209461276

Wright, I. 2015. *Platydema violaceum* (Fabricius) (Tenebrionidae) new to Oxfordshire (VC 23). *The Coleopterist* 24: 17.

Wright, I.R. and S.J. Gregory. 2006. The aculeate Hymenoptera of Shotover Hill, Oxfordshire. *British Journal of Entomology and Natural History* 19: 65–76.

Wright, I.R. and J.A. Wright. 2018. *Shotover: The life of an Oxfordshire hill*. Newbury: Pisces Publications.

Wychwood Project. 2008. *Blenheim High Park. An Assessment of Tree Size Class*. Woodstock: Wychwood Project.

Wychwood Project. 2014. *A management plan for High Park, the Blenheim Estate*. Eynsham: The Wychwood Project with Thames Valley Environmental Records Centre.

Yardley, M. 2015. The history of the Pheasant. *The Field*, 9 October. https://www.thefield.co.uk/shooting/the-history-of-the-pheasant-22364